Alternative Energy: Oil and Gas

Alternative Energy: Oil and Gas

Edited by Oliver Haghi

SYRAWOOD
PUBLISHING HOUSE
New York

Published by Syrawood Publishing House,
750 Third Avenue, 9th Floor,
New York, NY 10017, USA
www.syrawoodpublishinghouse.com

Alternative Energy: Oil and Gas
Edited by Oliver Haghi

International Standard Book Number: 978-1-68286-458-6 (Hardback)

Cataloging-in-publication Data

Alternative energy : oil and gas / edited by Oliver Haghi.
p. cm.
Includes bibliographical references and index.
ISBN 978-1-68286-458-6
1. Renewable energy sources. 2. Energy harvesting. 3. Petroleum. 4. Gas. I. Haghi, Oliver.
TJ808 .A48 2017
621.042--dc23

Printed in China.

TABLE OF CONTENTS

PREFACE

Biofuels and biogas are sources of alternative energy that are being extensively studied and explored. This book on alternative energy deal with renewable sources of oil and gas, their manufacture and their viability as long-term solutions to the damaging effects caused by the burning of fossil fuels. Topics that are explained in the book include mass production techniques of alternate energy and the ecological and commercial aspects of their production. Some of the diverse chapters in this text address the varied of alternatives that have been developed in the place of the burning of fossil fuels. The various advancements in this discipline are glanced at and their applications and ramifications are looked at in detail. This book will be very beneficial to students, experts, researchers, scientists and professionals engaged in the fields of conservation engineering, energy engineering and green technology development.

This book is a comprehensive compilation of works of different researchers from varied parts of the world. It includes valuable experiences of the researchers with the sole objective of providing the readers (learners) with a proper knowledge of the concerned field. This book will be beneficial in evoking inspiration and enhancing the knowledge of the interested readers.

In the end, I would like to extend my heartiest thanks to the authors who worked with great determination on their chapters. I also appreciate the publisher's support in the course of the book. I would also like to deeply acknowledge my family who stood by me as a source of inspiration during the project.

Editor

1

Experimental Study and Mathematical Modeling of Asphaltene Deposition Mechanism in Core Samples

T. Jafari Behbahani[1,2], C. Ghotbi[1]*, V. Taghikhani[1] and A. Shahrabadi[2]

[1] Department of Chemical and Petroleum Engineering, Sharif University of Technology, Tehran - Iran
[2] Research Institute of Petroleum Industry (RIPI), Tehran - Iran
e-mail: jafaribehbahani@che.sharif.ir - ghotbi@sharif.edu - taghikhani@sharif.edu - sharabadia@ripi.ir

* Corresponding author

Résumé — Étude expérimentale et modélisation mathématique du mécanisme de déposition d'asphaltène dans des carottes de forage — Les études expérimentales présentées dans ce document ont été effectuées afin de déterminer les effets de la déposition d'asphaltène sur la réduction de la perméabilité et de la porosité d'échantillons rocheux de carbonate, de grès et de dolomite en utilisant un échantillon de pétrole brut de fond de puits proche des conditions du réservoir, tandis que dans la majorité des travaux précédents un mélange d'huile recombinée (mélange de pétrole dégazé et des gaz associés) est injecté dans un échantillon de forage dans des conditions très éloignées de celles du réservoir. Les effets du taux d'injection de pétrole sur la déposition d'asphaltène et la réduction de la perméabilité ont été étudiés. Les résultats expérimentaux montrent qu'une augmentation du débit d'injection de pétrole peut causer une augmentation du dépôt d'asphaltène et une réduction de la perméabilité. Il est aussi observé que, pour des débits d'injection plus faibles, une diminution monotone de la perméabilité des échantillons rocheux peut être obtenue en augmentant le débit d'injection, alors que pour des débits d'injection plus élevés, après une diminution de la perméabilité de la roche, une tendance à l'augmentation est observée avant d'atteindre un état stationnaire. Les résultats expérimentaux montrent aussi que le type de roche peut affecter la quantité d'asphaltène déposée, et que la déposition d'asphaltène est liée à différents mécanismes dans les carottes de forage de grès et de carbonate. Il est mis en évidence que les mécanismes d'adsorption et de colmatage jouent un rôle plus important dans les carottes de carbonate que dans ceux de grès. Ces résultats indiquent que les volumes de pore du pétrole brut injecté sont plus élevés dans les échantillons de grès que dans ceux de carbonate. On peut ainsi en déduire que trois types de dépôts peuvent avoir lieu durant l'injection de pétrole brut : un dépôt continu pour les forages à faible perméabilité, un colmatage lent et progressif pour les forages à haute perméabilité et un dépôt progressif pour les forages à moyenne perméabilité. Les résultats expérimentaux indiquent ainsi qu'une augmentation de la pression de production provoque une augmentation des endommagements des carottes de forage. Ceux-ci montrent aussi que la quantité d'asphaltène restante dans les carottes de carbonate est plus élevée que celle restante dans les carottes de grès. De plus, des micrographies MEB (Microscopie Électronique à Balayage) des carottes de carbonate indiquent que la formation d'agglomérats d'asphaltène est plus importante dans ces échantillons que dans ceux en grès durant l'épuisement naturel. Il peut être conclu à partir des résultats de modélisation que les modèles proposés basés sur le mécanisme d'absorption à l'équilibre multicouches et quatre équations de bilan de matière sont plus précis que ceux obtenus à partir du mécanisme d'absorption à l'équilibre monocouche et deux équations de bilan de matière : ces résultats sont en accord avec les données expérimentales d'épuisement naturel présentées dans cette étude et dans la littérature.

Abstract — Experimental Study and Mathematical Modeling of Asphaltene Deposition Mechanism in Core Samples — *In this work, experimental studies were conducted to determine the effect of asphaltene deposition on the permeability reduction and porosity reduction of carbonate, sandstone and dolomite rock samples using an Iranian bottom hole live oil sample which is close to reservoir conditions, whereas in the majority of previous work, a mixture of recombined oil (a mixture of dead oil and associated gas) was injected into a core sample which is far from reservoir conditions. The effect of the oil injection rate on asphaltene deposition and permeability reduction was studied. The experimental results showed that an increase in the oil injection flow rate can result in an increase in asphaltene deposition and permeability reduction. Also, it can be observed that at lower injection flow rates, a monotonic decrease in permeability of the rock samples can be attained upon increasing the injection flow rate, while at higher injection rates, after a decrease in rock permeability, an increasing trend is observed before a steady-state condition can be reached. The experimental results also showed that the rock type can affect the amount of asphaltene deposition, and the asphaltene deposition has different mechanisms in sandstone and carbonate core samples. It can be seen that the adsorption and plugging mechanisms have a more important role in asphaltene deposition in carbonate core samples than sandstone core samples. From the results, it can be observed that the pore volumes of the injected crude oil are higher for sandstone cores compared with the carbonate cores. Also, it can be inferred that three depositional types may take place during the crude oil injection, i.e., continuous deposition for low-permeability cores, slow, steady plugging for high-permeability cores and steady deposition for medium-permeability cores. It can be seen from the experimental results that damage to the core samples was found to increase when the production pressures were increased. The experimental results show that the amount of remaining asphaltene in carbonate core samples is higher than those in sandstone core samples. Also, SEM (Scanning Electron Microscopy) micrographs of carbonate core samples showed the formation of large clusters of asphaltene in comparison with sandstone core samples during natural depletion. It can be seen from the modeling results that the proposed model based on the multilayer adsorption equilibrium mechanism and four material balance equations is more accurate than those obtained from the monolayer adsorption equilibrium adsorption mechanism and two material balance equations, and is in agreement with the experimental data of natural depletion reported in this work and with those reported in the literature.*

NOMENCLATURE

C_{sf}	Mass of suspended asphaltenes per mass of the oil phase
C	Experimental concentration (in this case, in mol/L)
C_A	Volume fraction of the suspended asphaltene precipitates in the liquid phase
D_{pt}	Average pore throat diameter
D_{ptcr}	Critical pore throat diameter, assumed constant
E_A	Volume fraction of the deposited asphaltene in the bulk volume of the porous media
E_{ad}	Fractional pore volume of asphaltene adsorbed
f	Fugacity
K_a	Ratio of adsorption/desorption rate constants
K	Permeability
k_1	First adsorption step parameter (in our case, this step is taken to be adsorption of asphaltenes in solution to the surface of the rock)
k_2	Second adsorption step parameter (in our case, this step is taken to be the adsorption of asphaltenes in solution to those asphaltenes already adsorbed to the rock)
R	Gas constant
S_{oi}	Initial oil saturation
S_{wi}	Initial water saturation
n	Mean aggregation number of the adsorbed asphaltenes (in the modified model)
u	Internal energy
u_L	Flux of the liquid phase
v	Molar volume
v_L	Interstitial velocity of the liquid phase

v_c	Critical interstitial velocity of the liquid phase (cm/s)
w_{AL}	Dissolved asphaltene in the liquid phase
W_{asp}	Asphaltene content of the bottom hole live oil
w_G	Mass fraction of gas
w_{OL}	Mass fraction of the oil in the liquid phase
W_{sa}	Mass of adsorbed asphaltene per mass of rock
w_{SAL}	Mass fractions of the suspended asphaltene precipitates
$W_{sa,\ max}$	Maximum adsorbed mass fraction, mg/g
w_{WL}	Mass fraction of water in the liquid phase

Greek Symbols

Ø	Porosity
M	Viscosity
ρ	Density
Γ	Amount of adsorbed substance in mol/L
Γ_∞	Maximum possible adsorption for the whole isotherm
α	Surface deposition rate coefficient, sec^{-1}
β	Entrainment rate coefficient, cm^{-1}
γ	Plugging deposition rate coefficient, cm^{-1}
σ	Snowball-effect deposition constant
μ_p	Chemical potential
δ_p	Solubility parameter of a liquid
Φ_s	Volume fraction of solvent (liquid oil)

Subscripts

AAD	Average absolute deviation
cal	Calculated
exp	Exponential function
O	Oil phase
P_{res}	Reservoir pressure
Q	Flow rate
T_{res}	Reservoir temperature

INTRODUCTION

Asphaltenes and resins are the polar fractions of crude oil that can be separated by addition of low-molecular-weight *n*-alkanes. By definition, asphaltenes are the fraction of a crude oil that is soluble in toluene and insoluble in an *n*-alkane, typically pentane or heptane. These fractions are formed by molecules with a polyaromatic structure containing paraffinic and naphthenic chains, as well as oxygen, nitrogen and sulfur as functional groups or heteroatoms. On the other hand, asphaltenes are heavy hydrocarbon molecules that are in colloidal suspension in the oil, stabilized by resins adsorbed on their surface [1]. Asphaltenes can reduce the hydrocarbon effective mobility by plugging pore throats, thus reducing the rock permeability, altering the formation wettability from water-wet to oil-wet and increasing the reservoir fluid viscosity by forming a colloidal solution in the oil phase [2]. The processes of asphaltene precipitation and deposition in porous media have various effects on oil flow during primary oil production and enhanced oil recovery processes. The natural depletion of a reservoir may cause the precipitation and deposition of asphaltenes, decreasing the productivity of wells during primary oil recovery. The majority of existing work studies the asphaltene deposition during core flooding using the injection of recombined oil (a mixture of dead oil and associated gas) into the core and/or with static systems and in the absence of reservoir rocks [3-5]. Therefore, such results are not applicable to the dynamic conditions which are related to the complex interactions between the asphaltenes and the rock surface. Asphaltene precipitation and deposition are two different processes and are governed by different mechanisms. The asphaltene deposition is governed by four mechanisms; surface deposition, entrainment, plugging and adsorption (surface excess or Langmuir theory) [6-12]. When precipitation occurs, asphaltene particles are aggregated and a new solid phase is generated in the oil flow, whereas when deposition takes place on the rock, this can change the wettability of the reservoir formation and reduce oil production [13]. The asphaltene deposition phenomenon is described using several mathematical models that have been proposed for the prediction of porosity and permeability reduction of porous media [14-19]. These models are based on two different parallel pathways with small and large pore sizes [20]. Most models are capable of correlating the permeability reduction experimental data using adjustable parameters and include the flow continuity equation, a mass balance equation and a kinetic equation. Other models based on network modeling theory have also been developed to describe the deposition of asphaltene in reservoir rock [21, 22].

In short, due to the complexity of asphaltene deposition on the rock surface, very few experimental results currently exist on asphaltene deposition under dynamic conditions using live reservoir fluid in porous media. The majority of existing work addresses the deposition of asphaltenes using dead residual oil or recombined oil [6, 23-25], while it is well-known that the asphaltene in bottom hole live oil is different than that of recombined oil [26]. In this work, an Iranian bottom hole live oil sample was used under dynamic conditions in porous

TABLE 1
Studied live reservoir fluid compositions

Components	Dead residual oil (mol%)	Associated gas (mol%)	Live reservoir fluid (mol%)	Components	Dead residual oil (mol%)	Associated gas (mol%)	Live reservoir fluid (mol%)
H_2S	0	0	0	n_C_5	1.35	1.88	1.59
N_2	0	0.67	0.3	C_6	11.19	1.62	6.95
CO_2	0	4.12	1.83	C_7	6.87	0.63	4.1
C_1	0	51.19	22.7	C_8	6.84	0.15	3.88
C_2	0.51	17.94	8.24	C_9	4.47	0.01	2.49
C_3	1.17	12.38	6.14	C_{10}	7.24	0	4.03
i_C_4	0.52	2.02	1.19	C_{11}	5.13	0	2.85
n_C_4	1.94	5.7	3.61	C_{12}^{+}	51.64	0	28.74
i_C_5	1.14	1.68	1.38	Total	100	100	100

media for investigation of the effect of several parameters and different production conditions such as crude oil type, the flow rate of the oil, rock type and production pressure on asphaltene deposition. Also, the performance of the proposed model based on the multilayer adsorption equilibrium mechanism was studied in more detail and compared with the Wang and Civan [7] model based on the mechanical plugging mechanism and the asphaltene deposition model based on the monolayer adsorption equilibrium mechanism.

1 EXPERIMENTAL SECTION

1.1 Materials

In the majority of previous work, a mixture of recombined oil (a mixture of dead oil and associated gas) was injected into a core sample which is far from reservoir conditions. In this work, a bottom hole live oil sample from an Iranian reservoir in the Southeast region of Iran which is close to reservoir conditions was used. This reservoir has a serious asphaltene deposition problem during production conditions. The oil density and viscosity of the bottom hole live oil sample were measured to be 0.7674 g/cc and 4.35 cP at the reservoir pressure, respectively. The asphaltene content of the live reservoir fluid was measured to be wasp = 16.3 wt% using the SARA analyses. The compositional analysis result of this crude oil was obtained using the gas chromatography method and is given in Table 1.

A reservoir brine sample was collected from the same oilfield, cleaned, and analyzed. Its detailed physical and chemical properties are listed in Table 2.

TABLE 2
Physical and chemical properties of reservoir brine sample

Viscosity (CST at 40°C)	0.92
pH at 20°C	6.5
Oil (mg/L)	4
Chemical oxygen demand as O_2 (mg/L)	1 216
Biological oxygen demand as O_2 (mg/L)	530
Total suspended solids (mg/L)	352
Chloride (mg/L)	123 100
$CaCO_3$ (mg/L)	177
Sulfate (mg/L)	291
Total dissolved solids (mg/L)	202 050
Potassium (g/L)	1.1
Sodium (g/L)	60
Calcium (mg/L)	13 280
Magnesium (mg/L)	1 262
Iron (mg/L)	42
Strontium (g/L)	0.58
Lithium (mg/L)	13
Sulfide (mg/L)	7.2
Barium (mg/L)	1.2

TABLE 3
Characteristics of the various studied cores

Sample No.	Type of core	Length (cm)	Diameter (cm)	Porosity (%)	Permeability (mD)	Grain density (g/cm^3)	Weight (g)	PV
1	Carbonate	4.5	3.7	26	2.7	—	103.8	13.5
2	Dolomite	5.1	3.8	22.5	106.6	2.68	118.5	12.3
3	Sandstone	5.96	3.7	13.15	22.8	2.65	150.8	8.5

Various carbonate and sandstone cores with a cylindrical shape with lengths ranging from 4 to 6 cm and constant diameter equal to 4 cm were extracted at various depths of the Iranian wells, then characterized and tested in the EOR Research Center laboratory, Research Institute of Petroleum Industry (RIPI). All the cores investigated were cleaned prior to their use, by using various solvents (Soxhlet extraction with xylene, methanol and chloroform) according to the ASTM D2172 procedure. Their characteristics, including their porosities, permeabilities and pore volumes, were measured and are shown in Table 3.

1.2 Methods

1.2.1 Gas and Liquid Chromatography

When the live reservoir fluid sample was flashed in atmospheric conditions, vapor and liquid phases were analyzed by means of the High-Temperature Gas Chromatography (HTGC) technique. To analyze the lighter fraction up to C9 (including CO_2, H_2S and N_2) a high-temperature gas chromatograph (three packed columns and one capillary column) with thermal conductivity and flame ionization detectors was used, while the residual liquid was analyzed up to C12 with another high-temperature gas chromatograph (a capillary column) that included a flame ionization detector.

1.2.2 SARA Analyses of the Crude Oil

The constituents of crude oils are typically classified by solubility: Saturates, Aromatics, Resins and Asphaltenes (SARA) [27]. For SARA fraction analysis, the asphaltene fraction was extracted from crude oil by precipitation with *n*-heptane as described by the ASTM D3279-97 procedure. Subsequent elution with a series of increasingly polar solvents as the mobile phase yields saturates (eluted with a nonpolar solvent such as hexane), followed by the elution of aromatics with toluene, and the resins are separated with a more polar solvent.

TABLE 4
SARA test results of the studied reservoir fluid

Type of group	Live reservoir fluid	Residual oil
Saturate (wt%)	32.61	37.8
Aromatic (wt%)	43.48	46.2
Resin (wt%)	7.61	5.9
Asphaltene (wt%)	16.3	10.1

The SARA analysis results of live reservoir fluid are shown in Table 4.

1.2.3 Live Reservoir Fluid Viscosity

To determine the viscosity behavior of the reservoir fluid, at the studied temperature and different pressures, a constant force electromagnetic viscometer (maximum working pressure of 137.9 MPa, maximum temperature of 463.2 K) was used. The viscosity behavior of the live reservoir fluid is shown in Figure 1.

1.2.4 The PVT and Phase Behavior of the Live Reservoir Fluid

To determine the PVT and phase behavior of the live reservoir fluid, various experiments were conducted on the reservoir fluid, including constant mass expansion, differential vaporization and a separator test (DBR, VINCI PVT Cell). In these experiments, saturation pressure, relative volume, oil compressibility, the solution gas-oil ratio, the liberated gas-oil ratio, the oil formation volume factor and oil density were determined. In Tables 5 to 7, the PVT characteristics and phase behavior (constant mass expansion and differential vaporization results) of the studied live reservoir fluid are shown.

Also, the variation of relative volume, oil compressibility, the solution gas-oil ratio, the liberated gas-oil ratio, the oil formation volume factor and oil density with pressure are shown in Figures 2-4.

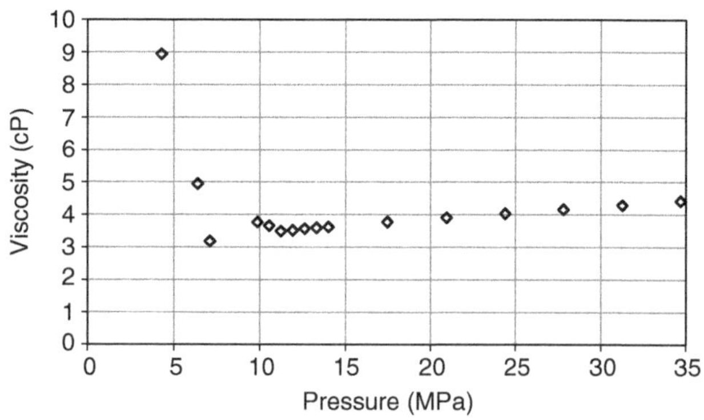

Figure 1

The viscosity behavior of the studied live reservoir fluid.

TABLE 5

PVT characteristics and phase behavior of the studied reservoir fluid

Reservoir pressure, P_{res} (MPa)	33
Reservoir temperature, T_{res} (°C)	96
Molar weight of the live reservoir fluid (g/mol)	182
Specific gravity of the live reservoir fluid	0.77
Molar weight of the heavy group (live reservoir fluid), C_{12}^{+} (g/mol)	491
Specific gravity of the heavy group (live reservoir fluid), C_{12}^{+}	0.99
Molar weight of the heavy group (dead residual oil), C_{12}^{+} (g/mol)	395

TABLE 6

Summary of constant mass expansion data for the studied live reservoir fluid

Saturation pressure	MPa	9.85
Solution GOR	SCF/STB	341.63
Density of total gas evolved	g/L	1.38
API gravity of residual oil	°API	20.29
Specific gravity of residual oil	60/60°F	0.93

TABLE 7

Summary of differential vaporization data for the studied live reservoir fluid

Solution GOR	SCF/STB	360.34
Density of total gas evolved	g/L	1.34
Oil density @ sat. pressure	g/cc	0.75
Oil density @ reservoir pressure	g/cc	0.77
Oil gravity of residual oil	°API	20.03
Specific gravity of residual oil	15/15°C	0.93
Oil viscosity @ sat. pressure	cP	3.50
Oil viscosity @ reservoir pressure	cP	4.36

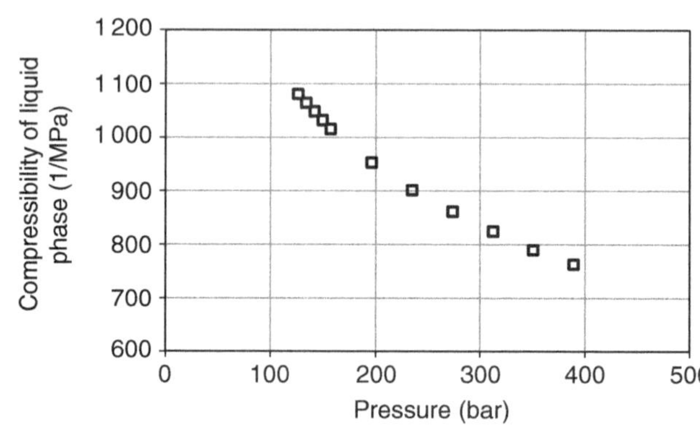

Figure 2

Variation of compressibility of liquid phase *versus* pressure for the studied live reservoir fluid.

1.3 Experimental Apparatus

Figure 5 shows the schematic of the experimental setup used in this work in order to determine the effect of asphaltene deposition on the properties of a reservoir rock.

This setup consists mainly of the following devices:

- an automatic displacement pump (*Vinci*, p = 70 MPa, accuracy = 0.000687 MPa) was used to displace the crude oil and reservoir brine through the composite reservoir core plugs inside a core holder (*Temco*);
- three high-pressure stainless steel cylinders (500-10-P-316-2, *DBR*, Canada) were used to store and deliver

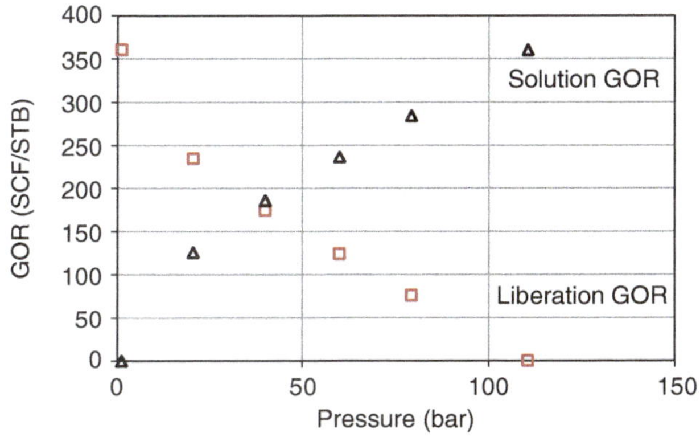

Figure 3

GOR results for the studied live reservoir fluid.

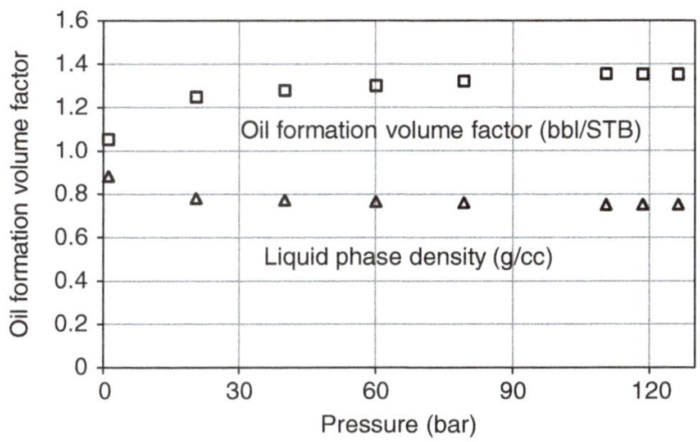

Figure 4

Differential liberation results *versus* pressure for the studied live reservoir fluid.

the live reservoir fluid, dead residual oil and reservoir brine, respectively (*DBR*, capacity of 500 mL);

- the core holder (*Temco*, p = 70 MPa, T = 150°C) holds a core through a sleeve. A constant overburden pressure was applied around this sleeve, which was always kept 5 MPa higher than the inlet pressure of the core holder. The inlet and outlet ports of the core holder were connected to the pressure transducer. The inlet ports of the core holder were also connected to the positive displacement pump. Two cylinders containing the live reservoir fluid sample allowed injection of samples inside the core holder. The positive displacement pump was operated at a constant rate or in pressure mode. The outlet port of the core holder was connected to the fractional collector through a back-pressure regulator for collection of produced fluids from the core.

The pressure drop during the core flood tests was measured with a digital pressure indicator. The core flooding system was thermoregulated by means of an air-bath oven. An important feature was the possibility of continuous measurement of the pressure drop along sections of the porous medium. The annular space between the sleeve and the body was filled with paraffin oil through which the confining pressure is applied on the external surface of the sleeve. The advantage of this setup is the uniform pressure in both axial and radial directions:

- a pressure transducer (*Jumo*, accuracy = 0.05);
- a back-pressure regulator (*Jeafer DBR*, p = 70 MPa, T = 250°C). The regulator was used to maintain the prespecified injection pressure inside the core holder during each flooding test;
- an overburden pump (*Enerpac*, p = 70 MPa);
- an air-bath oven (*Vinci*, T = 200°C);
- rigid valves;
- a data acquisition system (Logger screen 500 *Jumo*: pressure, temperature, volume).

1.4 Experimental Procedure

The reservoir core plugs were cleaned with toluene, methanol and chloroform in sequence to remove hydrocarbons, salts and clays, respectively. Then, the reservoir core plugs were dried and vacuumed at 0.7 bar for one hour, and the reservoir brine was imbibed to measure the pore volume of the reservoir core plugs. Afterward, the reservoir brine was injected at different rates (1-30 cc/h) to determine the water permeability of the core plugs. As listed in Table 3, the measured porosity was in the range of φ = 13.15-26% and the measured absolute permeability was in the range of k = 2.7-106.6 mD.

Then, reservoir brine was displaced by the live reservoir fluid until irreducible water saturation. After the core plugs were fully saturated with the crude oil, the live reservoir fluid was injected into the core plugs and several flooding tests were conducted. Then, the differential pressure (ΔP) between the inlet and outlet of the core holder was measured by a digital pressure indicator and was indicated by the data acquisition system at preset time intervals.

To study the effect of production pressure, the production pump was adjusted to 15 MPa and the injected pump to various pressures. The live reservoir fluid was injected into the core and attained stabilized conditions within a few minutes. Then, the procedure of gradual pressure reduction was followed throughout the experiment. The injection pressure used in this experiment was in the range of 50-41 MPa with temperature at 70°C using live reservoir fluid. There was no flow of

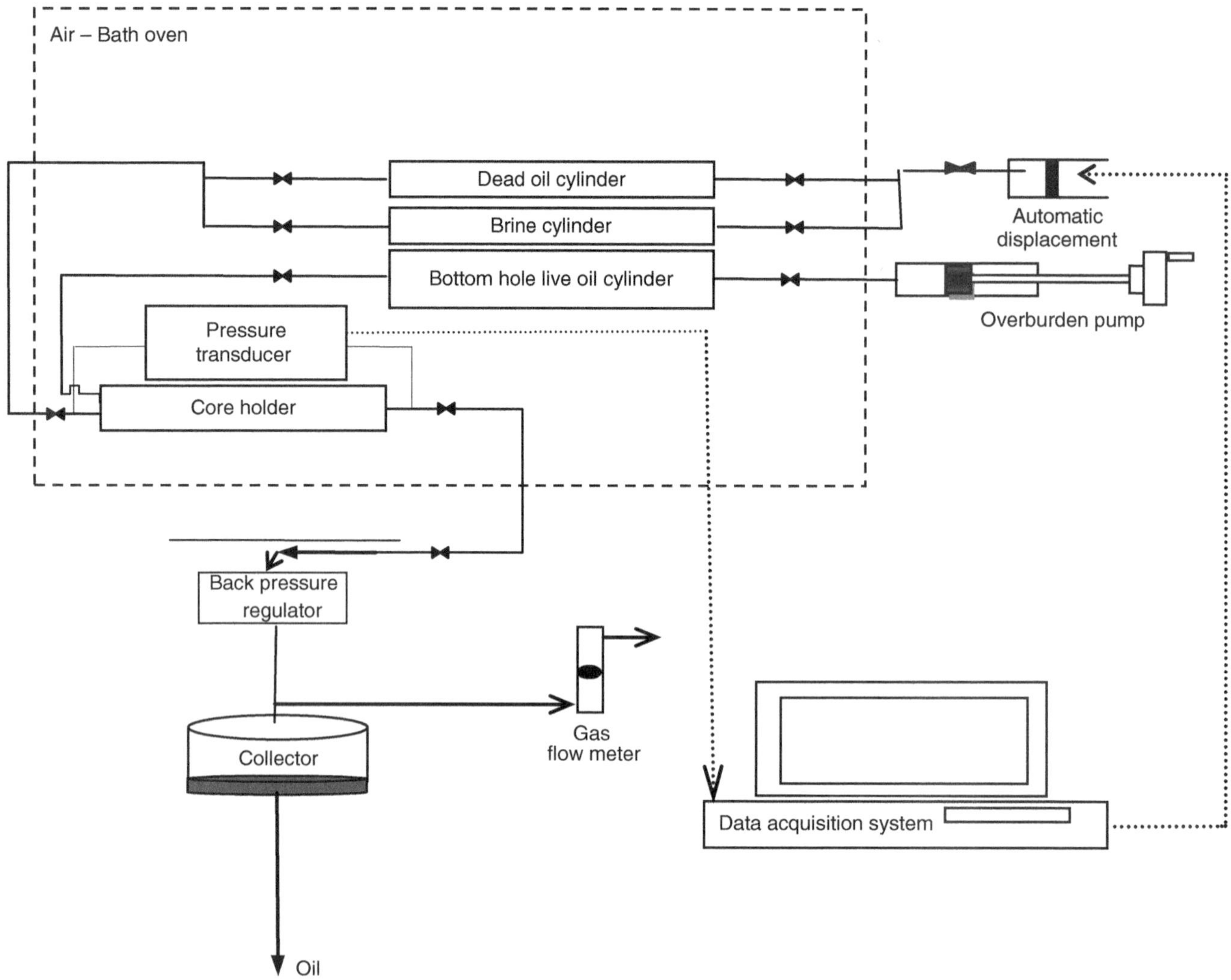

Figure 5

Schematic of core flooding apparatus.

crude oil at injection pressures lower than 41 MPa. In total, 13 sets of flooding tests were performed. Table 8 shows the experimental conditions.

Samples of the core, after the flooding, were inspected by Scanning Electron Microscopy (SEM) (Scanning Electron Microscope, VEGA, *TESCAN*). The core samples that were damaged in the process mentioned previously would develop a layer on their surface. A thin section of the damaged core was then cut at a thickness of one centimeter. These samples were then dried and viewed under the electron microscope. The SEM image enabled a three-dimensional view of the core sample at a microscopic level.

During this experiment, permeability was calculated from the measured pressure drop along the core through Darcy's law. According to Darcy's equation (*Eq. 1*), the permeability is dependent on the interfacial surface area, the viscosity, etc.:

$$Q = \frac{kA\Delta p}{\mu L} \tag{1}$$

2 MATHEMATICAL MODELING OF ASPHALTENE DEPOSITION IN POROUS MEDIA

The asphaltene deposition modeling is not only according to the mechanical plugging mechanism but also according to the asphaltene adsorption mechanism. In the majority of previous models, the asphaltene

TABLE 8

Experimental conditions for natural depletion tests

Test No.	Core sample	SWi (%)	SOi (%)	Q (cc/h)	Injection pressure (MPa)
1	3	36.7	63.3	1	–
2	3	36.7	63.3	3	–
3	3	36.7	63.3	10	–
4	3	36.7	63.3	30	–
5	3	36.7	63.3	40	–
6	4	37	63	10	–
7	1	43.7	56.3	1	–
8	1	43.7	56.3	5	–
9	2	43.7	56.3	3	–
10	1	43.7	56.3	–	41
11	1	43.7	56.3	–	44
12	1	43.7	56.3	–	48
13	1	43.7	56.3	–	50

TABLE 9

Considerations and parameters of the studied models

No.	Model	Ref.	Numbers of phases	Mechanisms of asphaltene deposition	Parameters
1	Proposed model (2012)	[28]	4	Multilayer adsorption equilibrium, surface deposition, entrainment of deposits and mechanical plugging	The first adsorption step parameter (this step is taken to be adsorption of asphaltenes in solution to the surface of the rock, the second adsorption step parameter (this step is taken to be the adsorption of asphaltenes in solution to those asphaltenes already adsorbed to the rock, the mean aggregation number of the adsorbed asphaltenes (in developed model), the maximum possible adsorption for the whole isotherm, surface deposition, entrainment of deposits, mechanical plugging, snowball-effect deposition constant and critical interstitial velocity of liquid phase parameters
2	Almehaideb (2004)	[8]	4	Monolayer adsorption equilibrium, surface deposition, entrainment of deposits and pore throat plugging	The adsorption parameter, the maximum possible adsorption for the whole isotherm, surface deposition, entrainment of deposits, mechanical plugging, snowball-effect deposition constant and critical interstitial velocity of liquid phase parameters
3	Wang and Civan (2001)	[7]	2	Surface deposition, entrainment of deposits and pore throat plugging	The surface deposition, entrainment of deposits and mechanical plugging, snowball-effect deposition constant and critical interstitial velocity of liquid phase parameters

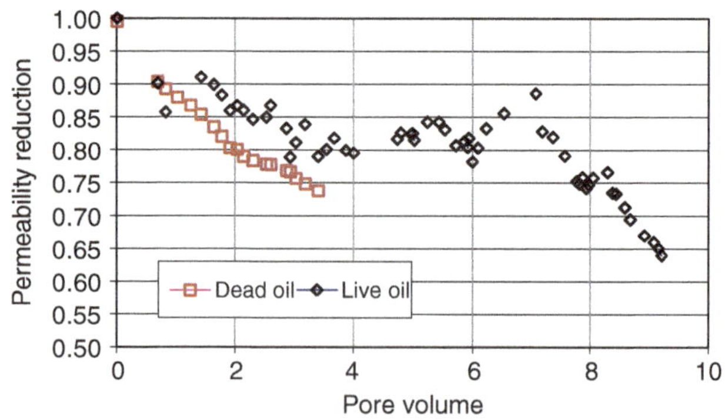

Figure 6

Permeability reduction for studied live reservoir fluid and dead residual oil *versus* the injected pore volume for a sandstone core sample.

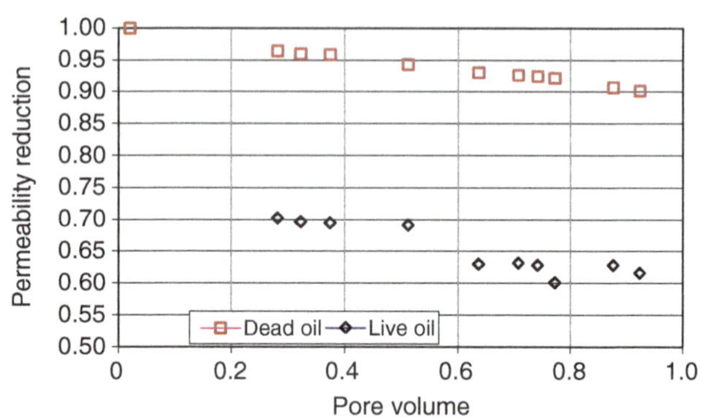

Figure 7

Permeability reduction for studied live reservoir fluid and dead residual oil *versus* the injected pore volume for a carbonate core sample.

deposition is based only on the mechanical trapping mechanism and it is different from the asphaltene adsorption mechanism. Also, the majority of previous work used two material balance equations based on the oil and asphaltene phases for core samples. In order to develop asphaltene deposition modeling including the asphaltene adsorption mechanism, a new model based on multilayer adsorption equilibrium theory and four material balance equations (oil, asphaltene, gas and water phases) was developed to account for asphaltene multilayer adsorption on core samples, as presented in our previous publication [28]. In this work, the results of asphaltene deposition in core samples during natural depletion were predicted using our proposed

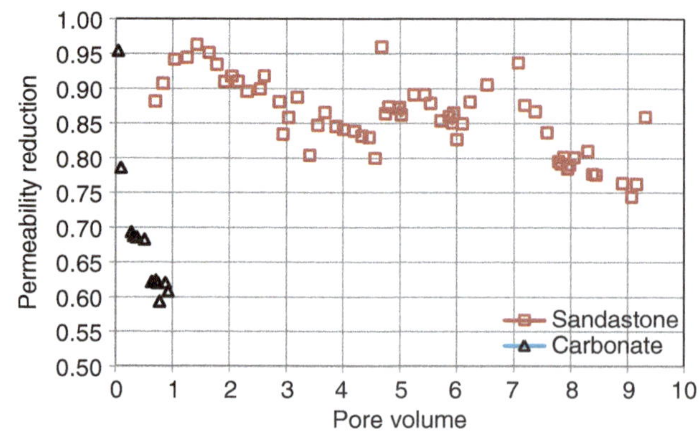

Figure 8

Permeability reduction *versus* the injected pore volume for studied live reservoir fluid.

model based on the multilayer adsorption equilibrium mechanism and compared with those obtained using asphaltene deposition modeling based on the monolayer adsorption equilibrium mechanism [8], and obtained by Wang and Civan's [7] model based only on the mechanical plugging mechanism.

2.1 Asphaltene Precipitation Model

Solid modeling treats the precipitating asphaltene as a single component residing in the solid phase, while the oil and gas phases are modeled with a cubic EOS. Solid models may require many empirical parameters and excessive tuning to match experimental data. In this study, the Nghiem model [29] was used; the fugacity of asphaltene in the solid phase is given by:

$$Inf_a = Inf_a^* + \frac{v_a(P - P^*)}{RT} \quad (2)$$

In a mixture, when the vapor, liquid and asphaltene phases coexist, the following thermodynamic equilibrium equations should be satisfied [29]:

$$Inf_{iv} = Inf_{il} \quad (3)$$

A three-phase flash algorithm is used to solve the above equations by using SRK EOS for the liquid and vapor phases. The existence of a solid phase satisfies the following criterion [29]:

$$f_{as} \geq f_{al} \quad (4)$$

In this model, the only parameter that should be adjusted from the experimental data is the referenced pressure, P^*. Also, according to our previous publication [30],

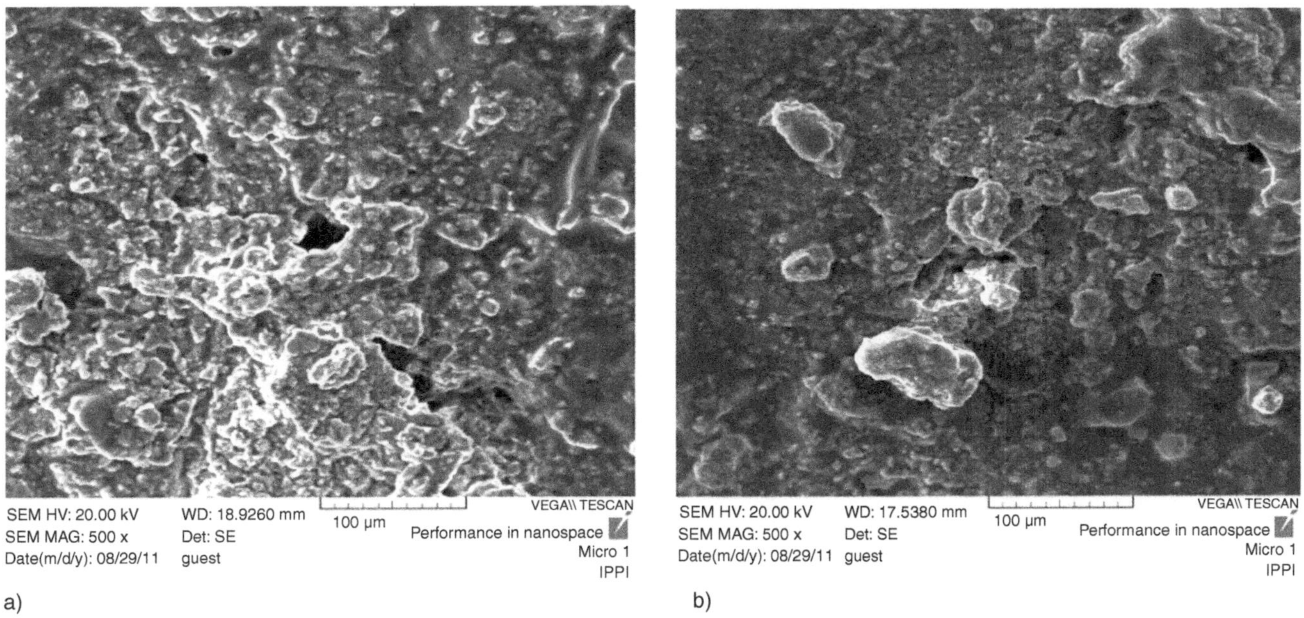

Figure 9

SEM images in studied a) carbonate and b) dolomite core samples.

the asphaltene precipitation in this bottom hole live oil was calculated based on the Flory-Huggins theory.

2.2 Asphaltene Adsorption and Deposition Models

There are two asphaltene deposition mechanisms on porous media, adsorption and mechanical plugging.

The asphaltene adsorption mechanism, which is related to the interactions between the asphaltene functional groups and the rock surface, involves surface polarity, affinity or other attractive forces. This process is reversible, with asphaltene desorbing from porous media when the concentration of suspended asphaltene in crude oil decreases. Asphaltene is a polar component; therefore, formations containing clay and carbonate formation have greater ability than sandstone formations to adsorb asphaltene.

In order to develop a new model for asphaltene deposition in porous media, a multilayer adsorption equilibrium term was included in the asphaltene flow rate and porosity relations, as presented in our previous publication [28]. In this model, two steps are considered for asphaltene adsorption. The first adsorption step is taken as the adsorption of asphaltenes in solution to the surface of the rock, and the second adsorption step is taken as the adsorption of asphaltenes in solution to those asphaltenes already adsorbed to the rock, as follows [31]:

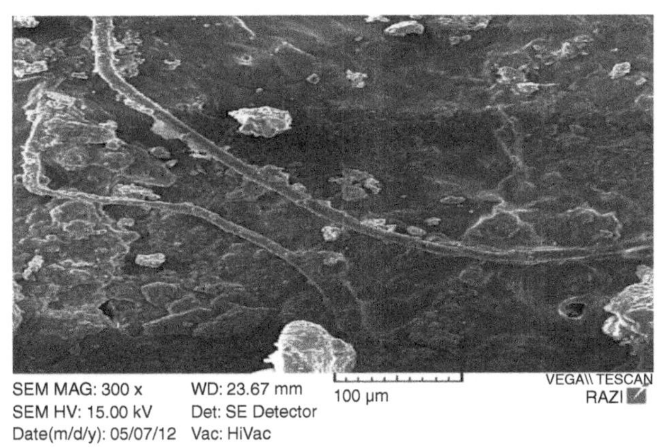

Figure 10

SEM image in studied sandstone core sample.

$$\Gamma = \frac{\Gamma_{\infty} k_1\, C(n^{-1} + k_2 C^{n-1})}{1 + k_1\, C(1 + k_2 C^{n-1})} \tag{5}$$

Also, by comparison, the asphaltene adsorption was modeled using a monolayer adsorption equilibrium term (Langmuir isotherm equation), as follows [8]:

$$W_{sa} = \frac{W_{sa,max} \cdot K_a \cdot C_{sf}}{K_a \cdot C_{sf} + 1} \tag{6}$$

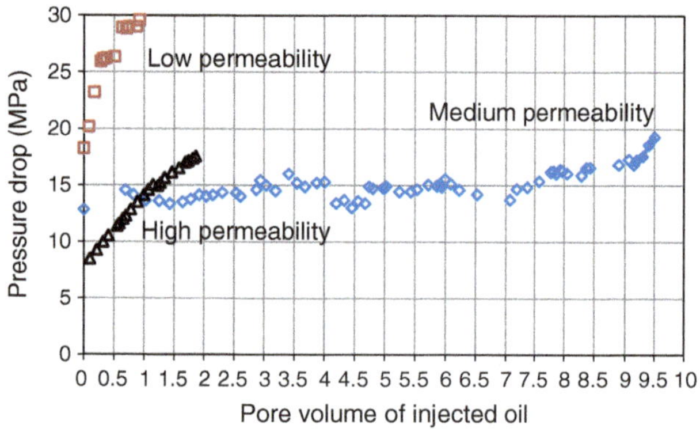

Figure 11

Variation of pressure drop *versus* injected pore volume for studied cores.

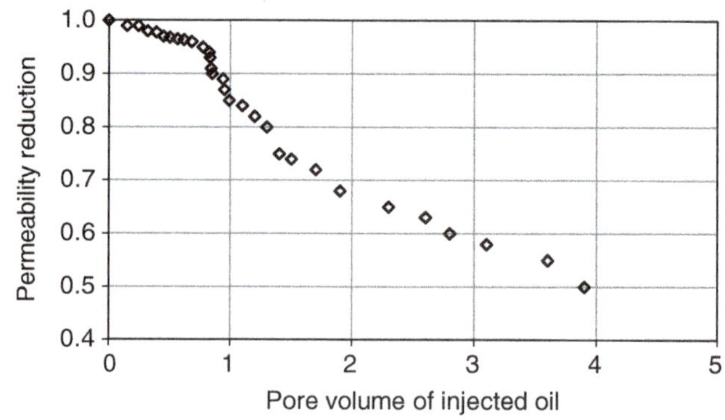

Figure 14

Effect of flow rate on permeability reduction in carbonate core sample (live oil).

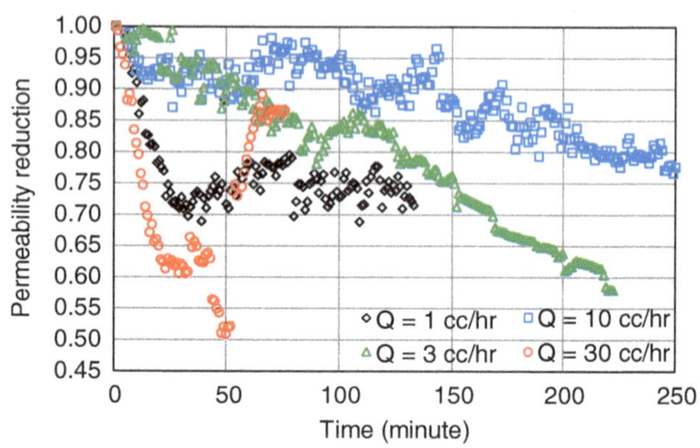

Figure 12

Effect of flow rate on permeability reduction in sandstone core sample (live reservoir fluid).

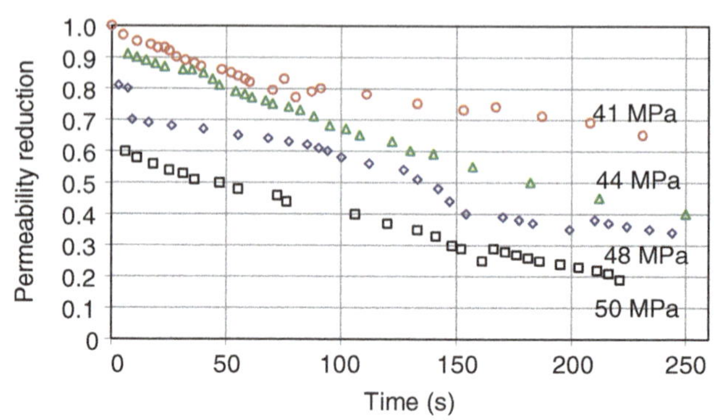

Figure 15

Effect of injection pressure on formation damage.

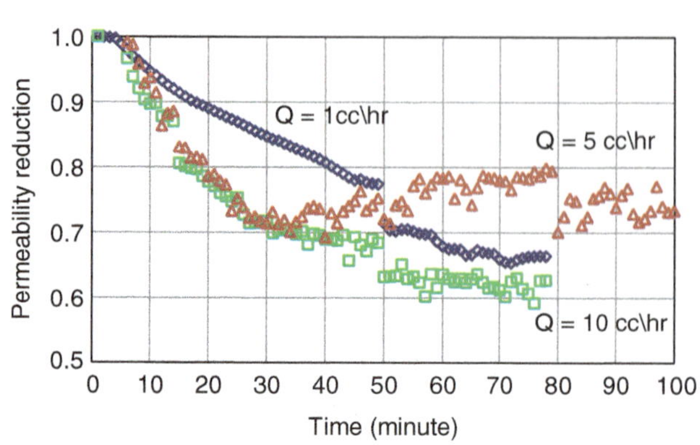

Figure 13

Effect of flow rate on permeability reduction in carbonate core sample (live oil).

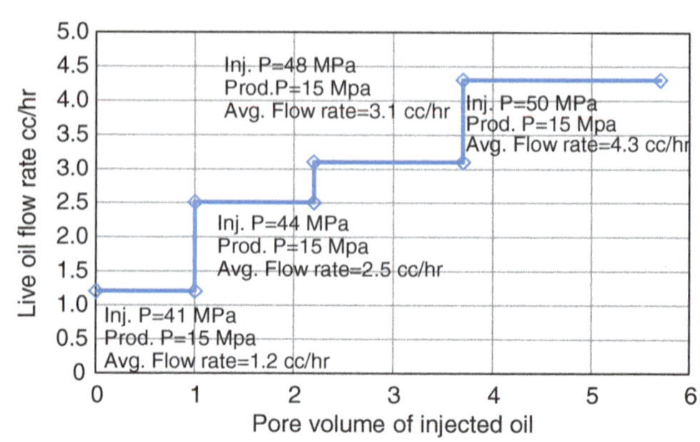

Figure 16

Variation of flow rates of live reservoir fluid recorded *versus* pore volume injected during flooding.

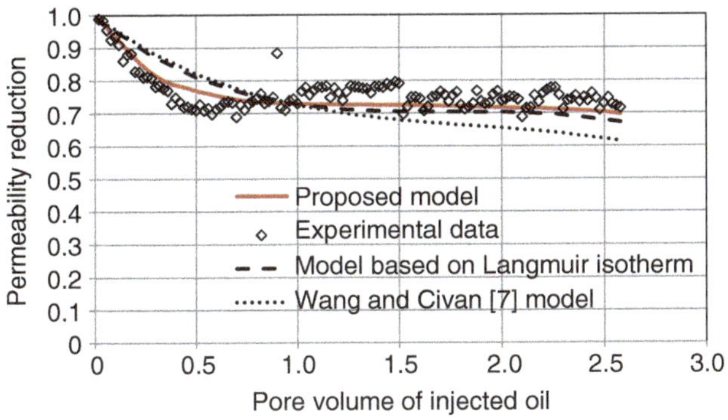

Figure 17

Experimental and simulated permeability reduction during primary deposition ($Q = 1$ cc/h) in sandstone core sample (live reservoir fluid).

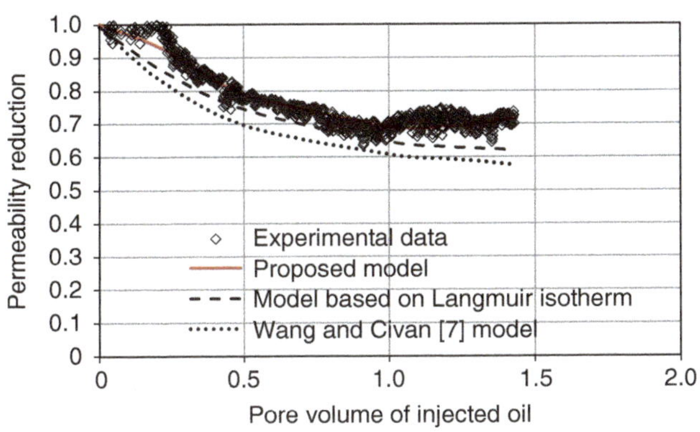

Figure 18

Experimental and simulated permeability reduction during primary deposition ($Q = 1$ cc/h) in carbonate core sample (live reservoir fluid).

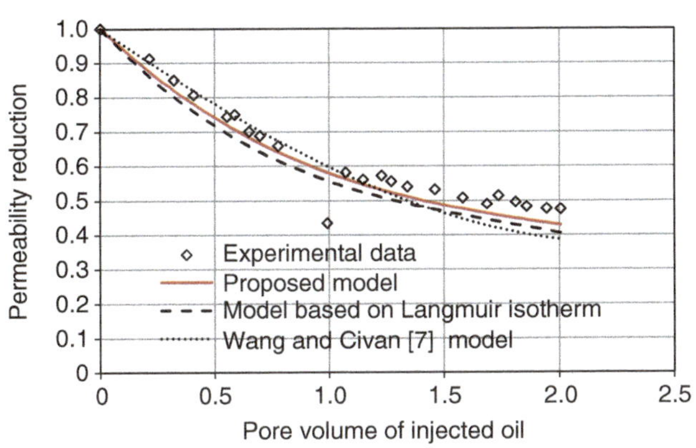

Figure 19

Experimental and simulated permeability reduction during primary deposition ($Q = 3$ cc/h) in dolomite core sample (live reservoir fluid).

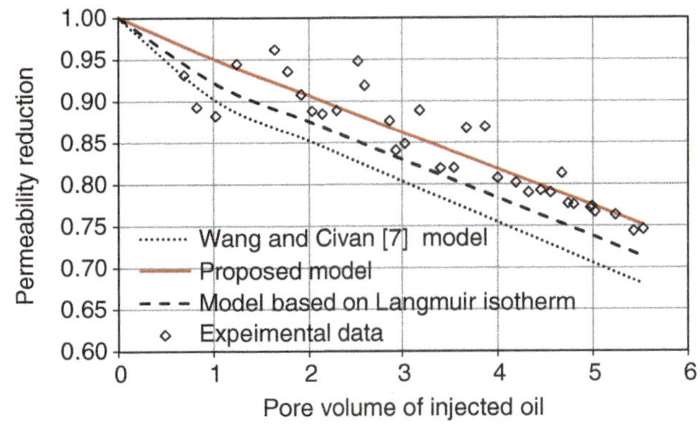

Figure 20

Experimental and simulated permeability reduction during primary deposition ($Q = 10$ cc/h) in sandstone core sample (live reservoir fluid).

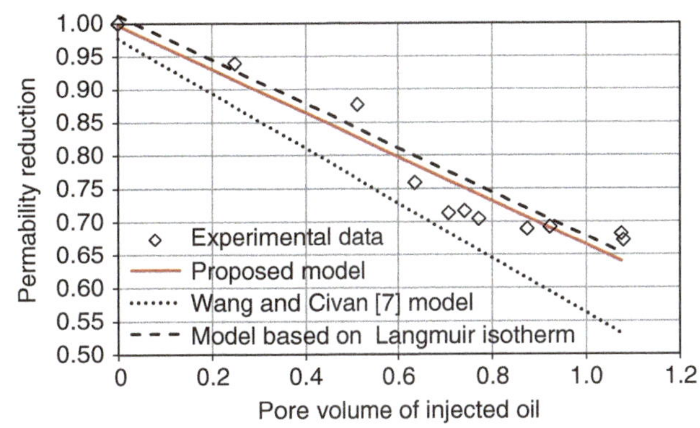

Figure 21

Experimental and simulated permeability reduction during primary deposition ($Q = 10$ cc/h) in carbonate core sample (live reservoir fluid).

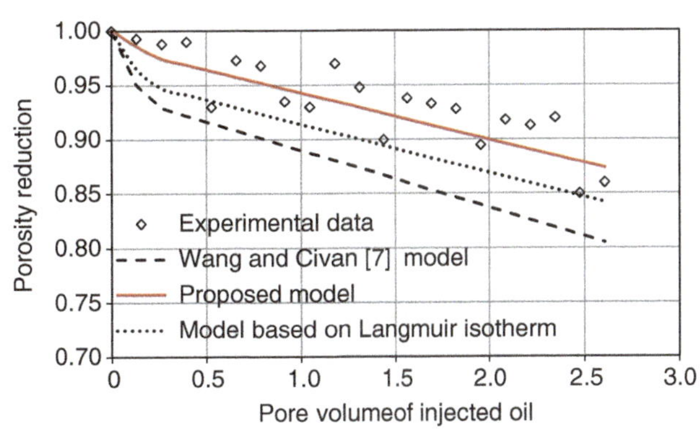

Figure 22

Experimental and simulated porosity reduction during primary deposition ($Q = 1$ cc/h) in sandstone core sample (live reservoir fluid).

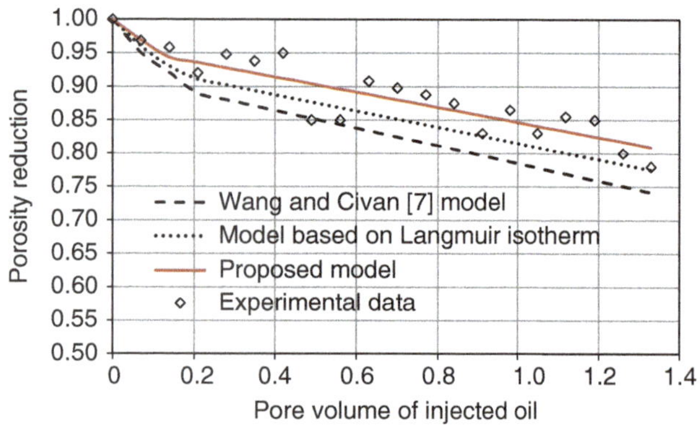

Figure 23

Experimental and simulated porosity reduction during primary deposition (Q = 1 cc/h) in carbonate core sample (live reservoir fluid).

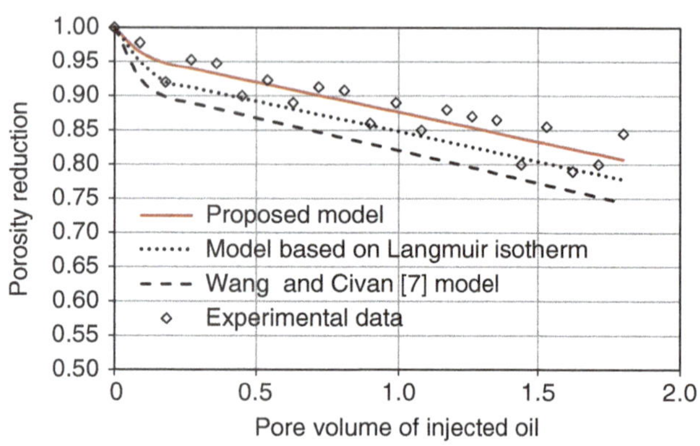

Figure 24

Experimental and simulated porosity reduction during primary deposition (Q = 3 cc/h) in dolomite core sample (live reservoir fluid).

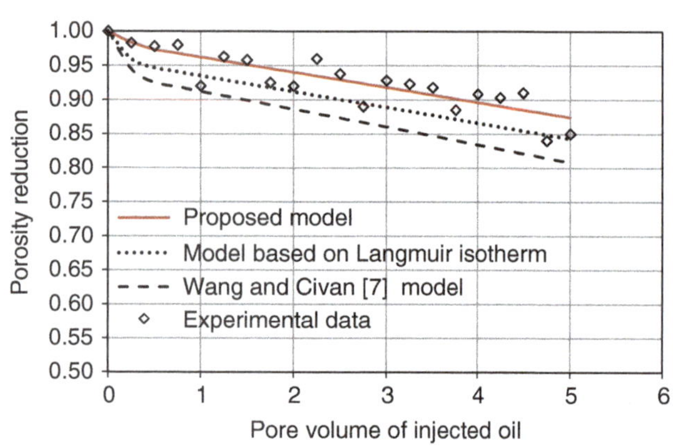

Figure 25

Experimental and simulated porosity reduction during primary deposition (Q = 10 cc/h) in sandstone core sample (live reservoir fluid).

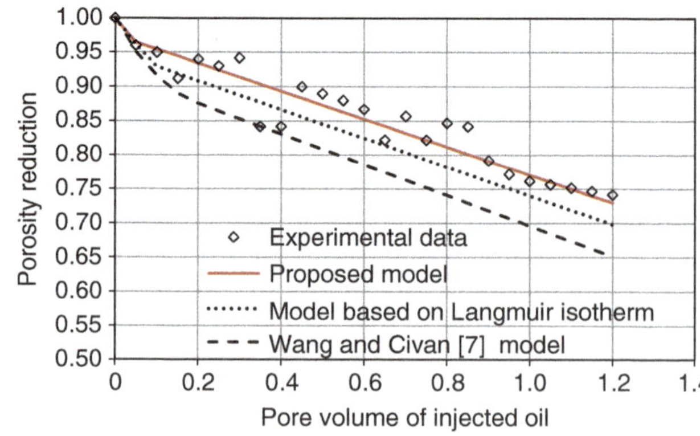

Figure 26

Experimental and simulated porosity reduction during primary deposition (Q = 10 cc/h) in carbonate core sample (live reservoir fluid).

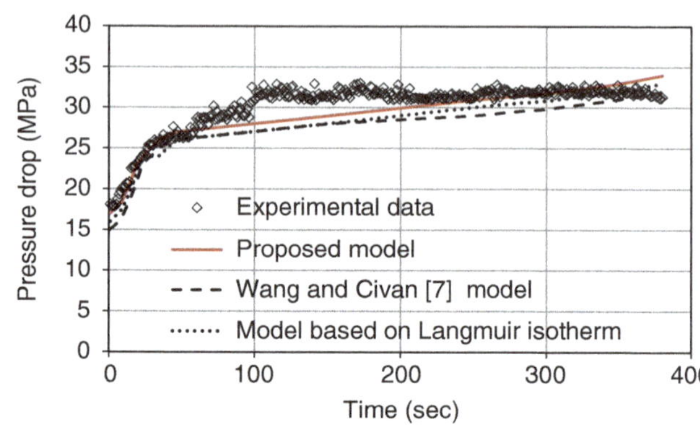

Figure 27

Experimental and simulated pressure drop during primary deposition (Q = 10 cc/h) in carbonate core sample (live reservoir fluid).

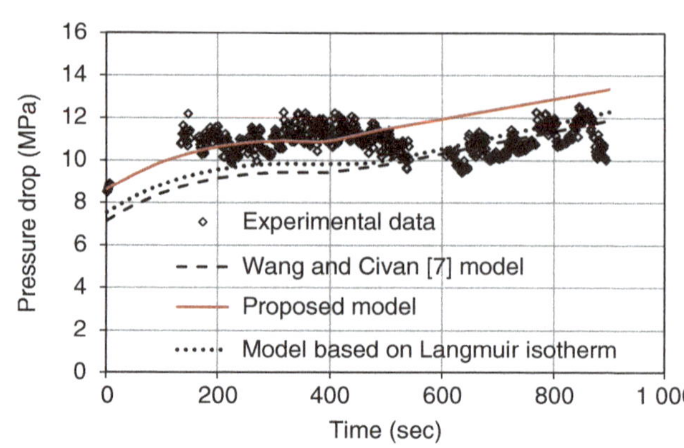

Figure 28

Experimental and simulated pressure drop during primary deposition (Q = 1 cc/h) in sandstone core sample (live reservoir fluid).

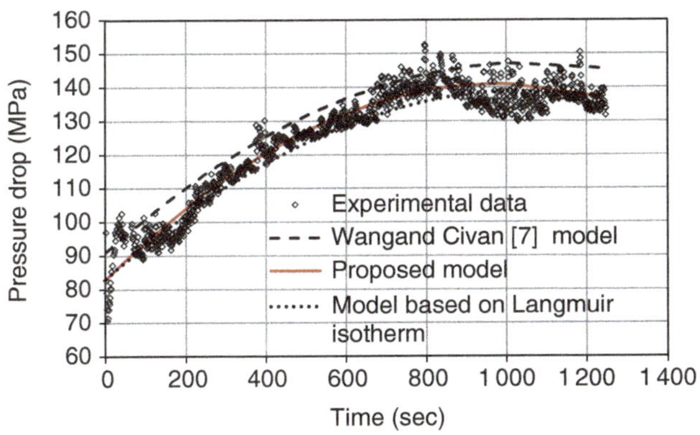

Figure 29

Experimental and simulated pressure drop during primary deposition (Q = 1 cc/h) in carbonate core sample (live reservoir fluid).

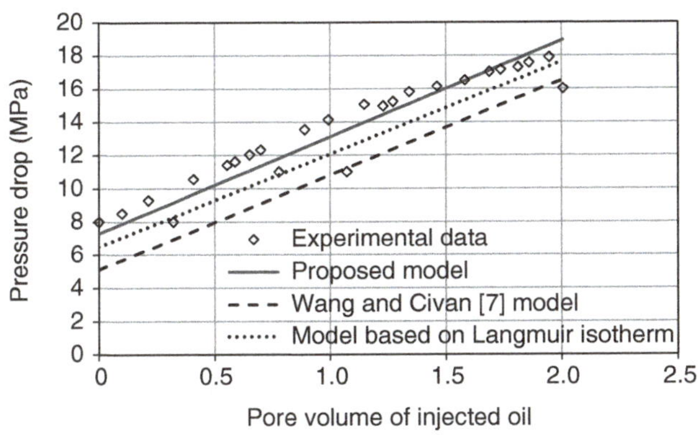

Figure 30

Experimental and simulated pressure drop during primary deposition (Q = 3 cc/h) in dolomite core sample (live reservoir fluid).

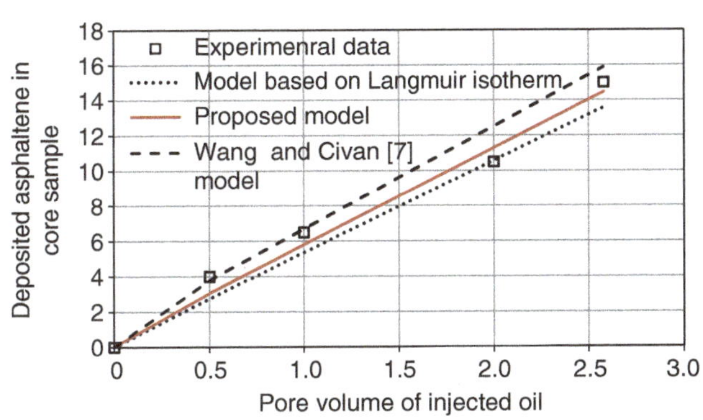

Figure 31

Experimental and simulated deposited asphaltene during primary deposition (Q = 3 cc/h) in dolomite core sample (live reservoir fluid).

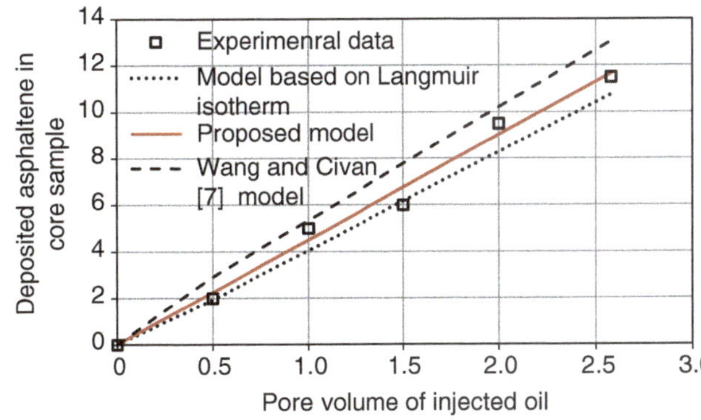

Figure 32

Experimental and simulated deposited asphaltene during primary deposition (Q = 1 cc/h) in sandstone core sample (live reservoir fluid).

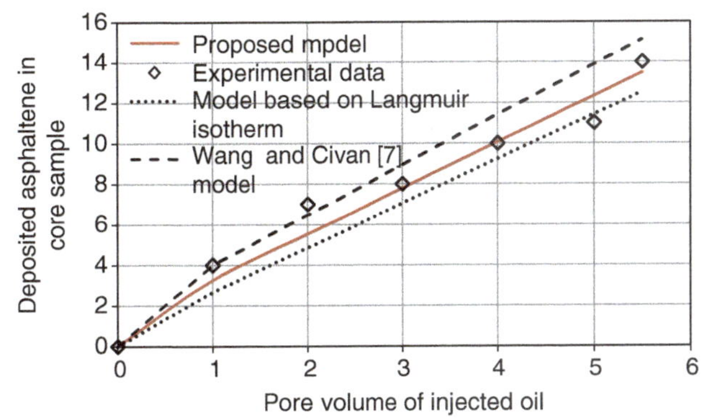

Figure 33

Experimental and simulated deposited asphaltene during primary deposition (Q = 10 cc/h) in sandstone core sample (live reservoir fluid).

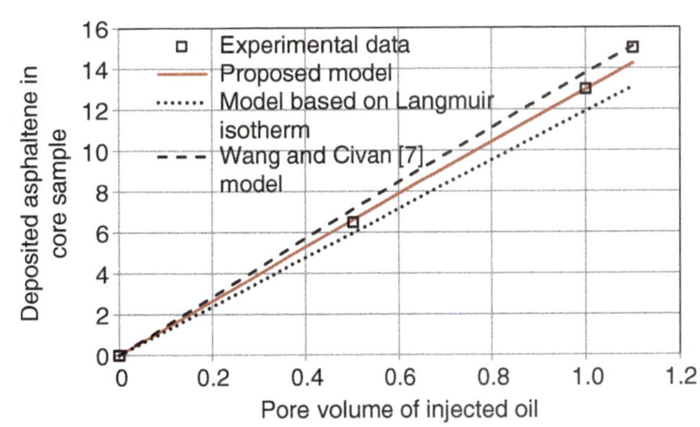

Figure 34

Experimental and simulated deposited asphaltene during primary deposition (Q = 10 cc/h) in carbonate core sample (live reservoir fluid).

The deposition rate for asphaltene is given by:

$$\frac{\partial E_A}{\partial t} = \alpha C_A \O - \beta E_A (v_L - v_{cr,L}) + \gamma u_L C_A \quad (7)$$

where the first term represents the surface deposition, the second term represents the entrainment of asphaltene deposits by the flowing phase when the interstitial velocity is larger than a critical interstitial velocity, and the last term represents the pore throat plugging rate. The value of β is set as [7]:

$$\beta = \beta_i,\ \text{when}\ v_L > v_{cr,L} \qquad \beta = 0,\ \text{otherwise} \quad (8)$$

where v_L is equal to u_L/ϕ The value of γ is described as [7]:

$$\gamma = \gamma_i(1 + \sigma E_A),\ \text{when}\ D_{pt} < D_{ptcr} \qquad \gamma = 0,\ \text{otherwise} \quad (9)$$

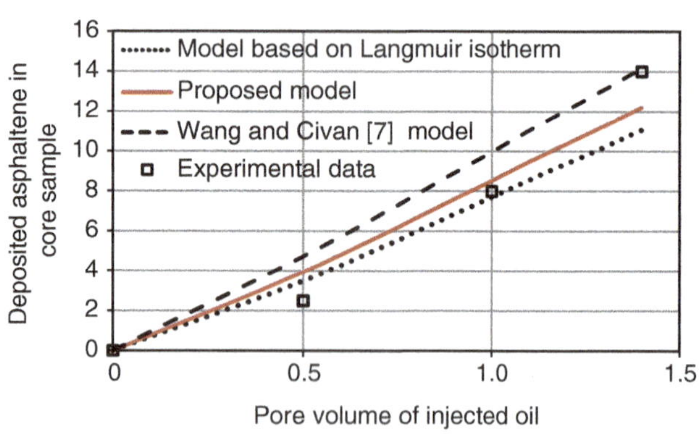

Figure 35

Experimental and simulated deposited asphaltene during primary deposition ($Q = 1$ cc/h) in carbonate core sample (live reservoir fluid).

Thus, the pore throat plugging deposition rate increases proportionally with the total deposits. When D_{pt} is less than D_{ptcr}, pore throat plugging deposition will occur. The details of these models are described else where [7].

2.3 Mass Balance Equations

The asphaltene precipitation and deposition problems are represented by mass balance equations for the liquid and asphaltene, the momentum balance equation, the asphaltene precipitation and deposition models, and the porosity and permeability reduction models [7].

The mass balance for asphaltene is given by [7]:

$$\frac{\partial}{\partial t}(\O\rho_A C_A + \O\rho_L W_{AL}) + \frac{\partial}{\partial x}(\rho_L u_L w_{SAL} + \rho_L u_L w_{AL}) = -\rho_A \frac{\partial}{\partial t} E_A \quad (10)$$

The mass balance equation for the live oil phase is given by [7]:

$$\frac{\partial}{\partial t}(\O\rho_L w_{OL}) + \frac{\partial}{\partial x}(\rho_L u_L w_{OL}) = 0 \quad (11)$$

where O represents the oil component and L represents the liquid phase. The details of these models are described elsewhere [7].

The majority of previous work used two material balance equations based on the oil and asphaltene phases for core samples. The material balance equations are included for the water and gas components during flooding processes, and therefore the material balance equations are proposed as follows:

Gas phase [28]:

$$\frac{\partial}{\partial t}(\O S_V \rho_G + \O S_L \rho_L W_G) + \frac{\partial}{\partial x}(\rho_L u_L w_G + \rho_L u_G) = 0 \quad (12)$$

TABLE 10

Absolute Average Deviation of the correlated permeability reduction from the experimental results by the proposed model and other models

	Absolute Average Deviation of asphaltene deposition modeling (%)								
Oil injection flow rate (cc/h)	Proposed model	Model based on monolayer adsorption equilibrium mechanism	Wang and Civan [7] model	Proposed model	Model based on monolayer adsorption equilibrium mechanism	Wang and Civan [7] model	Proposed model	Model based on monolayer adsorption equilibrium mechanism	Wang and Civan [7] model
	Sandstone core sample (3)			Carbonate core sample			Dolomite core sample		
1	2.8	3.5	4.2	3.1	3.5	5.6			
3	–	–	—	—		—	2.4	3.2	5.1
10	2.9	3.8	4.5	2.6	3.1	4.8	—		–

TABLE 11

Asphaltene deposition adjusted parameters of the proposed model in core samples

Adjusted parameter	Oil injection rate (cc/h)				
	Sandstone core sample		Carbonate core sample		Dolomite core sample
	1	10	1	10	3
α (1/s)	0.02	0.05	0.07	0.09	0.06
β (1/cm)	0.05	0.06	0.42	0.58	0.08
v_c (cm/s)	0.04	0.02	0.03	0.02	0.02
Γ (1/cm)	0.03	0.05	0.24	0.36	0.07
Γ_∞ (mg/g)	159.5	159.5	159.5	159.5	159.5
k_1	1.1×10^{-4}	1.1×10^{-4}	6.1×10^{-3}	9.1×10^{-3}	4.6×10^{-3}
k_2	4.6×10^{-9}	5.9×10^{-9}	8.9×10^{-8}	9.6×10^{-8}	4.3×10^{-8}
n	17	18	11	14	10
f_p	1	1	1	1	1
σ	32	36	41	45	38

Water phase [28]:

$$\frac{\partial}{\partial t}(\O S_W \rho_W w_{WL}) + \frac{\partial}{\partial x}(\rho_W u_W w_{WL}) = 0 \quad (13)$$

The momentum balance equation is given by Darcy's law [7]:

$$u_L = -\frac{k}{\mu_L}\frac{\partial P}{\partial x} \quad (14)$$

The instantaneous local porosity during asphaltene deposition is equal to the difference between the initial porosity and the fractional pore volume occupied by the deposited asphaltene and adsorbed asphaltene [28]:

$$\O = \O_0 - E_A - E_{ad} \quad (15)$$

The instantaneous, local permeability, k, is calculated by [7]:

$$k = f_p k_0 \left(\frac{\O}{\O_0}\right)^m \quad (16)$$

Table 9 summarizes the considerations and parameters for all the studied models.

Experimental permeability damage due to asphaltene deposition in cores was correlated using the above model of asphaltene deposition and their coefficients were adjusted to achieve the best match with the experimental data. The partial differential equations were coupled and solved using a finite-difference method, backward both in time and space, in MATLAB software to determine pressure and asphaltene concentration in the core samples. The Runge-Kutta fourth-order scheme was applied to calculate the volume fraction of asphaltene deposition and adsorption. A fully implicit numerical model was performed and solved by iterations. Numerical simulation runs were conducted to obtain the best match between experimental and numerical results. Numerical simulation was carried out in a linear grid system of 90 grid blocks. Time was discretized with a Δt of 30 second. For optimization and determination of the model parameters, history matching was used. In this study, square roots of the summation of the differences between measured and calculated porosity data are defined as the objective function [28]:

$$\text{Objective Function} = \sqrt{\sum_{i=1}^{n}\left(\left(\frac{k}{k_0}\right)_{measured} - \left(\frac{k}{k_0}\right)_{calculated}\right)^2} \quad (17)$$

As a result, the model parameters obtained by the optimization procedure are Γ_∞, k_1, k_2, m, n, α, β, v_c, γ, K_a, σ and f_p.

The simulated data are not appreciably sensitive to time and space steps.

TABLE 12
Asphaltene deposition adjusted parameters of the model based on the Langmuir isotherm in core samples

Adjusted parameter	Oil injection rate (cc/h)				
	Sandstone core sample		Carbonate core sample		Dolomite core sample
	1	10	1	10	3
α (1/s)	0.02	0.04	0.06	0.08	0.06
β (1/cm)	0.04	0.05	0.38	0.62	0.09
v_c (cm/s)	0.04	0.02	0.03	0.02	0.02
γ (1/cm)	0.02	0.04	0.31	0.56	0.07
$w_{sa,max}$ (mg/g)	0.2	0.2	0.2	0.2	0.2
K_a	2 500	2 500	2 500	2 500	2 500
f_p	1	1	1	1	1
σ	31	34	43	48	37

TABLE 13
Asphaltene deposition adjusted parameters of Wang and Civan's [7] model in core samples

Adjusted parameter	Oil injection rate (cc/h)				
	Sandstone core sample		Carbonate core sample		Dolomite core sample
	1	10	1	10	3
α (1/s)	0.01	0.01	0.02	0.01	0.02
β (1/cm)	0.06	0.07	0.32	0.48	0.09
v_c (cm/s)	0.04	0.02	0.03	0.02	0.02
γ (1/cm)	0.018	0.03	0.54	0.75	0.05
f_p	1	1	1	1	1
σ	32	38	43	48	41

3 RESULTS AND DISCUSSION

3.1 Effect of Crude Oil Type on Asphaltene Deposition

Figures 6 and 7 show the permeability reduction of two sandstone and carbonate core samples *versus* the pore volume of injected live reservoir fluid and dead residual oil samples.

As observed, higher permeability reduction is caused when injecting the live reservoir fluid samples in comparison with the results obtained by injecting the dead residual oil samples for the studied sandstone and carbonate cores. This observation can be justified by the fact that the C_{12}^{+} fraction in live reservoir fluid samples is heavier than the dead residual oil and contains more asphaltene content. Also, this difference may be from the state of asphaltene in live crude oil that is quite different from that in dead residual oil. It is clear that live and dead residual oil asphaltene show clear compositional differences, including [26]:

- the dead residual oil asphaltene sample contains a lower amount of the nitrogen-containing species but is higher in rings plus double bonds in comparison with the live reservoir asphaltene sample;
- the live reservoir asphaltenes are rich in sulfur and oxygen compared with the dead residual oil asphaltene;
- the live reservoir asphaltene has less aromatic compounds compared with the dead residual oil asphaltene;
- the live oil asphaltene contains more functional groups and less saturated groups in comparison with the dead oil asphaltene.

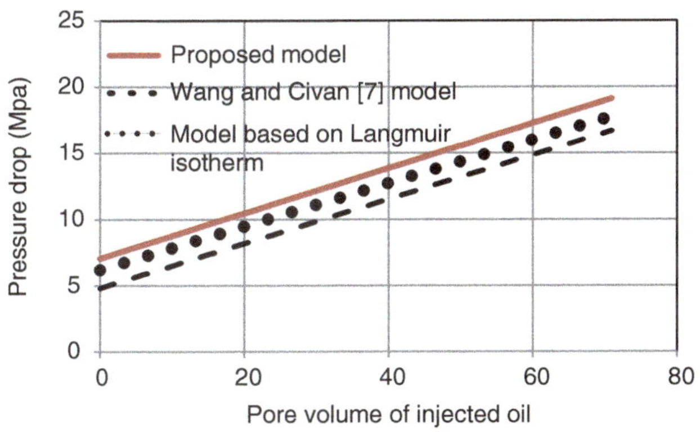

Figure 36

Pressure drop *versus* pore volume oil injected for oil GF3.

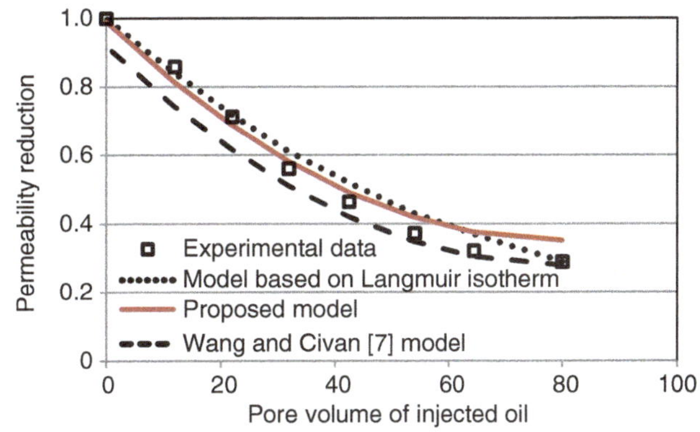

Figure 39

Permeability reduction *versus* pore volume oil injected for oil GF3.

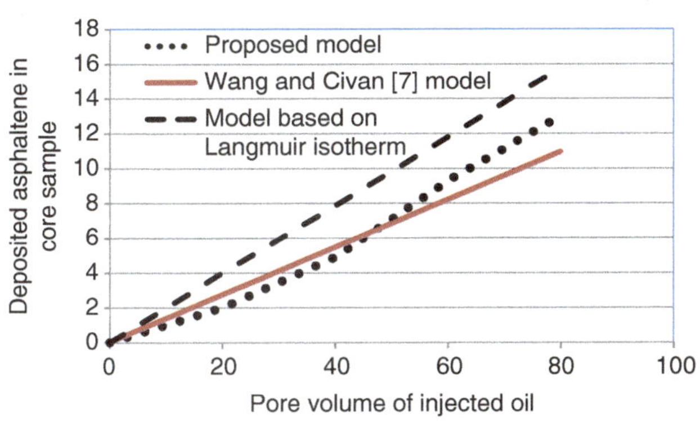

Figure 37

Deposited asphaltene in core sample *versus* pore volume of injected oil.

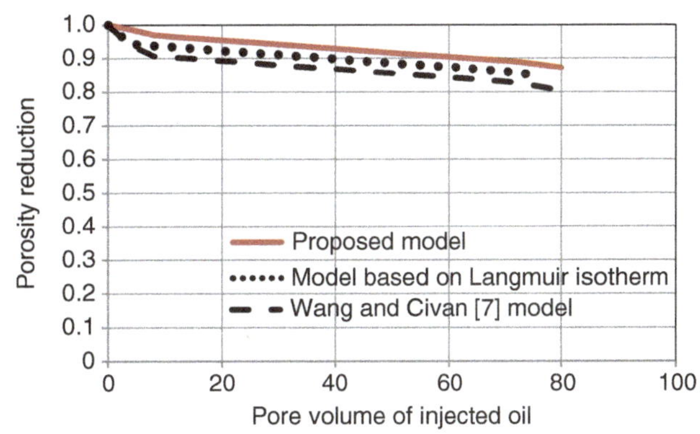

Figure 40

Porosity reduction *versus* pore volume oil injected for oil GF3.

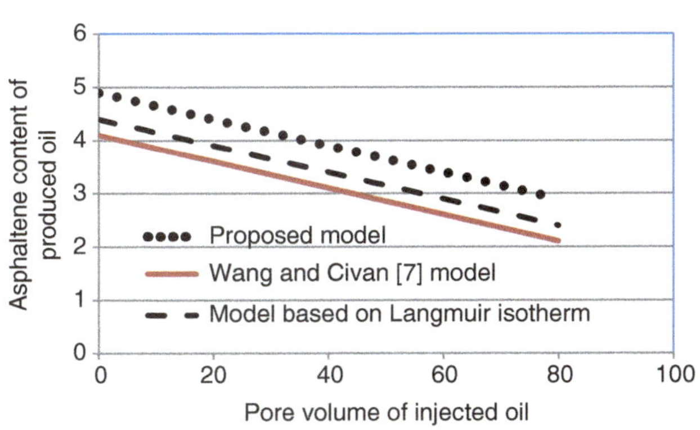

Figure 38

Asphaltene content of produced oil *versus* pore volume of injected oil.

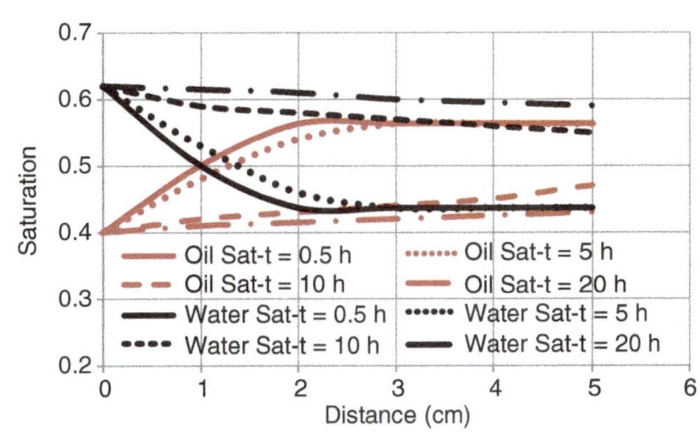

Figure 41

Saturation profiles at different time steps in carbonate core sample (Q = 1 cc/h).

Based on these results, it was concluded that the propensity for the asphaltenes of live reservoir fluid to aggregate into larger particles and to deposit on the rock was higher than for asphaltenes of dead residual oil. Therefore, permeability reduction and formation damage by asphaltene deposition during live reservoir fluid flooding were higher than for the dead residual oil flooding.

3.2 Effect of Rock Type on Asphaltene Deposition

Figure 8 shows the variation of the permeability for different types of core samples at various pore volumes of the injected live reservoir fluid.

As can be seen from this figure, the permeability reduction in carbonate rock types is steeper than that observed in sandstone rock cores over the whole range of pore volume numbers. Also, it can be observed that the carbonate core samples can be plugged at lower pore volume numbers due to the higher rate of asphaltene deposition and permeability reduction. Thus, it can be found that the plugging deposition mechanism is very significant in asphaltene deposition in carbonate core samples.

The SEM images for both carbonate and sandstone rock types are given in Figures 9 and 10. As shown in these figures, a comparison of SEM images revealed that the carbonate core samples would develop a layer on their surface. Thus, the carbonate core samples experienced more particle plugging compared with sandstone core samples, and the carbonate core samples were severely plugged by aspaltene deposition. It can be seen that asphaltene deposition has different mechanisms in carbonate and sandstone core samples. Therefore, it can be concluded that the adsorption mechanism in asphaltene deposition in carbonate core samples is more significant than that in sandstone core samples.

The difference between the carbonate and the sandstone cores is related to the core structure, mineral composition and consolidation. Hence, the carbonate core's inner surface may contain polar groups, which can exert polar interactions with asphaltene surface groups. Therefore, it is likely that asphaltene deposition on the carbonate core's surface can involve π-interaction and hydrogen bonding between the asphaltene groups and core surface polar groups. Asphaltene is a polar component: therefore, formations containing clay and carbonate have greater ability than sandstone formations to adsorb asphaltene [23]. In fact, the existence of polar components makes the asphaltene molecules adhere more strongly to the surface, and hence increases the amount of asphaltene deposition on the surface of the carbonate rock samples.

Figure 11 indicates the pressure drop *versus* pore volume for three rocks with different permeability.

It can be concluded that the formation damage caused by asphaltene deposition in low-permeability rock was more severe than that in higher-permeability rock for the same asphaltene content of the crude oil and under the same dynamic potential force. For low-permeability core samples, there may be a continuous deposition until complete plugging. In low-permeability cores, the plugging mechanism acts like a snowball growth that can be inferred from the sharp slopes of Figure 11 once plugging deposition is activated, leading to the first significant pressure variation. The results indicated that in low-permeability cores, the region of the asphaltene deposition was spread. There may also be a slow steady plugging for high-permeability cores. Furthermore, the deposition of asphaltene in the core was lower. In the high-permeability core, the region of the asphaltene deposition was limited in the inlet end of the core. Finally, a steady deposition can occur until interstitial velocity is high enough to not allow further mechanical plugging for medium permeability cores.

3.3 Effect of Oil Injection Flow Rate on Asphaltene Deposition

Figures 12-14 show the plot of the permeability reduction as a function of time for the cases of 1 to 30 cc/h injection rates for sandstone cores and 1 to 10 cc/h injection rates for carbonate cores.

One of the most important results from this experiment is the permeability reduction of the core samples at the beginning of the injection, which shows that the plugging of the pore throats occurs immediately after asphaltene deposition. The results also show that the asphaltene deposition on the covered pore surface with asphaltene layers is less than that on the clean surface.

These plots clearly show that an increase in flow rate is followed by an increase in asphaltene deposition at low flow rates. At these flow rates, the permeability reduction took place in a steady manner. The reduction continued until a steady state was reached. At this point, the relatively low deposition rate leads to the development of a steady state in pores. Also, these figures show that reduction in permeability until a minimum value was observed, and then a reversal in permeability reduction took place at higher flow rates. At this point, the permeability increased and reached a steady state. For this case, flow begins with a speed higher than the critical speed but it declines steadily below the critical velocity and the asphaltene deposits. Because of asphaltene deposition, the pores are blocked and the speed increases beyond the critical speed and the entrainment phenomenon begins, resulting in an increasing permeability.

Figure 14 shows that at a higher flow rate, 40 cc/h, the asphaltene plugging process appeared to begin late.

However, as the flow continued, permeability reduction took place very quickly. For this case, a steady-state condition was never reached, and the injection pressure continued to rise until the injection had to be stopped. For this case, local speed values remain above the critical speed. Consequently, permeability reduction remains steady. This also means that most asphaltenes trapped would be in the vicinity of the wellbore, an area that can be treated with production stimulation operations [6].

3.4 Effect of Production Pressure on Permeability Reduction

Figure 15 shows the effect of production pressure on permeability reduction as a function of time.

It can be seen from this figure that the highest permeability reduction occurs at the highest production pressure at 50 MPa. The damage of the core sample was found to increase when the production pressures were increased. At higher production pressures, the asphaltene deposition in the core sample increased due to the increase in the driving force acting on the asphaltene particles. Greater asphaltene deposition caused severe blockage of pore throats, and thus greater reduction in permeability of the core sample induced.

Also, at high pressures, the asphaltenes are soluble in oil. However, during pressure depletion, the oil expands, reducing the oil solubility parameter, and it becomes a poor solvent for asphaltene. At a low pressure, asphaltenes begin to precipitate. Pressure depletion is the main reason for asphaltene deposition due to the destabilization of asphaltenes. This phenomenon occurred due to the decrease in the crude oil density, the vaporization of the light component and the decrease in the solubility of asphaltenes in the oil [32].

The flow rates of live reservoir fluid recorded for each injection pressure during the live reservoir fluid flooding experiment are shown in Figure 16.

It is seen that the recorded flow rates were in continuous response to the imposed pressure drops. In other words, a decrease in the system pressure was followed almost immediately or instantaneously, by a resulting increase in the system flow rate. This behavior indicates that an increase in injection pressure is followed by a constant production pressure.

3.5 Modeling Approach

In the majority of previous models, the asphaltene deposition is based on the mechanical plugging mechanism [7] and also, only two material balance equations were applied. Therefore, the previous models in the literature do not consider the adsorption mechanism, which is related to interactions between the positive polar component of asphaltenes and the rock surface.

Therefore, in the proposed asphaltene deposition model [28], in addition to the mechanical plugging mechanism term, a multilayer adsorption equilibrium mechanism term for asphaltene was included. Also, application of four material balance equations (oil, asphaltene, gas and water phases) instead of two material balance equations was considered. In this section, the results of asphaltene deposition in core samples during natural depletion were predicted using our proposed model based on the multilayer adsorption equilibrium mechanism and compared with those obtained using asphaltene deposition modeling based on the monolayer adsorption equilibrium mechanism [8] and obtained by Wang and Civan's [7] model based only on the mechanical plugging mechanism.

The proposed model was correlated using the natural depletion experimental data obtained in this work which are denoted as 1 to 12 in Table 8.

Figures 17 to 35 show the results of permeability reduction, porosity reduction, pressure drop and deposited asphaltene in the sandstone, carbonate and dolomite core samples due to asphaltene deposition using the proposed model and compared with those obtained using asphaltene deposition modeling based on the monolayer adsorption equilibrium mechanism [8] and obtained by Wang and Civan's [7] model based only on the mechanical plugging mechanism.

Figures 17 to 35 show that the proposed model can correlate the experimental data more accurately in comparison with the Wang and Civan [7] model and the model based on the monolayer adsorption equilibrium mechanism. These figures show that reduction in permeability until a minimum value was observed and then permeability increased in the experimental data. Because of asphaltene deposition, the pores are blocked and the speed increase beyond the critical speed and the entrainment phenomenon begins, resulting in increasing permeability.

Figures 22 to 26 compare the results of porosity reduction due to asphaltene deposition in core samples using the proposed model with those obtained from the Wang and Civan [7] model and the model based on the monolayer adsorption equilibrium mechanism. It should be noted that the reduction of the porosity ratio is much lower than that of the permeability ratio.

Figures 27 to 30 compare the results of pressure drop in core samples during oil flooding due to asphaltene deposition using the proposed model with those obtained from the Wang and Civan [7] model and the model based on the monolayer adsorption equilibrium mechanism.

Figures 31 to 35 compare the results of deposited asphaltene in core samples during oil flooding using the proposed model with those obtained from the Wang and Civan [7] model and the model based on the monolayer adsorption equilibrium mechanism.

Table 10 shows the Average Absolute Deviation (AAD) of the predicted permeability reduction from the experimental data due to asphaltene deposition using the proposed model, that of the Wang and Civan [7] model, and that of the model based on the monolayer adsorption equilibrium mechanism. These results also confirm that the proposed model based on the multilayer adsorption equilibrium mechanism of asphaltene can correlate the permeability reduction experimental data with AAD of 2.4-3.1%, whereas the Wang and Civan [7] model correlates asphaltene deposition less accurately with AAD of 4.2-5.6%. This suggests that the multilayer adsorption mechanism of asphaltene has an important role in asphaltene deposition on core samples.

The results show that modeling based on the monolayer adsorption equilibrium mechanism is less accurate in the modeling of asphaltene deposition during natural depletion with AAD of 3.1-3.8%. Therefore, the asphaltene adsorption behavior on core samples is far from the monolayer adsorption equilibrium mechanism and is closer to the modified model based on multilayer adsorption equilibrium behavior.

This phenomenon can be explained by the fact that the asphaltene adsorption mechanism, which is related to the interactions between the asphaltene functional groups and the core surface, involves surface polarity, affinity or other attractive forces. It is also known that asphaltene surface groups may have acidic (carboxylic, benzoic, phenolic), and/or basic (pyridine, pyrazine, dimethylsulfoxide) functionality [23].

Tables 11 to 13 show the adjusted parameter values of the proposed model based on the multilayer adsorption equilibrium mechanism of asphaltene, the model based on the Langmuir isotherm and Wang and Civan's [7] model during natural depletion in core samples. It should be noted that the adjusted parameters are affected by the nature of core samples (rock morphology, rock-surface characteristics, mineral composition, etc.). For example, the adjusted parameters in the proposed model such as the first adsorption step parameter (k_1) and the second adsorption step parameter (k_2) have different quantities in carbonate and sandstone core samples, which is related to different mineral composition and rock morphology. Also, the results show that the second adsorption step parameter (k_2) is much lower than the first adsorption step parameter (k_1). It was found that the tendency for asphaltene deposition on the pore surface that was covered with asphaltene is less than that for the clean surface of core samples.

3.6 Case Study

In order to verify the accuracy of the proposed asphaltene deposition model, the model was run using the data given by Minssieux *et al.* [10], as denoted GF3. Figures 36 to 40 compare the results of the pressure drop, deposited asphaltene in the porous rock, remaining asphaltene concentration in the flowing oil, and permeability and porosity profiles at various times using the proposed model with those obtained from the Wang and Civan [7] model and the model based on the monolayer adsorption equilibrium mechanism. It should be noted that the deposition model constants given in Tables 11 to 13 are used in the above-mentioned models without using the GF3 experimental data.

The water and live oil saturation profiles were studied and the results are shown in Figure 41. The water saturation increases through the core sample. The water saturation increases from an initial value of 0.44 toward the value of 0.62. The oil saturation decreases through the core sample. The oil saturation reduces from an initial value of 0.56 toward the value of 0.4.

CONCLUSION

In this work, a set of experiments was conducted using bottom hole live oil samples under dynamic conditions in porous media with the purpose of investigation of asphaltene deposition and adsorption mechanisms involved in the interfacial interaction of the asphaltene/mineral rock. The effective parameters of porous media on asphaltene deposition, such as morphology of the surface, were studied in this work using SEM micrographs. Also, the experimental data obtained in this work were correlated by an asphaltene deposition model based on the multilayer adsorption equilibrium mechanism and four material balance equations (oil, asphaltene, gas and water phases) and the correlation results were compared with other models based on the mechanical plugging mechanism and the monolayer adsorption equilibrium mechanism to account for permeability reduction in porous media in dynamic conditions. While none of the studied models cover the experimental results with satisfaction, the proposed model correlates the experimental formation damage data due to asphaltene deposition better in comparison with the other studied models.

The experimental and modeling results of asphaltene deposition and adsorption during core flooding led to the following conclusions:

- the proposed model, based on the multilayer adsorption equilibrium mechanism of asphaltene and four material balance equations, is found to be more accurate than Wang and Civan's [7] model based on the mechanical plugging mechanism with AAD of less than 3.2. Hence, the multilayer adsorption equilibrium mechanism of asphaltene has an important role in asphaltene deposition, permeability reduction and wettability alteration of reservoir rock;
- the experimental results showed that an increase in flow rate is followed by an increase in asphaltene deposition; permeability reduction and porosity reduction took place in a monotonous manner until a steady-state condition was reached at low flow rates. At higher flow rates, initial permeability was reduced until a minimum value and then an increase in permeability was observed until a steady-state condition was reached;
- the effect of rock type on asphaltene deposition was also studied. It can be seen that the asphaltene deposition has different mechanisms in carbonate and sandstone core samples. SEM micrographs of asphaltene deposition on the surface of carbonate core samples show the formation of large clusters and propose multilayer adsorption of asphaltene;
- the experimental results showed that the permeability reduction rate using the live reservoir fluid samples obtained from wells can be different from that of the same dead oil samples in the same conditions due to higher asphaltene content;
- as inferred, the experimental results indicated that three depositional types, continuous deposition for low-permeability cores, slow, steady plugging for high-permeability cores and steady deposition for medium-permeability cores, can be observed. The results showed that the values of pore volumes of the injected crude oil are higher for sandstone cores compared with carbonate cores;
- also, the experimental results showed that the damage of the core sample was found to increase when the injection pressure was increased;
- the proposed model was verified using experimental data reported in the literature.

ACKNOWLEDGMENTS

This research is sponsored by the Research Institute of Petroleum Industry (RIPI). This is gratefully acknowledged.

REFERENCES

1 Mullins O.C., Sheu E.Y., Hammami A., Marshal A.G. (2007) Asphaltene Self-Association and Precipitation in Solvents-AC Conductivity Measurements, in *Asphaltenes, Heavy Oils and Petroleomics*, Springer, New York.

2 Civan F., Knapp R.M., Ohen H.A. (1989) Alteration of Permeability by Fine Particles Processes, *J. Petrol. Sci. Eng.* **3**, 65-79.

3 Rassamdana H., Dabir B., Nematy M., Farhani M., Sahimi M. (1996) Asphalt Flocculation and Deposition: I, *The Onset of Precipitation, AICHE J.* **42**, 1, 10-21.

4 Andersen S.I., Speight J.G. (1999) Thermodynamic models for asphaltene solubility and precipitation, *J. Petrol. Sci. Eng.* **22**, 53-66.

5 Hu Y.F., Li S., Liu N., Chu Y.P., Park S.J., Mansoori G. A., Guo T.M. (2004) Measurement and corresponding states modeling of asphaltene precipitation in Jilin reservoir oils, *J. Petrol. Sci. Eng.* **41**, 169-182.

6 Ali M.A., Islam M.R. (1997) The effect of asphaltene precipitation on carbonate rock permeability: An experimental and numerical approach, *SPE Annual Technical Conference and Exhibition*, San Antonio, 5-6 Oct.

7 Wang S., Civan F. (2001) Productivity decline of vertical and horizontal wells by asphaltene deposition in petroleum reservoirs, *SPE International Symposium on Oilfield Chemistry*, Houston, Texas, 13-16 Feb., *SPE paper* 64991.

8 Almehaideb R.A. (2004) Asphaltene precipitation and deposition in the near wellbore region: a modeling approach, *J. Petrol. Sci. Eng.* **42**, 157-170.

9 Minssieux L. (1997) Core Damage from Crude Asphaltene Deposition, *SPE International Symposium on Oilfield Chemistry*, Houston, Texas, 18-21 Feb., *SPE* 37250.

10 Minssieux L., Nabzar L., Chauveteau G., Longeron D., Bensalem R. (1998) Permeability damage due to asphaltene deposition experimental and modeling aspects, *Oil Gas Sci. Technol. – Rev. IFP* **53**, 3, 313-327.

11 Papadimitriou N.I., Romanos G.E., Charalambopoulou G.C., Kainourgiakis M.E., Katsaros F.K., Stubos A.K. (2007) Experimental investigation of asphaltene deposition mechanism during oil flow in core samples, *J. Petrol. Sci. Eng.* **57**, 281-293.

12 Mendoza de la Cruz J.L., Castellanos-Ramirez I.V., Ortiz-Tapia A., Buenrostro-Gonzalez E., Duran-Valencia C., de los A., Lopez-Ramirez S. (2009) Study of monolayer to multilayer adsorption of asphaltenes on reservoir rock minerals, *Colloids Surf. A: Physicochem. Eng. Aspects* **340**, 149-154.

13 Buckley J.S. (1998) Wetting Alternation of Solid Surfaces by Crude Oils and their Asphaltenes, *Oil Gas Sci. Technol. – Rev. IFP* **53**, 3, 303-312.

14 Leontaritis K.J. (1989) Asphaltene deposition: A comprehensive description of problem manifestations and modeling approaches, *SPE Production Operations Symposium*, Oklahoma, City, 13-14 March.

15 Srivastava R.K., Huang S.S., Mingzhe D. (1999) Asphaltene Deposition during CO_2 Flooding, *SPE Prod. Facil.* **14**, 4, 235-245.

16 Kocabas I., Islam M.R., Modarress H. (2000) A wellbore model for field-scale modeling of asphaltene plugging, *J. Petrol. Sci. Eng.* **26**, 19-30.

17 Mendoza de la Cruz J.L., Argüelles-Vivas F.J., Matias-Perez V., Duran-Valencia C., de los A., Lopez-Ramirez S. (2009) Asphaltene-Induced Precipitation and Deposition During Pressure Depletion on a Porous Medium: An Experimental Investigation and Modeling Approach, *Energy Fuels* **23**, 5611-5625.

18 Civan F. (1995) Modeling and Simulation of Formation Damage By Organic Deposition, *First International Symposium in Colloid Chemistry in Oil Production: Asphaltenes and Wax Deposition, JSCOP'95*, Rio de Janeiro, Ri, Brazil, 26-29 Nov.

19 Jafari Behbahani T., Ghotbi C., Taghikhani V., Shahrabadi A. (2011) Investigation of Asphaltene Deposition Mechanisms During Primary Depletion and CO_2 Injection, *European Formation Damage Conference*, The Netherlands, 7-10 June.

20 Gruesbeck C., Collins R.E. (1982) Entrainment and deposition of fine particles in porous Media, *Soc. Pet. Eng. J.* **22**, 847-856.

21 Monteagudo J.E.P., Rajagopal K., Lage P.L.C. (2002) Simulating oil flow in porous media under asphaltene deposition, *Chem. Eng. Sci.* **57**, 323-337.

22 Nasri Z., Dabir B. (2010) Modeling of carbonate oil reservoirs in imbibition displacement with considering asphaltene deposition, *J. Petrol. Sci. Eng.* **70**, 273-281.

23 Hamadou R., Khodja M., Kartout M., Jada A. (2008) Permeability reduction by asphaltenes and resins deposition in porous media, *Fuel* **87**, 2178-2185.

24 Leontaritis K., Mansoori G. (1988) Asphaltene Deposition: A Survey of Field Experiences and Research Approaches, *J. Petrol. Sci. Eng.* **1**, 229-239.

25 Bagheri M.B., Kharrat R., Ghotbi C. (2011) Experimental Investigation of the Asphaltene Deposition Process during Different Production Schemes, *Oil Gas Sci. Technol. – Rev. IFP* **66**, 3, 507-519.

26 Klein G.C., Kim S., Rodgers R.P., Marshall A.G. (2006) Mass Spectral Analysis of Asphaltenes. I. Compositional Differences between Pressure-Drop and Solvent-Drop Asphaltenes Determined by Electrospray Ionization Fourier Transform Ion Cyclotron Resonance Mass Spectrometry, *Energy Fuels* **20**, 1965-1972.

27 Carbognani L., Izquierdo A. (1989) Preparative and automated compound class separation of venezuelan vacuum residua by high performance liquid chromatography, *J. chromatography* **484**, 399-405.

28 Jafari Behbahani T., Ghotbi C., Taghikhani V., Shahrabadi A. (2012) Investigation on Asphaltene Deposition mechanisms during CO_2 Flooding processes in porous media: A novel Experimental study and a modified model based on multilayer theory for asphaltene adsorption, *Energy Fuels* **26**, 5080-5091.

29 Nghiem L., Coombe D.A. (1997) Modeling Asphaltene Precipitation during Primary Depletion, *SPE J.* **2**, 170-176.

30 Jafari Behbahani T., Ghotbi C., Taghikhani V., Shahrabadi A. (2011) Experimental investigation and thermodynamic modeling of asphaltene precipitation, *Scientia Iranica C.* **18**, 6, 1384-1390.

31 Zhu B.-Y., Gu T. (1990) Reverse hemimicelle formation of 1-decanol from heptane at the solution/graphite interface, *Colloids Surf.* **46**, 339-345.

32 Vargas F.M., Gonzalez D.L., Hirasaki G.J., Chapman W. G. (2009) Modeling Asphaltene Phase Behavior in Crude Oil Systems Using the Perturbed Chain Form of the Statistical Associating Fluid Theory (PC-SAFT) Equation of State, *Energy Fuels* **23**, 1140-1146.

2

Ligand Control of the Metal Coordination Sphere: Structures, Reactivity and Catalysis

Andreas A. Danopoulos[1,2*] and Pierre Braunstein[1*]

[1] Laboratoire de Chimie de Coordination, Institut de Chimie (UMR 7177 CNRS), Université de Strasbourg, 4 rue Blaise Pascal, 67081 Strasbourg Cedex - France

[2] Institute for Advanced Study (USIAS), Université de Strasbourg, 5 allée du Général Rouvillois, 67083 Strasbourg Cedex - France

e-mail: danopoulos@unistra.fr - braunstein@unistra.fr

* Corresponding author

Dedicated to the memory of Yves Chauvin, a most remarkable and visionary man and scientist, with gratitude for our scientific and non-scientific conversations and a very fruitful and enjoyable collaboration.

Abstract — *Two major aspects of coordination/organometallic chemistry are discussed in this article: (i) the use of functional chelating ligands to stabilize metal complexes while allowing easy stereodifferentiation within the coordination sphere and (ii) the choice of suitable ligands to stabilize challenging 'underligated' metal complexes with electronically highly unsaturated metal centres, thus potentially displaying unusual reactivity. In both cases, the relevance to homogeneous catalysis will be discussed.*

Résumé — **Contrôle de la sphère de coordination des métaux par les ligands : structures, réactivité et catalyse** — Deux aspects majeurs de la chimie de coordination/organométallique sont discutés dans cet article : d'une part, l'utilisation de ligands chélatants fonctionnels pour stabiliser des complexes métalliques en permettant une différenciation stéréoélectronique aisée au sein de la sphère de coordination et, d'autre part, la sélection de ligands adaptés à la stabilisation de complexes métalliques sous-coordinés dont les centres métalliques sont hautement insaturés électroniquement et donc potentiellement le siège d'une réactivité inhabituelle. Dans les deux cas, nous évoquerons la pertinence de ces approches en catalyse homogène.

INTRODUCTION

The synthesis, coordination chemistry and catalytic applications of functional phosphine-type ligands have represented a very fast growing field of research during the past 30 years and the diversity of donor groups that can thus be associated with the phosphorus donor opens unlimited possibilities. The chemically different nature of the functions available for coordination to metal centres immediately leads to consider the intrinsic properties of the resulting bonds and their consequences on the nature of the interactions between the metal(s) and other ligands present in the coordination sphere. This is illustrated in Scheme 1 with the simple case of a square-planar metal complex. Even if the ligands A and B are chemically the same (*e.g.* two chlorido or two methyl ligands), their reactivity will not be identical because of the stereochemically-different environment generated by the chelating functional phosphine ligand. This would not be the case in the presence of a symmetrical *P,P* or a *Z,Z* chelate with identical donor groups. Hybrid ligands thus represent a straightforward stereoelectronic differentiation tool within the metal coordination sphere, allowing selective reactivity and fine-tuning of the catalytic properties of metal complexes, as exemplified in the case of ethylene oligomerization [1, 2].

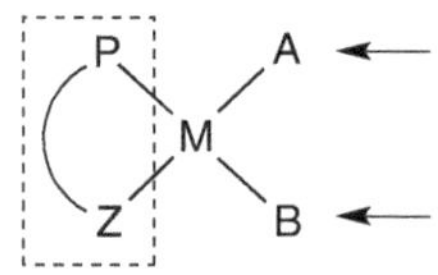

Functional phosphine

Scheme 1

Stereochemical differentiation in a metal coordination sphere resulting from chelation of a functional phosphine.

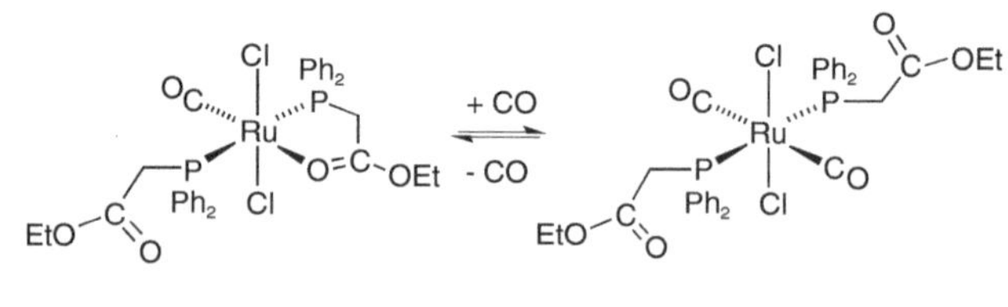

Scheme 2

Reversible binding of CO resulting from the opening and closing of the metal-oxygen dative bond involving the P,O chelate [24].

Ethylene, propylene and butenes are in increasing demand [3-13]. Ethylene alone is at the origin of *ca.* 30% of the products originating from the petrochemical industry. Among the α-olefins in the range C_4-C_{20}, those with shorter chain lengths (C_4-C_8) enjoy a rapidly expanding market owing to their applications as comonomers for the production of high density polyethylene and linear low density polyethylene and as intermediates in the production of oxo alcohols, hexyl- and octyl-mercaptans, amines, etc. Other linear α-olefins find applications *e.g.* for the fabrication of lubricants (C_{10}), biodegradable detergents (C_{10}, C_{12}, C_{14}), surfactants and lubricant additives (C_{12}-C_{20}) and many other useful chemicals. These features explain the continuing interest in both academia and industry for this research area [14-19].

The first part of this contribution will illustrate some of the work we carried out with metal complexes containing P,O-, P,N-, and N,O-type ligands relevant to ethylene oligomerization. This topic was selected because it corresponds to the starting points of our collaboration with Yves Chauvin and his group in the mid-90s. The second part deals with very recent developments concerning the stereoelectronic stabilization of 'underligated' metal complexes, a concept that illustrates a change of paradigm with respect to the classical 18 valence electron rule. Underligated complexes attract a rapidly increasing interest because of their often unusual features and reactivity and relevance to catalysis owing to the electronic unsaturation of their coordination sphere.

1 METAL COMPLEXES WITH FUNCTIONAL CHELATING LIGANDS

1.1 Metal Complexes with P,O-Type Ligands

In our first publication on transition metal complexes with functional phosphines, we described the coordination of the phosphino-ester ligand $Ph_2PCH_2C(O)OEt$ to various transition metal centres [20, 21]. In the case of Rh(III) and Ir(III), octahedral trihalide complexes containing two such ligands were characterized, in which one of them is terminally P-bound to the metal whereas the other acts as a P,O chelate. It was found by ^{1}H and $^{31}P\{^{1}H\}$ variable-temperature NMR spectroscopy that a dynamic behaviour was taking place in the Rh(III) complex, but not in the Ir(III) complex, that led the two ligands to exchange their role above coalescence temperature. This dynamic behaviour, which involves opening and closing of the metal-oxygen bond of the P,O chelate, became shortly after known as hemilabile behaviour [22, 23]. Related observations were made with Ru(II) complexes and applied to the reversible binding of carbon monoxide under ambient conditions (Scheme 2) [24].

Similarly to the phosphino-ester ligand $Ph_2PCH_2C(O)OEt$, the phosphino-ketone $Ph_2PCH_2C(O)Ph$ was found to chelate transition metals and its easy and modular synthesis prompted numerous studies in coordination chemistry [25]. It was rapidly found that when this ligand is coordinated to a transition metal, either κ^1-*P* or κ^2-*P,O*, its deprotonation readily affords complexes containing a P,O-chelating phosphino-enolate ligand $[Ph_2PCH{\cdots}C({\cdots}O)Ph]^-$, weak bases being sufficient as a result of the enhanced acidity of the PCH_2 protons upon coordination [26]. This observation appeared very promising, in particular in view of the remarkable properties of nickel complexes containing such phosphino-enolate and related ligands in catalytic ethylene oligomerization, which form the basis of the industrial Shell Higher Olefin Process (SHOP) [27-49]. This process does not require any cocatalyst and leads to linear α-olefins with remarkable selectivity. Furthermore, access to a diversity of transition metal complexes containing phosphino-enolates and related ligands opens interesting perspectives since the reaction pathway leading to the SHOP-type model catalyst $[NiPh\{Ph_2PCH{\cdots}C({\cdots}O)Ph\}(PPh_3)]$ (**1**), involving oxidative-addition of a P-Ph bond of the corresponding β-keto phosphorus ylid $Ph_3PCH{=}C(O)Ph$ across Ni(0), only works with Ni(0) (Scheme 3). These considerations formed the basis of stimulating discussions with Yves Chauvin at the *Institut Français du Pétrole*, Rueil-Malmaison, which initiated a

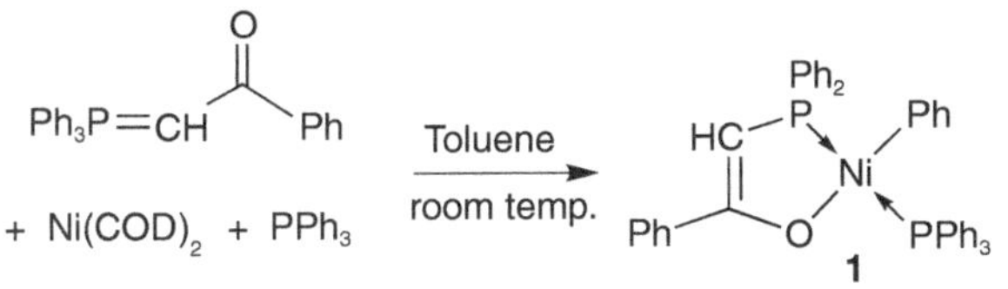
1

Scheme 3

Typical synthesis of a SHOP-type ethylene oligomerization pre-catalyst.

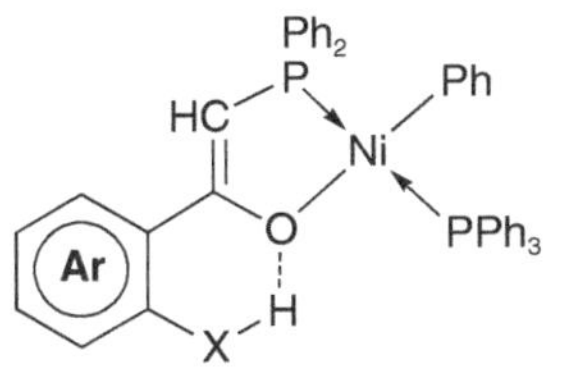
X = O, NMe, NPh

2

Scheme 4

Intramolecular hydrogen bonding between the *ortho*-substituent in the aryl group and the oxygen atom of the P,O-chelate.

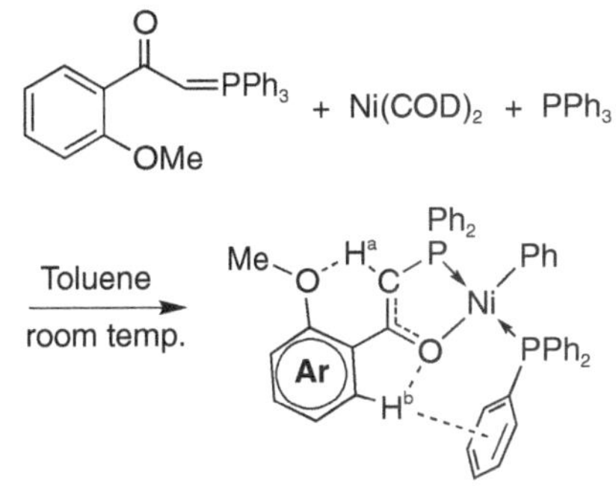
3

Scheme 5

Non-classical intramolecular hydrogen bonding and C-H-π interactions involving the P,O-chelate.

fruitful collaboration, still very active more than 20 years later. In our first joint paper, we examined the influence of a OH or NHR *ortho*-substituent in the aryl group **Ar** of Ni(II) complexes of type **2** (Scheme 4) on their catalytic properties toward ethylene oligomerization [47]. It was found that H-bonding interactions between this substituent and the enolate O atom resulted in a remarkable shift of the molecular mass distribution towards the highly desirable shorter chains oligomers.

These very encouraging results on the influence of aryl substituents on catalytically relevant intramolecular H-bonding interactions prompted further studies. Thus, when the *ortho*-OH group on the ring **Ar** was replaced with a methoxy substituent in:

$$[NiPh\{Ph_2PCH\cdots C(\cdots O)(o\text{-}MeO\text{-}C_6H_4)\}(PPh_3)]$$

it was found by X-ray diffraction that the oxygen atom of the ether group is involved in a non-classical interaction with the hydrogen atom H^a of the C-*H* group α to the phosphorus atom (Scheme 5). Furthermore, the hydrogen atom H^b in the *o*'- position in **Ar** is involved in another non-classical intramolecular hydrogen bonding interaction with the enolate oxygen atom (2.34 Å) and in a C-H-π interaction with one of the phenyl rings of the PPh_3 ligand (2.74 Å from its centroid) [48]. The contrasting orientations of the *o*-C_6H_4XH and *o*-C_6H_4OMe aromatic cycles in **2** and **3**, respectively, and the associated H-bonding behaviour of the P,O ligand, which functions as acceptor in the former and both a donor and an acceptor in the latter, are noteworthy and emphasize the need for detailed structural investigations to identify the occurrence and consequences of subtle substituent effects.

These results triggered a theoretical structure-reactivity investigation of ethylene insertion into nickel-alkyl bonds [49]. The activation energy of the olefin insertion into nickel-alkyl bonds (*i.e.* alkyl migration) in a series of square-planar Ni(II) complexes of formula [Ni(X^Y)(ethyl)(ethylene)], where X^Y is a anionic bidentate ligand, was calculated by DFT methods. Where X $\neq$ Y, the reactions of the two possible isomers have been compared and it was found that when one of the coordinating groups exerts a Weak Trans Influence (WTI) and the other a Strong Trans Influence (STI), as in the case of a phosphino-enolate P^O ligand, one of the two isomers has an activation energy for olefin insertion significantly lower than that of the reaction of complexes with symmetrical bidentate ligands, either "WTI^WTI" or "STI^STI" [49].

The coordinated anionic P,O-ligands $[Ph_2PCH\cdots C(\cdots O)Ph]^-$ and $[Ph_2PCH\cdots C(\cdots O)OEt]^-$ were shown to carry significant electronic density at the α-PCH carbon and this was evidenced by reactions with a number of organic and inorganic electrophiles [26]. A remarkably facile reversible binding reaction of CO_2 by a Pd(II) complex under ambient conditions was discovered and its application to the catalytic telomerization of butadiene with CO_2 [50, 51], and carbon-carbon bond formation by insertion of organic isocyanates RN=C=O into the α-P*C*-*H* bond [52] were some of the outcome of these investigations. Extensions to isoelectronic or closely related metalloligands demonstrated the potential of these often air-stable but reactive systems [26, 53]. Furthermore, reactions with metal electrophiles provided novel rational entries into homo- and heterodinuclear chemistry [26].

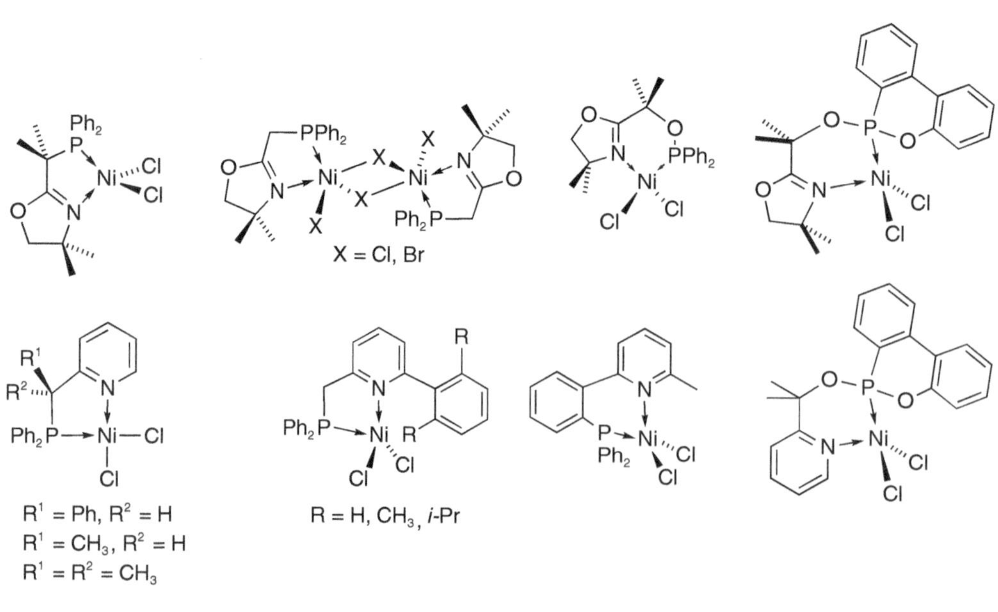

Scheme 6

Various oxazoline- and pyridine-based P,N-type Ni(II) complexes, with chelate ring sizes ranging from 5 to 7, applied to catalytic ethylene oligomerization [2].

1.2 Metal Complexes with P,N-Type Ligands

In the course of our studies – of which many were performed in collaboration with *IFP Energies nouvelles* – on the coordination properties of heterobidentate P,N-type ligands in relation to the catalytic properties of their nickel(II) complexes [2], the structures of numerous diamagnetic and paramagnetic complexes have been established by X-ray diffraction, which were often unpredictable in terms of metal coordination geometry and nuclearity (Scheme 6).

These investigations led *i.a.* to the finding that in the solid state, a microcrystalline sample of the green tetranuclear complex **4** could be transformed under pressure into its red mononuclear component **5** (Scheme 7) [54].

In other cases, different crystallisation temperatures could result in the formation of either mono- (**6**) or di-nuclear (**7**) complexes (Scheme 8) [55, 56]. The influence of the chelating ligands on activity, selectivity and catalyst lifetime was investigated while keeping the amount of cocatalysts to the minimum required. Complex **6** has been evaluated as precatalyst in the oligomerization of ethylene, with $AlEtCl_2$ or MAO as cocatalyst, or propylene, with MAO as cocatalyst. With only 2 equiv. of $AlEtCl_2$, it was highly selective for the formation of ethylene dimers (up to 96%) and with propylene, the selectivity for C6 products was higher than 98.5% [55].

These examples serve again to illustrate the importance of a full structural characterization of pre-catalysts in order to be in a position to draw meaningful structure-properties relationships.

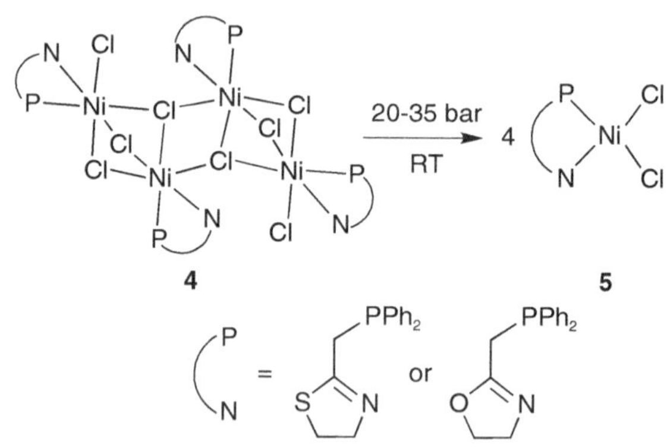

Scheme 7

Effect of pressure on complex nuclearity: conversion of a microcrystalline sample of the Ni(II) complex **4** into **5** under pressure in the solid state [54].

1.3 Metal Complexes with N,O-Type Ligands

Extending our studies of Ni(II) complexes with P,O- or P,N- to N,O-type heterofunctional chelates also provided interesting findings. We focused on oxazolinyl- and pyridinyl-alcohols and – alcoholates as ligands and a range of polynuclear complexes with unexpected nuclearities and structures could be characterized (Scheme 9). The ability of the alcoholate oxygen to bridge between two or three metal centres plays a critical role and accounts for the isolation of a number of coordination clusters.

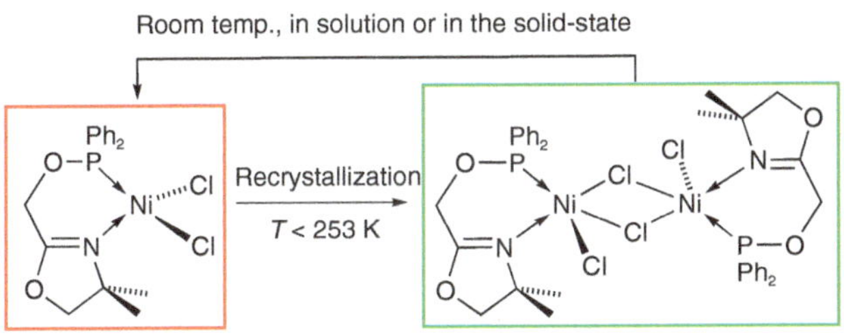

6
Red

7
Green

Scheme 8

Influence of the crystallization temperature on the nuclearity of the complex [55, 56].

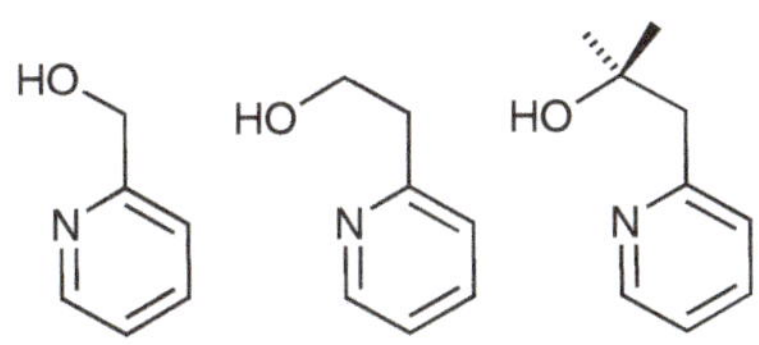

Scheme 9

Examples of pyridinyl-alcohols with different spacers between the heterocycle and the alcohol function investigated as ligands in Ni(II) complexes.

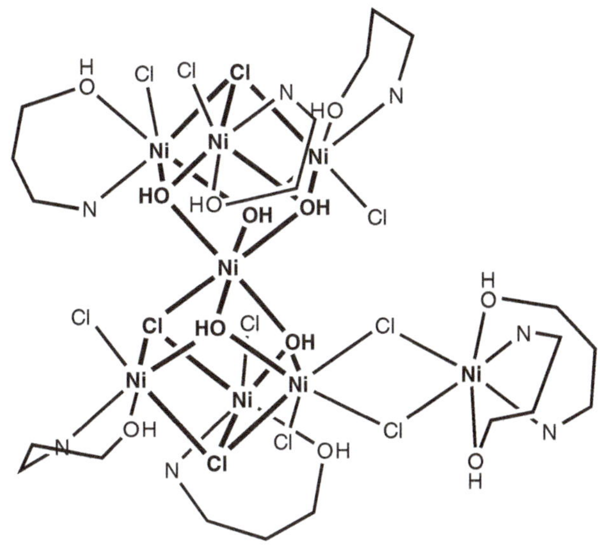

8

Figure 1

The Ni_8 complex **8** containing cubane-type structural subunits [61].

In particular, complexes with cubane-type structures could be structurally characterized, as well as higher nuclearity complexes constituted of assembled (incomplete) cubane units such as the octanuclear complex $[Ni(PyCH_2O)]_3(\mu_2\text{-}OCH_2Py)_3(\mu_3\text{-}OH)(\mu_5\text{-}Na)_3[Ni(PyCH_2O)_3]Cl$ **8** (Fig. 1) [57-61].

Many of these polynuclear complexes turned out to have interesting magnetic properties and some behave as single molecule magnets [57-61].

Polynuclear metal architectures held together by coordination bonds or by direct metal-metal bonds may be considered as models to heterogeneous catalytic sites but can also serve as molecular precursors to nanosized heterogeneous particles. This is particularly useful in the case of heterometallic systems, where cooperativity and synergistic effects may arise [62]. Such compounds may also serve as reservoirs of lower nuclearity, or even underligated (see below), mononuclear, catalytically active fragments. These fundamental questions clearly deserve further investigations.

In conclusion to this section, it is clear that the diversity of heterobidentate ligands available offers an unlimited playground for transition metal coordination chemistry. In the course of systematic and rational approaches to fine-tune the coordination sphere of a transition metal ion and thus the chemical and/or physical properties of its complexes, unexpected results may occur which open the way to new ideas and unanticipated developments; the role of serendipity should not be underestimated [63].

2 UNDERLIGATED METAL COMPLEXES

The renewed interest in the catalytic applications of 'base metals' (*e.g.* Co, Fe, Ni), which fulfil to a certain extent green environmental criteria, drives the re-investigation of their coordination chemistry and the catalytic potential of well defined reactive complexes. In this respect, the paradigm that accompanied and guided organometallic chemistry with strong-field σ-donor and π-acceptor ligands since the Wilkinson years, *viz.* the 18 electron rule, predicting high stability in complexes with 18 valence electrons has to be abandoned to reach the realm of coordinative and electronic unsaturation in 'underligated', reactive complexes. With the expectation that a number of ligands and thus a valence electron count lower than usual will be associated with enhanced reactivity and catalytic properties, we have investigated, since 2010, the synthesis and properties of 'underligated' metal complexes of Fe, Co and Cr stabilized by N-Heterocyclic Carbene ligands (NHC).

The approach was based on the use of bulky NHC as a means of supporting homoleptic and heteroleptic

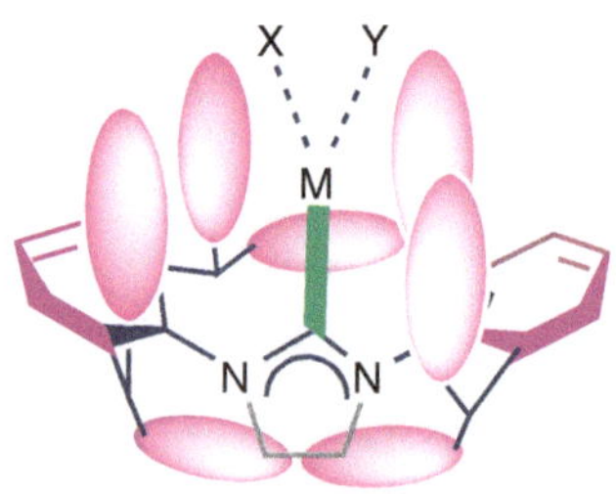

Figure 2

The NHC 'cavity' targeted to support 'underligation'.

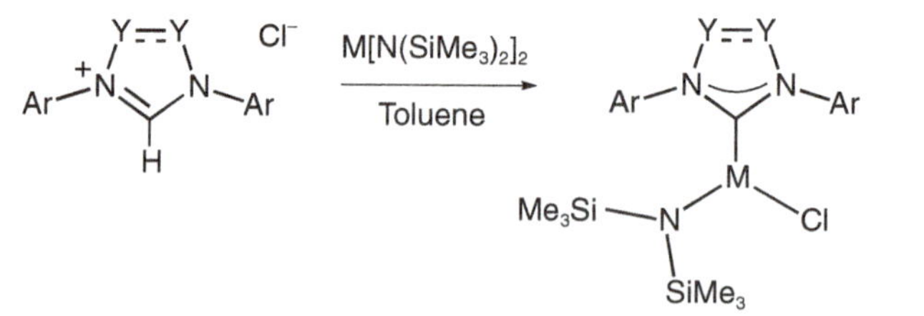

Ar = DiPP: Y = CH, **9**; Y = CH_2, **10**
Ar = Mes: Y = CH, **11**; Y = CH_2, **12**

Scheme 10

Aminolysis of $[Fe\{N(SiMe_3)_2\}_2]_2$ by imidazol(in)ium chlorides to give three-coordinate complexes; $[Co\{N(SiMe_3)_2\}_2]_2$ reacts analogously.

underligated structures. The NHC have versatile and tuneable electronic properties (strong σ-donors, but occasionally accepting or donating π-electron density from or to the metal). Hence, compared to the strongly σ- and π-donating amido-, alkoxo- and nacnac- (anionic bidentate ligands derived from 1,3-diketimines) co-ligands, NHC should lead to *'genuine' electronically unsaturated* metal centres, the stability, bonding behaviour and reactivity of which under these conditions are not predictable. Compared to other bulky ligands (amido, terphenyl *etc*.), NHC offer easier and diverse steric tuning options leading to concave metal nesting 'pockets' (Fig. 2).

In this respect, the bulky ligands IPr (di*iso*propylphenyl substituents on the imidazole N atoms) and IMes (with mesityl substituents at N), and the sterically relevant SIPr and SIMes (with a saturated heterocycle backbone), provide similar bulk and space organization to the bulky terphenyls developed by Ni and Power [64], although the latter are anionic and the NHC are neutral ligands.

Furthermore, underligated 3d metal complexes are expected to show weak ligand fields with the following implications:

- open-shell energetically close-lying electronic states;
- non-covalent interactions frequently becoming functional in the stabilization of structural motifs;
- common occurrence of unusual oxidation states;
- control of the stoichiometric or catalytic reactivity by metal-centred or -generated paramagnetic intermediates;
- stabilization of metal-ligand multiple bonds with $3d^n$ $2 < n < 10$ electron counts;
- spin crossovers during chemical reactions;
- unusual magnetic properties related to spin-orbit coupling and anisotropy.

All these characteristics set the scene for novel structures and reactivities in functional complexes.

A convenient entry into the synthetically useful 3-coordinate complexes of Fe^{II} and Co^{II} with high potential for mild and selective derivatization was the discovery of the aminolysis of $[M\{N(SiMe_3)_2\}_2]_2$ M = Fe, Co, with bulky imidazol(in)ium halides [(NHC)H]Cl,

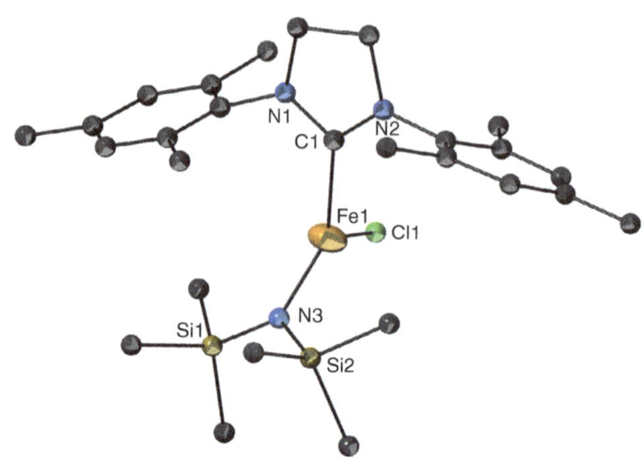

Figure 3

The structure of the Fe^{II} complex **12**.

i.e. [(S)IPrH]Cl, [(S)IMesH]Cl (S for saturated) and the isolation of $[M(NHC)\{N(SiMe_3)_2\}Cl]$ complexes (Scheme 10) [65, 66]. The method was recently extended to $[M\{N(SiMe_2Ph)_2\}_2]_2$, M = Fe, Co [67]. However, a coordinating counteranion *i.e.* halide, should accompany the imidazole (in)um salt.

The structure of **12** is shown in Figure 3.

The presence of three types of ligands that can be replaced using different reaction types (*e.g.* chloride by salt metathesis, bis(silylamide) by further aminolysis and NHC by substitution with neutral 2 electron donors) adds to the versatility and value of the three-coordinate complexes in accessing new 3- or higher-coordinate species of Fe and Co under mild conditions. Selected aminolysis possibilities for the iron complexes are shown in Scheme 11 [64]; the structures of complexes **13**, **14**, **17-19** are given in Figures 4-8, respectively; analogous reactivity has been described for Co [65, 67].

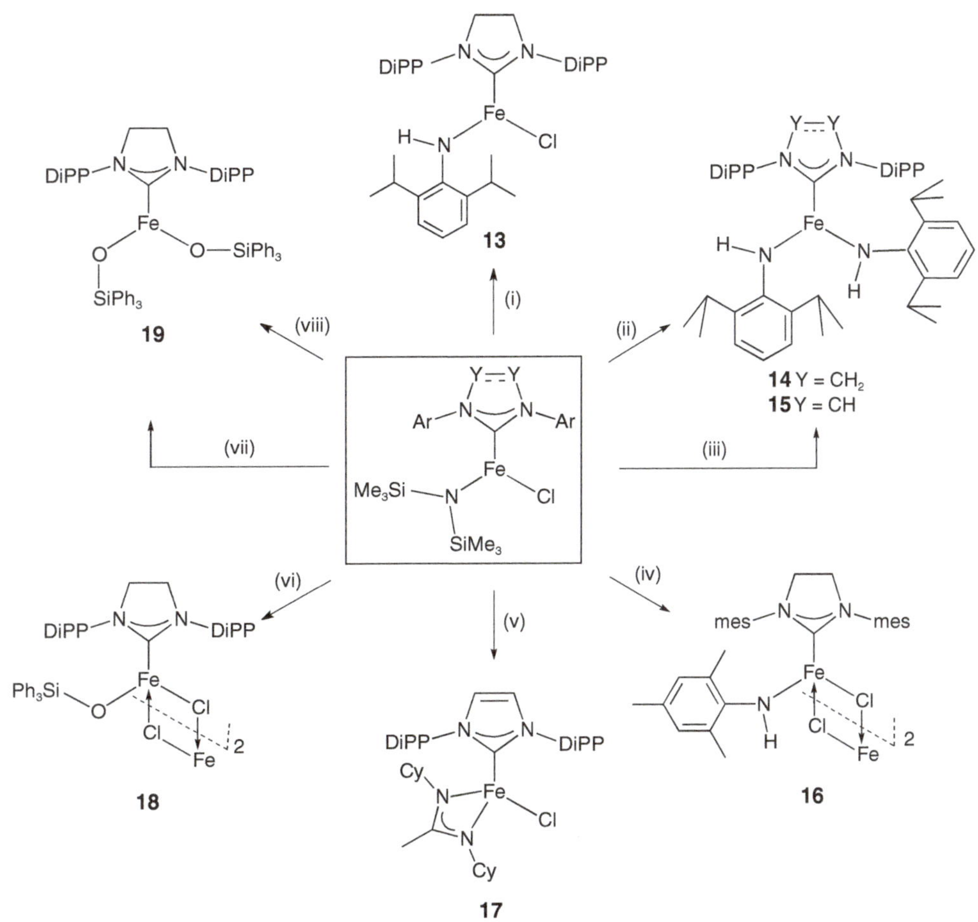

Scheme 11

Representative aminolysis reactions of 3-coordinate [Fe(NHC){N(SiMe$_3$)$_2$}Cl] complexes. Reagents and conditions: (i) 1 equiv. DiPPNH$_2$ in octane, reflux for 5 min; (ii) 1 equiv. DiPPNH$_2$ in toluene, room temp., 5 h; (iii) 1 or 2 equiv. DiPPNHLi in ether, −78°C to room temp., 5 h or 2 equiv. DiPPNH$_2$ in ether, room temp., 12 h; (iv) 1 equiv. mesNH$_2$ in toluene; (v) *N,N'*-di-cyclohexyl-acetamidine, toluene 100°C, 1 h; (vi) 1 equiv. Ph$_3$SiOH in toluene, room temp.; (vii) 2 equiv. Ph$_3$SiOTl in toluene, room temp., 24 h; (viii) 2 equiv. Ph$_3$SiOH in toluene 100°C, 1 h.

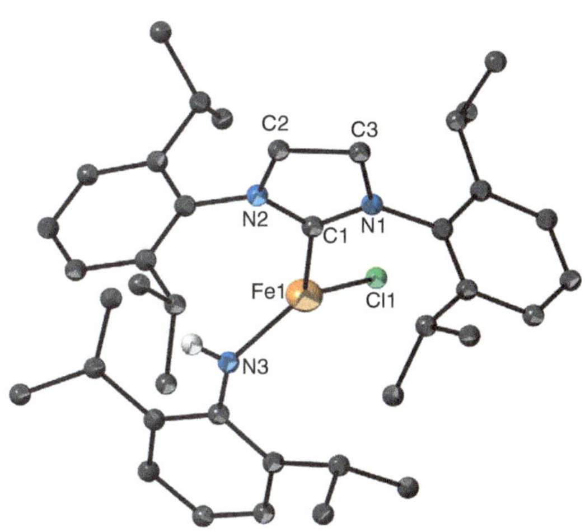

Figure 4

The structure of the Fe^{II} complex **13**. H-atoms (except the one at the anilido N) are omitted.

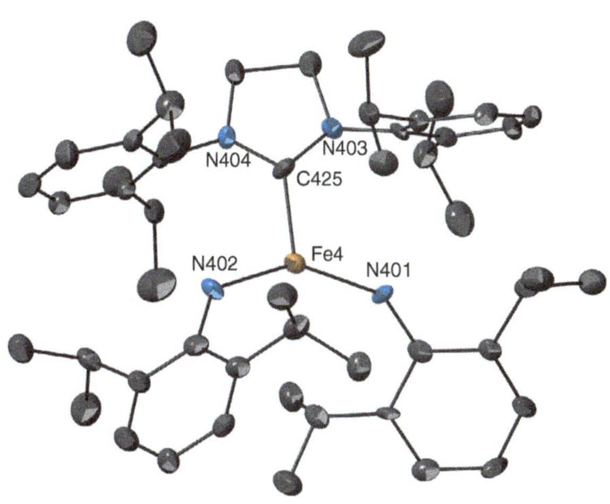

Figure 5

The structure of the Fe^{II} complex **14**. One of the four molecules found in the asymmetric unit is shown. All H atoms are omitted.

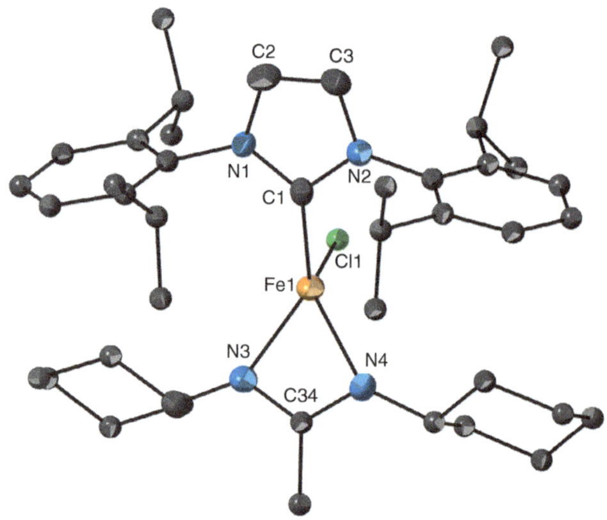

Figure 6

The structure of the Fe^{II} complex **17**. All H atoms are omitted.

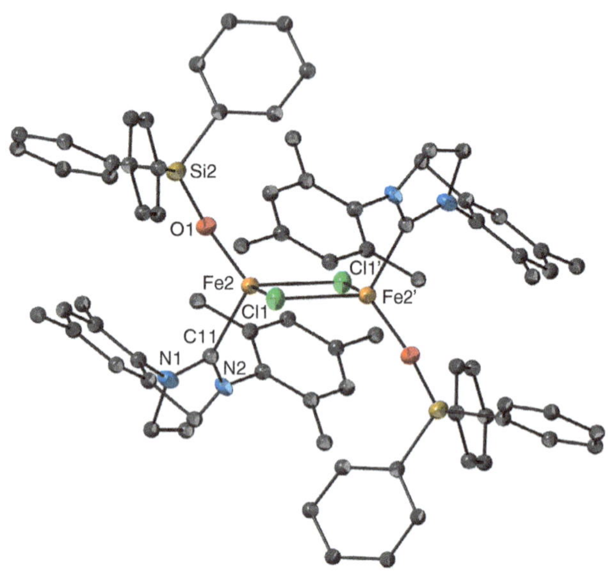

Figure 7

The dinuclear Fe^{II} complex **18**. All H atoms are omitted.

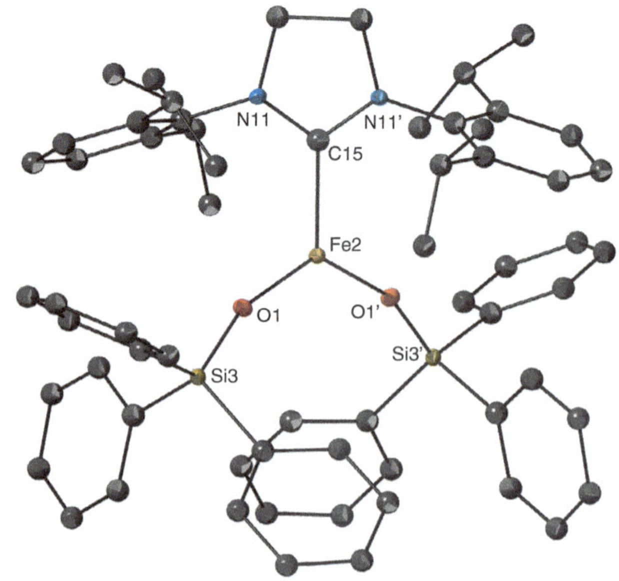

Figure 8

The structure of the Fe^{II} complex **19**. One of the two molecules found in the asymmetric unit is shown.

It can be concluded that the nature of the products obtained is dependent on the bulk of the ligands at Fe. In this respect, the sterics of the bis(trimethylsilylamide) ligand promote mononuclear 3-coordinate structures (**9-12**). Mesitylanilido- and triphenylsilyloxo-chlorido species **16** and **18**, respectively, dimerize *via* chlorido bridge formation, while the 3-coordinate DiPPanilido-chlorido derivative **13**, that was prepared under carefully controlled conditions in octane, easily conproportionates in polar solvents by ligand redistribution to form the thermodynamically more stable mononuclear 3-coordinate bis(anilido) species **14** and **15**. Conproportionation has been observed in the reaction of **10** with thallium triphenylsiloxide. It may be that conproportionation is favoured with the combination on the metal of the bulkiest co-ligands; $LiN(SiMe_3)_2$ and $TlN(SiMe_3)_2$ may be leaving groups in these cases.

The isoelectronic relationship between amido and alkyl ligands guided efforts to prepare 3-coordinate di-alkyls and chlorido-alkyls with the bulky (S)IPr and (S)IMes as co-ligands. For Fe, simple salt metathesis reactions using $FeCl_2$ and $MgR_2(THF)_2$ (R = CH_2SiMe_3, CH_2CMe_2Ph, CH_2Ph, mesityl), in the presence of IPr or SIPr in THF/dioxane was the method of choice. The alkylation was sequential (except for R = Mes); thus, adjustment of the stoichiometry led to either chlorido-alkyls or di-alkyls [68]. Remarkably, the chlorido-alkyls are mononuclear species (Scheme 12). The underligated complexes **20**, **23**, **26**, **31** and **32** have been previously reported by us [66-68]; all other complexes shown in Scheme 12 are new. The complex [Fe(NHC)Mes_2], where NHC = *N,N'*-di*iso*propyl-imidazol-2-ylidene, was reported simultaneously with our work [69]. The thermally stable 12 valence electron dialkyl complexes undergo IPr or (S)IPr substitution by (S)IMes to give the analogous 3-coordinate complexes (one example, **29**, is shown in Scheme 12). The mechanism of the

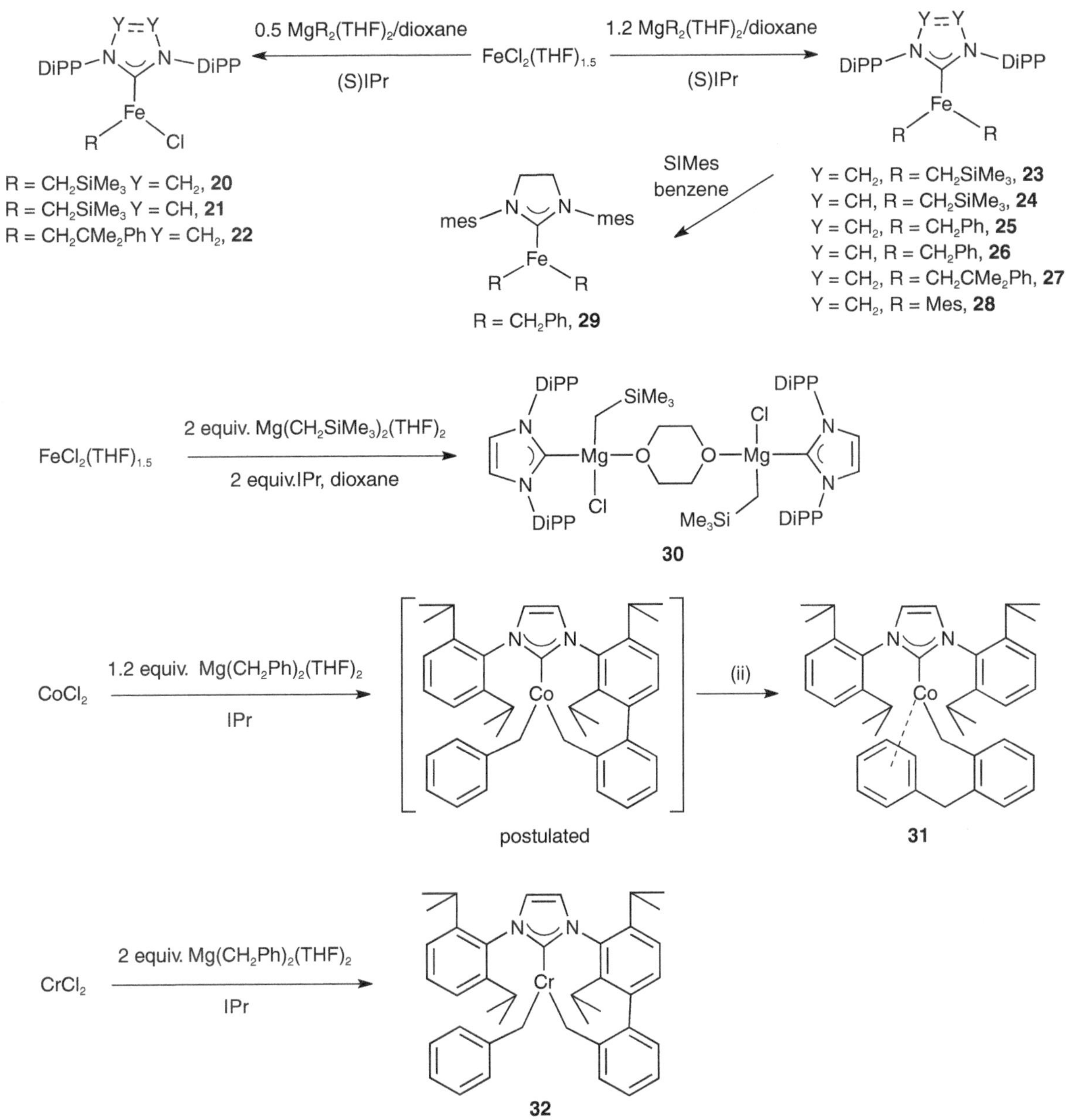

Scheme 12

Synthetic transformations leading to 3-coordinate, underligated chlorido-alkyl and di-alkyl iron, chlorido-alkyl magnesium, alkyl cobalt and dibenzyl chromium complexes.

substitution (associative *versus* dissociative) has not yet been clarified.

An insight into the speciation and the nature of the Mg alkylating agent was gained by carrying out the alkylation of $FeCl_2$ with 2 equiv. of $Mg(CH_2SiMe_3)_2(THF)_2$ and 2 equiv. IPr, and fractionally crystallizing the reaction products. After **24**, the magnesium complex **30** was isolated as colourless crystals from pentane containing a few drops of dioxane. It is plausible that **31** could be accessible from the reaction of the corresponding Grignard reagent with IPr in the presence of dioxane. The structures of the underligated complexes **21**, **22**, **24**, **25**, **27-30** are shown in Figures 9-16, respectively.

On attempting the preparation of the Co analogue of **26**, *i.e.* $[Co(IPr)(CH_2Ph)_2]$, an unprecedented C-H activation/coupling reaction occurred between the two benzyl ligands at the coordination sphere of the postulated intermediate $[Co(IPr)(CH_2Ph)_2]$ to give eventually **31** that features the new aryl-substituted, anionic, benzyl $8e^-$ donor ligand

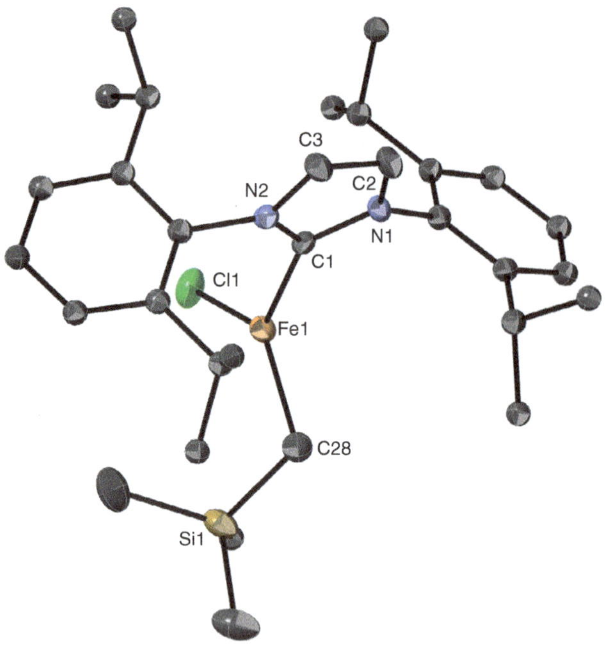

Figure 9

The structure of the Fe^{II} complex **21**. All H atoms are omitted.

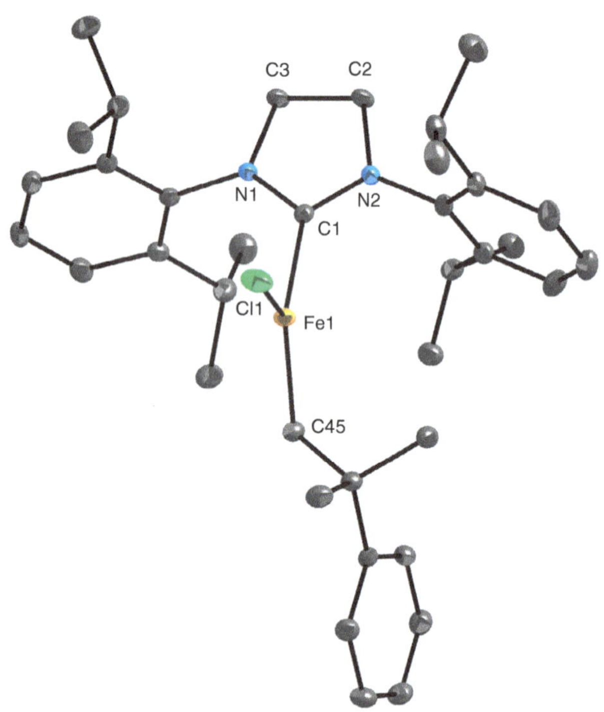

Figure 10

The structure of the Fe^{II} complex **22**. All H atoms are omitted.

(arene 6 + benzyl 2) [66]. In contrast, the introduction of CH_2SiMe_3 occurred smoothly, giving Co dialkyl complexes analogous to **24** [70].

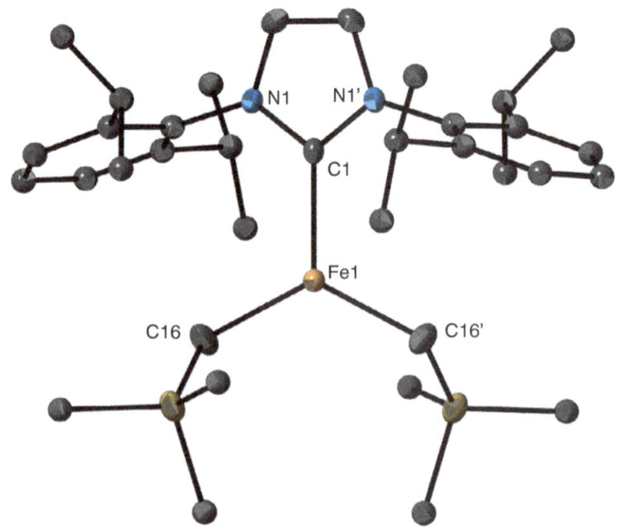

Figure 11

The structure of the Fe^{II} complex **24**. All H atoms are omitted.

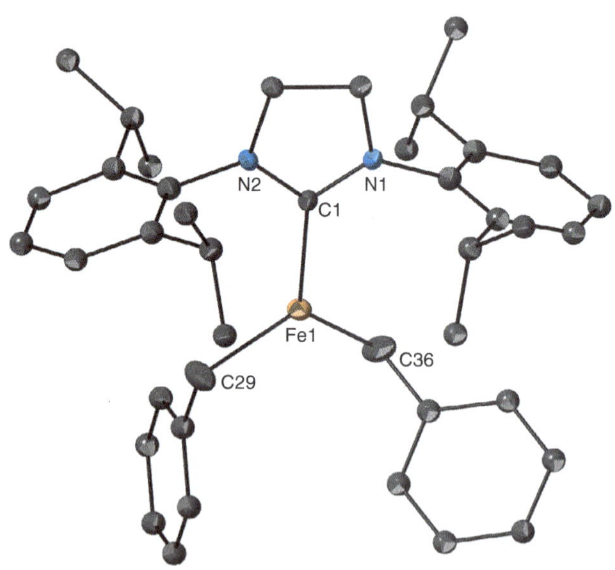

Figure 12

The structure of the bis(benzyl) Fe^{II} complex **25**. All H atoms are omitted.

The activation of C-H bonds can proceed by mechanistically distinct pathways with metals from across the periodic table. Low oxidation state 'underligated' species may participate in C-H activation *via* oxidative-addition, in higher oxidation state centres *via* σ-bond metathesis and in multiple-bonded species *via* 1,2-concerted additions or hydrogen atom transfers. One distinctive aspect of the

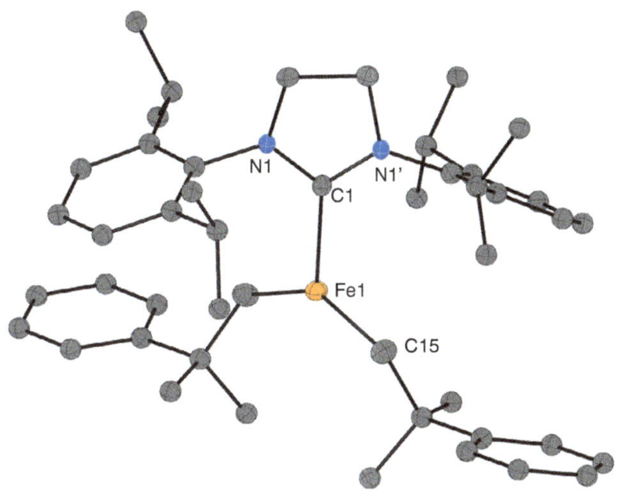

Figure 13

The structure of the Fe^{II} complex **27**. All H atoms are omitted.

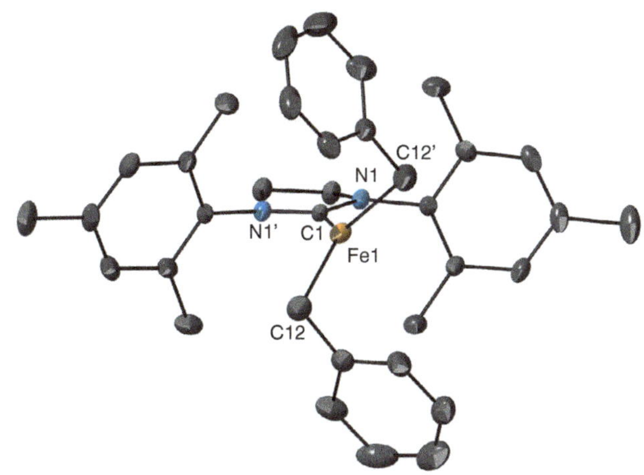

Figure 15

The structure of the bis(benzyl) Fe^{II} complex **29**. All H atoms are omitted.

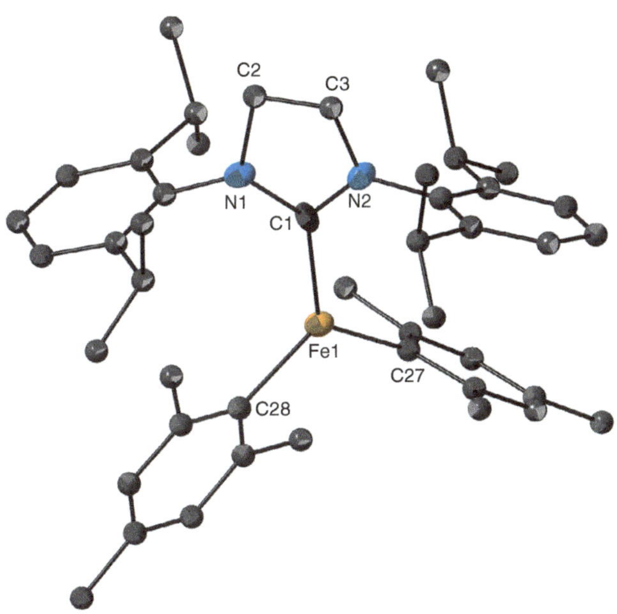

Figure 14

The structure of the Fe^{II} complex **28**. All H atoms are omitted.

'underligated' 3*d*-complexes is the presence of energetically close, usually open-shell electronic structure(s) with different spin states that may be variable during reactions, potentially promoting diverse C-H activation mechanisms, some of them of bioinorganic relevance.

The synthesis of 3-coordinate Cr organometallics *e.g.* **32** (Scheme 12) may provide models for organometallic species involved in Cr-catalyzed C-C bond formation reactions [71]. The remarkable structural feature of the 10 valence electron complex **32** is the non-sterically induced severe angular distorsion away form the usual sp^3 angle at the benzyl carbon atoms (97.9(2)° and 76.6(2)°), which led to theoretical investigations and the conclusion that "Polarization-Induced Metal–Arene" interactions of a new type were at work (Fig. 17). The energy of this interaction has been calculated to be *ca.* 50 kJ/mol [71].

The observation of structural effects attributable to weak energetic interactions may be possible due to the sterically open coordination sphere and the weak 3-coordinate ligand field. With respect to ethylene and alkene oligomerization, organometallic species like $[Cr(IPr)(CH_2Ph)_2]$ raise interest for the following reasons:

- they can be considered as models of precursors to catalytically active species;
- they can show potential for activator-free oligomerization (since both Cr-C carbon bonds and coordinative unsaturation are already present features);
- they may provide useful reagents for the Cr immobilization and creation of novel organometallics of catalytic relevance on supports.

The study of the organometallic chemistry of these simple systems in solution under catalytic conditions by XAS techniques should be particularly interesting. The strategy of using bulky NHC for the stabilization of 'underligated' 3d metal centres has been further developed recently by using the strongly σ-donating and π-accepting cAAC ligands [72, 73].

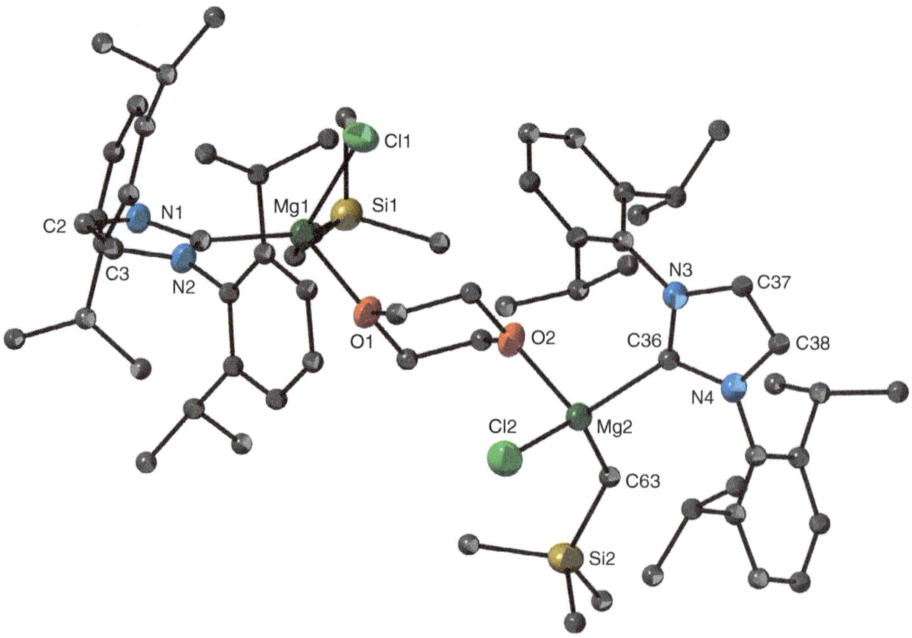

Figure 16

The structure of the dinuclear Mg complex **30**. All H atoms are omitted.

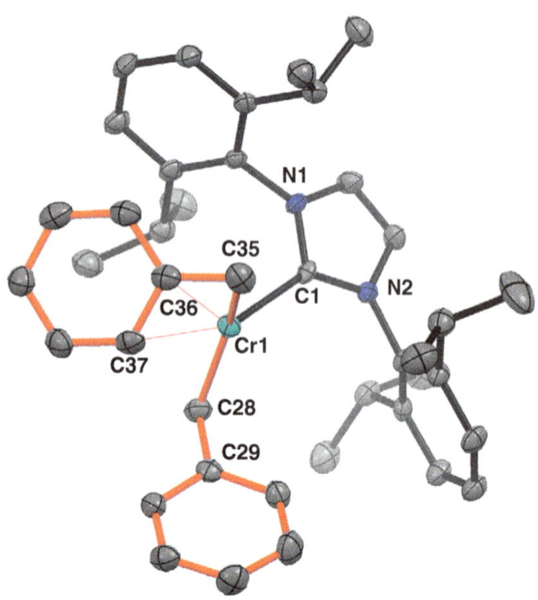

Figure 17

Polarization-induced metal-arene interactions operating in the 3-coordinate complex [Cr(IPr)(CH_2Ph)$_2$] (**32**). Reprinted from [71]. Copyright American Chemical Society.

In conclusion, research on 'underligated' transition metal complexes opens new dimensions in molecular inorganic chemistry, leading:

- to fundamental advances in the understanding of bonding and reactivity theory,
- to applications in homogeneous and heterogeneous catalysis,
- to the exploitation of electronic and physical properties for the manufacture of useful devices [74].

It is expected to further broaden the scope and applications of 'base metals' [75].

ACKNOWLEDGMENTS

We are grateful to the coworkers and collaborators cited in the references and we thank Dr N. Stylianides and Dr R. Pattacini for assistance in solving the crystal structures of **12**, **18** and **13**, **24**, respectively, and Dr. L. Karmazin and Miss C. Bailly (*Service de Radiocristallographie*, UdS) for the determination of the crystal structures of **17**, **21**, **24**, **25**, **29** and **30**. The USIAS, CNRS, *Université de Strasbourg*, *Région Alsace* and *Communauté Urbaine de Strasbourg* are acknowledged for the award of fellowships and a Gutenberg Excellence Chair (2010–11) to AAD. We gratefully thank the CNRS, the MESR (Paris) and IFP Energies nouvelles for support.

REFERENCES

1 Braunstein P. (2004) *J. Organomet. Chem.* **689**, 3953.

2 Speiser F., Braunstein P., Saussine L. (2005) *Acc. Chem. Res.* **38**, 784

3 Burridge E. (2008) *Chemical Profile: Alpha Olefins, ICIS Chemical Business.*

4 Lappin G.R., Sauer J.D. (eds) (1989) *Alphaolefins Applications Handbook*, Marcel Decker Inc., Berkeley, CA.

5 Vogt D. (2000) *Oligomerization of ethylene to higher linear alpha olefins*, in *Applied homogeneous catalysis with organometallic compounds*, Cornils B., Hermann W.A. (eds), Wiley-VCH, Weinheim, pp. 245-258.

6 Weissermel K., Arpe H.J. (2003) *Olefins in Industrial Organic Chemistry*, 4th edition, Weissermel K., Arpe H.J. (eds), Wiley VCH, Weinheim, pp. 63-89.

7 Bjacek P. (2003) Petrochemical prospects, *Oil & Gas Journal*, **Dec 15**, p. 50.

8 Nakamura D.N. (2007) Ethylene report, *Oil & Gas Journal*, **July 16**, p. 46.

9 Forestière A., Olivier-Bourbigou H., Saussine L. (2009) *Oil Gas Sci. Technol.* **64**, 649.

10 Olivier-Bourbigou H., Forestière A., Saussine L., Magna L., Favre F., Hugues F. (2010) *Oil Gas Eur. Mag.* **36**, 97.

11 Belov G.P., Matkovsky P.E. (2010) *Pet. Chem.* **50**, 283.

12 Keim W. (2013) *Angew. Chem. Int. Ed.* **52**, 12492.

13 Breuil P.-A.R., Magna L., Olivier-Bourbigou H. (2015) *Catal. Lett.* **145**, 173.

14 Britovsek G.J.P., McGuinness D.S., Wierenga T., Young C. (2015) *ACS Catal.* **5**, 7, 4152-4166.

15 Delferro M., Marks T.J. (2011) *Chem. Rev.* **111**, 2450.

16 McGuinness D.S. (2011) *Chem. Rev.* **111**, 2321.

17 Agapie T. (2011) *Coord. Chem. Rev.* **255**, 861.

18 van Leeuwen P.W.N.M., Clément N.D., Tschan M.J.-L. (2011) *Coord. Chem. Rev.* **255**, 1499.

19 Bianchini C., Giambastiani G., Luconi L., Meli A. (2010) *Coord. Chem. Rev.* **254**, 431, and references cited therein.

20 Braunstein P., Matt D., Mathey F., Thavard D. (1978) *J. Chem. Res. Synop.* **232**.

21 Braunstein P., Matt D., Mathey F., Thavard D. (1978) *J. Chem. Res. Miniprint* **3041**.

22 Jeffrey J.C., Rauchfuss T.B. (1979) *Inorg. Chem.* **18**, 2658.

23 Braunstein P., Naud F. (2001) *Angew. Chem. Int. Ed.* **40**, 680.

24 Braunstein P., Matt D., Dusausoy Y. (1983) *Inorg. Chem.* **22**, 2043.

25 Bouaoud S.-E., Braunstein P., Grandjean D., Matt D., Nobel D. (1986) *Inorg. Chem.* **25**, 3765.

26 Braunstein P. (2006) *Chem. Rev.* **106**, 134.

27 Keim W., Kowaldt F.H., Goddard R., Krüger C. (1978) *Angew. Chem. Int. Ed.* **17**, 466.

28 Beach D.L., Harrison J.J. (1980) US Patent 4,293,502, (to Gulf Research).

29 Keim W., Behr A., Gruber B., Hoffmann B., Kowaldt F.H., Kürschner U., Limbäcker B., Sistig F.P. (1986) *Organometallics* **5**, 2356.

30 Keim W. (1990) *Angew. Chem. Int. Ed. Engl.* **29**, 235.

31 Pietsch J., Braunstein P., Chauvin Y. (1998) *New J. Chem.* **22**, 467.

32 Klabunde U., Ittel S.D. (1987) *J. Mol. Catal.* **41**, 123.

33 Klabunde U., Mühlhaupt R., Herskowitz T., Janowicz A.H., Calabrese J., Ittel S.D. (1987) *J. Polym. Sci. Polym. Chem.* **25**, 1989.

34 Keim W. (1987) *New J. Chem.* **11**, 531.

35 Soula R., Broyer J.P., Llauro M.F., Tomov A., Spitz R., Claverie J., Drujon X., Malinge J., Saudemont T. (2001) *Macromolecules* **8**, 2438.

36 Keim W. (1989) *J. Mol. Catal.* **52**, 19.

37 Kuhn P., Sémeril D., Matt D., Chetcuti M.J., Lutz P. (2007) *Dalton Trans.* 515.

38 Pietsch J., Braunstein P., Chauvin Y. (1998) *New J. Chem* 467.

39 Heinicke J., Koesling M., Brüll R., Keim W., Pritzkow H. (2000) *Eur. J. Inorg. Chem.* 299.

40 Heinicke J., He M., Dal A., Klein H.F., Hetche O., Keim W., Flörke U., Haupt H.J. (2000) *Eur. J. Inorg. Chem.* 431.

41 Heinicke J., Köhler M., Peulecke N., He M., Kindermann M.K., Keim W., Fink G. (2003) *Chem.–Eur. J.* **9**, 6093.

42 Heinicke J., Köhler M., Peulecke N., Keim W. (2004) *J. Catal.* **225**, 16.

43 Heinicke J., Köhler M., Peulecke N., Keim W., Jones P.G. (2004) *Z. Anorg. Allg. Chem.* **630**, 1181.

44 Heinicke J., Köhler M., Peulecke N., Kindermann M.K., Keim W., Köckerling M. (2005) *Organometallics* **24**, 344.

45 Braunstein P., Chauvin Y., Mercier S., Saussine L. (2005) *C. R. Chim.* **8**, 31.

46 Keim W. (2013) *Angew. Chem. Int. Ed.* **52**, 12492.

47 Braunstein P., Chauvin Y., Mercier S., Saussine L., De Cian A., Fischer J. (1994) *J. Chem. Soc. Chem. Commun.* 2203.

48 Kermagoret A., Braunstein P. (2008) *Dalton Trans.* 822.

49 Hasanayn F., Achord P., Braunstein P., Magnier H.J., Krogh-Jespersen K., Goldman A.S. (2012) *Organometallics* **31**, 4680.

50 Braunstein P., Matt D., Nobel D. (1988) *Chem. Rev.* **88**, 747.

51 Braunstein P., Matt D., Nobel D. (1988) *J. Am. Chem. Soc.* **110**, 3207.

52 Braunstein P., Nobel D. (1989) *Chem. Rev.* **89**, 1927.

53 Zhang S., Pattacini R., Braunstein P. (2014) in *Advances in Organometallic Chemistry and Catalysis: The Silver/Gold Jubilee International Conference on Organometallic Chemistry Celebratory Book*, Pombeiro A.J.L. (ed.), John Wiley & Sons, Chapter 14.

54 Kermagoret A., Pattacini R., Chavez Vasquez P., Rogez G., Welter R., Braunstein P. (2007) *Angew. Chem. Int. Ed.* **46**, 6438. In this paper, a pressure much higher than actually required was indicated.

55 Chavez Vasquez P., Guerrero Rios I., Kermagoret A., Pattacini R., Meli A., Bianchini C., Giambastiani G., Braunstein P. (2009) *Organometallics* **28**, 1776.

56 Flapper J., Kooijman H., Lutz M., Spek A.L., van Leeuwen P. W.N.M., Elsevier C.J., Kamer P.C.J. (2009) *Organometallics* **28**, 3272.

57 Zhang J., Teo P., Pattacini R., Kermagoret A., Welter R., Rogez G., Hor T.S.A., Braunstein P. (2010) *Angew. Chem. Int. Ed.* **49**, 4443.

58 Kayser L., Pattacini R., Rogez G., Braunstein P. (2010) *Chem. Commun.* **46**, 6461.

59 Hameury S., Kayser L., Pattacini R., Rogez G., Wernsdorfer W., Braunstein P. (2013) *Dalton Trans.* **42**, 5013.

60 Massard A., Rogez G., Braunstein P. (2014) *Dalton Trans.* **43**, 42.

61 Hameury S., Kayser L., Pattacini R., Rosa P., Barra A.-L., Braunstein P. (2015) *ChemPlusChem.* **80**, 1312.

62 Buchwalter P., Rosé J., Braunstein P. (2015) *Chem. Rev.* **115**, 28.

63 Braunstein P. (2015) *Chem. Rec.* **15**, 1132-1136.

64 Ni C., Power P. (2010) *Struct. Bond.* **136**, 59-111.

65 Danopoulos A.A., Braunstein P., Stylianides N., Wesolek M. (2011) *Organometallics* **30**, 6514.

66 Danopoulos A.A., Braunstein P. (2013) *Dalton Trans.* **42**, 7276.

67 Danopoulos A.A., manuscript in preparation.

68 Danopoulos A.A., Braunstein P., Wesolek M., Monakhov K.Yu., Rabu P., Robert V. (2012) *Organometallics* **31**, 4102.

69 Xiang L., Xiao J., Deng L. (2011) *Organometallics* **30**, 2018.

70 Przyojski J.A., Arman H.D., Tonzetich Z.J. (2012) *Organometallics* **32**, 723.

71 Danopoulos A.A., Monakhov K.Yu., Robert V., Braunstein P., Pattacini R., Conde-Guadaño S., Hanton M., Tooze R.P. (2013) *Organometallics* **32**, 1842.

72 Samuel P.P., Mondal K.C., Amin Sk N., Roesky H.W., Carl E., Neufeld R., Stalke D., Demeshko S., Meyer F., Ungur L., Chibotaru L.F., Christian J., Ramachandran V., van Tol J., Dalal N.S. (2014) *J. Am. Chem. Soc.* **136**, 11964.

73 Ung G., Rittle J., Soleilhavoup M., Bertrand G., Peters J.C. (2014) *Angew. Chem. Int. Ed.* 53, 8427.

74 Power P.P. (2012) *Chem. Rev.* **112**, 3482.

75 75 Electronic Supplementary Information (ESI) available: X-ray crystallographic data in CIF format for CCDC reference numbers: 1413910, **12**; 1413905, **13**; 1413906, **14**; 1413907, **17**; 1413908, **18**; 1413909, **19**; 1413835, **21**; 1413836, **22**; 1413837, **24**; 1413838, **25**; 1413839, **27**; 1413840, **28**; 1413841, **29**; 1413842, **30**.

3

Design of Compact Photoelectrochemical Cells for Water Splitting

Tom Bosserez[1]*, Jan Rongé[1], Jan van Humbeeck[2], Sophia Haussener[3] and Johan Martens[1]

[1] *KU Leuven, Centre for Surface Chemistry and Catalysis, Kasteelpark Arenberg 23, Post box 2461, 3001 Leuven - Belgium*

[2] *KU Leuven, Materials Engineering Department (MTM), Kasteelpark Arenberg 44, Post box 2450, 3001 Leuven - Belgium*

[3] *École Polytechnique Fédérale de Lausanne (EPFL), Laboratory of Renewable Energy Science and Engineering, Station 9, 1015 Lausanne - Switzerland*

e-mail: tom.bosserez@kuleuven.be - jan.rongé@kuleuven.be - johan.martens@kuleuven.be jan.vanhumbeeck@kuleuven.be - sophia.haussener@epfl.ch

* Corresponding author

Abstract *— Solar driven water splitting can be achieved by coupling electrolyzers with PhotoVoltaics (PV). Integration of both functions in a compact PhotoElectroChemical (PEC) cell is an attractive option but presents significant scientific challenges. In this work, the design of single- and dual-compartment PEC cells for research purposes is discussed. The fabrication of separator-electrode assemblies is an important aspect, and upscaling of these architectures even to centimeter scale is not trivial. The layout of a new dual-compartment compact PEC cell with* in-situ *monitoring of pH, temperatures, and oxygen and hydrogen evolution for research purposes is presented. Finally, a prospect of future PEC cells for practical applications is presented.*

Résumé — Conception de cellules photoélectrochimiques compactes pour la décomposition de l'eau — La décomposition de l'eau en utilisant la lumière du soleil s'effectue par couplage d'un électrolyseur aux cellules PhotoVoltaïques (PV). L'intégration des deux fonctions dans une seule cellule PhotoElectroChimique (PEC) compacte est envisagée, mais présente un très grand défi scientifique. Cet article traite la conception des cellules PEC de recherche avec ou sans compartimentation. L'assemblage de séparateurs et d'électrodes est un aspect important, et la fabrication de ces architectures, même à l'échelle centimétrique et n'est pas évidente. La disposition d'une nouvelle cellule compacte PEC à deux compartiments permettant de contrôler *in situ* l'évolution du pH, de la température, et les concentrations d'oxygène et d'hydrogène est présentée. Enfin, une perspective de futures cellules PEC pour des applications pratiques est présentée.

INTRODUCTION

In a not so distant future, our society will need to be CO_2 neutral by limiting dependence on fossil energy sources. Solar energy can cover a substantial share of our energy needs, but the intermittency and diluteness of sunlight must be alleviated through storage in energy-dense commodities (Mathiesen and Lund, 2009). Solar production of hydrogen is such a storage option. Solar hydrogen can be produced using an electrolyzer powered by PhotoVoltaics (PV) and this technology is readily being applied worldwide. PhotoElectroChemical (PEC) cells present an alternative to the PV-electrolyzer combination by integration of photoabsorption and photocatalysis in a single device. Depending on the application, PEC cells have potential advantages over the coupled PV/electrolysis system as they can provide a more compact and integrated design. Additionally, they can be operated at lower current density exhibiting lower overpotentials and higher efficiency, and enable the use of earth-abundant, cheap electrode materials instead of expensive platinum metal for the electrolyzer. PV/electrolysis systems, need an additional DC/DC converter and thus three devices in total. Coupling these units causes energy losses. Both technologies have their merits and the reader is referred to literature for a more detailed scientific comparison (Rongé *et al.*, 2014a; Conibeer and Richards, 2007; Jacobsson *et al.*, 2014).

A PEC cell is a device that achieves water splitting and production of a solar commodity or fuel, *e.g.* hydrogen, with sunlight as the only energy source. The electrode assembly is the heart of the PEC cell. It consists of photoactive materials such as semiconductors or photosensitizers that absorb light and produce charge carriers. These are separated and consumed in redox reactions at the surface of the photocatalyst or the (electro)catalysts attached to the photoactive material. The electrode assembly is submerged in an electrolyte for ion conductivity and the whole is encased within the reactor or cell, which also offers the channeling for providing the educts to the reaction centers and evacuating the products. A PEC cell hereby simultaneously combines light absorption, charge generation and separation, hole, electron, ionic and molecular transport in a single device (Rongé *et al.*, 2014b). Hydrogen produced in PEC cells is a useful fuel as such. It can be compressed or converted downstream together with CO or CO_2 to liquid hydrocarbon fuel.

Steady progress is being made in the development of light absorbers and (photo)electrocatalysts for photoelectrochemical water splitting. These energy materials are conveniently tested in a three-electrode electrochemical setup to assess their water splitting potential. As PEC cells are reaching maturity, effort should be put into exploring the potential of new energy materials in assemblies, and separation of oxygen and hydrogen by separators dividing anode and cathode compartments. Separator-Electrode Assemblies (SEA) are being implemented in research programs (Rongé *et al.*, 2014a). This paper provides an overview of the development of PEC cells for research purposes, denoted as exPEC cells, and presents some ideas on the design of PEC cells for an ultimate practical application.

1 STATE-OF-THE ART OF EXPERIMENTAL PEC CELLS

The multitude of physical and (photo-electro)chemical processes in a PEC cell can only be achieved by a combination of different materials. Many configurations of functional assemblies are possible and different terminologies have been proposed (Rongé *et al.*, 2014b; Nielander *et al.*, 2014). After the successful research efforts on photoactive materials, electrocatalysts, and their combinations, the design of assemblies and the development of cells for operating assemblies are the next research challenge. Clever engineering of PEC cells and optimization at the different length scales are essential with respect to overall efficiency and cost.

It is important to make a distinction between wired and wireless electrode assembly configurations (Fig. 1), (Newman, 2013). In a wired configuration, electrical connection between anode and cathode is established by external wires. These connections enable measurement of (photo)currents and potentials using voltmeter,

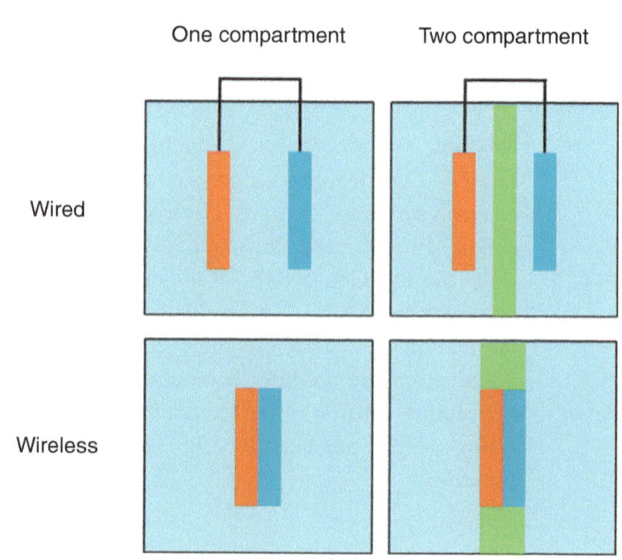

Figure 1

Schematic representation of four possible configurations of electrodes and molecular separators in exPEC cells. Red, blue and green colors represent anode, cathode and separator, respectively.

ammeter or potentiostat, and detailed (photo)electrochemical characterization of the assembly. The electron pathlength in wired configurations is considerably larger compared to a wireless configuration, possibly taking advantage of the high electrical conductivity of metal wires. At the same time, the electrodes can be placed very closely, reducing the ionic pathlength and enabling the use of low ionic conductivity electrolyte. The opposite holds for the wireless configuration, in which the electron pathlength is minimized while the ionic pathlength is enhanced. A further distinction can be made between wireless and monolithic setup. Wireless means that anode and cathode (and possibly separator) are part of the same physical assembly. Wireless systems are always monolithic. Wired system can be monolithic or not (Rongé *et al.*, 2014b, c).

PEC cells for research purposes denoted exPEC cells can be classified according to compartmentalization. ExPEC cells either have one or two compartments (Fig. 1). To avoid confusion, the term dual compartment will be used here when two compartments are molecularly separated[(1)] by using an ion exchange separator that allows ionic transport between both compartments but limits the mixing of molecular oxygen and hydrogen. Many different separators can be used such as glass frits, salt bridge or ion exchange membranes. The use of a molecular separator in dual compartment PEC cells has important advantage over single-compartment exPEC cells. It provides a means for product separation and facilitates analysis of reaction product formation and determination of Faradaic and quantum efficiencies.

1.1 Single Compartment Experimental PEC Cells

Single-compartment exPEC cells are most common in water splitting research. Their purpose is to characterize individual (photo)electrodes in a wired, three-electrode configuration (Chen Z. *et al.*, 2011). Here, the working electrode is the study object. The counter electrode typically is a Pt electrode with large surface area to limit its influence on the working electrode. A reference electrode is implemented. Electrochemical techniques for characterizing photoelectrodes include cyclic- or linear voltammetry, Mott-Schottky analysis, impedance spectroscopy, chronoamperometry, etc. (Chen Z. *et al.*, 2013).

Both wired and wireless configurations can be used to characterize an electrode assembly. For wired configurations in single exPEC cells, it is appropriate to use two-electrode measurements. For wireless samples, characterization is limited to product analysis and current cannot be measured. Depending on the sample and the properties to be

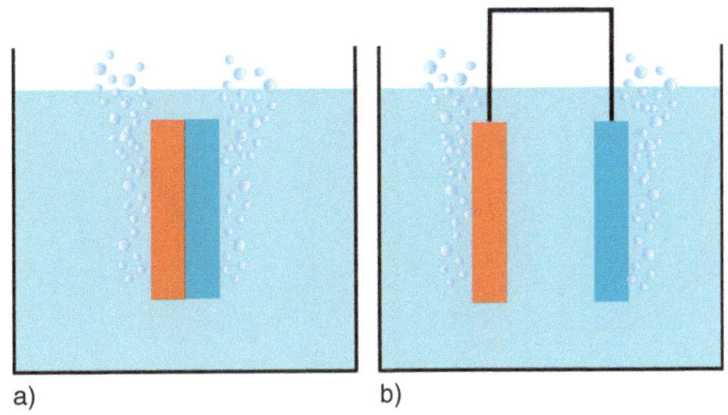

Figure 2

Single compartment exPEC cell with liquid filled container for a) wireless and b) wired electrode configuration.

investigated, many variations in exPEC cell design exist (Minggu *et al.*, 2010).

The most basic exPEC cell is an open beaker filled with a suitable electrolyte (Fig. 2) in which the assembly is submerged and illuminated. This type of setup has been used to demonstrate the working of both wired and wireless cells (Reece *et al.*, 2011; Kelly and Gibson, 2006). Obviously, this design is less convenient for reaction product quantification and it is often used for demonstration purposes only. A more advanced exPEC cell can be sealed from the environment in an air-tight manner to prevent electrolyte leaking and air diffusion into the cell. This is especially important as oxygen at the anode can act as an electron acceptor and its penetration into the cell leads to an overestimation of the oxygen production. Similarly, oxygen at the cathode will recombine with hydrogen to form water and thus underestimate the hydrogen production. In- and outlets can be foreseen to flush the volume with reactive or inert gases to create a controlled environment and enable quantitative product analysis by *e.g.* gas chromatography or mass spectrometry.

The exPEC cell can be made out of any type of glass such as Pyrex® and quartz for visible and UV-light illumination, respectively. However, depending on design, manufacturing these types of cells is not trivial and the implementation of quartz windows is expensive. Chemically inert polymer materials such as Teflon® or PolyEther Ether Ketone (PEEK) are often used as support materials. Insertion of smaller quartz or glass windows can be achieved by sealing with epoxy resin, although generally the use of O-rings is more convenient. ExPEC cells are commercially not available and researchers often design their own exPEC cells. Researchers design their own exPEC cells. Examples are the so called 'Cappuccino' cell designed at EPFL (Fig. 3a), (Lopes *et al.*, 2010), Lausanne, and the exPEC cell designed at TU Delft (van de Krol and Grätzel, 2012).

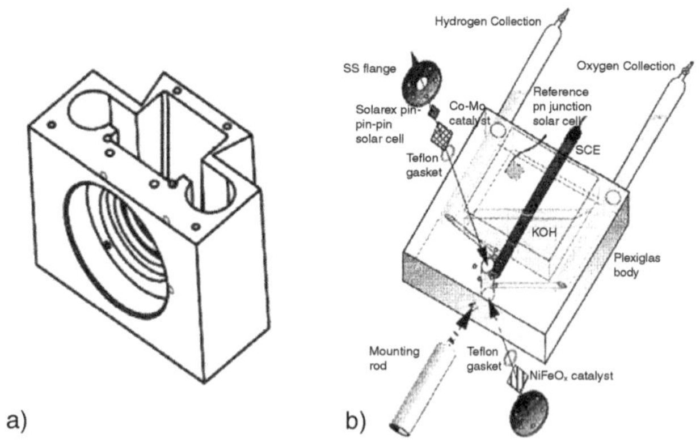

Figure 3

a)[(2)] Example of the Cappuccino cell designed at EPFL, Switerland, (Lopes *et al.*, 2010); b)[(3)] exPEC cell reported by Rocheleau *et al.* (1998).

The latter is an example of a gas tight exPEC cell made out of PEEK. It is handy for characterization of photoelectrodes in a 2- or 3-electrode setup in dark or illuminated conditions. The small characteristic distances between the electrodes minimize ohmic losses. Both the sample and fused silica window can be removed by a threaded insert with an O-ring for gas tightness.

The major drawback of single compartment cells is the mixing of the evolving oxygen and hydrogen gas which entails the complication of recombination reactions. There is a safety aspect if the concentration of hydrogen would reach the explosion limit (Schröder *et al.*, 2004). Hydrogen can be collected separately by proper design of the exPEC cell. In a wired configuration, compartmentalization can be achieved in an H-cell in which a glass frit reduces crossover of bubbles and products recombination. A similar approach was demonstrated by Rocheleau *et al.* (1998) with a wired configuration (Fig. 3b), (Miller *et al.*, 2003). This experimental setup is suited for characterization of samples working at high current densities as in this case readily detectable amounts of oxygen and hydrogen are produced. Spatial separation of the electrodes, however, leads to larger characteristics distances between the electrodes and correspondingly higher ohmic losses in the electrolyte which can be critical when working at high current densities.

It is obvious that alternative exPEC cell designs present advantages and disadvantages, and the choice will depend on the research focus. While most exPEC cells are suited for screening of (photo)electrodes, they are less convenient for characterizing wired or wireless separator-based electrode assemblies in which anode, cathode and separator are contributing to the overall performance of the system.

1.2 Dual Compartment Experimental PEC Cells

Dual compartment exPEC cells have the advantage of producing oxygen and hydrogen gas separately in anode and cathode compartments, respectively. Most designs are based on a Proton Exchange Membrane (PEM) fuel cell, with the water formation reaction occurring in reverse electrochemical direction. In dual compartment exPEC cells two- or four-electrode measurements are relevant. In a two-electrode setup, as opposed to a three-electrode setup, the counter electrode contributes to the photoelectrochemical characteristics and thus gives a more realistic representation of the complete PEC cell. In a four-electrode setup, two additional reference electrodes are added and their potential difference is measured while applying a current between the work- and counter electrode. This approach gives insight in potential losses between electrodes while discarding the electrodes themselves. It makes it possible to measure the electrolyte resistance (dependent on cell geometry and the resistance of the ionic separator).

The H-cell is a typical dual-compartment cell for wired configurations (Fig. 4), (Hernández-Pagán *et al.*, 2012; Lopes *et al.*, 2014). Anodic and cathodic reactions occur in two separate compartments, connected by a small tube in which the ionic separator is placed with appropriate sealing to prevent leaking. Product gases can be collected vertically while ionic transport occurs in horizontal direction. By implementing pH electrodes a buildup of pH gradient across the membrane can be observed (Hernández-Pagán *et al.*, 2012). The phenomenon of pH gradient buildup occurs even in concentrated buffers. Hernández-Pagán *et al.* (2012) revealed that the choice of electrolyte, electrolyte volume, and the ionic path length are key parameters for cell performance.

A more advanced type of cell was recently reported by Lopes *et al.* (2014), (Fig. 4b). The design has additional fused silica windows on both sides of the compact cell. Two compartments can be separated by a Teflon® separator. The illuminated sample surface area can be up to 100 cm^2. These properties enable detailed characterization of wired, non-monolithic PEC configurations.

Monolithic electrode assemblies offer opportunities for reducing ionic transport distances (Rongé *et al.*, 2014b). Ichikawa and Doi (1996) reported one of the first investigations of a monolithic assembly. They used a dual-compartment cell to investigate the water splitting reaction in a wired configuration with a thin film anatase anode and

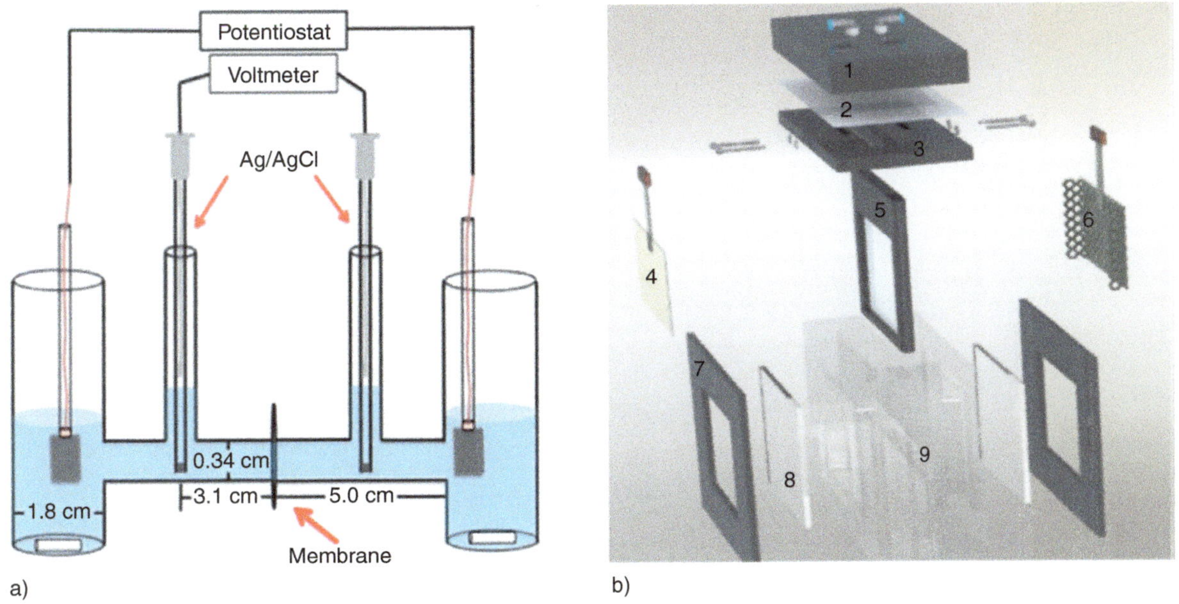

Figure 4

a)[4] Typical H-cell with four-electrode setup (Hernández-Pagán *et al.*, 2012); b)[5] dual compartment exPEC cell designed at LEPABE, Portugal (Lopes *et al.*, 2014).

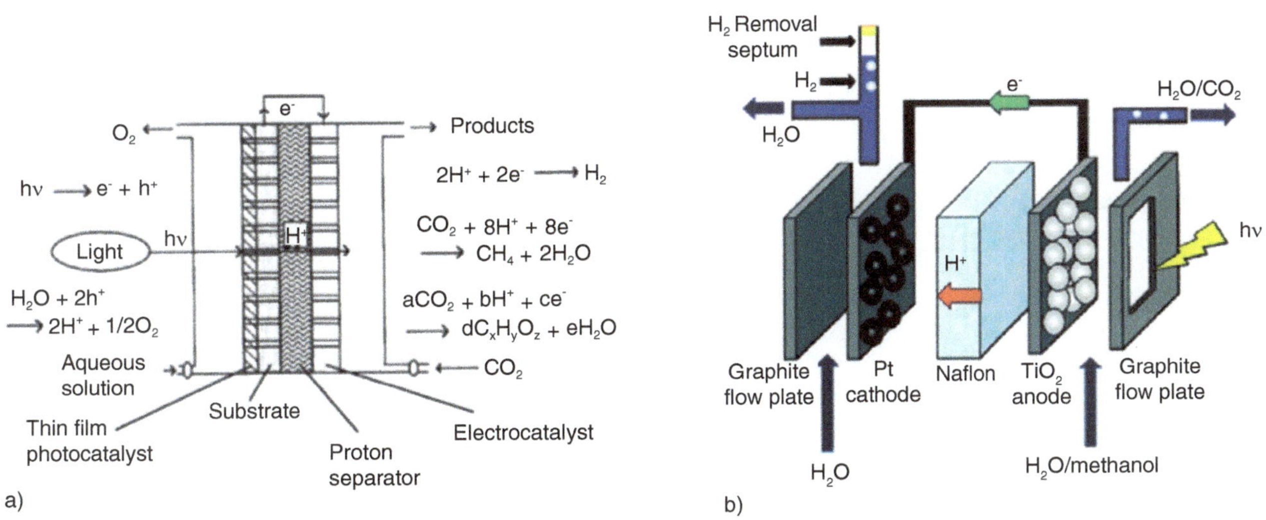

Figure 5

Schemes of dual compartment exPEC cells presented by a)[6] Ichikawa and Doi (1996) and b)[7] Seger and Kamat (2009).

---

a Pt cathode, separated by a Nafion membrane (Fig. 5a) (Ichikawa and Doi, 1996). Later on, similar cells were designed and used by others (Seger and Kamat, 2009; Ampelli *et al.*, 2010; Qian *et al.*, 2010).

Recently, some of us reported a monolithic assembly in a dual compartment exPEC cell that performs solar splitting of water molecules from water vapor collected from ambient air (Fig. 6), (Rongé *et al.*, 2013, 2014c; Schröder *et al.*, 2004). The use of carbon nanotubes coated with a conformal TiO_2 coating deposited by Atomic Layer Deposition (ALD) enabled protection of the sample against the detrimental effects of atmospheric oxygen. The same PEC cell was also

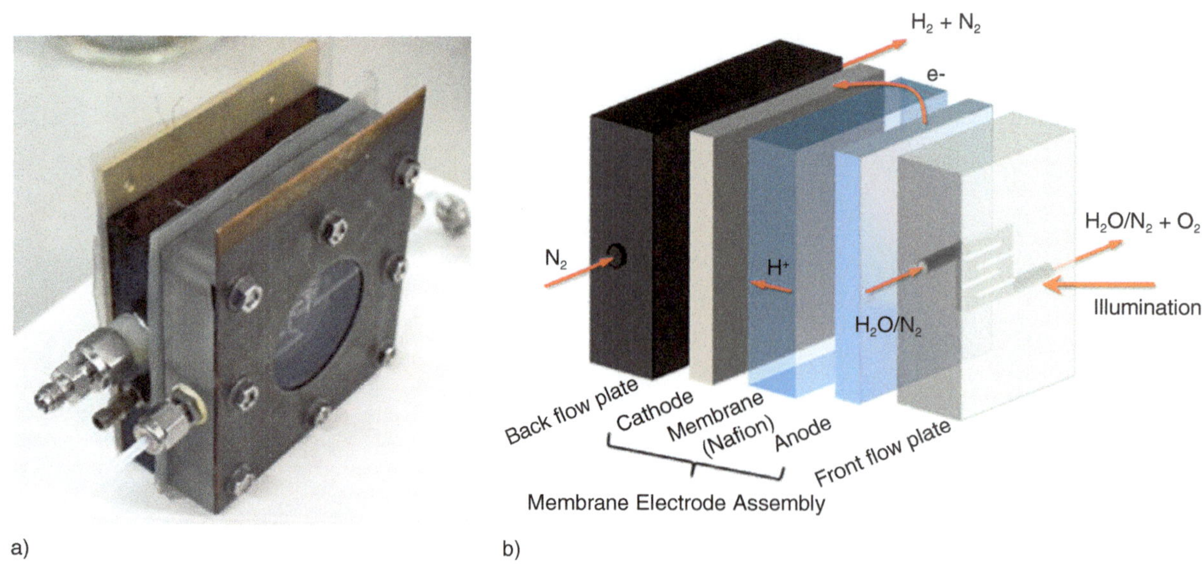

Figure 6

a) Fused silica/graphite dual-compartment exPEC cell from previous work (Schröder *et al.*, 2004) with illumination area of 0.785 cm^2; b) schematic representation of the cell and its components[(8)], (Rongé *et al.*, 2013).

used to investigate scalability of the separator electrode assembly as will be discussed below.

2 UPSCALING OF MONOLITHIC ASSEMBLIES AND EXPERIMENTAL PEC CELLS

Increasing surface area of Separator-Electrode Assemblies (SEA) is not obvious. In this section, the upscaling of a recently presented SEA (Rongé *et al.*, 2013) based on carbon nanotubes, carbon paper, Nafion membrane, anatase and platinum electrodes, tested in a dual compartment exPEC cell, is presented.

As mentioned previously, penetration of atmospheric oxygen has detrimental effects (Hagfeldt and Lindström, 1995) and, therefore, it is important for PEC cells to be air tight. Water splitting experiments were performed in a dual compartment exPEC cell with a surface area of 9 cm^2, roughly 10 times larger than in previous work (Rongé *et al.*, 2013). This exPEC cell (Fig. 6) is a dual-compartment cell with in- and outlets for reactant feeding and product collection. The cathode compartment was made out of graphite, has serpentine flow channels for optimal fluid flow over the sample surface while providing electrical contact with the electrode. The anode compartment was made entirely out of quartz and is provided with serpentine flow channels. The SEA was mounted in between two silicone gaskets which sealed off the sample from the environment and prevent leaking. Subsequently the mounted SEA was introduced between both compartments and screwed tight by means of 8 insulated screws. UV-illumination of the anode compartment was provided by means of a Xe lamp with IR filter (*Oriel 66984*, Newport, Irvine, CA) with an incident intensity of 100 $mW.cm^{-2}$. In- and outlets allow the exPEC cell to be flushed with either liquid water or water vapor at the anode while the cathode was fed with water saturated N_2. The outlets were connected to a mass spectrometer or gas chromatograph for analysis.

Chronoamperometry is convenient for SEA characterization. The photocurrent between anode and cathode upon illumination at short circuit (0 V) in a two-electrode setup was measured using a potentiostat (*Versastat 4*, Princeton Applied Research, Oak Ridge, TN with Versastudio electrochemical analysis software). The electrical connection was achieved either by direct contact with the graphite frameworks or if needed by placing Ti wires and stainless steel wires on the surface of the anode and cathode, respectively. The gas composition at the outlet of the cathode compartment was monitored during illumination with a Quantitative Gas Analyzer (*Hiden Analytical*, Warrington, UK). The hydrogen signal was corrected for interference by water and the N_2 signal was used as internal standard.

Anode and cathode electrodes consisted of carbon fiber paper coated with TiO_2 and Pt on Carbon Black (Pt/CB), respectively. SEA measuring ca. 1 cm^2 previously were prepared using a hot press procedure (Rongé *et al.*, 2013, 2014c).

[8] Reproduced from reference Rongé *et al.* (2013) with permission from The Royal Society of Chemistry.

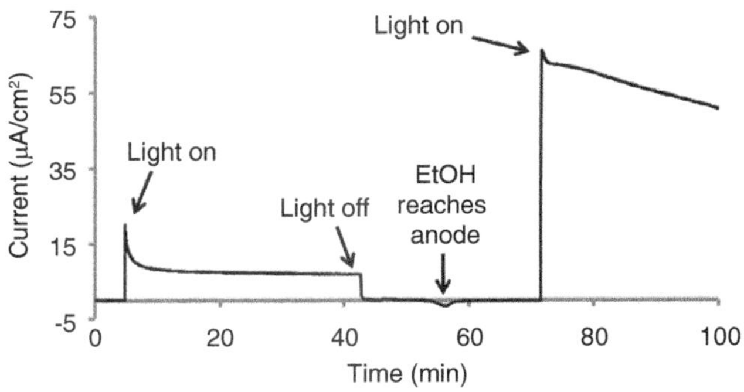

Figure 7

Influence of ethanol addition (after 55 min) on observed photocurrent in dual compartment exPEC cell. Ethanol acts as a hole scavenger and is oxidized providing electrons and protons for H_2 formation at the cathode.

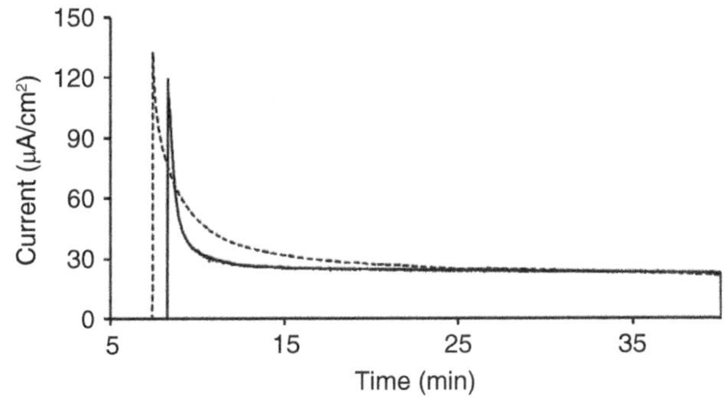

Figure 8

Photocurrent profiles in exPEC cells with illuminated surface area of 1 cm^2 (solid line) and 9 cm^2 (dotted line).

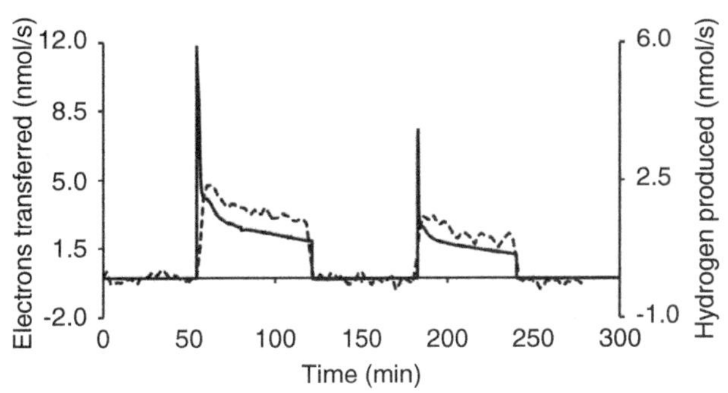

Figure 9

Photocurrent profile (solid) and continuous monitoring of hydrogen production by mass spectrometry (dotted).

The Nafion® membrane was sandwiched between the anode and cathode with a few drops of Nafion® suspension, which acted as an adhesive, added on each side. The resulting assembly was then placed between two Teflon® plates and pressed at 50 bar and 130°C. The SEA was then allowed to cool and was stored at room temperature before use.

Unfortunately, practical problems were encountered during fabrication of 9 cm^2 samples. Samples showed cracks in the carbon fiber substrate. These cracks locally interrupted the electrical contact. It was concluded that this cracking was due to a combination of high pressures and temperatures present in the hot press method causing curling of the Nafion membrane after pressing. To overcome the problem a thinner membrane Nafion® XL (27.5 μm thickness) was used. This more flexible variety of Nafion® membrane did not exhibit the curling effect. Additionally, the hot press method was adapted to lower pressures and temperatures.

The Nafion suspension used to prepare SEA contained residual ethanol solvent. EtOH is a known hole scavenger and readily oxidizes on TiO_2 surfaces to form acetaldehyde (Araña *et al.*, 2004).

$$Ethanol \leftrightarrow Acethaldehyde + 2e^- + 2H^+ \left(E^0 = +0.197\,\text{V}\right)$$

Thus, the presence of ethanol residues may have dramatic effects on PEC performance, leading to overestimation of the quantum efficiency. To assess the influence of ethanol impurities on PEC performance, the anode was flushed with N_2 purged through a water vessel containing 10% ethanol. Figure 7 shows the photocurrent profile in the presence of EtOH vapor. A short standard measurement was performed prior to the EtOH test to indicate the SEA and exPEC cell were working properly. After introduction of EtOH to the anode compartment, the photocurrent increased significantly. This current does not contribute to the water splitting reaction. This experience revealed that when working at low currents it is important to make sure no residual organic compounds are present. However, this observation also shows that PEC cells potentially can be used for photoelectrochemical degradation of volatile organic compounds to generate hydrogen.

Reproducible behavior was obtained with the 9 cm^2 SEA as shown in Figure 8. When using 1 cm^2 SEA samples the concentration of hydrogen at the cathode was often below the detection limit of the mass spectrometer (1×10^{-11} torr) and a continuous measurement was not possible. However, when upscaling, a continuous measurement of the produced hydrogen of H_2 formations during an unbiased experiment was possible, as shown in Figure 9.

In the first illumination period, a large anodic overshoot was observed and the production of hydrogen lagged the photocurrent. In the second illumination period, the anodic overshoot was smaller and the hydrogen production was detected faster after illumination. The change of behavior

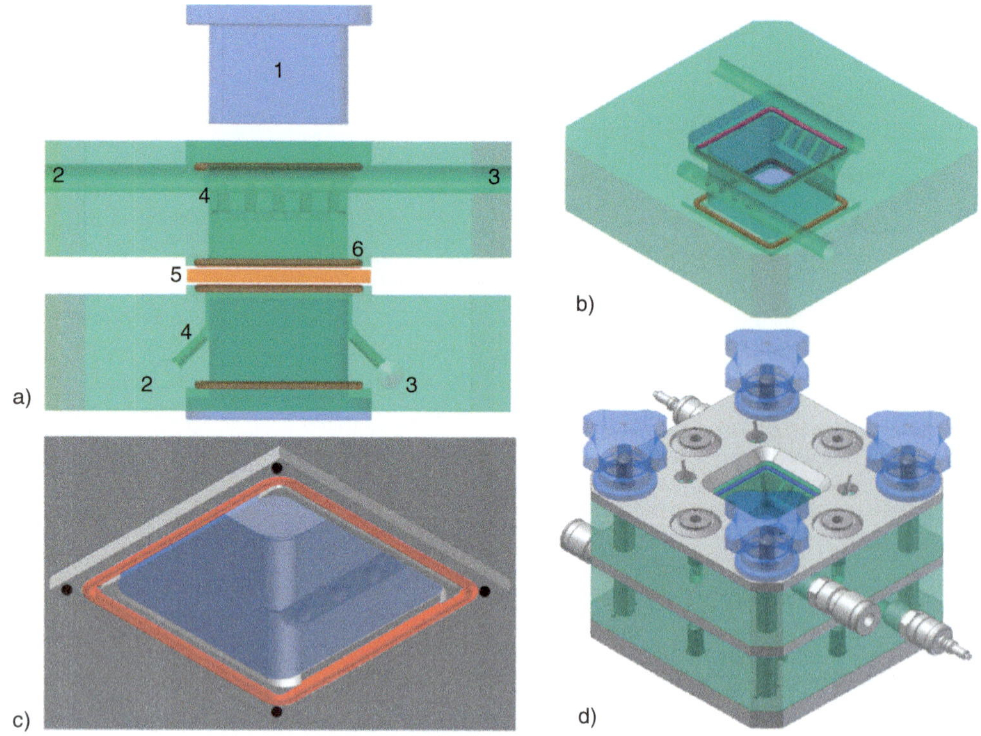

Figure 10

Advanced exPEC cell a) side view of the half cells. 1: Quartz window; 2: Reagent inlet; 3: Product outlet; 4: Inlet distribution or outlet collection; 5: Sample; 6: O-ring. b) Top-side view of the half-cell. c) Bottom-side view of the half-cell. The black dots indicate the spring contacts for electrical contact with the substrate. d) Assembled exPEC cell.

can be explained as follows. A fresh anode has a high concentration of trap states which can be saturated by the sudden production of photoexcited holes. Additionally, the cathode forms hydrides at the Pt surface until saturation only after which the formation of hydrogen molecules was started. In the second illumination period, the surface states at the anode were not yet fully discharged and the Pt surface at the cathode presumably remained hydride saturated. This explains the lower anodic overshoot and the immediate detection of hydrogen production. The lack of a hydrogen overshoot could also be explained by the presence of this charge accumulation at the interface and by the temporal resolution of 2 minutes of the mass spectrometer. Additionally, it is important to mention that the length of the transfer line from cathode outlet to the mass spectrometer needs to be taken into account when designing an exPEC cell to avoid lagging artifacts.

3 ADVANCED RESEARCH PEC CELLS

State-of-the-art exPEC cells offers access to a variety of relevant experimental parameters, namely illumination intensity and composition, electrolyte composition (concentration and pH), reactant flow rates, and gas composition. There is ample room for refinement and sophistication. Future exPEC cells should enable investigation of the influence of assembly configuration, geometry, temperature, humidity, light incident angle, etc. For this aim, a new dual-compartment exPEC cell with additional functionalities was designed (Fig. 10). The cell is divided in two compartments and consists of two modular half-cells which can be fabricated and optimized separately. Each half-cell is fabricated out of PEEK and is equipped with a quartz window (1) which is pressed into the PEEK half-cell and sealed with a square-shaped polymer sealing. The illuminated surface area is 3 × 3 cm extendable by using a larger window.

Each half-cell was manufactured from a solid PEEK block through mechanical milling. The in- and outflow of the cell are conceived for an even flow distribution. At the inlet (2), the flow enters a cylindrical channel after which it is equally divided into 5 smaller cylindrical channels (4). Before reaching the SEA surface, these 5 channels gather into one common channel to redistribute the flow. The flow channel over the sample chamber is approximately 1 mm

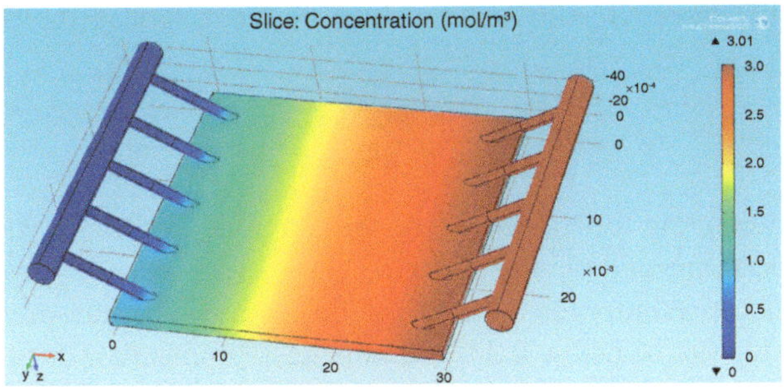

Figure 11

Simulation of the new exPEC cell operation showing the homogeneous buildup of the hydrogen concentration through water splitting in vapor phase.

high to avoid product buildup in dead volumes. The path for reactant and product flow in gas or liquid phase is well defined. The SEA (5) can be mounted in between these two half-cells, each of which seal off the sample from the environment by a square-shaped sealing (6). An electrical contact with the sample is obtained by a connection with four spring contacts located at the four corners of the sample for an equal current or potential distribution across the sample surface area.

Numerical simulations of the 3-D design were performed to evaluate the uniformity of the flow. Width and height of the different channels were varied as well as the flow velocity. A constant hydrogen production activity across the sample surface was assumed. The resulting pressure loss, concentration gradient across the sample surface and velocity profile were calculated using numerical software. The results showed no significant impact of the variables on the observables and thus it was concluded that an equal distribution of the flow over the sample surface could be obtained with this design. Additionally, the model showed that the products can be evacuated from the flow channel without creating a significant concentration gradient over the sample surface (Fig. 11).

By adjusting the height of the quartz window, it is possible to implement sensors close to the sample surface. Sealable insertion holes for the implementation of various sensors is envisioned. Reference electrodes can be inserted for (photo)electrochemical testing in 3- or 4-electrode setup. pH sensors can be mounted for *in situ* investigation of pH gradients. Such pH gradients over SEA recently have been simulated and experimentally observed (Hernández-Pagán *et al.*, 2012; Jin *et al.*, 2014). Additionally, O_2- and H_2-microsensors fitting in the flow channel can be mounted for time resolved reaction monitoring, close to the active SEA surface. Temperatures can be monitored by insertion of thermocouples or by infrared detection directly on the surface.

4 DESIGN OF PRACTICAL SMALL TO LARGE SCALE PEC CELLS

PEC cells can be envisioned at a small scale for domestic applications. A PEC cell with an efficiency of 20% would need about 50 m² of surface area to cover the daily energy needs of an average family with a consumption of 20 $kWh.day^{-1}$. This energy corresponds to a volume of about 3 m^3 of hydrogen at atmospheric pressure which can be highly reduced when pressurized. PEC cells can also be employed on a large scale for industrial applications where the produced hydrogen gas should be pressurized to ensure practical transportation. Additionally hydrogen could be coupled to carbon dioxide mitigation for storage as hydrocarbons or used for downstream processes such as ammonia production.

Anyhow, practical PEC cells will differ from exPEC cells in many respects. Commercial PEC cells should have a design that is simple, cost effective, sustainable, safe, and durable. As the scale of PEC-based hydrogen processing increases more technical difficulties concerning both the SEA and the cell assembly will arise. Many designs are possible and have already been investigated by techno-economic and life cycle analysis (Pinaud *et al.*, 2013; James *et al.*, 2009; Sathre *et al.*, 2014; Zhai *et al.* 2013). In general, two designs seem feasible for large-scale hydrogen production:

- fixed panel PEC cells,
- concentrator PEC cells.

Systems based on suspended nanoparticles have been proposed as well, but those are in an earlier development state. Storage of hydrogen is still a highly disputed topic within the community of PEC cells, the storage *versus* downstream transformation will not be considered here.

Both wired and wireless configurations have been proposed for large scale fixed panel PEC cells (Pinaud *et al.*, 2013; James *et al.*, 2009; Sathre *et al.*, 2014; Zhai *et al.*, 2013). In wired design, the importance of correct electrical contacts of the large electrodes should not be overlooked. Although ohmic losses in electrical systems are generally negligible compared to losses due to ionic transport, in practice it is not evident to manufacture effective electrical contacts and they can be quite demanding in the overall cell design. A low resistance ohmic contact is imperative for current collection and is not always straightforward to obtain as it depends on the involved materials[(9)].

[9] The choice of contact material is based on the work functions of both materials.

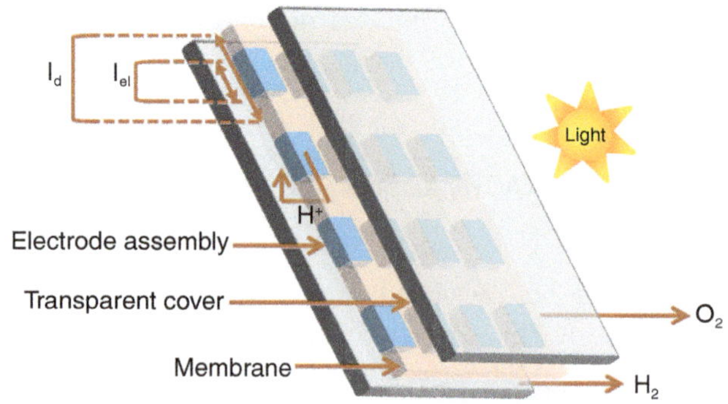

Figure 12

Separator embedded electrode assembly which combines a wireless PEC cell assembly with short ionic pathways.

Additionally, the contact must be protected from the electrolyte and the oxidative impact of the environment. Like in the PV manufacturing and scale up, there will be a trade-off between shading effects and ohmic losses, and a small detrimental effect on overall efficiency cannot be avoided. The use of Transparent Conductive Oxides (TCO) can minimize active surface area losses at the expense of the introduction of larger ohmic losses by the electron transport as TCO show typically larger resistances compared to metals. This disadvantage is slightly reduced by the fact that a smaller current density is experienced in the TCO compared to the metallic current collector due to the larger area available for electron transport in the TCO.

For a wireless configuration, the ionic transport and its associated losses are dominating (Newman, 2013). Here, the electrical contact is made within the SEA and the need for complex contact methods is absent. However, as most efficient PEC cells use dense silicon technology, an ionic transportation problem arises. As the PEC cell surface area increases, ionic transport distance around the SEA increases accordingly (Haussener *et al.*, 2012). In proposed large PEC cell designs, these ohmic losses due to long ion transportation distances are neglected. For devices exceeding centimeter scale dimensions, overpotentials increase significantly with size and will eventually hinder the system operation when not enough driving potential is delivered by the SEA. This problem is gaining awareness and could possibly be overcome by embedding the assemblies as an array inside a membrane as depicted in Figure 12. However, from a manufacturing point of view, this would be difficult to produce as the dimensions of this array should be in the cm- to µm-scale. It is currently unknown if it would be technically and economically possible to produce such a device. Either way, it is obvious that for wireless systems there is a trade-off between the illuminated area and the area required for ionic transport. This trade-off can be characterized by the ratio f:

$$f = \frac{l_{el}}{l_d} \tag{1}$$

representing the ratio between the electrode length, l_{el}, and the illuminated device length, l_d (Fig. 12). f must be close to 1 to maximize the light absorption potential but must preferably not be below 0.9 to avoid large overpotentials due to high ionic current densities (Haussener *et al.*, 2012). Another possible way to alleviate this trade-off is the use of carefully designed membrane-based PEC cells in which the membrane is placed vertically to minimize its share in the aforementioned ratio (Walczak *et al.*, 2015). Another approach is the use of porous assemblies that simultaneously and perpendicular to the surface allow electronic and ionic transport. One example of such a system are the membrane embedded silicon microwires as reported by Spurgeon *et al.* (2011). Here, silicon microwires with additional co-catalysts significantly increase active surface area and are embedded in a Nafion® membrane for proton transport with an additional layer of PEDOT:PSS[(10)] for electrical connection between the anodic and cathodic wires.

In fixed panel PEC cells, the SEA ideally will be placed between two plates (*e.g.* Pyrex®) that serve as flow plates to provide water flow to the anode and collect hydrogen at the cathode. Single compartment and dual compartment systems have been proposed (Pinaud *et al.*, 2013; Sathre *et al.*, 2014; Zhai *et al.*, 2013). Compared to exPEC cells, the requirements will be less stringent, but cost and energy requirements for production become important issues. The use of expensive quartz glass is no option. Unless needed, no complicated flow designs and expensive fabrication methods should be used in the simplest possible designs.

In single compartment systems, inclination of the panel has been proposed for the separation of hydrogen and oxygen, and inclination simultaneously ensures optimal illumination intensity (Pinaud *et al.*, 2013).

The choice of electrolyte will have a large impact on the final design, practicability, and cost of a PEC cell. To allow sufficient ionic transport when no separators are present, often highly acidic or basic electrolytes are used (Seger and Kamat, 2009; Chen Q. *et al.*, 2013; Currao *et al.*, 2004). These have the disadvantage of being corrosive and expensive which makes them difficult to apply on a large scale as highly resistant reactor materials must be used throughout the system. In this respect, the development of suitable photoactive materials and catalysts, and device designs, that continuously work under neutral pH conditions is a much investigated aspect of PEC water splitting.

10 PEDOT:PSS = Poly(3,4-EthyleneDiOxyThiophene) Polystyrene Sulfonate.

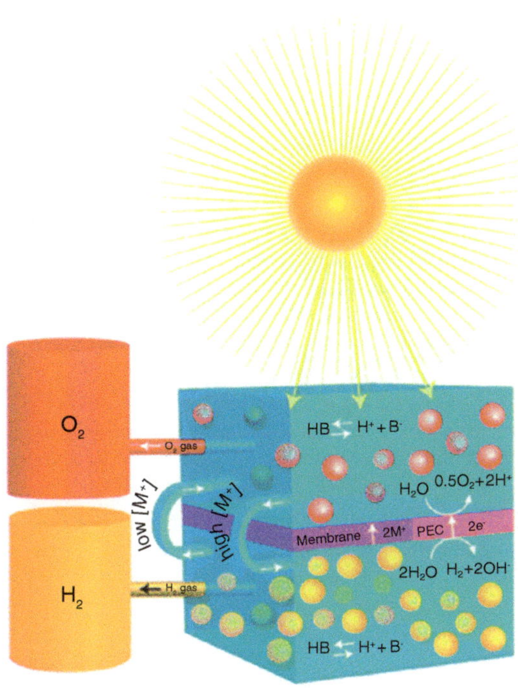

Figure 13

Self-regulating, continuously operating solar fuel generator reported by Modestino *et al.* (2014).

Additionally, there will be a trade-off concerning the amount of electrolyte inside the PEC cell which needs a certain volume above the electrode to minimize ohmic losses in the solution (Haussener *et al.*, 2012), but must not exceed an upper limit because the added weight and light absorption can become performance limiting factors (Döscher *et al.*, 2014).

However, even when working in concentrated buffered solutions at neutral pH, the formation of a pH gradient is observed which makes it impossible to keep working in a neutral pH. Additionally, the buffered electrolyte is depleted at the cathode side and a concentration gradient is developed during operation which poses a threat for long term operation of such systems. In the future, adequate membrane materials could allow selective transport of neutral buffer species back to the anode side to alleviate this problem. Here, reactor design can contribute to solutions as well. As reported recently by Modestino *et al.* (2014), by introducing a controlled recirculating system it is possible to prevent the formation of a significant concentration gradient in buffered near neutral electrolytes, while maintaining sufficient gaseous product separation (Fig. 13). Another option is the use of ambient air as a feedstock. It has been proposed that even in remote, dry areas, air contains enough water to sustain PEC operation. This would make the reactor design a lot easier as water in vapor phase reduces the weight of the PEC cell significantly, is non-corrosive and poses no risk of freezing in winter. However, a cover would still be needed to protect the surface from heavy winds and debris (Rongé *et al.*, 2014a, c).

Concentrating PEC cells, depending on the solar concentration ratio, use between 10 to 1 000 times less surface area of the most expensive part of the PEC cell namely the electrode assembly. Here, the ratio *f* will become even more vital as the small surface area must be used optimally and overpotentials become more severe as the current densities are significantly higher. Concentration PEC cells need to be operated under pressure to take maximum profit from the concentrated electric currents. Concentrator PEC cells have been reported to be technically feasible and economically attractive for large scale hydrogen production. However, publications in literature concerning this technology are scarce (Kelly and Gibson, 2008). Nonetheless, it is certain that multijunction materials will have to be used to maximally take advantage of the solar spectrum and an optimization of bandgap combinations will be needed for maximal efficiency (Haussener *et al.*, 2013). Designing and building the reactor will come at a higher cost as the requirements will be more severe. Elevated temperatures and pressures require more resistant materials both for the electrode assembly as for the reactor, and possibly require an advanced thermal management. Additionally, for wired configurations, electrical contacts will be more resistant (Gessert and Coutts, 1990) and without the use of noble metals, current collection will become problematic as the state-of-the-art reported catalysts for both redox reactions are limited to currents of about 1 $A.cm^{-2}$ which would limit the concentration to about 100 suns.

CONCLUSIONS

An advanced understanding of PEC cell operation is required for practical and large-scale implementation. This understanding can be gained by the design and use of advanced, custom-made experimental PEC cells (exPEC cells). Currently, exPEC cells are mostly single- or dual-compartment cells and are used for evaluation of individual energy materials and their integration in SEA and photoactive assemblies. However, if these systems are to be applied, more insight is to be gained using dual compartments cells with uniform flow patterns and equipped with reference electrodes, and temperature- and pH sensors as this will reflect more realistic operating conditions. An advanced exPEC cell design meeting these requirements was proposed. Additionally, it is emphasized upscaling of sample fabrication processes plays a crucial role in the applicability of large-scale PEC cells. Care must be taken that no other side reactions due to the presence of impurities can occur at the surface caused as this will overestimate the performance. As size increases and samples are often fragile, technical problems arise which have a detrimental effects on the performance of the PEC cell. Large-scale manufacturing of photoactive assemblies such as spatial ALD and inktjet printing have been proposed and these techniques could provide a suitable method for large scale production of PEC cells (Hiller *et al.*, 2002; Poodt *et al.*, 2012).

Wired or small-scale wireless dual-compartment membrane-based flow-type PEC cells are expected to dominate the future PEC market for solar hydrogen production as they provide means for the continuous production and safe separation of hydrogen and oxygen. Therefore a focus on these PEC cells must be put in future research, additionally focusing on scalability in realistic conditions. The potential for concentrated PEC cells for large-scale solar hydrogen production must be investigated as it holds great promise as an alternative for non-concentrated PEC cells which seem to be more suitable for small-scale applications.

ACKNOWLEDGMENTS

The authors thank Marc Peeters and Joop Vandeurzen for technical drawings and building of the new experimental PEC cell, Philip van Hoof (MSc) for performing PEC experiments and Louis Huguenin (MSc) for performing modeling simulations. The Flemish government is acknowledged for Methusalem long-term structural funding of JAM. JR is grateful to Research Foundation – Flanders (FWO) for a research fellowship.

REFERENCES

Ampelli C., Centi G., Passalacqua R., Perathoner S. (2010) Synthesis of solar fuels by a novel photoelectrocatalytic approach, *Energy & Environmental Science* **3**, 3, 253.

Araña J., Doña-Rodríguez J.M., González-Díaz O., Tello Rendón E., Herrera Meliá J.a., Colón G., Navío J.a., Pérez Peña J. (2004) Gas-phase ethanol photocatalytic degradation study with TiO_2 doped with Fe, Pd and Cu, *Journal of Molecular Catalysis A: Chemical* **215**, 1-2, 153-160.

Chen Q., Li J., Li X., Huang K., Zhou B., Shangguan W. (2013) Self-biasing photoelectrochemical cell for spontaneous overall water splitting under visible-light illumination, *ChemSusChem* **6**, 7, 1276-1281.

Chen Z., Jaramillo T.F., Deutsch T.G., Kleiman-Shwarsctein A., Forman A.J., Gaillard N., Garland R., Takanabe K., Heske C., Sunkara M., McFarland E.W., Domen K., Miller E.L., Turner J.a., Dinh H.N. (2011) Accelerating materials development for photoelectrochemical hydrogen production: Standards for methods, definitions, and reporting protocols, *Journal of Materials Research* **25**, 01, 3-16.

Chen Z., Dinh H.N., Miller E. (2013) *Photoelectrochemical water splitting Standards, Experimental Methods, and Protocols*, Springer, New York.

Conibeer G., Richards B. (2007) A comparison of PV/electrolyser and photoelectrolytic technologies for use in solar to hydrogen energy storage systems, *International Journal of Hydrogen Energy* **32**, 14, 2703-2711.

Currao A., Reddy V.R., van Veen M.K., Schropp R.E.I., Calzaferri G. (2004) Water splitting with silver chloride photoanodes and amorphous silicon solar cells, *Photochemical & Photobiological Sciences: Official Journal of the European Photochemistry Association and the European Society for Photobiology* **3**, 11-12, 1017-1025.

Döscher H., Geisz J.F., Deutsch T.G., Turner J.A. (2014) Sunlight absorption in water – efficiency and design implications for photoelectrochemical devices, *Energy & Environmental Science* **7**, 2951-2956.

Gessert T., Coutts T. (1990) Requirements of Electrical Contacts to Photovoltaic Solar Cells, *MRS Proceedings*.

Hagfeldt A., Lindström H. (1995) Photoelectrochemical studies of colloidal TiO 2 films: The effect of oxygen studied by photocurrent transients, *Journal of Electroanalytical Chemistry* **381**, 39-46.

Haussener S., Xiang C., Spurgeon J.M., Ardo S., Lewis N.S., Weber A.Z. (2012) Modeling, simulation, and design criteria for photoelectrochemical water-splitting systems, *Energy & Environmental Science* **5**, 12, 9922.

Haussener S., Hu S., Xiang C., Weber A.Z., Lewis N.S. (2013) Simulations of the irradiation and temperature dependence of the efficiency of tandem photoelectrochemical water-splitting systems, *Energy & Environmental Science* **6**, 12, 3605.

Hernández-Pagán E.a., Vargas-Barbosa N.M., Wang T., Zhao Y., Smotkin E.S., Mallouk T.E. (2012) Resistance and polarization losses in aqueous buffer–membrane electrolytes for water-splitting photoelectrochemical cells, *Energy & Environmental Science* **5**, 6, 7582.

Hiller J., Mendelsohn J.D., Rubner M.F. (2002) Reversibly erasable nanoporous anti-reflection coatings from polyelectrolyte multilayers, *Nature Materials* **1**, 1, 59-63.

Ichikawa S., Doi R. (1996), Hydrogen production from water and conversion of carbon dioxide to useful chemicals by room temperature photoelectrocatalysis, *Catalysis Today* **27**, 12, 271-277

Jacobsson T.J., Fjällström V., Edoff M., Edvinsson T. (2014) Sustainable Solar Hydrogen Production: From Photoelectrochemical Cells to PV-Electrolysis and Back Again, *Energy & Environmental Science* **7**, 2056-2070.

James B., Baum G.N., Perez J., Baum K.N. (2009), *Technoeconomic analysis of photoelectrochemical (PEC) hydrogen production.*

Jin J., Walczak K., Singh M.R., Karp C., Lewis N.S., Xiang C. (2014) An experimental and modeling/simulation-based evaluation of the efficiency and operational performance characteristics of an integrated, membrane-free, neutral pH solar-driven water-splitting system, *Energy & Environmental Science* **7**, 10, 3371-3380.

Kelly N., Gibson T. (2006) Design and characterization of a robust photoelectrochemical device to generate hydrogen using solar water splitting, *International Journal of Hydrogen Energy* **31**, 12, 1658-1673.

Kelly N., Gibson T. (2008) Solar energy concentrating reactors for hydrogen production by photoelectrochemical water splitting, *International Journal of Hydrogen Energy* **33**, 22, 6420-6431.

Lopes T., Andrade L., Ribeiro H.A., Mendes A. (2010) Characterization of photoelectrochemical cells for water splitting by electrochemical impedance spectroscopy, *International Journal of Hydrogen Energy* **35**, 20, 11601-11608.

Lopes T., Dias P., Andrade L., Mendes A. (2014) An innovative photoelectrochemical lab device for solar water splitting, *Solar Energy Materials and Solar Cells* **128**, 399-410.

Mathiesen B., Lund H. (2009) Comparative analyses of seven technologies to facilitate the integration of fluctuating renewable energy sources, *IET Renewable Power Generation* **3**, August 2008, 190-204.

Miller E.L., Rocheleau R.E., Deng X.M. (2003) Design considerations for a hybrid amorphous silicon/photoelectrochemical multijunction cell for hydrogen production, *International Journal of Hydrogen Energy* **28**, 615-623.

Minggu L.J., Wan Daud W.R., Kassim M.B. (2010) An overview of photocells and photoreactors for photoelectrochemical water splitting, *International Journal of Hydrogen Energy* **35**, 11, 5233-5244.

Modestino M.a., Walczak K., Berger A.D., Evans C.M., Haussener S., Koval C., Newman J., Ager J.W., Segalman R.a. (2014) Robust production of purified H_2 in a stable, self-regulating, and continuously operating solar fuel generator, *Energy & Environmental Science* **7**, 1, 297-301.

Newman J. (2013) Scaling with Ohm's Law; Wired *vs.* Wireless Photoelectrochemical Cells, *Journal of the Electrochemical Society* **160**, 3, F309-F311.

Nielander A.C., Shaner M.R., Papadantonakis K.M., Francis S.a., Lewis N.S., (2014) A taxonomy for solar fuels generators, *Energy Environ. Sci.* **8**, c, 16-25.

Pinaud B.a, Benck J.D., Seitz L.C., Forman A.J., Chen Z., Deutsch T.G., James B.D., Baum K.N., Baum G.N., Ardo S., Wang H., Miller E.L., Jaramillo T.F. (2013) Technical and economic feasibility of centralized facilities for solar hydrogen production *via* photocatalysis and photoelectrochemistry, *Energy & Environmental Science* **6**, 7, 1983.

Poodt P., Cameron D.C., Dickey E., George S.M., Kuznetsov V., Parsons G.N., Roozeboom F., Sundaram G., Vermeer A. (2012) Spatial atomic layer deposition: A route towards further industrialization of atomic layer deposition, *Journal of Vacuum Science & Technology A: Vacuum, Surfaces, and Films* **30**, 1, 010802.

Qian F., Wang G., Li Y. (2010) Solar-driven microbial photoelectrochemical cells with a nanowire photocathode, *Nano Letters* **10**, 11, 4686-4691.

Reece S., Hamel J., Sung K., Jarvi T., Esswein A.J., Pijpers J.J.H., Nocera D.G. (2011) Wireless solar water splitting using silicon-based semiconductors and earth-abundant catalysts, *Science* **334**, 645-648.

Rocheleau R.E., Miller E.L., Misra A. (1998) High-efficiency photoelectrochemical hydrogen production using multijunction amorphous silicon photoelectrodes, *Energy & Fuels* **12**, 1, 3-10.

Rongé J., Nijs D., Kerkhofs S., Masschaele K., Martens J. (2013) Chronoamperometric study of membrane electrode assembly operation in continuous flow photoelectrochemical water splitting, *Physical Chemistry Chemical Physics* **15**, 23, 9315-9325.

Rongé J., Bosserez T., Huguenin L., Dumortier M., Haussener S., Martens J. (2014a) Solar hydrogen reaching maturity, *OGST - Revue d'IFP Energies Nouvelles* (this issue). DOI: 10.2516/ogst/201406.

Rongé J., Bosserez T., Martel D., Nervi C., Boarino L., Taulelle F., Decher G., Bordiga S., Martens J.a. (2014b) Monolithic cells for solar fuels, *Chemical Society Reviews* **43**, 7963-7981.

Rongé J., Deng S., Pulinthanathu Sree S., Bosserez T., Verbruggen S.W., Kumar Singh N., Dendooven J., Roeffaers M.B.J., Taulelle F., De Volder M., Detavernier C., Martens J.A. (2014c) Air-based photoelectrochemical cell capturing water molecules from ambient air for hydrogen production, *RSC Advances* **4**, 55, 29286-29290.

Sathre R., Scown C.D., Morrow W.R., Stevens J.C., Sharp I.D., Ager J.W., Walczak K., Houle F.a., Greenblatt J.B. (2014) Life-cycle net energy assessment of large-scale hydrogen production *via* photoelectrochemical water splitting, *Energy & Environmental Science* **7**, 10, 3264-3278.

Schröder V., Emonts B., Janßen H., Schulze H.P. (2004) Explosion Limits of Hydrogen/Oxygen Mixtures at Initial Pressures up to 200 bar, *Chemical Engineering & Technology* **27**, 8, 847-851.

Seger B., Kamat P. (2009) Fuel cell geared in reverse: photocatalytic hydrogen production using a TiO2/Nafion/Pt membrane assembly with no applied bias, *The Journal of Physical Chemistry C* **113**, 43, 18946-18952.

Spurgeon J.M., Walter M.G., Zhou J., Kohl P. a., Lewis N.S. (2011) Electrical conductivity, ionic conductivity, optical absorption, and gas separation properties of ionically conductive polymer membranes embedded with Si microwire arrays, *Energy & Environmental Science* **4**, 5, 1772-1780.

van de Krol R., Grätzel M. (2012) *Photoelectrochemical hydrogen production*, Springer, New York.

Walczak K., Chen Y., Karp C., Beeman J.W., Shaner M., Spurgeon J., Sharp I.D., Amashukeli X., West W., Jin J., Lewis N.S., Xiang C. (2015) Modeling, Simulation, and Fabrication of a Fully Integrated, Acid-stable, Scalable Solar-Driven Water-Splitting System, *Chem-SusChem* **8**, 3, 544-551.

Zhai P., Haussener S., Ager J., Sathre R., Walczak K., Greenblatt J., McKone T. (2013) Net primary energy balance of a solar-driven photoelectrochemical water-splitting device, *Energy & Environmental Science* **6**, 8, 2380.

4

Zn(II), Mn(II) and Sr(II) Behavior in a Natural Carbonate Reservoir System. Part I: Impact of Salinity, Initial pH and Initial Zn(II) Concentration in Atmospheric Conditions

B. Auffray[1], B. Garcia[1]*, C.-P. Lienemann[2], L. Sorbier[2] and A. Cerepi[3]

[1] IFP Energies nouvelles, 1 et 4 avenue de Bois Préau, 92852 Rueil-Malmaison - France
[2] IFP Energies nouvelles, Rond-point de l'échangeur de Solaize, BP 3, 69360 Solaize - France
[3] EA 4592 G&E, ENSEGID, Université de Bordeaux, 1 allée F. Daguin, 33607 Pessac Cedex - France
e-mail: baptiste.auffray@beicip.com - bruno.garcia@ifpen.fr
charles.lienemann@ifpen.fr - loic.sorbier@ifpen.fr - adrian.cerepi@ipb.fr

* Corresponding author

Abstract — *The sorption of inorganic elements on carbonate minerals is well known in strictly controlled conditions which limit the impact of other phenomena such as dissolution and/or precipitation. In this study, we evidence the behavior of Zn(II) (initially in solution) and two trace elements, Mn(II) and Sr(II) (released by carbonate dissolution) in the context of a leakage from a CO_2 storage site. The initial pH chosen are either equal to the pH of the water-CO_2 equilibrium (~ 2.98) or equal to the pH of the water-CO_2-calcite system (~ 4.8) in CO_2 storage conditions. From this initial influx of liquid, saturated or not with respect to calcite, the batch experiments evolve freely to their equilibrium, as it would occur in a natural context after a perturbation.*
The batch experiments are carried out on two natural carbonates (from Lavoux and St-Emilion) with $P_{CO_2} = 10^{-3.5}$ bar, with different initial conditions ($[Zn(II)]_i$ from 10^{-4} to 10^{-6} M, either with pure water or 100 g/L NaCl brine). The equilibrium regarding calcite dissolution is confirmed in all experiments, while the zinc sorption evidenced does not always correspond to the two-step mechanism described in the literature. A preferential sorption of about 10% of the concentration is evidenced for Mn(II) in aqueous experiments, while Sr(II) is more sorbed in saline conditions. This study also shows that this preferential sorption, depending on the salinity, is independent of the natural carbonate considered. Then, the simulations carried out with PHREEQC show that experiments and simulations match well concerning the equilibrium of dissolution and the sole zinc sorption, with log $K_{Zn(II)}$ ~ 2 in pure water and close to 4 in high salinity conditions. When the simulations were possible, the log K values for Mn(II) and Sr(II) were much different from those in the literature obtained by sorption in controlled conditions. It is shown that a new conceptual model regarding multiple Trace Elements (TE) sorption is required, to enable us to better understand the fate of contaminants in natural systems.

Résumé — Comportement du Zn(II), Mn(II) et Sr(II) au sein d'un système réservoir carbonate naturel. Partie 1 : impact de la salinité, du pH initial et de la concentration en Zn(II) en conditions atmosphériques — La sorption d'éléments inorganiques sur les minéraux carbonatés est bien connue

dans des conditions strictement contrôlées, limitant l'impact d'autres phénomènes tels que la dissolution et/ou précipitation. Dans cette étude, nous mettons en évidence le comportement de Zn(II) (initialement en solution) et deux éléments « traces », Mn(II) et Sr(II) (relargués au cours de la dissolution des carbonates) dans un contexte d'une fuite d'un site de stockage de CO_2. Le pH initial choisi est, soit égal au pH de l'équilibre eau-CO_2 (~ 2,98), soit égal au pH du système eau-CO_2-calcite (~ 4,8) et ce, dans des conditions de stockage de CO_2. De cet afflux initial de liquide saturé ou non par rapport à la calcite, les expériences en réacteurs fermés évoluent librement jusqu'à équilibre, telles que des conditions de contexte naturel après une perturbation.
Les expériences en réacteurs fermés sont effectuées sur deux carbonates naturels (Lavoux et St-Émilion) avec une pression $P_{CO_2} = 10^{-3,5}$ bar, et des conditions initiales différentes ($[Zn(II)]_i$ variant de 10^{-4} à 10^{-6} M, soit avec de l'eau pure ou en présence de 100 g/L de NaCl). L'équilibre en ce qui concerne la dissolution de la calcite est confirmé dans toutes les expériences, tandis que la sorption de zinc mise en évidence ne correspond pas toujours au mécanisme en deux étapes décrit dans la littérature. Une sorption préférentielle d'environ 10 % de la concentration est mise en évidence pour Mn(II) dans des expériences aqueuses, tandis que le Sr(II) est plus « sorbé » dans des conditions salines. Cette étude montre également que cette sorption préférentielle, en fonction de la salinité, est indépendante du carbonate naturel considéré.
Les simulations effectuées avec PHREEQC montrent une bonne adéquation entre les expériences et les simulations concernant l'équilibre de dissolution et la seule sorption de zinc, avec un log $K_{Zn(II)} \sim 2$ dans l'eau pure et proche de 4 dans des conditions de haute salinité. Lorsque les simulations étaient possibles, les valeurs obtenues de log K de Mn(II) et Sr(II) sont très différentes de celles de la littérature obtenues dans des conditions contrôlées. La nécessité d'un nouveau modèle conceptuel concernant plusieurs éléments traces et leur sorption est mis en avant, et ce, afin de mieux comprendre le devenir des contaminants dans les systèmes naturels.

INTRODUCTION

Carbonate minerals are widely present in natural environments, be they in a surface, subsurface or marine context (Reeder, 1983). Their high reactivity is also well established and implies several phenomena such as precipitation/dissolution and sorption/desorption, which control the availability and transport cycles of many chemical elements. A lot of studies have been carried into the surface behavior of calcite regarding divalent metallic cations to obtain the surface exchange properties of this mineral. At the macro scale, calcite powder equilibrated with water at various salinities is exposed to a given Me(II) concentration, and the evolution of the metal concentration in solution is measured along time or a pH path (Lakshtanov and Stipp, 2007; Tertre *et al.*, 2010; Zachara *et al.*, 1991, 1988). At the micro scale, spectroscopic, atomic force microscopy and synchrotron investigations are carried out to provide a better mechanistic representation of the processes happening at the calcite-liquid interface (Elzinga and Reeder, 2002; Freij *et al.*, 2005; Pokrovsky *et al.*, 2000; Stipp, 1999; Villegas-Jiménez *et al.*, 2009a, b). The data obtained from those experiments allowed the authors to constrain Surface Complexation Models (SCM) for several carbonate minerals, but uncertainties remain regarding the mechanistic description used in those models. For example, the two-site scheme based on the 1014 plan classically used for SCM reactions (Stipp and Hochella, 1991; Wright *et al.*, 2001) was discussed recently and a new one-site scheme finally proposed (Villegas-Jiménez *et al.*, 2009b). This new scheme represents much better the results of spectroscopic investigations and also allows a simplification of the numerical models used for SCM.

Still, those studies present surface properties of calcite regarding a maximum of two divalent metallic cations: Ca(II) from equilibration of solid calcite with the liquid phase, and another Me(II) cation added in solution. As the aim of sorption experiments is to get data to constrain models in the natural context, the study of the behavior of several Me(II) seems necessary. More specifically, the effect of the competitive sorption of major cations (naturally present in solution or released by dissolution processes) and trace elements (from the same possible sources) should be investigated. The most studied trace elements regarding sorption on calcite are divalent transition metals (Cd, Co, Mn, Ni, Pb, Zn) and earth-alkaline elements (Ba, Sr). A batch or a stirred flow-through reactor is filled with calcite powder and a water of variable salinity and pH, and a given amount of the trace element is inserted (Lakshtanov and Stipp, 2007; Martin-Garin *et al.*, 2003; Zachara *et al.*, 1991). All the publications on this subject agree that the time dependant plots of Me(II) concentration present a two step behavior: a rapid uptake of the trace element (adsorption), and a second slower phase (coprecipitation, recrystallization).

In this study, the aim was to obtain sorption data analogous to what would happen in a natural surface context, but also to set up a methodology that could be applied to subsurface conditions. An example of a use for this methodology could be the storage of nuclear waste in deep subsurface layers, the behavior of a radionuclide in such a context and its possible long term fate in association with either the mineral or the liquid phase (Bruno *et al.*, 1998; Curti, 1999; Pérez del Villar *et al.*, 2002). A second example concerns the case of geological CO_2 storage, where the acidification of water due to CO_2 injection dissolves a part of the rock, which can lead to the mobilization of significant amounts of trace elements (Kharaka *et al.*, 2009; Little and Jackson, 2010; Pauwels *et al.*, 2007). To replicate such behaviors, no equilibration of the liquid phase with the natural carbonate is achieved, to reproduce the response of a carbonate system to a flux of water coming from the reservoir. The initial pH is used to reproduce the arrival of two solutions coming from the reservoir storage: water in equilibrium with the water-CO_2 system and the water-CO_2-calcite system. In both cases, the mobilization of TE (Mn(II) and Sr(II)) is studied in addition to that of Zn(II), to consider the multi-component aspect in the system. The impact of other factors such as salinity and initial Zn(II) concentration is also investigated. The results obtained in the experiments are then used as input data for simulations with PHREEQC (Parkhurst and Appelo, 1999) with several aims:

- to check that equilibrium is reached regarding calcite dissolution,
- to obtain surface complexation constants,
- to discuss those values in the light of the available literature.

1 MATERIALS AND METHODS

1.1 Samples Origin and Preparation

Two carbonates are considered in this study: the Lavoux carbonate and the St-Emilion carbonate. The first one is dated from the Callovian period (164.7-161.2 MY) and is produced in a carrier located at Lavoux, France. It comes from a 20 m thick oolithic white carbonate which presents fine grains cemented with calcite (Bourgueil and Gabilly, 1971). The St-Emilion carbonate is located in the upper Oligocene 1B layer (Stampian, 33.9-23.03 MY, 30 m thick), in the north of the Aquitaine basin, France (Cerepi, 1997). More information concerning the St-Emilion site and this carbonate are available in the articles concerning the CO_2-Vadose project (Loisy *et al.*, 2013). Previous to all analyses and experiments, the carbonates are crushed and sieved to recover the 10-100 μm fraction with an average grain diameter of 40 μm. The powders are then ultrasonically cleaned five times in pure ethanol over 10 min in order to eliminate the fraction inferior to 10 μm (the supernatant was removed and replaced until it remained clear). Each powder is then rinsed with ultrapure deionized water (18.2 MΩ cm^{-1}) for 5 min and dried for 12 hours at 80°C. The efficiency of this procedure is checked with Scanning Electron Microscopy (SEM) and a laser scattering particle size distribution analyzer.

1.2 Solid Characterization

1.2.1 Chemical Composition

The crystallographic phases (more than 1% weight) are evidenced by X-Ray Diffraction (XRD) with a *PANalytical* X'Pert Pro diffractometer using the $CoK\alpha_1$ radiation ($\lambda = 0.1789$ nm). Each sample is analyzed in reflection mode with a 50 kV tension and a 30 mA intensity. The diffraction pattern is acquired over the 2θ range from 2 to 80° in continuous mode with a step size equal to 0.0167°, a time/step equal to 200 s and a scan speed about 0.0106°/s. A Soller slit of 0.04 radian and a divergence slit of 1/4° are set up for the incident beam optics; while for the diffracted beam optics, a Soller slit of 0.04 radian, an anti-scatter slit of 5 mm and a Fe filter are used. This qualitative measure is completed with a quantitative Rock-Eval 6 analysis. In this device, 10 mg of each sample are submitted to combustion in a closed system at temperatures from 300°C to 650°C with a heating rate of 25°C/min in the presence of an inert gas (pyrolysis cycle), and in an open system under synthetic air flow at temperatures from 300°C to 850°C with a heating rate of 20°C/min (oxidation cycle). The CO and CO_2 produced by the decarbonation are measured as a function of the temperature for each cycle and allow us to distinguish organic carbon from mineral carbon, and even to determine the nature of the cations related to CO_3^{2-} groups (Behar *et al.*, 2001; Garcia *et al.*, 2010; Lafargue *et al.*, 1998). The mass balance obtained, coupled with the XRD analysis, allows the calculation of the weight percentage of each evidenced phase.

In order to analyze the TE content of the samples, each one is completely dissolved according to the procedure used by Le Pape *et al.* (2012). Half of the final solution obtained is diluted 10 times while the other half is diluted 100 times. Those two solutions are then analyzed with an ELAN 6100 (*Perkin-Elmer*) ICP-MS, calibrated from 0 to 100 μg/L for Mn, Zn and Sr (*Spex Certiprep*, Stanmore, UK). An external quality control is also analyzed to assess a 5% accuracy of the measured concentration.

1.2.2 Morphology and Petrophysics

The microscopic analyses of initial and reacted minerals are performed on a *Zeiss* EVO MA10 scanning electron microscope on Au coated samples. The sample morphology is imaged in classical SEM mode with a beam energy of 15 kV, the chemical contrast is monitored in backscattered electron mode with beam energy of 15 kV and the operating conditions for punctual analysis (EDS) are HT = 15 kV, I_{probe} = 700 pA, the dead time of detection being below 25%. The Specific Surface Area (SSA) of each powder is measured, after 12 hours of drying at 110°C, with an ASAP 2000 (*Micromeritics*) using the three point BET method with Kr as the adsorbent gas, as calcite often presents low values of SSA (below 1 m^2/g). The estimated accuracy based on repeated measurements is below 10% for 1 g of carbonate.

1.3 Experimental Data

1.3.1 Apparatus and Analysis

The different sorption experiments are performed in 50 mL polypropylene beakers at $P_{CO_2} = 10^{-3.5}$ atm and consist of one gram of carbonate powder with 40 mL of one of the solutions (Tab. 1), thus 25 g/L of sorbent. The solid samples are dried at 90°C for ~ 30 min before each experiment to avoid over-estimation of the sample mass due to adsorbed water (Daval *et al.*, 2009). The initial pH is adjusted by addition of NaOH (0.01 M) or HCl (0.01 M). All the experiments are run at ambient temperature (~ 20°C), with an initial stirring. Four milliliters of solution are sampled after 24, 95, 196 and 363 hours of the experiment with the Lavoux carbonate and after 22.5, 88.5, 211 and 383 hours of the experiment with the St-Emilion carbonate. All these samplings are realized with polypropylene syringes, filtered through a 0.2 μm membrane, and immediately diluted (two times) in an acidified (0.2% HNO_3) solution for subsequent analysis. At the end of the experiments, 2 mL of solution is sampled for Ca(II) analysis and the solid phase is recovered by Büchner filtration using a 0.2 μm mesh size filter.

Before each sampling, the pH of the solution is measured with a *Mettler Toledo* SevenGo pro pH-meter calibrated with three NIST solutions (pH 4.01, pH 6.00 and pH 9.21) at 20°C. The trace element, Zn(II), Mn(II) and Sr(II), concentrations in solution are measured by ICP-AES (IRIS Advantage, *Thermo*), after an optimization of the injection and plasma conditions. An external calibration from 0 to 1 000 μg/L of each element (*Spex Certiprep*, Stanmore, UK) is done using NaCl 50 g/L solutions (*VWR*, GPR Rectapur 99%) as matrix matched solutions. Preliminary analyses were carried out in order to test the repeatability and the sensitivity of the instrument with the 50 g/L NaCl matrix. As for the ICP-MS analyses, an external quality control is also analyzed to assess a 5% accuracy of the measured concentration. The Ca(II) concentration at the end of the experiments is measured with a IonPac® CG12A Guard Column and CS12A Analytical Column (*Dionex*) after calibrations with both deionized and saline solutions. It is performed for an aqueous and for a saline solution (100 g/L) from 50 to 1 000 μmol/L of each element. Every ten samples analyzed, a blank and a control (200 μmol/L) are introduced to check that no drift occurs.

To complement the aqueous analyses, the reacted solid phases are studied with all the analytical methods presented in Section 1.2, but also by microprobe. To do so, reacted powders are prepared with epoxy resin in the shape of a pastille, then polished with heptane to obtain sections of grains and finally metalized with carbon. Those preparations are analyzed with a *JEOL* 8100 electron probe microanalyzer fitted with 5 wavelength dispersive spectrometers in order to investigate the zinc content of the reacted grains and more specifically the repartition profile of the zinc. To carry out this investigation, punctual analysis by steps of about 5 μm across the length of grains are realized, with a 25 kV acceleration voltage and a 240 nA beam current. Characteristic lines are measured with spectrometers: Si Kα on TAP, Ca Kβ on LiFH and Zn Kα on LiFH. C and O content are calculated by stoichiometry assuming Ca as $CaCO_3$, Si as SiO_2 and Zn as metal. The procedure developed by Armstrong (1988) is used for matrix correction. The position of background measurements is chosen at both sides of the peaks to avoid interferences, and subtraction is performed assuming a linear background. The counting time is 40 s on peaks and background, leading to a detection limit of 75 ppm for Zn. Standards are acquired just before measurement on albite for Si, on apatite for Ca and zinc sulfide for Zn.

1.3.2 Supplementary Calculations/Data Processing

The proportion of adsorbed zinc for each experiment is calculated with the formula:

$$\mathrm{Zn(II)}_{\mathrm{sorbed}} = \frac{\left([\mathrm{Zn(II)}]_i - [\mathrm{Zn(II)}]_f\right) \times 100}{[\mathrm{Zn(II)}]_i} \tag{1}$$

where $Zn(II)_{sorbed}$ is expressed in percent of the initial Zn(II) concentration, and $[Zn(II)]_i$ and $[Zn(II)]_f$ are the initial and final Zn(II) concentration in the solution in mol/L, respectively.

The calcite mass dissolved is calculated with the formula:

$$M_{diss} = \frac{[\mathrm{Ca(II)}]_f \times V_i}{M_{\mathrm{CaCO_3}}} \tag{2}$$

TABLE 1
Initial conditions of sorption experiments

Experiment ID	Carbonate	Salinity (g/L NaCl)	M_i, initial mass of powder (g)	$[Zn(II)]_i$ (mol/L)	pH_i	Equivalent equilibrium
1	Lavoux	/	1.0002	10^{-4}	4.12	Water CO_2 calcite
2	Lavoux	/	1.0001	10^{-5}	3.69	
3	Lavoux	/	1.0006	10^{-6}	3.53	
4	Lavoux	100	0.9996	10^{-4}	4.11	
5	Lavoux	100	1.0000	10^{-5}	3.51	
6	Lavoux	100	1.0008	10^{-6}	3.36	
7	Lavoux	/	0.9998	10^{-4}	2.97	Water CO_2
8	Lavoux	/	0.9998	10^{-5}	2.61	
9	Lavoux	/	1.0004	10^{-6}	2.54	
10	Lavoux	100	1.0005	10^{-4}	3.08	
11	Lavoux	100	1.0000	10^{-5}	3.00	
12	Lavoux	100	1.0009	10^{-6}	2.91	
13	St-Emilion	/	1.0003	10^{-4}	4.11	Water CO_2 calcite
14	St-Emilion	/	1.0005	10^{-5}	3.68	
15	St-Emilion	/	0.9997	10^{-6}	3.60	
16	St-Emilion	100	1.0004	10^{-4}	4.11	
17	St-Emilion	100	1.0001	10^{-5}	3.51	
18	St-Emilion	100	1.0001	10^{-6}	3.70	
19	St-Emilion	/	1.0001	10^{-4}	2.85	Water CO_2
20	St-Emilion	/	1.0003	10^{-5}	2.83	
21	St-Emilion	/	1.0000	10^{-6}	2.71	
22	St-Emilion	100	1.0001	10^{-4}	2.97	
23	St-Emilion	100	1.0000	10^{-5}	2.83	
24	St-Emilion	100	1.0003	10^{-6}	2.83	

where M_{diss} is the dissolved calcite mass in g, $[Ca(II)]_f$ is the Ca(II) concentration measured at the end of the experiment in mol/L, V_i is the initial volume used in the experiment in L and M_{CaCO_3} is the molar weight of calcium carbonate in g/mol. This calculation implies that the equilibrium of calcite dissolution is achieved within the first 25 hours of the experiments (before the first sampling). This hypothesis is considered valid as calcite is well known for its minute-scale dissolution (Villegas-Jiménez *et al.*, 2009a). From this calcite mass dissolved, the theoretical amount of TE released in solution is estimated considering an homogeneous repartition of the TE in the grains. To do so, the following equation is used:

$$n_{Me(II)} = \frac{[Ca(II)]_f \times V_i \times M_{CaCO_3} \times A_{TE}}{10^6 \times M_{TE}} \tag{3}$$

where $[Ca(II)]_f$ is the calcium concentration at the end of the experiment in mol/L, V_i is the volume of solution at the beginning of the experiment in L, M_{CaCO_3} is the molar weight of the calcium carbonate in g/mol, A_{TE} is the natural carbonate's content of TE (mg/kg) and M_{TE} is the molar weight of the TE considered in the calculation.

1.4 Simulations

The geochemical code PHREEQC (Parkhurst and Appelo, 2013) is used in order to simulate both dissolution and sorption processes under the experimental conditions described in Table 1. The following section presents the methodology used to carry out those simulations.

1.4.1 Database Selection

The question of the database is very important concerning simulations, especially when half of the experiments are performed in high salinity solutions (100 g/L of NaCl, *i.e.* 1.71 M). Indeed, the extended Debye-Hückel model used in PHREEQC for the calculation of the activity coefficient is considered valid for solutions with ionic strengths below 1 M (Parkhurst and Appelo, 1999), which is about half of the current salinity. Simulations were performed with both a Debye-Hückel and a Pitzer model (minteq.v4.dat and pitzer.dat databases respectively), and a maximum of 20% of the Ca(II) concentration underestimation was observed for simulations run with the classic model at this salinity (see Appendix). More precisely, this underestimation was comprised between 16% and 20% for basic initial conditions and around 10% for acid initial conditions. In the end, the lack of values for the elements studied in this paper forbade using Pitzer's database. The database minteq.v4.dat with the classic Debye-Hückel model for the calculation of the activity coefficient is thus used for all the subsequent simulations, despite its inadequacy for high salinity conditions. This is a recurrent issue for people working in high salinity systems, as stated in other studies (De Lucia *et al.*, 2012).

1.4.2 Equilibrium in the System

1.4.2.1 Calcite Dissolution Equilibrium

To ensure that our experiments match the equilibrium hypothesis, simulations are run using initial experimental conditions as input data: temperature, zinc concentration, pH, salinity, quantity of water, number of moles of calcite and partial CO_2 pressure. The initial pH is calculated as the quantity protons coming from the zinc solution and from the HCl or NaOH added to prepare the experimental aqueous and saline solutions: the results of those calculations are given in Table 1, as "pH_i". A maximum 10% difference is calculated between the measured pH in the remaining initial solutions and calculated initial pH values (data not shown).

1.4.2.2 Thermodynamic Precipitation of Secondary Phases

The selection of the database is also justified by the amount of data available concerning equilibrium of other mineral phases. Indeed, several chemical elements (Zn(II), Mn(II) and Sr(II)) at various concentrations are considered in this study and could lead to precipitation of other mineral phases. For example, in the phase diagram of the system $\{Zn - H_2O - CO_2\}$ the point corresponding to the conditions of this study belongs to the hydrozincite ($Zn_5(OH)_6(CO_3)_2$) predominance area (Preis and Gamsjäger, 2001). The possible precipitation of this phase is often discussed in the literature, especially for zinc concentrations above 10^{-5} M (Zachara *et al.*, 1991), thus its dissolution equilibrium value was taken from the database 'thermo.com.V8.R6.230' (Johnson *et al.*, 1992) and added to the database used in this study. The solubilities of phases related to Mn(II) and Sr(II) mineral species are numerous in the database, and as for hydrozincite, its saturation index is checked at the end of the simulations.

1.4.3 Surface Complexation Model (SCM)

The experimental results are used as an input dataset for surface complexation model simulations. The simulation of sorption is designed as a two-site SCM, according to the large literature available (Lakshtanov and Stipp, 2007; Martin-Garin *et al.*, 2003; Pokrovsky *et al.*, 2000; Stipp, 1999; Villegas-Jiménez *et al.*, 2009b). The equilibrium equations accounting for surface complexation in the context of the study are presented in Table 2 with the related log K values at 25°C and $I = 0$. The equilibrium constants concerning the three elements of interest in this study are initially taken to be equal to values from the literature: Glasner and Weiss (1980), Tsusue and Holland (1966) and Zachara *et al.* (1991). Two types of SCM simulation are run: in the first, only Zn(II) ions interact with the surface. For each experiment, the log K value for equation 07 (Tab. 2) is adjusted until the aqueous zinc concentration obtained and the final zinc concentration measured experimentally differs by less than 1%. In the second run, the whole system (Zn(II), Mn(II) and Sr(II)) is considered by adding the concentrations of the TE as an input data when a sorption calculation is possible. The fitting procedure is a bit more complex: the log K value for Zn(II) is fitted first, and then corrected as the log K values for Mn(II) and Sr (II) are also fitted. Several iterations of this procedure are necessary to obtain a difference lower than 1% between experimental and simulated concentrations for all elements.

2 RESULTS

2.1 Unreacted Samples

2.1.1 Mineral Phases, Carbon and TE

The XRD investigation of the natural carbonates (Fig. 1a) identified two main phases present in each carbonate by comparison to the reference diffractograms of those pure phases (Morris *et al.*, 1981; Swanson *et al.*, 1954).

TABLE 2

Surface complexation reactions relevant for the present system at 25°C and $I = 0$, used in PHREEQC. Data for reactions 01 to 06 from (Pokrovsky *et al.*, 2000); log K for reactions 07, 08 and 09 are determined from experimental results and presented in Table 5

	Reactions	Surface complexation equilibrium reactions	log K°_{int} (25°C, $I = 0$)
Data from literature and thermodynamic databases	01	$CaOH^{\circ} - H^{+} =>> CaO^{-}$	−12
	02	$CaOH^{\circ} + H^{+} => CaOH_2^{+}$	11.5
	03	$CaOH^{\circ} + CO_3^{2-} + 2H^{+} => CaHCO_3^{\circ} + H_2O$	23.5
	04	$CaOH^{\circ} + CO_3^{2-} + H^{+} => CaCO_3^{-} + H_2O$	17.1
	05	$CO_3H^{\circ} => CO_3^{-} + H^{+}$	−5.1
	06	$CO_3H^{\circ} + Ca^{2+} => CO_3Ca^{+} + H^{+}$	−1.7
Fit from experimental data	07	$CO_3Ca^{+} + Zn^{2+} => CO_3Zn^{+} + Ca^{2+}$	
	08	$CO_3Ca^{+} + Sr^{2+} => CO_3Sr^{+} + Ca^{2+}$	
	09	$CO_3Ca^{+} + Mn^{2+} => CO_3Mn^{+} + Ca^{2+}$	

The presence of calcite is shown by the presence in both samples of the peak located at 34.3° (main calcite peak) and several others peaks (27, 42, 46, 51, 56 and 57°). For quartz, only the main peak (located at 31°) is detected clearly in the St-Emilion carbonate, while it is barely visible for the Lavoux carbonate. The results of the Rock-Eval 6 analyses are plotted in Figure 1b, and represent the quantity of CO_2 emitted during the pyrolysis cycle (dotted line) and during the oxidation cycle (full line). For both initial natural carbonates, the intensity of the peak obtained during the pyrolysis cycle is about 45 times less important than the peak obtained during the oxidation cycle. This information, combined with the temperature of the maximum CO_2 production, confirm that:

- no organic carbon (*i.e.* organic matter) is present in those samples, and
- that mineral carbon is contained by $CaCO_3$ (Lafargue *et al.*, 1998).

Those values allow the calculation of the initial proportion of $CaCO_3$ in the samples, according to Behar's methodology (Behar *et al.*, 2001). The compositions obtained are presented in Table 3, assuming that quartz is the only other phase present in the samples (a hypothesis which is confirmed by the SEM-EDXS analyses on initial carbonates, data not shown). The results of the ICP-MS analysis for TE are also given in Table 3 and complete the geochemical initial characterization of the samples. Both carbonates present a similar content of Zn and Sr but there is close to 10 times as much Mn in the St-Emilion carbonate than in the Lavoux carbonate.

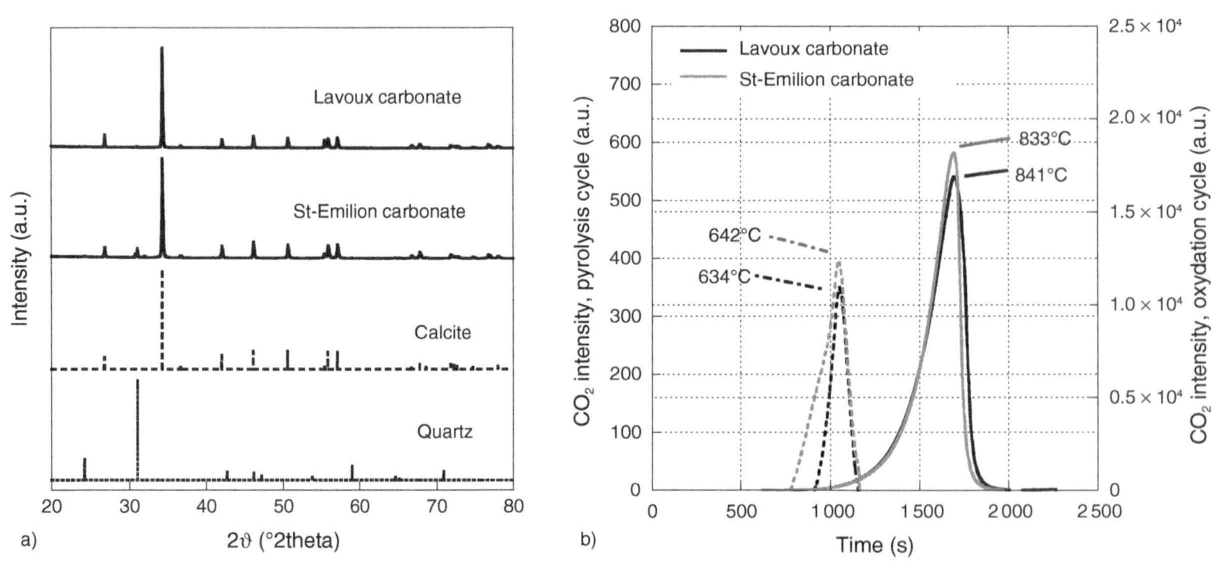

Figure 1

a) Plots of the Lavoux and the St-Emilion natural carbonates XRD analysis and those of pure calcite and quartz. Diffractograms are normalized with respect to their highest peak. b) Comparative Rock-Eval 6 analysis of the Lavoux and the St-Emilion carbonates. Dotted lines represent the pyrolysis cycle (intensity scale on the left axis) whereas full lines represent the oxidation cycle (intensity scale on the right axis).

2.2 Experimental Results

2.2.1 pH, Calcite Dissolution and TE Release

After about 200 hours of reaction the pH reaches a stable value in all the experiments: for pure water its mean value is 8.11, while it is 8.20 for brine experiments. The results for the final Ca(II) concentration (Tab. 4) show that salinity enhances dissolution by a factor of two for an initial basic pH, as seen in a paper from Coto *et al.* (2012), while it has no noticeable impact on dissolution when the initial pH is acid. The corresponding amount of calcite dissolved in each experiment is then calculated: this value ranges from 1.9 (experiment 14) to 8.4 mg (experiment 9); with a mean value of ~ 4 mg of calcite dissolved. This represents less than 1% of the initial solid phase present in the experiments, a value which is commonly used in the literature (Martin-Garin *et al.*, 2003; Tertre *et al.*, 2010; Villegas-Jiménez *et al.*, 2009b) to ensure that sorption is dominant regarding dissolution. In addition, the quantity of TE released in solution is calculated for each experiment. The released amount of Zn(II) is estimated to a concentration of 10^{-8} M, which is neglectible by comparison with the initial Zn(II) present in solution. Unfortunately, the results obtained for the maximum theoretical concentrations of Mn(II) and Sr(II) are in most cases lower than the experimental data. Thus, the sorption of those elements has to be approached differently, as discussed in the appropriate section.

TABLE 3

Composition of the initial rock samples and the parameters necessary for the simulation of sorption

		Lavoux	St-Emilion
Composition	Calcite (%wt)	100 ± 2	96 ± 2
	Quartz (%wt)	0 ± 2	4 ± 2
	A_{Mn} (ppm)	33.4	282.8
	A_{Zn} (ppm)	29.0	16.6
	A_{Sr} (ppm)	160.6	180.9
	Initial SSA (m²/g)	0.60	1.20
Sorption simulation parameters	Average reacted SSA (m²/g)	0.53	1.10
	Density of surface sites (moles/m²)	8.22×10^{-6}	8.22×10^{-6}
	Number of available surface complexation sites (moles)	4.357×10^{-6}	9.042×10^{-6}

2.2.2 Sorption

The behaviors of Zn(II), Mn(II) and Sr(II) are presented considering the evolution of their aqueous concentrations with respect to time (Fig. 2, 3).

2.2.2.1 pH Equivalent to Calcite-Water-CO_2 Partial Equilibrium in Geological Storage Conditions

Independently of the $[Zn(II)]_I$, of the salinity and of the natural carbonate (experiments 1-6;13-18, Tab. 1), a rapid decrease of [Zn(II)] occurs between 0 and 24 hours (Fig. 2a, b). This decrease is followed by either a stable [Zn(II)] value or a much slower sorption rate. This behavior is well-known in the literature (Zachara *et al.*, 1991), and is discussed in the appropriate section. The Zn(II) sorption calculations (Tab. 4) show that the presence of NaCl is the main control of the amount of Zn(II) sorbed, followed by the initial Zn(II) concentration.

In Figure 2c, the concentration of Mn(II) after 24 hours of reaction is the same (around 6×10^{-7} M), independent of the salinity and the initial Zn(II) concentration. Without NaCl, the Mn(II) concentration decreases more than in the presence of salt, and two different behaviors are evidenced at the end of the batch experiments. In brine systems, the end concentration of Mn(II) is approximately 2×10^{-7} M; while without salt it is ten times lower. In Figure 2d, the Mn(II) concentration is close to 7×10^{-6} for the first sample (24 hours of reactions), given the high uncertainty on the two lower points. The end concentration without salt is around 5×10^{-8} M and 5 times higher in brine. Independently of the natural carbonate considered, salinity clearly inhibits sorption of Mn(II). The behavior of Sr(II) released by the dissolution of the Lavoux carbonate is presented in Figure 2e. The first measures of Sr(II) concentration are higher in non-saline systems (between 1×10^{-6} and 3×10^{-6} M) than with NaCl (~ 6×10^{-7} M). In both cases, Sr(II) concentration decreases and there is evidence that the sorption of this element is enhanced by a factor 5 in the presence of NaCl. In Figure 2f, the Sr(II) concentration in pure water decreases from 8×10^{-6} to 8×10^{-5} M, divided by ten; while in the presence of salt, it decreases from 3-4$\times10^{-6}$ to 1.6×10^{-7} M, divided by a factor close to twenty. Contrary to Mn(II), salinity enhances the sorption of Sr(II), independently of the natural carbonate considered.

To perform sorption calculations in the subsequent parts of this study, the higher concentration measured in an experiment is considered as an initial concentration. This is a consequence of the invalidity of the calculations relying on the hypothesis of homogeneous distribution of the TE in the powders (Sect. 2.2.1).

2.2.2.2 pH Equivalent to Water-CO_2 Equilibrium in Geological Storage Conditions

A different trend is observed for zinc sorption with respect to time (Fig. 3a, b). The apparent sorption rate is much slower: it

TABLE 4

Results of calcite dissolution measurements, final Zn(II) concentrations and the related amount sorbed, initial and final concentrations for Mn(II) and Sr(II) estimated and measured in sorption experiments along with the estimated amount of TE sorbed

Experiment ID	Final pH	$[Ca(II)]_{final}$ measured (μmol/L)	Dissolved calcite, calculated (mg)	$[Zn(II)]_{final}$ (mol/L)	$Zn(II)_{sorbed}$ (%)	$[Mn(II)]_i$ (mol/L)	$[Mn(II)]_{final}$ (mol/L)	$Mn(II)_{sorbed}$ (%)	$[Sr(II)]_i$ (mol/L)	$[Sr(II)]_{final}$ (mol/L)	$Sr(II)_{sorbed}$ (%)
1	7.89	571.7	2.3	1.1955×10^{-5}	88.0	5.95×10^{-7}	2.9×10^{-8}	95.20	8.48×10^{-7}	5.25×10^{-7}	38.07
2	8.07	502.4	2.0	1.058×10^{-6}	89.4	7.49×10^{-7}	1.7×10^{-8}	97.67	1.001×10^{-6}	4.11×10^{-7}	58.92
3	8.13	553.9	2.2	1.23×10^{-7}	87.7	6.80×10^{-7}	1.9×10^{-8}	97.18	2.758×10^{-6}	4.76×10^{-7}	82.76
4	8.17	1154.0	4.6	1.8203×10^{-5}	81.8	5.21×10^{-7}	1.73×10^{-7}	66.82	5.47×10^{-7}	1.08×10^{-7}	80.19
5	8.24	1093.8	4.4	6.45×10^{-7}	93.6	4.98×10^{-7}	1.86×10^{-7}	62.62	6.21×10^{-7}	1.17×10^{-7}	81.21
6	8.21	1060.7	4.2	3.1×10^{-8}	96.9	/	2.52×10^{-7}	/	/	1.58×10^{-7}	/
7	8.03	1083.3	4.3	1.2522×10^{-5}	87.5	8.6×10^{-8}	1.6×10^{-8}	82.00	/	7.23×10^{-7}	/
8	8.12	1815.7	7.3	4.48×10^{-7}	95.5	8.2×10^{-8}	1.0×10^{-8}	87.57	/	1.087×10^{-6}	/
9	8.07	2101.3	8.4	1.52×10^{-7}	84.7	2.23×10^{-7}	4.0×10^{-8}	81.98	2.053×10^{-6}	1.308×10^{-6}	36.21
10	8.16	1263.8	5.1	2.1040×10^{-5}	78.9	/	1.91×10^{-7}	/	/	1.20×10^{-7}	/
11	8.25	1154.8	4.6	7.35×10^{-7}	92.7	/	1.72×10^{-7}	/	/	1.08×10^{-7}	/
12	8.24	1138.2	4.6	3.8×10^{-8}	96.1	/	2.20×10^{-7}	/	/	1.38×10^{-7}	/
13	7.96	539.2	2.2	1.0078×10^{-5}	89.9	9.4×10^{-8}	6.6×10^{-8}	29.95	/	7.20×10^{-7}	/
14	8.09	462.7	1.9	4.55×10^{-7}	95.4	1.0401×10^{-5}	5.3×10^{-8}	99.49	7.448×10^{-6}	7.61×10^{-7}	89.79
15	8.14	516.2	2.1	1.14×10^{-7}	88.6	7.682×10^{-6}	4.8×10^{-8}	99.38	7.921×10^{-6}	8.53×10^{-7}	89.23
16	8.15	1267.7	5.1	2.9616×10^{-5}	70.4	7.579×10^{-6}	2.51×10^{-7}	96.69	3.502×10^{-6}	1.57×10^{-7}	95.51
17	8.24	1312.3	5.3	9.76×10^{-7}	90.2	5.551×10^{-6}	2.83×10^{-7}	94.90	3.709×10^{-6}	1.77×10^{-7}	95.22
18	8.21	1599.0	6.4	5.4×10^{-8}	94.5	/	2.74×10^{-7}	/	/	1.72×10^{-7}	/
19	8.18	1121.6	4.5	1.1518×10^{-5}	88.5	1.34×10^{-7}	1.23×10^{-7}	8.20	/	1.08×10^{-6}	/
20	8.30	1161.1	4.6	7.35×10^{-7}	92.6	1.38×10^{-7}	1.09×10^{-7}	21.08	/	1.236×10^{-6}	/
21	8.32	1414.7	5.7	1.35×10^{-7}	86.5	3.30×10^{-7}	1.49×10^{-7}	54.74	2.240×10^{-6}	1.420×10^{-6}	36.60
22	8.18	1251.5	5.0	3.1825×10^{-5}	68.2	/	2.74×10^{-7}	/	2.47×10^{-7}	1.72×10^{-7}	30.39
23	8.14	1198.3	4.8	1.298×10^{-6}	87.0	/	2.70×10^{-7}	/	2.89×10^{-7}	1.69×10^{-7}	41.41
24	8.17	1168.9	4.7	9.6×10^{-8}	90.4	4.62×10^{-7}	2.86×10^{-7}	38.10	/	1.79×10^{-7}	/

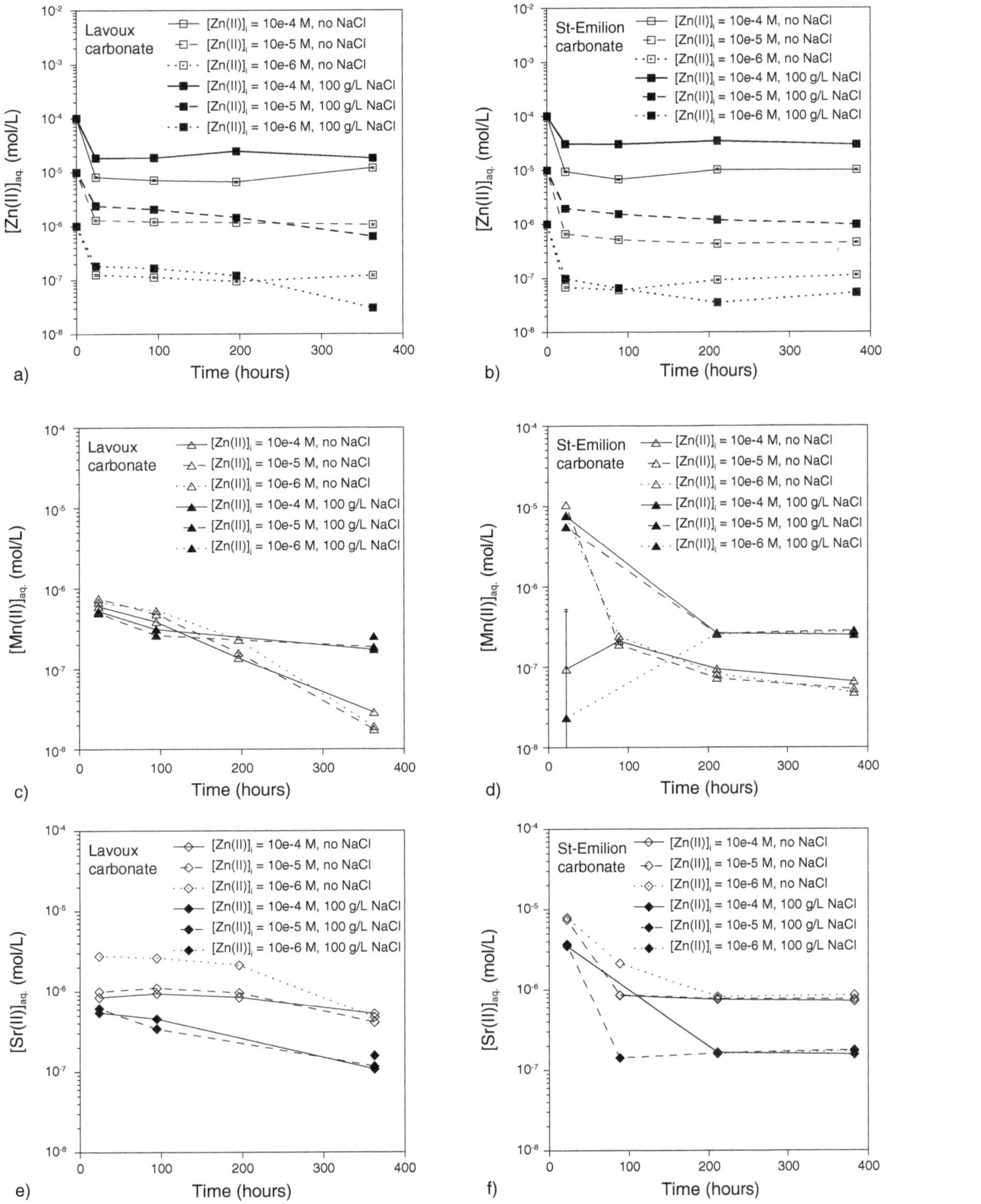

Figure 2

Results of sorption experiments considering an initial water-CO_2-calcite equilibrium. a,b) Zinc concentration evolution as a function of time for experiments performed with the Lavoux and the St-Emilion carbonate in aqueous (empty symbols) and saline solution (black symbols): $[Zn(II)]_i$ equal to 10^{-4} M (full line), 10^{-5} M (dotted line), 10^{-6} M (dashed line). c,d) Evolution with time of [Mn(II)]. The same code as for plots a) and b) is used. When not visible, error bars are comprised within the symbol size. e,f) Evolution with time of [Sr(II)]. The same code as for plots a) and b) is used. When not visible, error bars are comprised within the symbol size.

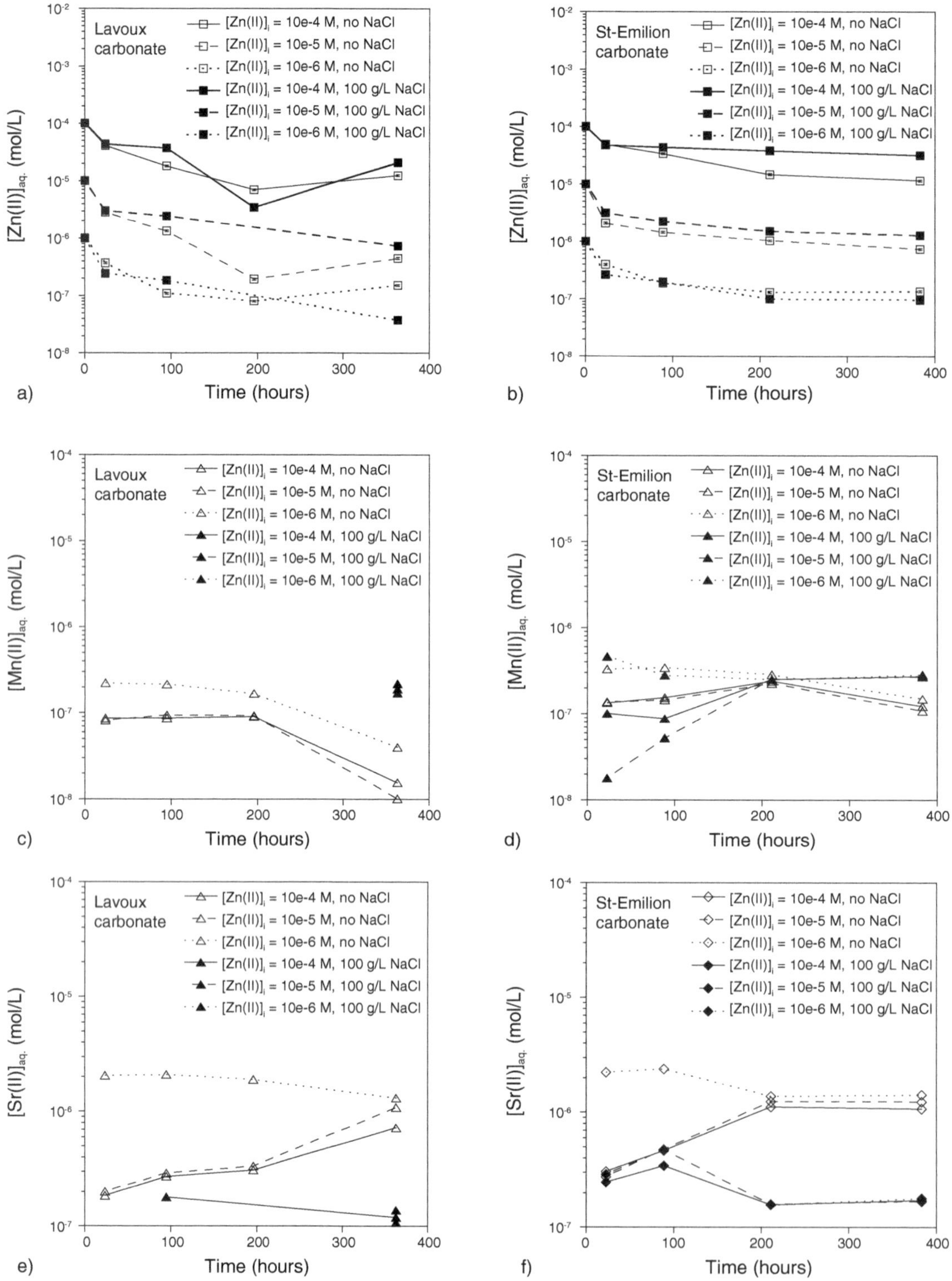

Figure 3

Results of sorption experiments considering an initial water-CO_2 equilibrium . a,b) Zinc concentration evolution as a function of time for experiments performed with the Lavoux and the St-Emilion carbonate in aqueous (empty symbols) and saline solution (full black symbols): $[Zn(II)]_i$ equal to 10^{-4} M (full line), 10^{-5} M (dotted line), 10^{-6} M (dashed line). c,d) Evolution with time of [Mn(II)]. The same code as for plots a) and b) is used. When not visible, error bars are comprised within the symbol size. e,f) Evolution with time of [Sr(II)]. The same code as for plots a) and b) is used. When not visible, error bars are comprised within the symbol size.

takes about 200 hours to get to the concentration corresponding to 24 hours of reaction in the condition previously investigated (about 8 times as much time). Though, if we consider the zinc concentration values at the end (~380 hours) of the experiments, the values are nearly the same. Translated in terms of zinc sorbed (experiments 7-12 and 19-24, Tab. 4), a slight reduction (between 1% and 3%) of that quantity is observed, except for experiment 8, which shows a 7% higher zinc sorption, by comparison with experiment 2. Not taking into account this value, the same conclusion as for basic experiments can be stated: in pure water the initial Zn(II) concentration does not impact the amount of zinc sorbed, whereas in brine the amount of zinc sorbed increases when $[Zn(II)]_i$ decreases. The measured aqueous concentrations of Mn(II) and Sr(II) (Fig. 3c-f) are also higher than the maximum theoretical concentration, which confirms that the hypothesis of homogeneous repartition of TE in the sample is not proper, or that the TE content of initial samples was underestimated. Again, the same behavior is observed for both carbonates: there is less Mn(II) sorbed in high salinity experiments (7-12 and 19-24, Tab. 4), and the amount of Sr(II) sorbed is reduced by the salinity. This shows that initial pH only impacts the kinetics of the sorption phenomenon (factor 8 in time) for zinc. Though, the stronger dissolution that occurs in those conditions seems to impact the sorption kinetics of TE released in the system: in many of the experiments their concentrations present far fewer variations than in the basic conditions, and can even increase continuously (experiments 7, 8, 9, 19, 20, 23). The final concentrations of TE confirm that salinity is the major parameter controlling the amount of sorbed element.

2.2.3 Newly-Formed Phases and Surface Changes

Despite a careful study of the reacted samples with all the techniques used for initial characterization (Sect. 1.2), the presence of newly formed solid phases could not be evidenced, except locally with the presence of NaCl crystals in high salinity experiments (data not shown), which are not relevant in the present study. Enrichment in zinc on the periphery of the grains could only be evidenced after microprobe analysis, around 150 ppm for the Lavoux carbonate, and up to 700 ppm for the St-Emilion carbonate, as clearly shown in Figure 4. The SSA measurements of the reacted powders showed a decrease of 0.07 and 0.1 m²/g for the Lavoux and the St-Emilion carbonate powder respectively by comparison with the non-reacted measurements.

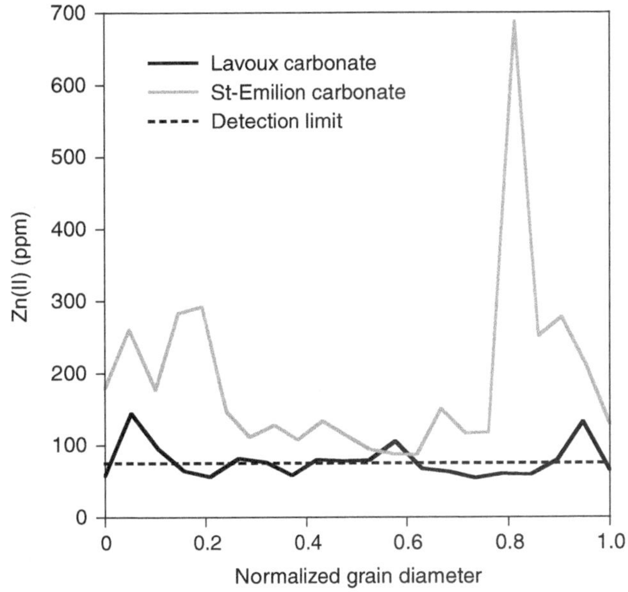

Figure 4

Microprobe analyses of reacted Lavoux and St-Emilion powder. Each profile represents the mean zinc content (ppm) of the cross-sections for 15 grains, in function of the grain radius normalized to 1. The dotted line represents the detection limit (75 ppm).

2.3 Modeling Results

2.3.1 Simulation of Calcite Dissolution

The experimental values are then compared to the simulated concentrations obtained with PHREEQC (Tab. 5). The final pH values obtained by simulation present a good match with the experimental measures, with a maximum difference of 5%. Concerning the concentration of Ca(II), the experiments in aqueous solution match well with the simulated data (only 1 value with over 20% difference between result and simulation); but as expected, due to the non adequacy of the database in high salinity solutions, dissolution is overestimated in simulations (up to 45% in terms of concentration).

2.3.2 Sorption Modeling Results

2.3.2.1 Zn(II) as the Only Metal Considered

The result of the fitting procedure for each experiment is given in Table 5 under the form of a log K value. All the log K values obtained are between 1.7 and 4.62, and can be specified according to the carbonate considered and to the salinity of the aqueous phase. The initial zinc concentration seems to control the log K value in non saline experiments: log K decreases when $[Zn(II)]_i$ decreases (experiments 1-3, 7-8 and 13-15). The absence of impact from the initial pH of the solution is observed again in the simulations: this leads the Lavoux carbonate to record log K = 2.56 +/− 0.52 in pure water and log K = 4.50 +/− 0.14 in brine; and for the St-Emilion

TABLE 5

PHREEQC simulation results of [Ca(II)] for calcite-solution-CO_2 equilibrium and log K values obtained for Zn(II) sole sorption, and multiple Zn(II), Mn(II) and Sr(II) sorption

			Zn(II) sorption	Zn(II), Mn(II) and Sr(II) sorption		
Experiment ID	$[Ca(II)]_{final}$ simulated (μmol/L)	Error (%)	log K_{Zn}	log K_{Zn}	log K_{Mn}	log K_{Sr}
1	560.5	1.99	3.05	3.08	3.35	1.84
2	606.6	−17.17	2.09	2.10	2.63	1.16
3	640.3	−13.49	1.99	2.00	2.53	1.67
4	1440.0	−19.86	4.58	4.60	4.93	2.69
5	1541.0	−29.02	4.32	4.32	4.06	1.93
6	1599.0	−33.66	4.62	/	/	/
7	958.2	13.05	3.17	3.18	2.89	/
8	1602.0	13.34	2.83	2.83	2.27	/
9	1847.0	13.77	2.24	2.25	2.09	1.18
10	1785.0	−29.19	4.52	/	/	/
11	1867.0	−38.15	4.33	/	/	/
12	1983.0	−42.60	4.61	/	/	/
13	561.2	−3.92	2.05	2.05	0.57	/
14	608.3	−23.94	2.15	2.20	3.01	1.65
15	623.9	−17.26	1.7	1.74	2.89	1.60
16	1440.0	−11.96	3.36	3.41	5.18	2.49
17	1541.0	−14.84	3.78	3.80	4.78	2.27
18	1495.0	6.96	4.01	/	/	/
19	1110.0	1.04	2.07	2.07	0.06	/
20	1140.0	1.85	2.13	2.13	0.35	/
21	1360.0	4.02	1.87	1.88	1.07	0.73
22	1903.0	−34.24	3.42	3.42	/	0.86
23	2129.0	−43.72	3.77	3.76	/	0.92
24	2111.0	−44.63	3.89	3.89	3.39	/

carbonate to log K = 1.99 +/− 0.17 in pure water and log K = 3.70 +/− 0.26 in brine. Whatever the carbonate considered, the log K value is higher in high salinity conditions.

2.3.2.2 Zn(II), Mn(II) and Sr(II) Considered

As for the simulation considering only zinc sorption, the results are given as log K values in Table 5. Firstly, the log K values for zinc sorption are slightly impacted when the sorption of several different species is considered: indeed the variations of those values are below 0.05 units whatever the experiment considered. Regarding log K obtained for Mn(II) and Sr(II), the variability of the behaviors observed in the experiments did not allow the authors to simulate all experiments. Especially in high salinity conditions, the first concentration measured (after ~24 hours) was often below the final concentration, preventing analysis of the data as a classic sorption phenomenon. Nevertheless, some trends are clearly visible. First the relative importance of the log K values in experiments 1, 2, 3, 4, 14, 15, 16, and 17 (initial pH basic independent of the carbonate considered) is the same: $\log K_{Mn} > \log K_{Zn} > \log K_{Sr}$. When considering initial pH impact, the values obtained indicate that the log K values are lower when the initial pH is about 3, whether Mn(II) or Sr(II) are considered.

3 DISCUSSION

3.1 Equilibrium of the Dissolution/Precipitation Processes

The values obtained for equilibrium Ca(II) concentration in the simulations and the experiments (comparison in Tab. 5) are in good accordance for experiments in aqueous solution (only 1 value showed over 20% of difference between the result and simulation, experiment 14); but in high salinity solutions, the difference between experiment and simulation increases (between 12% and 45%). To explain those higher differences for high salinity, the main argument is the impact of the database used and especially of the classic Debye-Hückel model. Indeed, as presented in Section 1.3.1.1, a difference of up to 20% is introduced if we consider the Pitzer database as the reference (see Appendix). This variability can also be due to the interaction of calcite with solute species: Martin-Garin *et al.* (2003) evidenced that cadmium impacts the reactivity of the calcite crystal surface, as showed in previous studies (Dove and Hochella, 1993; Fenter *et al.*, 2000; Hillner *et al.*, 1992; Reeder, 1996). Also, some experiments present a Ca(II) concentration higher than the maximum value of the calibration range, which can lead to a drift in the measured value. Finally, the mean pH values obtained, 8.17 in pure water and 8.11 in brine, confirm the equilibrium hypothesis in the system, with only a few percent difference with the simulated value (8.27). Concerning the possible precipitation of newly formed phases, only hydrozincite presents, for some simulations, a positive Saturation Index (SI). This point is discussed specially for initial zinc concentrations above 10^{-5} M (Zachara *et al.*, 1991) and corresponds effectively to the predominance area of hydrozincite in the phase diagram of the system $\{Zn - H_2O - CO_2\}$ corresponding to the present conditions (Preis and Gamsjäger, 2001). Even though no hydrozincite presence was evidenced by all the techniques used in this study, thin deposits (~10 nm) could certainly have been distinguished with an atomic force microscope, as showed by Freij (Freij *et al.*, 2005). The solubility of phases related to Mn(II) and Sr(II) mineral species are numerous in the database used, and as for hydrozincite, their SI at the end of the simulations is sometimes slightly positive but again, it was never related to the observation of any newly formed phase. These discussions concerning dissolution, precipitation and equilibrium in the system ensure the validity of the major contribution of sorption in the interpretation of the results.

3.2 Sorption of Zn(II), Mn(II) and Sr(II)

The trends observed in Zn(II) concentration in the experiments are in good accordance with those observed in many studies (Freij *et al.*, 2005; Tertre *et al.*, 2010; Zachara *et al.*, 1988), as are the amounts of Zn(II) sorbed: whatever the initial conditions of the experiments the strong zinc retention by calcite is evidenced. But another aspect is emphasized here: the role of high salinity on the amount of zinc sorbed. Indeed, the 1.71 Meq ionic strength used in this study is much higher than in Zachara's experiments (maximum 0.03 M) and implies different amounts of Zn(II) sorbed, depending on the natural carbonate used and the initial concentration of Zn(II). This salinity impact could be due to the thickness of the diffuse double layer κ^{-1}, defined by $\kappa^{-1} \sim 3.09/(I^{1/2})$ where I is the ionic strength in mol/dm^3 in the Debye-Hückel theory (Appelo and Postma, 1993). From about 75 Å in pure water conditions, this thickness is reduced to ~2 Å in 100 g/L NaCl conditions, but such dimensions count only for a very small amount of the porous medium considered here. More probably, the surface speciation of calcite in such high salinities is not the same as in lower salinities, where it is much better understood and constrained (Pokrovsky *et al.*, 2000; Villegas-Jiménez *et al.*, 2009a). Despite this lack of knowledge regarding the surface speciation for all experimental conditions, the log K values obtained by simulation for zinc are in agreement with the values obtained by Zachara (Zachara *et al.*, 1988, 1991), even though they present a larger range due to the wider experimental conditions investigated. A supplementary point to discuss regarding those discrepancies is the log K dependence on standard states for activities of surface sites and species, as showed by Sverjensky (2003). The sorbent used (*i.e.* the natural carbonate), the number of surface sites, its SSA and the salinity of the solution strongly impact the log K values obtained. When the three elements are considered for sorption, we have showed (Fig. 2, 3) that an initial acid pH controls the appearance in solution of the TE by a slow increase of their concentration as the system tends towards a balance between the release of species in solution and the adsorption of these same released species. To the best of our knowledge, no such observation is mentioned in the literature, either because people who studied sorption did it in strictly controlled conditions with pre-equilibrated solutions (with regard to calcite SI) or because they were studying the sorption of one element on a specific sorbent. Despite the impossibility of interpreting all the data obtained as sorption data, log K values for Mn(II) and Sr(II) could be estimated. Those values are very different from those presented in the literature, where sorption of those elements regarding calcite is studied. Concerning manganese, the values reported lie between 0.6 and 1.39 (Lorens, 1981; Pingitore *et al.*, 1988; Zachara *et al.*, 1991). For strontium, the values of the literature are all negative, lying between −2.04 and −0.52 (Mucci, 1986; Pingitore and Eastman, 1986, 1984; Zachara *et al.*, 1991). These discrepancies indicate that the values obtained in the studies regarding mono-element sorption added in solution are not relevant when multiple elements are considered, and are even less so when some of those elements are not initially in solution but released by the dissolution of the sorbent. Despite their mutual coherence, the log K values obtained are good in the sense that they match well the experimental data and indicate a scaling of sorption regarding initial pH and salinity. The kinetic processes controlling combined calcite dissolution and zinc sorption coupled with TE release and re-sorption are yet to be put in equations to constrain rigorously those phenomena, either in pure water or under high salinity conditions. It is all the more clearly evidenced when considering the experiments with an initial pH close to 3: only half of the data allowed a sorptive interpretation of the experiments.

3.3 Implications in Case of CO_2 Leakage Up to a Shallow Aquifer

The impact of a leakage is still one of the key issues connected with CO_2 geological storage as a remedy to carbon concentration increase in the atmosphere (IPCC, 2005). Uncertainties concerning the displacement of the CO_2 plume and fluids from the underground storage formation extend also to uncertainties concerning the species that could be released by the reactivity induced by the CO_2 or acid fluid influx. Indeed, it was clearly evidenced by field studies (Kharaka *et al.*, 2006) that the concentration of iron in solution was very sensitive to CO_2 injection, for example. Though, in such field studies, it is not possible to calculate the quantity of iron that stays in the liquid phase after being transported in the porous media across tens of meters. Another study, from Viswanathan *et al.* (2012), showed that the fate of arsenic in an aquifer subjected to natural CO_2 flux is a very challenging field in which to study this kind of reactive-transport with field scale data and laboratory experiments. They explain the immediate increase in As following CO_2 injection observed in their experiments by either the dissolution of calcite or As desorption from oxides and/or clays. It is coherent with the process that occurs in the experiments presented in this paper: the Mn(II) and Sr(II) concentrations increase from nought to a given concentration, depending on the quantity of carbonate dissolved and on the initial TE content in the natural carbonate. In particular, the equilibrium concentration of As that they measure and simulate after 25 days of experiments is not zero, which means that even if this concentration drops by a factor of 10 in this time lapse, there is still As available in solution. In this paper, the amount of zinc still available after reaching the equilibrium corresponds well to the values from the literature, with between 70% and 97% of zinc concentration decrease, but no such statement can be made concerning TE. Mn(II) and Sr(II) seem to be captured well by the solid when the initial pH is close to 4.7, but this is much less the case with an initial pH of 3. The salinity is also a major parameter, as we showed variations of up to 10% of the sorption of Mn(II) and Sr(II) according to this parameter. Knowing that in such experiments, the sorption capacity is enhanced (higher SSA, and no transport phenomenon) regarding natural processes, this shows clearly that the amount of such elements available in solution will imply higher values.

At the same time, in a natural system, other mineralogical phases are present, such as clay minerals that show very interesting sorption properties. Viswanathan *et al.* (2012) discuss the impact of the dissolution of clay surfaces and of the precipitation of small volume fractions of kaolinite that modify the number of sorption sites in their simulations, along with the strong impact of pH on sorption. Though, the observations on rock samples taken from various aquifers in Texas and the Gulf Coast region by Smyth *et al.* (2009) are different from the experimental results of this article. Their "Type I" class of elements that contains Zn, Mn and Sr corresponds to elements that stay in solution with no sorption. In the conditions of this study, those elements should be contained by the "Type II" class as their concentration increases rapidly and then decreases over time. The results presented here imply that the differentiation between sorbed and non-sorbed elements is certainly related to the sorbent used, and the affinity between the elements considered and this sorbent. The "scavenging" effect considered shows that in a natural carbonate context, those trace elements are relevant and behave in accordance with the sorbent used, as evidenced at natural analogue sites (Aiuppa *et al.*, 2005).

CONCLUSIONS

The sorption experiments presented in this paper evidence the impact of different factors such as initial pH, initial Zn(II) concentration and salinity of the liquid phase on the dissolution/precipitation and sorption phenomena. A combination of multiple experimental techniques allowed us to obtain data which we then used to run simulations with the geochemical code PHREEQC. The results obtained both experimentally and by simulations intend to prove that the behavior of TE in such systems remains to be more deeply investigated. Indeed some behaviors could not be modeled with the classical methods, including for example the continuous release of Mn(II) and Sr(II) observed in experiments with an initial pH of about 3.

For all experiments, the equilibrium of the main kinetic phenomenon (fast Zn(II) uptake) was reached within 200 hours maximum, with either the end of sorption or a lower rate phenomenon occurring then. This sorption could neither be evidenced by SEM-EDX or XRD but it was possible to show an enrichment of the surface of the grains using the microprobe technique. Even though the dissolution was considered negligible (less than 1% of the initial powder mass was dissolved), the mobilization of trace elements was investigated. Mn(II) and Sr(II) concentrations were measured and their behavior regarding sorption established. The low amount of calcite dissolved implied strong variations in the composition of the solid-liquid interface. Concerning TE, not only the behaviors, but the equilibrium values observed were very dependent on the initial pH, showing that the sole sorption perspective is not adapted to consider the behavior of the gas-water-rock-TE assemblage. The concentration of the major element Ca(II) is constrained by calcite equilibrium, as is already well known, but there are two main factors controlling the evolution of elements in the system:

- the initial pH,
- the salinity of the water,

with a slight influence of initial Zn(II) concentration in brine. The simulations carried out with PHREEQC demonstrate that sorption of a relatively concentrated element (Zn(II) here) is well constrained by a simple log K value with the two site SCM scheme (Davis and Kent, 1990) used along with the equilibrium constants from the literature (Pokrovsky *et al.*, 2000; Villegas-Jiménez *et al.*, 2009b). However, these simulations remain dependent on the salinity and on the specific petrophysical and morphological properties of the natural carbonate (Sverjensky, 2003). The impossibility of carrying out several simulations showed that mathematical representation used is not relevant when one wants to consider the evolution of the TE released from the bulk solid in a natural carbonate system following dissolution, *i.e.* with an initial pH acid. In addition, the validity of the database used in high salinity conditions remains disputable, but no other possibility is currently available when one is interested in TE behavior.

In the context of a leakage from a CO_2 storage site into an aquifer closer to the surface, the potential consequences for water quality are clear: the strong reactivity of carbonate minerals, widely present on the surface of Earth (Reeder, 1983), could imply the release in solution of non-negligible amounts of potentially hazardous elements. The scavenging effect is thus a major center of interest in better understanding both the incorporation of TE in rocks and their mobility along the different geological reservoirs where gas-water-rock interactions occur.

These are the potential consequences for the hydrogeochemistry of subsurface water reservoirs, but a complementary study was also carried out to consider the scavenging effect under CO_2 geological storage conditions. Those conditions correspond indeed to a minimum of ~ 73 atm and 31°C, where the supercritical CO_2 injected in an aquifer is progressively dissolved and trapped in the reservoir formation, and imply many more dissolution reactions than in atmospheric conditions. Therefore, along with integrity studies and risk assessment investigations, the fate of TE mobilized in such conditions remains to be investigated in terms of the quantity of elements that will be definitively released in the aqueous phase and their competitive behavior regarding adsorption (Auffray *et al.*, 2016), and appropriate mathematical tools have yet to be built to simulate their migration along other natural compartments.

ACKNOWLEDGMENTS

B. Auffray is grateful to Frédéric Neyret-Martinez and Nathalie Texier for their presence and advice during the ICP-AES analyses campaign, to Anne-Sophie Gay and Yannick Blouet for their help concerning the microprobe analyses and to Véronique Bardin for their availability during the liquid chromatography analyses. The authors also thank the three anonymous reviewers who helped to substantially enlighten this article. This paper is part of the PhD Thesis of B. Auffray, funded by *IFP Energies nouvelles.*

REFERENCES

Aiuppa A., Federico C., Allard P., Gurrieri S., Valenza M. (2005) Trace metal modeling of groundwater-gas-rock interactions in a volcanic aquifer: Mount Vesuvius, Southern Italy, *Chemical Geology* **216**, 289-311.

Appelo C.A.J., Postma D. (1993) *Geochemistry, groundwater and pollution*, A.A. Balkema (ed.), Brookfield, VT.

Armstrong J.T. (1988) *Quantitative analysis of silicate and oxide materials: Comparison of Monte Carlo, ZAF, and $\Phi(\rho z)$ procedures, Microbeam Analysis*, San Francisco Press.

Auffray B., Garcia B., Lienemann C.-P., Sorbier L., Cerepi A. (2016) Zn(II), Mn(II) and Sr(II) behavior in natural carbonate reservoir system. Part II : impact of geological CO_2 storage conditions, submitted to *Oil & Gas Science and Technology.*

Behar F., Beaumont V., Penteado H.L.D.B. (2001) Rock-Eval 6 Technology: Performances and Developments, *Revue de l'Institut Français du Pétrole* **56**, 111-134.

Bourgueil B., Gabilly J. (1971) *Carte géologique de la France n° 590 à 1/50 000*, Chauvigny.

Bruno J., Duro L., de Pablo J., Casas I., Ayora C., Delgado J., Gimeno M.J., Peña J., Linklater C., Pérez del Villar L., Gomez P. (1998) Estimation of the concentrations of trace metals in natural systems: The application of codissolution and coprecipitation approaches to El Berrocal (Spain) and Poços de Caldas (Brazil), *Chemical Geology* **151**, 277-291.

Cerepi A. (1997) Milieux poreux matriciel, fractures et teneur en eau d'un calcaire en zone de diagenèse météorique : Calcaire à astéries "Pierre de Bordeaux", Oligocène (Bordeaux, France), *Ph.D. Thesis.*

Coto B., Martos C., Peña J.L., Rodríguez R., Pastor G. (2012) Fluid Phase Equilibria Effects in the solubility of $CaCO_3$: Experimental study and model description, *Fluid Phase Equilibria* **324**, 1-7.

Curti E. (1999) Coprecipitation of radionuclides with calcite: estimation of partition coefficients based on a review of laboratory investigations and geochemical data, *Applied Geochemistry* **14**, 433-445.

Daval D., Martinez I., Corvisier J., Findling N., Goffé B., Guyot F. (2009) Carbonation of Ca-bearing silicates, the case of wollastonite: Experimental investigations and kinetic modeling, *Chemical Geology* **265**, 63-78.

Davis J.A., Kent D.B. (1990) Surface Complexation Modeling in Aqueous Geochemistry, *Reviews in Mineralogy and Geochemistry* **23**.

De Lucia M., Bauer S., Beyer C., Kühn M., Nowak T., Pudlo D., Reitenbach V., Stadler S. (2012) Modelling CO_2-induced fluid-rock interactions in the Altensalzwedel gas reservoir. Part I: from experimental data to a reference geochemical model, *Environmental Earth Sciences* **67**, 563-572.

Dove P.M., Hochella Jr M.F. (1993) Calcite precipitation mechanisms and inhibition by orthophosphate: *In situ* observations by Scanning Force Microscopy, *Geochimica et Cosmochimica Acta* **57**, 705-714.

Elzinga E.J., Reeder R.J. (2002) X-ray absorption spectroscopy study of Cu^{2+} and Zn^{2+} adsorption complexes at the calcite surface: Implications for site-specific metal incorporation preferences during calcite crystal growth, *Geochimica et Cosmochimica Acta* **66**, 3943-3954.

Fenter P., Geissbühler P., DiMasi E., Srajer G., Sorensen L.B., Sturchio N.C. (2000) Surface speciation of calcite observed *in situ* by high-resolution X-ray reflectivity, *Geochimica et Cosmochimica Acta* **64**, 1221-1228.

Freij S.J., Godelitsas A., Putnis A. (2005) Crystal growth and dissolution processes at the calcite-water interface in the presence of zinc ions, *Journal of Crystal Growth* **273**, 535-545.

Garcia B., Beaumont V., Perfetti E., Rouchon V., Blanchet D., Oger P., Dromart G., Huc A., Haeseler F. (2010) Experiments and geochemical modelling of CO_2 sequestration by olivine: Potential, quantification, *Applied Geochemistry* **25**, 1383-1396.

Glasner A., Weiss D. (1980) The crystallization of calcite from aqueous solutions and the role of zinc and magnesium ions - I. Precipitation of calcite in the presence of Zn^{2+} ions, *Journal of Inorganic and Nuclear Chemistry* **42**, 655-663.

Hillner P.E., Manne S., Gratz A.J., Hansma P.K. (1992) AFM images of dissolution and growth on a calcite crystal, *Ultramicroscopy* **42-44**, 2, 1387-1393.

IPCC (2005) Special Report on Carbon Capture and Storage.

Johnson J.W., Oelkers E.H., Helgeson H.C. (1992) SUPCRT92: A software package for calculating the standard molal thermodynamic properties of mineral, gases, aqueous species, and reactions from 1 to 5000 bar and 0 to 1000°C, *Computer & Geosciences* **18**, 899-947.

Kharaka Y.K., Cole D.R., Hovorka S.D., Gunter W.D., Knauss K.G., Freifield B.M. (2006) Gas-water-rock interactions in Frio Formation following CO_2 injection: Implications for the storage of greenhouse gases in sedimentary basins, *Geology* **34**, 577-580.

Kharaka Y.K., Thordsen J.J., Hovorka S.D., Nance H.S., Cole D.R., Phelps T.J., Knauss K.G. (2009) Potential environmental issues of CO_2 storage in deep saline aquifers: Geochemical results from the Frio-I Brine Pilot test, Texas, USA, *Applied Geochemistry* **24**, 1106-1112.

Lafargue E., Marquis F., Pillot D. (1998) Rock-Eval 6 Applications in Hydrocarbon Exploration, Production, and Soil Contamination Studies, *Revue de l'Institut Français du Pétrole* **53**, 421-437.

Lakshtanov L.Z., Stipp S.L.S. (2007) Experimental study of nickel (II) interaction with calcite: Adsorption and coprecipitation, *Geochimica et Cosmochimica Acta* **71**, 3686-3697.

Le Pape P., Ayrault S., Quantin C. (2012) Trace element behavior and partition *versus* urbanization gradient in an urban river (Orge River, France), *Journal of Hydrology* **472-473**, 99-110.

Little M.G., Jackson R.B. (2010) Potential impacts of leakage from deep CO_2 geosequestration on overlying freshwater aquifers, *Environmental Science and Technology* **44**, 9225-32.

Loisy C., Cohen G., Laveuf C., Le Roux O., Delaplace P., Magnier C., Rouchon V., Cerepi A., Garcia B. (2013) The CO_2-Vadose Project: Dynamics of the natural CO_2 in a carbonate vadose zone, *International Journal of Greenhouse Gas Control* **14**, 97-112.

Lorens R.B. (1981) Sr, Cd, Mn and Co distribution coefficients in calcite as a function of calcite precipitation rate, *Geochimica et Cosmochimica Acta* **45**, 553-561.

Martin-Garin A., van Cappellen P., Charlet L. (2003) Aqueous cadmium uptake by calcite: A stirred flow-through reactor study, *Geochimica et Cosmochimica Acta* **67**, 2763-2774.

Morris M.C., McMurdie H.F., Evans E.H., Paretzkin B., Parker H.S., Panagiotopoulos N.C. (1981) Standard X-ray Diffraction Powder Patterns, *NBS Monograph* **25**, Section 18.

Mucci A. (1986) Growth kinetics and composition of magnesian calcite overgrowths precipitated from seawater: Quantitative influence of orthophosphate ions, *Geochimica et Cosmochimica Acta* **50**, 2255-2265.

Parkhurst D.L., Appelo C.A.J. (1999) User's guide to PHREEQC (version 2) - A Computer Program for Speciation, Batch-Reaction, One-Dimensional Transport, and Inverse Geochemical Calculations.

Parkhurst D.L., Appelo C.A.J. (2013) Description of Input and Examples for PHREEQC Version 3 - A Computer Program for Speciation, Batch-Reaction, One-Dimensional Transport, and Inverse Geochemical Calculations.

Pauwels H., Gaus I., Le Nindre Y.M., Pearce J., Czernichowski-Lauriol I. (2007) Chemistry of fluids from a natural analogue for a geological CO_2 storage site (Montmiral, France): Lessons for CO_2-water-rock interaction assessment and monitoring, *Applied Geochemistry* **22**, 2817-2833.

Pérez del Villar L., Bruno J., Campos R., Gomez P., Cozar J.S., Garralon A., Buil B., Arcos D., Carretero G., Ruiz Sanchez-Porro J., Hernan P. (2002) The uranium ore from Mina Fe (Salamanca, Spain) as a natural analogue of processes in a spent fuel repository, *Chemical Geology* **190**, 395-415.

Pingitore N.E.J., Eastman M.P. (1984) The experimental partitioning of Ba^{2+} into calcite, *Chemical Geology* **45**, 113-120.

Pingitore N.E.J., Eastman M.P. (1986) The coprecipitation of Sr^{2+} with calcite at 25°C and 1 atm, *Geochimica et Cosmochimica Acta* **50**, 2195-2203.

Pingitore N.E.J., Eastman M.P., Sandidge M., Oden K., Freiha B. (1988) The coprecipitation of manganese(II) with calcite: an experimental study, *Marine Chemistry* **25**, 107-120.

Pokrovsky O.S., Mielczarski J.A., Barres O., Schott J. (2000) Surface Speciation Models of Calcite and Dolomite/Aqueous Solution Interfaces and Their Spectroscopic Evaluation, *Langmuir* **16**, 2677-2688.

Preis W., Gamsjäger H. (2001) (Solid + solute) phase equilibria in aqueous solution. XIII. Thermodynamic properties of hydrozincite and predominance diagrams for (Zn^{2+} + H_2O + CO_2), *Journal of Chemical Thermodynamics* **33**, 803-819.

Reeder R.J. (1983) Carbonates: Mineralogy and Chemistry, Mineralogical Society of America.

Reeder R.J. (1996) Interaction of divalent cobalt, zinc, cadmium, and barium with the calcite surface during layer growth, *Geochimica et Cosmochimica Acta* **60**, 1543-1552.

Smyth R.C., Hovorka S.D., Lu J., Romanak K.D., Partin J.W., Wong C., Yang C. (2009) Assessing risk to fresh water resources from long term CO_2 injection - laboratory and field studies, *Energy Procedia* **1**, 1957-1964.

Stipp S.L., Hochella M.F. (1991) Structure and bonding environments at the calcite surface as observed with X-ray photoelectron spectroscopy (XPS) and low energy electron diffraction (LEED), *Geochimica et Cosmochimica Acta* **55**, 1723-1736.

Stipp S.L.S. (1999) Toward a conceptual model of the calcite surface: Hydration, hydrolysis, and surface potential, *Geochimica et Cosmochimica Acta* **63**, 3121-3131.

Sverjensky D.A. (2003) Standard states for the activities of mineral surface sites and species, *Geochimica et Cosmochimica Acta* **67**, 17-28.

Swanson H.E., Fuyat R.K., Ugrinic G.M. (1954) NBS Circular 539, Vol. 3 - Data for 34 inorganic substances.

Tertre E., Beaucaire C., Juery A., Ly J. (2010) Methodology to obtain exchange properties of the calcite surface - Application to major and trace elements: Ca(II), HCO_3^-, and Zn(II), *Journal of Colloid and Interface Science* **347**, 120-126.

Tsusue A., Holland H.D. (1966) The coprecipitation of cations with $CaCO_3$ - III. The coprecipitation of Zn^{2+} with calcite between 50 and 250°C, *Geochimica et Cosmochimica Acta* **30**, 439-440.

Villegas-Jiménez A., Mucci A., Paquette J. (2009a) Proton/calcium ion exchange behavior of calcite, *Physical Chemistry Chemical Physics* **11**, 8895-8912.

Villegas-Jiménez A., Mucci A., Pokrovsky O.S., Schott J. (2009b) Defining reactive sites on hydrated mineral surfaces: Rhombohedral carbonate minerals, *Geochimica et Cosmochimica Acta* **73**, 4326-4345.

Viswanathan H., Dai Z., Lopano C., Keating E., Hakala J.A., Scheckel K.G., Zheng L., Guthrie G.D., Pawar R. (2012) Developing a robust geochemical and reactive transport model to evaluate possible sources of arsenic at the CO_2 sequestration natural analog site in Chimayo, New Mexico, *International Journal of Greenhouse Gas Control* **10**, 199-214.

Wright K., Cygan R.T., Slater B. (2001) Structure of the $(10\bar{1}4)$ surfaces of calcite, dolomite and magnesite under wet and dry conditions, *Physical Chemistry Chemical Physics* **3**, 839-844.

Zachara J.M., Cowan C.E., Resch C.T. (1991) Sorption of divalent metals on calcite, *Geochimica et Cosmochimica Acta* **55**, 1549-1562.

Zachara J.M., Kittrick J.A., Harsh J.B. (1988) The mechanism of Zn^{2+} adsorption on calcite, *Geochimica et Cosmochimica Acta* **52**, 2281-2291.

APPENDIX

TABLE 1A

Impact of the database used for calculations of the $H_2O - CO_2 - CaCO_3$ system equilibrium in saline (100 g/L NaCl) conditions

Experiment ID	Minteq.v4.dat	Pitzer.dat	Minteq.v4.dat – pitzer.dat	
	Ca(II)eq (mol/L)	Ca(II)eq (mol/L)	Delta Ca(II) (mol/L)	% (relative to pitzer.dat)
4	1.15×10^{-3}	1.43×10^{-3}	-2.81×10^{-4}	−19.68
5	1.23×10^{-3}	1.49×10^{-3}	-2.63×10^{-4}	−17.66
6	1.27×10^{-3}	1.52×10^{-3}	-2.53×10^{-4}	−16.60
10	1.42×10^{-3}	1.63×10^{-3}	-2.16×10^{-4}	−13.23
11	1.48×10^{-3}	1.68×10^{-3}	-1.99×10^{-4}	−11.84
12	1.57×10^{-3}	1.75×10^{-3}	-1.75×10^{-4}	−10.02
16	1.15×10^{-3}	1.43×10^{-3}	-2.81×10^{-4}	−19.68
17	1.23×10^{-3}	1.49×10^{-3}	-2.63×10^{-4}	−17.66
18	1.19×10^{-3}	1.46×10^{-3}	-2.71×10^{-4}	−18.55
22	1.51×10^{-3}	1.70×10^{-3}	-1.93×10^{-4}	−11.34
23	1.69×10^{-3}	1.83×10^{-3}	-1.45×10^{-4}	-7.92
24	1.67×10^{-3}	1.82×10^{-3}	-1.49×10^{-4}	-8.18

5

Zn(II), Mn(II) and Sr(II) Behavior in a Natural Carbonate Reservoir System. Part II: Impact of Geological CO_2 Storage Conditions

B. Auffray[1], B. Garcia[1]*, C.-P. Lienemann[2], L. Sorbier[2] and A. Cerepi[3]

[1] IFP Energies nouvelles, 1-4 avenue de Bois Préau, 92852 Rueil-Malmaison - France
[2] IFP Energies nouvelles, Rond-point de l'échangeur de Solaize, BP 3, 69360 Solaize - France
[3] EA 4592 G&E, ENSEGID, Université de Bordeaux, 1 allée F. Daguin, 33607 Pessac Cedex - France
e-mail: baptiste.auffray@beicip.com; baptiste.auffray@gmail.com - bruno.garcia@ifpen.fr
charles.lienemann@ifpen.fr - loic.sorbier@ifpen.fr - adrian.cerepi@ipb.fr

* Corresponding author

Abstract *— Some key points still prevent the full development of geological carbon sequestration in underground formations, especially concerning the assessment of the integrity of such storage. Indeed, the consequences of gas injection on chemistry and petrophysical properties are still much discussed in the scientific community, and are still not well known at either laboratory or field scale. In this article, the results of an experimental study about the mobilization of Trace Elements (TE) during CO_2 injection in a reservoir are presented. The experimental conditions range from typical storage formation conditions (90 bar, supercritical CO_2) to shallower conditions (60 and 30 bar, CO_2 as gas phase), and consider the dissolution of the two carbonates, coupled with the sorption of an initial concentration of 10^{-5} M of Zn(II), and the consequent release in solution of Mn(II) and Sr(II). The investigation goes beyond the sole behavior of TE in the storage conditions: it presents the specific behavior of each element with respect to the pressure and the natural carbonate considered, showing that different equilibrium concentrations are to be expected if a fluid with a given concentration of TE leaks to an upper formation. Even though sorption is evidenced, it does not balance the amount of TE released by the dissolution process. The increase in porosity is clearly evidenced as a linear function of the CO_2 pressure imposed for the St-Emilion carbonate. For the Lavoux carbonate, this trend is not confirmed by the 90 bar experiment. A preferential dissolution of the bigger family of pores from the preexisting porosity is observed in one of the samples (Lavoux carbonate) while the second one (St-Emilion carbonate) presents a newly-formed family of pores. Both reacted samples evidence that the pore network evolves toward a tubular network type.*

Résumé — Comportement du Zn(II), Mn(II) et Sr(II) au sein d'un système réservoir carbonate naturel. Partie II : impact des conditions de stockage géologique de CO_2 — Aujourd'hui encore, quelques points essentiels limitent le développement industriel de la séquestration géologique du carbone dans des formations souterraines, en particulier celui concernant l'évaluation de l'intégrité du stockage. En effet, les conséquences de l'injection de gaz sur la chimie et les propriétés pétrophysiques sont encore beaucoup discutées au sein de la communauté scientifique, et doivent être mieux comprises tant à l'échelle de laboratoire que sur le terrain.
Dans cet article, le résultat d'une étude expérimentale de mobilisation d'éléments traces suite à une injection de CO_2 dans un réservoir est présenté. Les conditions expérimentales étudiées représentent

des conditions équivalentes à celles d'un stockage typique de formation (90 bar, CO_2 supercritique) jusqu'à des conditions peu profondes (60 et 30 bar, le CO_2 étant sous forme gazeuse), et considèrent la dissolution des deux carbonates, couplées avec la sorption de Zn(II) dont la concentration initiale est de 10^{-5} M, ainsi que la dissolution de Mn(II) et Sr(II).
L'étude présente le comportement spécifique de chaque élément par rapport à la pression et le carbonate naturel considéré, montrant que des concentrations différentes d'équilibre sont à prévoir si des fuites d'un fluide avec une concentration donnée en éléments traces sont présentes vers une formation supérieure. Et même si le phénomène de sorption est mis en évidence, celui-ci ne contrebalance pas la quantité d'éléments traces relargués par le processus de dissolution. L'augmentation de la porosité est clairement mise en évidence comme une fonction linéaire de la pression de CO_2 imposée pour le carbonate de St-Émilion. Pour le carbonate de Lavoux, cette tendance est confirmée par l'expérience de 90 bar. Une dissolution préférentielle de la plus grande famille de pores de la porosité préexistante est observée dans un des échantillons (carbonate de Lavoux) tandis que le second (carbonate de St-Émilion) présente une famille nouvellement formée de pores, avec une géométrie de type tubulaire.

INTRODUCTION

Some disagreements remain in the scientific community over the consequences of climate change and mean temperature increase (Albritton and Meira Filho, 2001). Even so, the role of CO_2 is now known and accepted: its increasing concentration in the atmosphere (from 280 to 380 ppm over the last two hundred years) is estimated to represent around two-thirds of the enhanced greenhouse effect (Bryant, 1997; IPCC, 2007). The major challenge is to mitigate the extent of climate change by reducing anthropic CO_2 emissions to the atmosphere by stabilizing the CO_2 concentration to twice its pre-industrial concentration (around 550 ppm). To do so, the geological storage of CO_2 has been recognized as a significant possibility (Gale *et al.*, 2004; Hepple and Benson, 2005; Holloway, 1997). Coal beds and flood basalts are considered potential storage sites (Garcia *et al.*, 2010; McGrail *et al.*, 2006; Schaef *et al.*, 2010), as are oil and gas fields which are well known and characterized, and which also possess the industrial facilities needed for CO_2 storage (Garcia *et al.*, 2012). But the structures that offer the largest potential for CO_2 geological storage are saline aquifers, as they are widely distributed throughout the globe in sedimentary basins.

Geological storage in saline aquifers involves the injection of SuperCritical CO_2 (SC-CO_2) into porous and permeable layers in the deep subsurface. The long-term storage of CO_2 implies multiple mechanisms:

① structural trapping (CO_2 is trapped by geological structures and buoyant rise);
② capillary trapping as a residual phase in small pores;
③ CO_2 dissolution into the brine (solubility trapping) and reaction with the host rock minerals to form ionic species (ionic trapping);
④ trapping as mineral phases as CO_2 reacts with the host rock and solution.

These reactions are slow but represent the most stable form of CO_2 trapping. The safety of the trapping increases with these different processes (Gaus, 2010). In this article, the focus is on steps ③ and ④ presented above, in the case of a natural carbonate reservoir system. The dissolution of the CO_2 in the water/brine implies many consequences for both the chemistry and the petrophysics of the host formation. The main process is the dissolution of calcite due to the pH decrease that follows the CO_2 dissolution in the aqueous phase. But this dissolution also means that all the species that were originally trapped in the mineral phase are solubilized and therefore able to be displaced by the natural or induced aquifer flux. Indeed, a large number of elements are present at variable concentrations in a carbonate reservoir rock and aquifer (Chadwick *et al.*, 2006).

As SC-CO_2 is injected into the aquifer, chemical interactions occur that lead to a pH drop of a few units and to the dissolution, mobilization and re-precipitation of metals and other rock-forming elements. This ability of the brine to leach metals from the aquifer rocks was evidenced in several articles (Czernichowski-Lauriol *et al.*, 2006; Emberley *et al.*, 2005; Fischer *et al.*, 2010; Gunter *et al.*, 1997; Kharaka *et al.*, 2006, 2009; Rempel *et al.*, 2011; Wigand *et al.*, 2008). This study aims to investigate the behavior of some divalent cations by examining dissolution, precipitation and sorption processes in conditions relevant for CO_2 storage. Indeed, most studies existing in literature are mainly simulation studies based on incomplete databases (Krupka *et al.*, 2010). It also aims to complete the chemical investigations with a petrophysical approach in order to estimate the consequences of CO_2 injection for the transport properties of a formation.

1 MATERIALS AND METHODS

1.1 Samples

The natural carbonates used are the Lavoux and St-Emilion carbonates. They were chosen for their composition (more than 95% of calcite and quartz) and for their homogeneity. More details on these samples and their preparation can be found in another paper (Auffray *et al.*, 2016).

1.2 Solid Characterization

The analytical methods and techniques used for the initial characterization of the carbonate powders are the same as those described in the previous article (Auffray *et al.*, 2016). The chemical composition is analyzed by coupled X-Ray Diffraction (XRD) and Rock-Eval 6 to obtain the phases present, and ICP-MS analysis for Trace Elements (TE) content. The morphology and petrophysical investigations of the samples are carried out with Scanning Election Microscopy (SEM), BET Specific Surface Area (SSA) and mercury injection. Local Zn enrichment into the carbonate grains is investigated by electron-probe microanalysis. For more details, see Auffray *et al.* (2016). Porosimetry data are obtained using a *Micromeritics* Model 9420 Autopore III Instrument. According to the theoretical basis defined by Washburn (1921) and developed by other authors (Li and Wardlaw, 1986; Lowell and Shields, 1984; Tsakiroglou and Payatakes, 1990; van Brakel, 1981), the porous medium is modeled as a group of cylindrical capillary tubes. The Washburn's equation: $P_c = 4\gamma|\cos\theta|/d_p$, (where P_c is the capillary pressure, d_p is the average pore diameter, γ is the interfacial tension and θ the contact angle between mercury and the pore wall); allows the calculation of petrophysical parameters such as total porosity, distribution of pore throat size, Hg surface area and cavity to throat size ratio. The total porosity is divided into three families according to their pore diameter: macroporosity A (Φ_{MA}: $d_p > 15$ µm), macroporosity B (Φ_{MB}: 0.1 µm $< d_p <$ 15 µm) and mesoporosity (Φ_{meso}: $d_p < 0.1$ µm). The free and trapped porosities (Φ_f and Φ_t respectively) are also calculated from a drainage-imbition cycle.

1.3 Experimental Settings

To investigate the impact of CO_2 geological storage conditions on the behavior of TE in the system $CaCO_3$ - H_2O - CO_2 - NaCl, an hastelloy autoclave (*PAAR*® autoclave) with a sampling feature was used (Fig. 1). To ensure the efficiency of the sampling, a greater quantity of reactants were used: 160 mL of brine along with 4 g of powder to respect the 25 g/L sorbent concentration used in the previous study (Auffray *et al.*, 2016). To ensure that the equilibrium was reached concerning sorption, the experiments lasted around 50 days, with several samplings. The experimental conditions used for those experiments are summarized in Table 1, and temperature and pressure were recorded continuously to detect an eventual leakage from the autoclave. The aqueous samples consisted of 5 mL of solution filtered through a 0.2 µm membrane, and were immediately diluted (factor 2) in an acidified (0.2% HNO_3) solution for subsequent ICP-OES analysis. The mass loss due to sampling all through the experiment did not exceed 16% of the initial liquid mass.

The Ca(II) concentration at the end of experiments was measured to check the equilibrium of calcite dissolution as no *in-situ* pH measure was possible; the solid phase was also recovered after a Büchner filtration with a 0.2 µm mesh size filter to establish the petrophysical modifications induced by either gas or supercritical CO_2 experimental conditions.

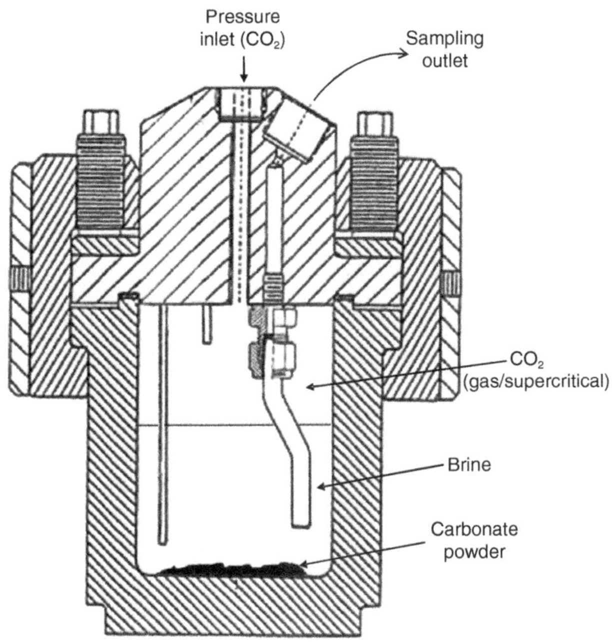

Figure 1

Design of the *PAAR*® autoclave used for the experiments, along with the sampling features.

1.4 Numerical Simulations

The geochemical code PHREEQC (Parkhurst and Appelo, 2013) was used with the pitzer.dat database to check the equilibrium of calcite dissolution in the system, as the salinity is around 1.71 M of NaCl (~100 g/L). However, the lack of reliable data on the amount of TE in the powders used, a consequence of the natural heterogeneity of the samples, did not allowed a complete sorption simulation study.

2 RESULTS

2.1 Unreacted Samples

The mineral and chemical composition of the two natural carbonates is available in another article from Auffray *et al.* (2016).

TABLE 1
Initial conditions of dissolution-sorption experiments

Experiment ID	Carbonate	Salinity (g/L NaCl)	Mass of powder (g)	$[Zn(II)]_{aq}$ (mol/L)	Temperature (°C)	Pressure CO_2 (bar)	CO_2 phase
Lav30	Lavoux	100	4.0020	10^{-5}	40	30	Gas
Lav60	Lavoux	100	4.0036	10^{-5}	40	60	Gas
Lav90	Lavoux	100	3.9997	10^{-5}	40	90	Supercritical
StEm30	St-Emilion	100	4.0021	10^{-5}	40	30	Gas
StEm60	St-Emilion	100	4.0012	10^{-5}	40	60	Gas
StEm90	St-Emilion	100	3.9926	10^{-5}	40	90	Supercritical

2.2 Chemistry

2.2.1 Equilibrium in the Systems

The steady state in the system was checked by measuring the concentration of Ca(II) in solution at the end of the experiments. The values obtained are presented in Table 2, along with the simulated values and the corresponding error. The Lavoux carbonate experiments present a good match with the simulation data, with less than 5% difference between the experimental and simulated values. Concerning the St-Emilion carbonate, the experimental data also fits very well with the simulated values, with a slightly high value measured for the 30 bar experiment (~7.3% of difference). Depending on the CO_2 pressure in the system, between 0.5 and 0.7 g of the initial calcite mass (~4 g) is dissolved, according to the final Ca(II) concentration measured in solution. The simulated values obtained with the database minteq.v4.dat are given in the last two columns of Table 2, to evidence the impact of the database on such simulations where high salinity experiments are concerned. The results show a strong overestimation of the dissolution process at such salinity and pressures with this last database.

2.2.2 Divalent Cation Behavior and Scavenging Effect

The behaviors of Zn(II), Mn(II), and Sr(II) are presented considering the evolution of their aqueous concentration with respect to time (Fig. 2-4).

In the experiments carried out with the Lavoux carbonate (Fig. 2a), the Zn(II) concentration shows an increasing trend when the pressure is fixed at 30 bar. At higher pressures (60 and 90 bar), a slight decrease of the Zn(II) concentration is observed, which means that the Zn(II) released by dissolution is adsorbed on the carbonate. The variations in concentration observed in those cases, correlated with the dissolution process, show that Zn(II) does not interact the same way with the Lavoux carbonate according to the pressure conditions. For the experiment carried out with the St-Emilion carbonate, three very different behaviors are evidenced. At 30 bar, sorption happens at a very slow

TABLE 2
Results of final Ca(II) concentration measurements and corresponding simulations with both databases. The error is calculated as the difference between measured and simulated concentration divided by the simulated value, and multiplied by 100

Experiment	Measured	Dissolved calcite, calculated	Pitzer.dat	Error, relative to pitzer simulations	Minteq.v4.dat	Error, relative to minteq.v4 simulations
	$[Ca(II)]_{eq}$ (mol/L)	(g)	$[Ca(II)]_{eq}$ (mol/L)	(%)	$[Ca(II)]_{eq}$ (mol/L)	(%)
Lav30	0.03579	0.57	0.0364	−1.7	0.04525	−20.9
Lav60	0.04227	0.68	0.0414	2.1	0.06127	−31.0
Lav90	0.04163	0.67	0.0432	−3.6	0.07335	−43.2
StEm30	0.03376	0.54	0.0364	−7.3	0.04525	−25.4
StEm60	0.04176	0.67	0.0414	0.9	0.06127	−31.8
StEm90	0.04136	0.66	0.0432	−4.3	0.07335	−43.6

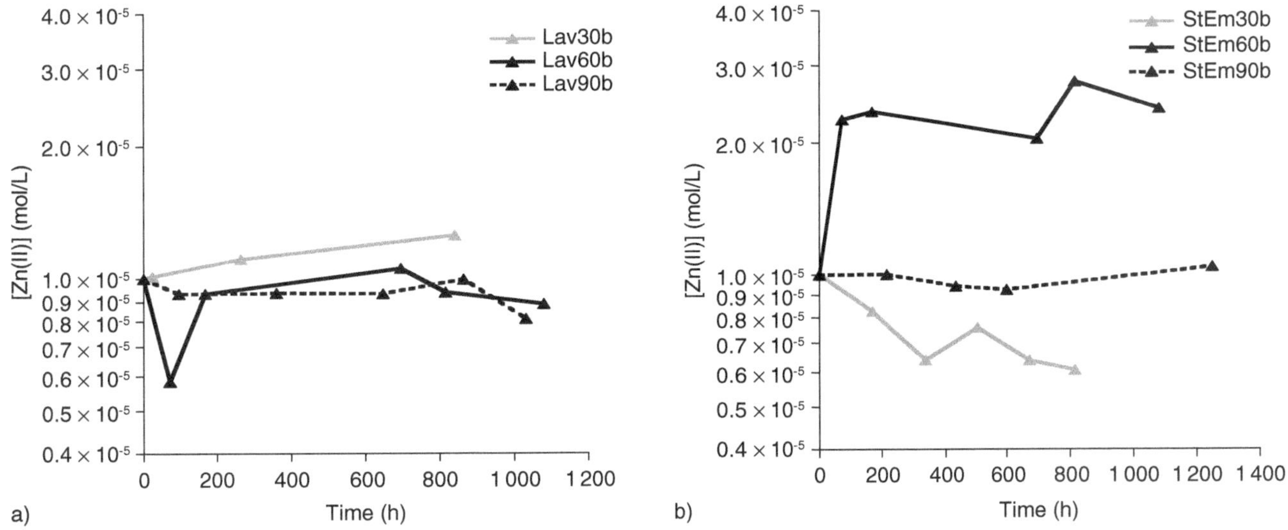

Figure 2

Zinc concentration as a function of time. a) Experiments run with the Lavoux carbonate at 30 bar (full line, grey), 60 bar (full line, black), and 90 bar (dotted line, black). b) Experiments run with the St-Emilion carbonate at 30 bar (full line, grey), 60 bar (full line, black), and 90 bar (dotted line, black). Error bars are comprised within symbol size.

rate with a decrease of about 40% of the Zn(II) concentration. At 60 bar, the end concentration measured is close to twice the initial Zn(II) concentration, implying a strong Zn (II) release from the St-Emilion carbonate due to calcite dissolution. In supercritical conditions, the Zn(II) concentration is nearly constant all through the experiment. As dissolution of the natural carbonate occurs at such pressure, there is a nearly perfect balance between the Zn(II) sorbed and the Zn(II) released due to calcite dissolution.

The study of Mn(II) concentration behavior (Fig. 3) shows that where the Lavoux carbonate is concerned, the difference between the three pressure conditions is very slight, as evidenced for Zn(II). However, as no Mn(II) is initially present in solution, the conclusions are not the same concerning the sorption of this element. Indeed the amount of Mn(II) released is supposed to be nearly equal at 30 and 90 bar, and higher at 60 bar, according to calcite dissolution data. However, the final Mn(II) concentrations at 60 and 90 bar are lower than the one at 30 bar, meaning that sorption is stronger when the CO_2 pressure increases. For the St-Emilion carbonate experiments, the Mn(II) concentration increase is more important, in agreement with the higher Mn(II) content in the St-Emilion carbonate, but the end concentrations indicate different behaviors according to pressure. At 30 bar, the end Zn(II) concentration is nearly the same as for experiment with the Lavoux carbonate. At higher pressures, the end concentrations indicate a lower amount of Mn(II) sorbed at 90 bar than at 60 bar. This trend is well correlated with the trend of Zn(II) in experiments with the same carbonate.

The behavior of Sr(II) concentration over time (Fig. 4), seems to evidence a pressure dependent sorption equilibrium, which can be correlated with the dissolution effect. Indeed, for both carbonates, the lower equilibrium concentration is observed at 30 bar, the next one is at 90 bar and the higher equilibrium value is at 60 bar: this observation is consistent with the amount of calcite dissolved given in Table 2. However this apparent pressure dependence for the final Sr(II) concentration has also to be correlated with the varying of the number of available sorption sites, which vary according to the dissolution of calcite. In all of these experiments, the time to reach the end concentrations is between 100 and 200 h, this variability relying on the sampling dates selected. An apparent organization according to pressure is evidenced for the three elements of interest in the St-Emilion carbonate experiments, observed also with Sr(II) for Lavoux carbonate experiments, which will be discussed later in this article.

2.3 Solid Phase Evolution

2.3.1 Petrophysical Consequences

The petrophysical changes following the different pressure conditions are presented in this section (Fig. 5, 6), based on mercury injection data (Tab. 3).

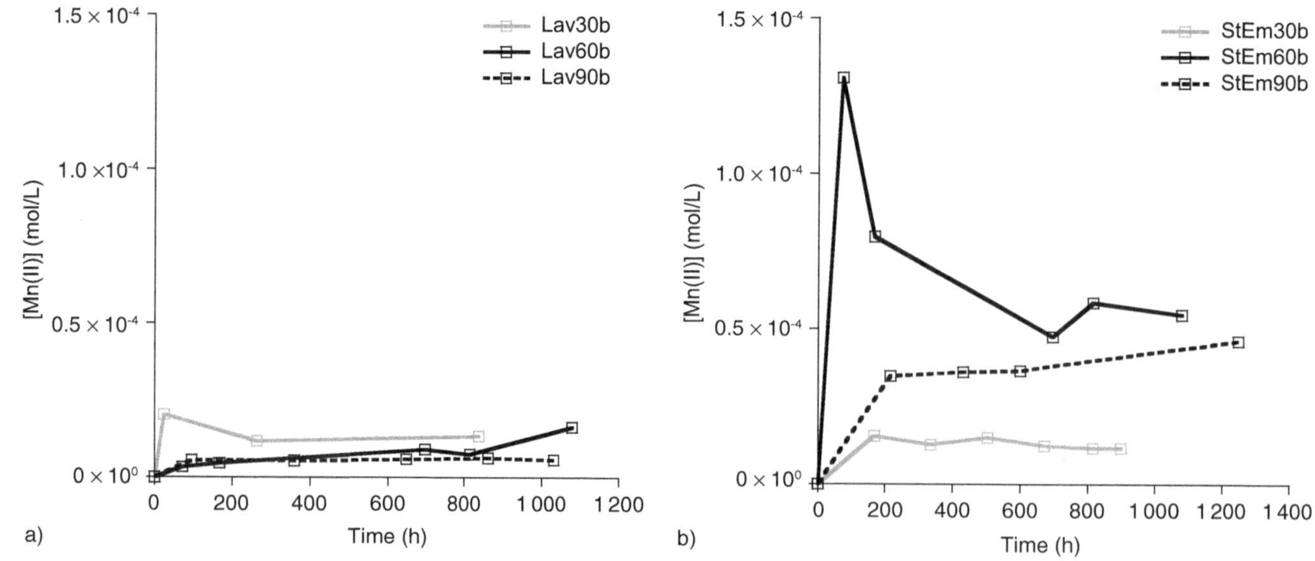

Figure 3

Manganese concentration as a function of time. a) Experiments run with the Lavoux carbonate at 30 bar (full line, grey), 60 bar (full line, black), and 90 bar (dotted line, black). b) Experiments run with the St-Emilion carbonate at 30 bar (full line, grey), 60 bar (full line, black), and 90 bar (dotted line, black). Error bars are comprised within symbol size.

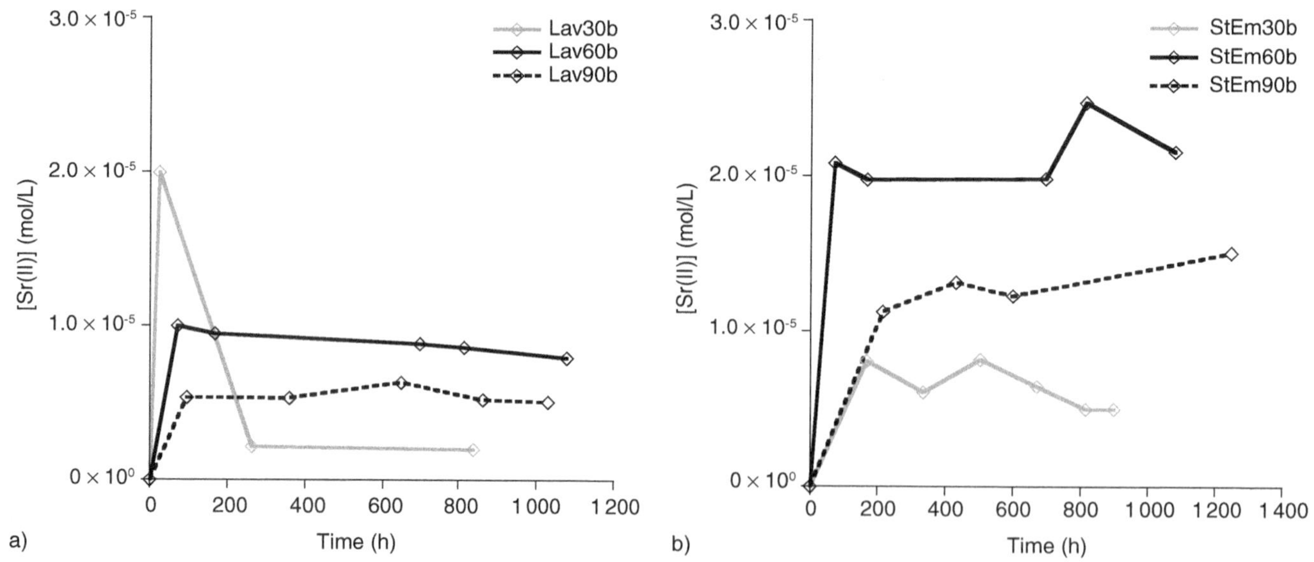

Figure 4

Strontium concentration as a function of time. a) Experiments run with the Lavoux carbonate at 30 bar (full line, grey), 60 bar (full line, black), and 90 bar (dotted line, black). b) Experiments run with the St-Emilion carbonate at 30 bar (full line, grey), 60 bar (full line, black), and 90 bar (dotted line, black). Error bars are comprised within symbol size.

2.3.1.1 The Lavoux Carbonate

On the Lavoux carbonate porosity distribution curves (Fig. 5a), the characterization of the initial powder is plotted with the reacted powders dataset. A progressive evolution is evidenced with the increasing pressure conditions of the experiments. The initial curve presents a double porosity characterized by two main families, FLav_1 (d_p below 10 μm) and FLav_2 (d_p over 10 μm). When reacted at

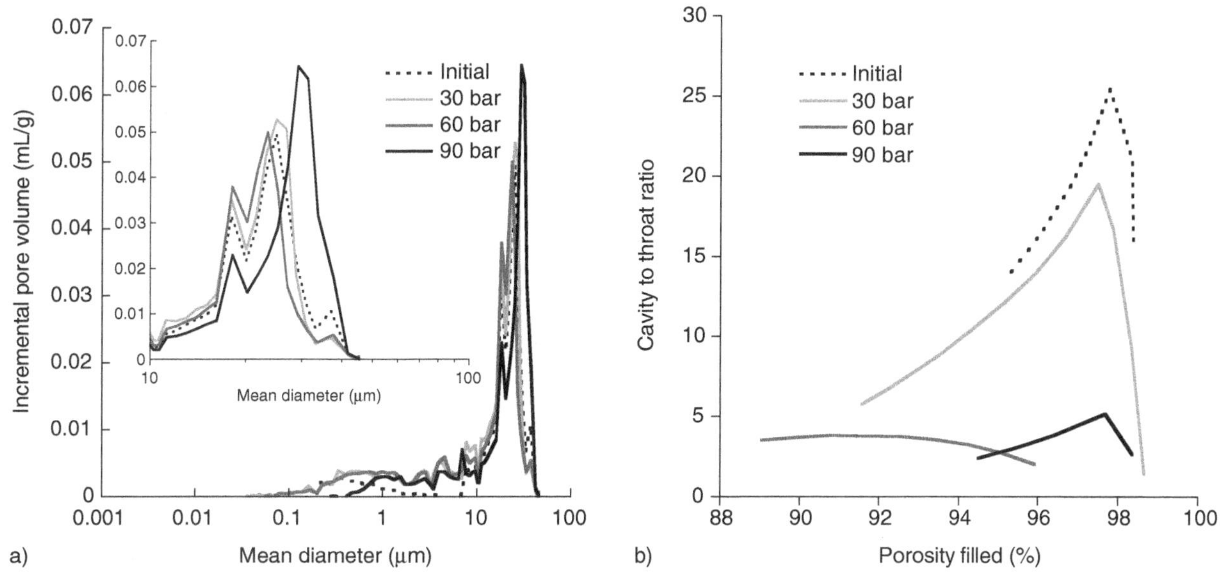

Figure 5

Lavoux carbonate mercury injection data comparison for initial carbonate, and reacted powder at 30, 60, and 90 bar of CO_2. a) Incremental pore volume as a function of the mean pore diameter. A zoom of the 10-100 μm range is added on this figure. b) Cavity to throat ratio evolution with respect to porosity filling.

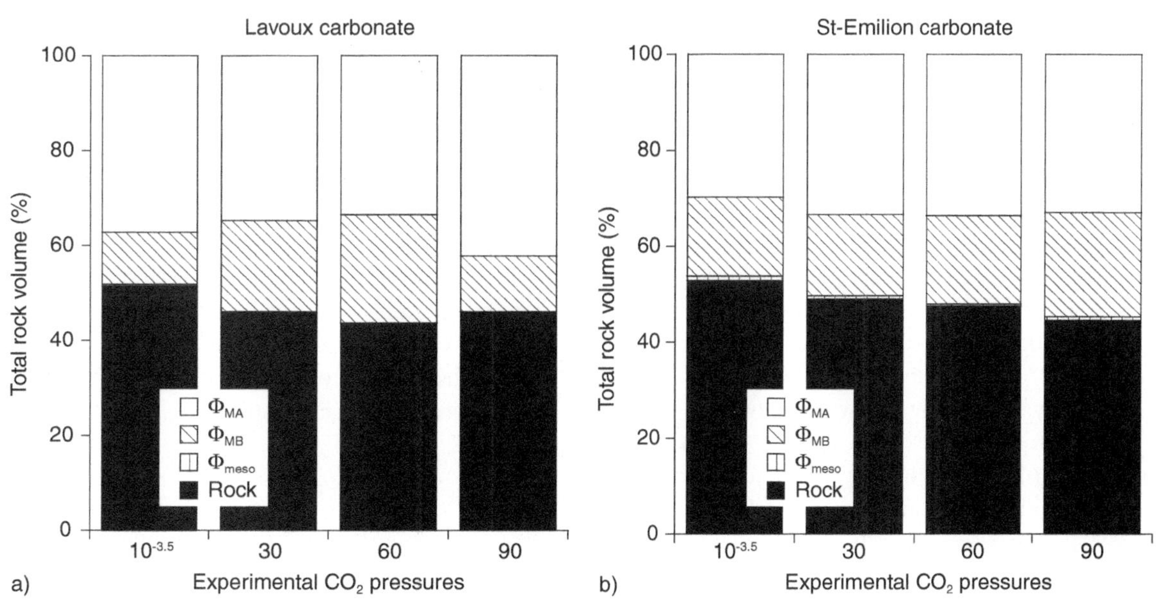

Figure 6

Comparative evolution of the families composing the total porosity of the Lavoux a) and the St-Emilion b) carbonates with the CO_2 pressure conditions of the experiments (atmospheric, 30, 60 and 90 bar). Three families of porosity are represented: macroporosity A (white), macroporosity B (diagonal lines) and mesoporosity (vertical lines), along with the rock fraction (black).

30 bar pressure, no major impact is evidenced on the FLav_1 family, but the FLav_2 family presents a larger peak, slightly shifted towards higher mean diameters. With a 60 bar pressure, the peak of the first family of pores is higher, while the second is slightly shifted towards smaller pore diameters. This shift to the left of the FLav_2 pore family is a side effect

TABLE 3
Results of SSA measurements and mercury injection

Experiment	SSA	Φ	Φ increase	Φ_{MA}	Φ_{MB}	Φ_{meso}	Φ_f	Φ_t	Asp	HD (4V/A)
	(m^2/g)	(%)	(%)	(%)	(%)	(%)	(%)	(%)	(m^2/g)	(nm)
Lav30	0.54	54.07	5.7	34.68	19.21	0.18	4.67	49.40	0.683	2.753
Lav60	0.75	64.63	16.3	41.17	23.32	0.14	7.49	57.14	0.578	2.858
Lav90	0.92	53.85	5.5	42.17	11.68	0.00	2.95	50.90	0.187	9.408
StEm30	1.30	50.89	3.8	33.43	16.80	0.66	4.06	46.83	0.912	1.737
StEm60	1.06	52.33	5.2	33.81	18.53	0.36	3.15	49.18	/	/
StEm90	1.53	55.47	8.4	34.95	19.75	0.78	3.05	52.42	0.974	1.703

of the global increase of the pore volume observed in the 20-25 μm range of mean diameter. Finally, at 90 bar pressure, the first family presents a much lower peak than in the three other conditions, correlated with a clear shift in the FLav_2 pore family towards 30 μm of mean diameter. The pressure increase effect on this carbonate can be summarized as: with CO_2 as gas phase, the dissolution increases the volume available behind the 20-25 μm mean pore diameter range; while the supercritical CO_2 mainly increases the diameter of the bigger pores. The curves plotted in Figure 5b illustrate this preferential dissolution effect by showing the evolution of the cavity to throat ratio with respect to porosity filling. As pressure increases, the trend shows that this ratio diminishes from a mean value of 20 down to a mean value of 5: dissolution tends to homogenize the porous media and to create tubular shaped pores.

The data in Table 3 indicates a 5% to 15% increase of the total porosity in the Lavoux carbonate (by comparison with the initial characterization data, Auffray *et al.*, 2016), depending on the pressure conditions. Concerning macroporosity A (Φ_{MA}: $d_p > 15$ μm), the contribution of this family to the total porosity value decreases with the increasing CO_2 pressure in the experiments (from about 77% to 60%). Macroporosity B (Φ_{MB}: 1 μm $< d_p <$ 15 μm) makes an increasing contribution to the total porosity when CO_2 pressure increases, and mesoporosity (Φ_{meso}: 1 μm $> d_p$) is very slightly impacted by the experimental conditions. The measure of the SSA indicates an increasing function with pressure. For a better representation of what happens for each family of porosity, Figure 6 summarize these results for both carbonates.

2.3.1.2 The St-Emilion Carbonate

The evolution of the pore size distribution of the St-Emilion carbonate (Fig. 7a) is more complicated. Initially it presents one pore family, referred as FStEm_1 (d_p below 10 μm), but the CO_2 pressure conditions and related dissolution lead to the creation of a second pore family, FStEm_2 (20 μm $< d_p <$ 30 μm). With a 30 bar CO_2 pressure, this second family has a peak as high as the one of FStEm_1. At 60 bar, the available pore volume after this range of pore size increases (higher peak) and shifts slightly towards lower diameters. Lastly, the 90 bar experiment evidences that the two main families of mean pore diameters represent almost the same pore volume. With this carbonate, the possible creation of a new pore family is evidenced. Considering the cavity to throat ratio (Fig. 7b), from the initial mean value of 17, the 30 and 60 bar of CO_2 conditions decrease this value to something around 8, and at 90 bar, this mean ratio is about 11. As for the Lavoux carbonate, the increasing CO_2 pressure seems to homogenize the porous media towards a cylindrical shape.

The total porosity (Tab. 3) increases almost linearly with pressure, from its initial value of 47% to 51% at 30 bar, 52.3% at 60 bar and 55.5% at 90 bar. The available volume behind macroporosity A increases about 3% of the total porosity regardless the pressure applied, while macroporosity B increases with CO_2 pressure. As for the Lavoux carbonate, the mesoporosity is only slightly impacted by the dissolution process (Fig. 6b).

2.3.2 Specific Surface Area

The variations of the SSA of each sample with respect to the imposed experimental conditions are presented in Table 3. The Lavoux carbonate evolves from its initial value of 0.60 m^2/g (Auffray *et al.*, 2016) to 0.54, 0.75 and 0.92 m^2/g for experiments carried out at 30, 60 and 90 bar of CO_2 respectively. The variation measured at 30 bar of CO_2 is comprised within the incertitude range, so no variation of the SSA is evidenced in this experiment. But clearly, at 60 and 90 bar, the dissolution of the carbonate is correlated with an increase of the specific surface area of 25% and 53% respectively. The SSA measures of the St-Emilion carbonate powders also evolved from the initial value of 1.20 m^2/g. The 30 bar CO_2 pressure leads to a 1.30 m^2/g SSA (8% more) and 1.06 m^2/g (about 12% less) for 60 bar of CO_2

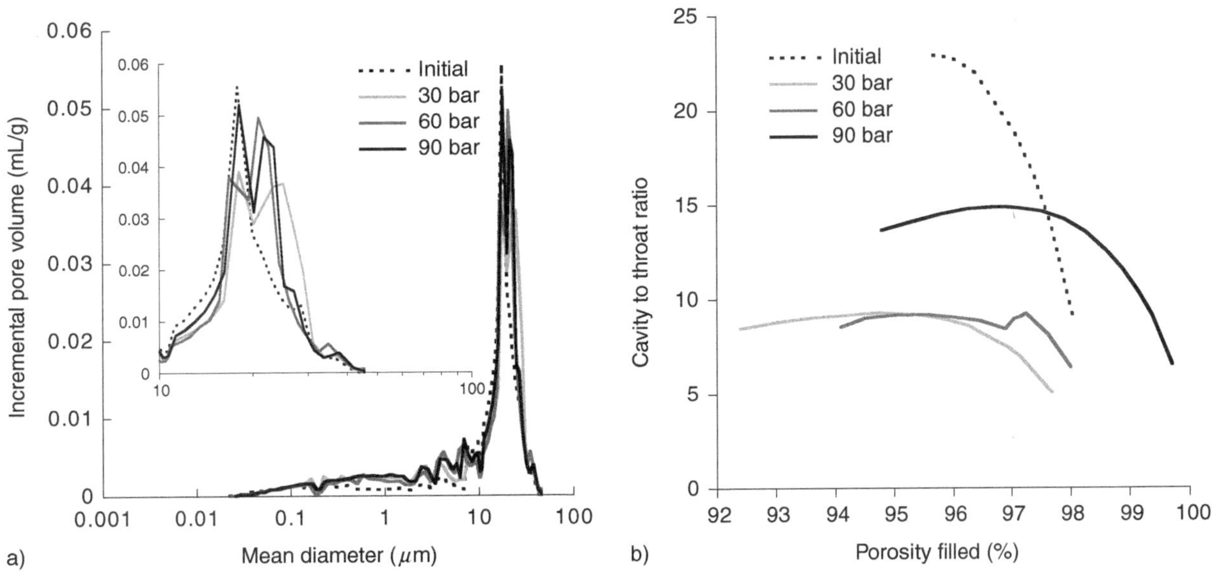

Figure 7

St-Emilion carbonate mercury injection data comparison for initial carbonate, and reacted powder at 30, 60, and 90 bar of CO_2. a) Incremental pore volume as a function of the mean pore diameter. A zoom of the 10-100 μm range is added on this figure. b) Cavity to throat ratio evolution with respect to porosity filling.

pressure: those values are closely within the 10% error range of the measure; but at 90 bar it increases up to 1.53 m^2/g (an increase of ~27%). The dissolution of the St-Emilion carbonate is not clearly correlated to an increase of the SSA, except at a CO_2 pressure of 90 bar.

2.3.3 Microprobe

The microprobe analysis of the reacted powders did not evidence any zinc enrichment of the periphery of the grains (data not shown). All the measures (about 15 grains for each experiment) indicated that the zinc content was always below the detection limit of about 75 ppm.

3 DISCUSSION

3.1 Equilibrium of the System and TE Behavior

CO_2 injection under the supercritical phase is considered the easiest way to inject this molecule underground, because in this state it has both the density of a liquid and the viscosity of a gas (IPCC, 2007). With these experiments (30, 60 and 90 bar at 40°C) coupled with PHREEQC simulations, we evidence that the gas and supercritical states of CO_2 in high salinity conditions (~100 g/L NaCl) are well managed by the numerical code, as stated by other studies (Krupka *et al.*, 2010). This is at least proved for CO_2 pressures up to 90 bar.

Thus, it is possible to simulate accurately the dissolution of calcite in conditions relevant for CO_2 geological storage (supercritical state, high salinity), but the system of interest in this article is more complex in terms of composition: the numerous TE that are solubilized and especially their fate are of major interest. As discussed in another article (Auffray *et al.*, 2016), the Pitzer database used for high salinity simulations does not contain the specific parameters for Zn(II), Mn(II), Sr(II) and their related species. This lack of thermodynamic data has already been pointed by several studies (Bateman *et al.*, 2005; Bentham and Kirby, 2005; Bethke, 2008; Gaus *et al.*, 2008; Kaszuba *et al.*, 2003; Xu *et al.*, 2004), but never concerning the TE released in the system by mineral dissolution. An accurate simulation tool for combined dissolution, precipitation and sorption phenomenon is still to be developed.

The importance of thermodynamic data concerning TE is emphasized in this article by focusing on the evolution with time of the TE in the aqueous phase. Another study (Viswanathan *et al.*, 2012) points out that in the specific context of their study, As is mobilized in the aqueous phase and that, despite sorption and/or reprecipitation, an amount of this As remains in solution. What we observe in the experiments presented here demonstrates that this mobilization effect also exists for elements such as zinc, manganese and strontium in the natural carbonate context studied here. After the initial increase of TE concentration following the dissolution of calcite, sorption occurs in the systems that lead to a reduced amount of element in solution. For example the steady-state concentration of Mn(II) in solution at 60 bar of CO_2 for the St-Emilion carbonate is a third of the initial concentration measured after the initial pulse (Fig. 3b). Another important result enlightened in this study is the specific behavior of the TE with respect to the natural carbonate used. Indeed, Zn(II) behavior is not the same if one considers the Lavoux or the St-Emilion carbonates experiments, whatever the pressure considered. A probable explanation could be the impact of the petrophysical properties of the different samples, coupled with their respective evolution to the CO_2 pressures.

The last result evidenced by this study is the specific sorption behavior of the TE with respect to pressure. Indeed, the Ca(II) concentration in solution, related to calcite dissolution, indicates nearly the same quantities of dissolved calcite at 60 and 90 bar. One would expect the same end concentration for each TE in solution, but this is absolutely not the case: the end concentration is always higher at 60 than at 90 bar, and this occurs whatever the TE considered. This result can be discussed in the light of the parameters relevant for the sorption phenomena, including the SSA (Dzombak and Morel, 1990; Sverjensky, 2003; Zachara *et al.*, 1991, 1988). The variation of this parameter could explain the difference observed by comparing values from Table 2 and Figures 2-4. Despite the same dissolved masses measured, the SSA increases much more for experiments carried out at 90 bar. To our knowledge, this was never reported in the literature, probably due to the fact that all the previous sorption studies were carried out only under atmospheric pressure and strictly controlled conditions to prevent the dissolution from being of major impact on the results (Martin-Garin *et al.*, 2003; Tertre *et al.*, 2010; Villegas-Jiménez *et al.*, 2009; Zachara *et al.*, 1988, 1991).

This discussion evidences well the actual lack of knowledge concerning the behavior of TE in a CO_2 geological storage context, especially if one wants to assess the safety of a storage site regarding the possible contamination by TE. It also points out the other subject still discussed in the community (Gaus *et al.*, 2008), which is the link between the injection of SC-CO_2, the chemical reactions that will follow and the evolution of the petrophysical properties of the host formation.

3.2 Linking Chemistry and Petrophysics

Indeed, as presented by several other studies (Bachaud *et al.*, 2011), a key point to better understand the consequences of SC-CO_2 injection in a given formation is to obtain more information about the linked evolution of chemistry and petrophysical properties in the system. To do so, a parallel between the calcium concentration in the aqueous phase and the petrophysical variations is discussed in this part, based on the values presented in Table 3 and Figures 5-7. No correlation is observed between the total porosity increase and the calcite mass dissolved in the experiments, and a first explanation of this discrepancy is the use of powders in these experiments. Indeed, the dissolution process occurs on both the external and internal (porosity) parts of the grains, while mercury injection gives only access to pores that have a diameter below 44 µm. This means that all the variations in the external structure of the grains are not taken into account via mercury injection analyses. Still, the impact of dissolution on porosity is specifically studied by considering the different families that compose the total porosity.

The results of the experiments show that dissolution impacts mainly the two pore families Φ_{MA} and Φ_{MB} as depicted by Figure 6. By comparison with the pore size distributions of the two initial carbonates (Fig. 5, 7), this observation is coherent. Indeed, the main variations correspond to the diameter range 10-100 µm. This observation agrees well with preferential dissolution zones evidenced by Bachaud *et al.* (2011) in cap-rock samples, particularly for the Lavoux carbonate. The St-Emilion carbonate, though, is useful in evidencing the fact that this preferential dissolution is not always valid. With the increasing CO_2 pressures (30, 60 and 90 bar) a new family of pores appears around 20 µm of pore diameter. To complete this discussion on the impact of dissolution on the porous media, the cavity to throat ratio indicates, coherently with the previous discussion, that the porous network is homogenized by dissolution. This is true for both carbonates, as their cavity to throat ratio decreases by a factor 4 approximately. Dissolution tends to dissolve throats and create a tube-like porous network. This result is expected in the case of flow through experiments, but it seems also to be the case for the powders used in these experiments.

3.3 Extrapolation to a CO_2 Storage Context and Possible Leakages from a Reservoir Formation

In the case of a real storage site, other specificities must be kept in mind: first, the fact that the rock formation does not have the same properties as powders relative to sorption and reactivity (Qafoku *et al.*, 2013), and then that the space (vertical and horizontal) and time scales considered are much wider. Indeed, in this article the focus is on impacts of a CO_2 pressure of 30, 60 and 90 bar, which corresponds roughly to 300, 600 and 900 m depths, depending on the regional conditions. At these depths, the mobilization of TE contained in host formations is evidenced, and potentially balanced by the sorption and/or reprecipitation phenomena, depending on the carbonate rock. The modifications of the petrophysical properties evidence a probable increase in the transport properties of the formation. But to complete the study of CO_2 impacted zones by a leakage, as discussed in other papers (Harvey *et al.*, 2013; Little and Jackson, 2010; Viswanathan *et al.*, 2012), the subsurface environment and shallow aquifers are also of importance. Results are presented on this particular aspect of the impact of the CO_2 underground storage on water quality in another paper from the authors (Auffray *et al.*, 2016), that deals with the influx of a brine loaded with a given concentration of Zn(II) in a carbonate reservoir. The last aspect to investigate is the impact on the vadose zone: a field study was carried out in an old quarry of the St-Emilion village (France) but with no aspect concerning the behavior of TE in those specific hydrologic conditions (Cohen *et al.*, 2013; Garcia *et al.*, 2013; Le Roux *et al.*, 2013; Loisy *et al.*, 2013).

CONCLUSIONS

A leakage of fluid (gas and/or brine) from the storage formation is a major uncertainty concerning the safety assessment of a CO_2 geological storage project. The consequences can vary a lot, all along the pathway that brine and gas follow as they leak from the storage formation to an upper one, as evidenced in this paper. Considering sorption and dissolution, pressure dependent behaviors for the TE considered, we evidence in this paper chemical variations in the fluid composition, but also modifications of the structure of the porous media, and consequently on the flow properties of the concerned formation. The increase of TE concentration in solution is evidenced, and sorption does not balance this release in solution in most cases, despite an available SSA that is much higher than the SSA that will be available in a real storage site. Even in a real storage context, the dissolution is supposed to be lower than what we observe in those experiments, due to aquifer flow, non negligible amounts are thought to be mobilized (Viswanathan *et al.*, 2012; Zheng *et al.*, 2009; Wigand *et al.*, 2008) when one considers the huge volume of injected CO_2 that is envisaged (Lu *et al.*, 2012; Rempel *et al.*, 2011). To fill this gap in our knowledge, more experimental studies must be carried out to understand more the real behavior of those TE in CO_2 storage conditions.

Where closer aquifers of potable water resources are concerned, the debate is still open to know whether a CO_2 leakage would be a benefit or an issue (Harvey *et al.*, 2013), but concerning at least the three species studied here, Zn(II) for example could become toxic with the concentrations observed, as this element reaches concentrations around 10^{-8} M (WHS, 2011). Nevertheless, the precipitation of metal carbonate minerals, an aspect not evidenced in this study, must be considered in a real storage site far from the injection well, as CO_2 density and concentration decrease, leading to possible incorporation of metals elements due to precipitation of secondary phases (Gunter *et al.*, 1997; Xu *et al.*, 2010, 2004).

ACKNOWLEDGMENTS

B. Auffray is thankful to Michel Chardin for his help in managing the *PAAR*® autoclave experimental settings and the associated procedures; Frederic Neyret-Martinez and Nathalie Texier for their availability during the ICP-AES analysis campaigns. The authors also thank the two anonymous reviewers for their contribution to improving this paper. This article is part of the PhD thesis of B. Auffray, funded by *IFP Energies nouvelles.*

REFERENCES

Albritton D.L., Meira Filho L.G. (2001) Contribution of Working Group I Third Assessment Report on Intergovernmental Panel on Climate Change, in *Climate Change 2001: the Scientific Basis*, Cambridge, Cambridge Univ. Press, UK and New York, USA.

Auffray B., Garcia B., Lienemann C.-P., Sorbier L., Cerepi A. (2016) Zn(II), Mn(II) and Sr(II) Behavior in a Natural Carbonate Reservoir System. Parti I: Impact of Salinity, Initial pH and Initial Zn(II) Concentration in Atmospheric Conditions, *Oil & Gas Science and Technology* **71**, 47.

Bachaud P., Berne P., Renard F., Sardin M., Leclerc J.P. (2011) Use of tracers to characterize the effects of a CO_2-saturated brine on the petrophysical properties of a low permeability carbonate caprock, *Chemical Engineering Research and Design* **89**, 1817-1826.

Bateman A.S., Kelly S.D., Jickells T.D. (2005) Nitrogen isotope relationships between crops and fertilizer: Implications for using nitrogen isotope analysis as an indicator of agricultural regime, *Journal of Agricultural and Food Chemistry* **53**, 5760-5765.

Bentham M., Kirby G. (2005) CO_2 Storage in Saline Aquifers, *Oil & Gas Science and Technology* **60**, 3, 559-567.

Bethke C.M. (2008) *Geochemical and Biogeochemical reaction modeling*, Cambridge University Press, New York.

Bryant E. (1997) *Climate Process and Change*, Cambridge University Press, Cambridge.

Chadwick A., Noy D., Lindeberg E., Arts R., Eiken O., Williams G. (2006) Calibrating reservoir performance with time-lapse seismic monitoring and flow simulations of the Sleipner CO_2 plume, *Proceedings of the 8th International Conference on Greenhouse Gas Control Technologies*, Trondheim, Norway, June 19-22.

Cohen G., Loisy C., Laveuf C., Le Roux O., Delaplace P., Magnier C., Rouchon V., Garcia B., Cerepi A. (2013) The CO_2-Vadose project: Experimental study and modelling of CO_2 induced leakage and tracers associated in the carbonate vadose zone, *International Journal of Greenhouse Gas Control* **14**, 128-140.

Czernichowski-Lauriol I., Rochelle C., Gaus I., Azaroual M., Pearce J., Durst J. (2006) Geochemical interactions between CO_2, pore-waters and reservoir rocks - Lessons learned from laboratory experiments, field studies and computer simulations, *NATO Science Series IV Earth and Environmental Sciences* **65**, 157-174.

Dzombak D.A., Morel F.M.M. (1990) *Surface Complexation Modeling: Hydrous Ferric Oxide*, Wiley-Interscience, New York.

Emberley S., Hutcheon I., Shevalier N., Durocher K., Mayer B., Gunter W.D., Perkins E.H. (2005) Monitoring of fluid-rock interaction and CO_2 storage through produced fluid sampling at the Weyburn CO_2-injection enhanced oil recovery site, Saskatchewan, Canada, *Applied Geochemistry* **20**, 1131-1157.

Fischer R., Lorenz M., Köhl M., Mues V., Granke O., Iost S., van Dobben H., Reinds G.J., de Vries W. (2010) The condition of forests in Europe, *ICP Forests, 2010 Executive Report*, Hamburg, Germany, and European Commission, Brussels, Belgium.

Gale J.F.W., Laubach S.E., Marrett R.A., Olson J.E., Holder J., Reed R.M. (2004) Predicting and characterizing fractures in dolostone reservoirs: Using the link between diagenesis and fracturing, in *The Geometry and Petrogenesis of Dolomite Hydrocarbon Reservoirs*, Braithwaite C.J.R., Rizzi G., Darke G. (eds), Geological Society (London) Special Publication 235.

Garcia B., Beaumont V., Perfetti E., Rouchon V., Blanchet D., Oger P., Dromart G., Huc A., Haeseler F. (2010) Experiments and geochemical modelling of CO_2 sequestration by olivine: Potential, quantification, *Applied Geochemistry* **25**, 1383-1396.

Garcia B., Hy Billiot J., Rouchon V., Mouronval G., Lescanne M., Lachet V., Aimard N. (2012) A Geochemical Approach for Monitoring a CO_2 Pilot Site: Rousse, France. A Major gases, CO_2-carbon isotopes and Noble Gases Combined Approach, *Oil Gas Science and Technology* **67**, 341-353.

Garcia B., Delaplace P., Rouchon V., Magnier C., Loisy C., Cohen G., Laveuf C., Le Roux O., Cerepi A. (2013) The CO_2-vadose project: Numerical modeling to perform a geochemical monitoring methodology and baseline performance assessment for various geochemical variables (gas flux, gas composition, stable isotopes and noble gases) in the carbonate vadose zone, *International Journal of Greenhouse Gas Control* **14**, 247-258.

Gaus I. (2010) Role and impact of CO_2-rock interactions during CO_2 storage in sedimentary rocks, *International Journal of Greenhouse Gas Control* **4**, 73-89.

Gaus I., Audigane P., André L., Lions J., Jacquemet N., Durst P., Czernichowski-Lauriol I., Azaroual, M. (2008) Geochemical and solute transport modelling for CO_2 storage, what to expect from it? *International Journal of Greenhouse Gas Control* **2**, 605-625.

Gunter W.D., Genetzis T., Rottenfuser Richardson R.J.H. (1997) Deep coalbed methane in Alberta Canada: A fuel resource with the potential of zero greenhouse gas emissions, *Energy Conservation and Management* **38S**, 217-222.

Harvey O.R., Qafoku N.P., Cantrell K.J., Lee G., Amonette J.E., Brown C.F. (2013) Geochemical Implications of Gas Leakage associated with Geologic CO_2 storage - A Qualitative Review, *Environmental Science and Technology* **47**, 23-26.

Hepple R.P., Benson S.M. (2005) Geologic storage of carbon dioxide as a climate change mitigation strategy: performance requirements and the implications of surface seepage, *Environmental Geology* **47**, 4, 576-585.

Holloway S. (1997) An overview of the underground disposal of carbon dioxide, *Energy Conversion and Management* **38**, 193-198.

IPCC (2007) Special Report on Carbon Capture and Storage.

Kaszuba J.P., Janecky D.R., Snow M.G. (2003) Carbon dioxide reaction processes in a model brine aquifer at 200°C and 200 bars: implications for geologic sequestration of carbon, *Applied Geochemistry* **18,** 1065-1080.

Kharaka Y.K., Cole D.R., Hovorka S.D., Gunter W.D., Knauss K.G., Freifield B.M. (2006) Gas-water-rock interactions in Frio Formation following CO_2 injection: Implications for the storage of greenhouse gases in sedimentary basinsn, *Geology* **34**, 577-580.

Kharaka Y.K., Thordsen J.J., Hovorka S.D., Nance H.S., Cole D.R., Phelps T.J., Knauss K.G. (2009) Potential environmental issues of CO_2 storage in deep saline aquifers: Geochemical results from the Frio-I Brine Pilot test, Texas, USA, *Applied Geochemistry* **24**, 1106-1112.

Krupka K.M., Cantrell K.J., McGrail B.P. (2010) Thermodynamic Data for Geochemical Modeling of Carbonate Reactions Associated with CO_2 Sequestration - Literature Review, in *Report PNNL-19766*.

Le Roux O., Cohen G., Loisy C., Laveuf C., Delaplace P., Magnier C., Rouchon V., Cerepi A., Garcia B. (2013) The CO_2-Vadose project: Time-lapse geoelectrical monitoring during CO_2 diffusion in the carbonate vadose zone, *International Journal of Grennhouse Gas Control* **16**, 156-166.

Li Y., Wardlaw N.C. (1986) Mechanisms of Nonwetting Phase Trapping during Imbibition at Slow Rates, *Journal of Colloid and Interface Science* **109**, 473-486.

Little M.G., Jackson R.B. (2010) Potential impacts of leakage from deep CO_2 geosequestration on overlying freshwater aquifers, *Environmental Science and Technology* **44**, 9225-9232.

Loisy C., Cohen G., Laveuf C., Le Roux O., Delaplace P., Magnier C., Rouchon V., Cerepi A., Garcia B. (2013) The CO_2-Vadose Project: Dynamics of the natural CO_2 in a carbonate vadose zone, *International Journal of Greenhouse Gas Control* **14**, 97-112.

Lowell S., Shields J.E. (1984) *Powder Surface Area and Porosity*, Chapman and Hall.

Lu J., Kharaka Y.K., Thordsen J.J., Horita J., Karamalidis A., Griffith C., Hakala J.A., Ambats G., Cole D.R., Phelps T.J., Manning M.A., Cook P.J., Hovorka S.D. (2012) CO_2-rock-brine interactions in Lower Tuscaloosa Formation at Cranfield CO_2 sequestration site, Mississippi, U.S.A., *Chemical Geology* **291**, 269-277.

Martin-Garin A., van Cappellen P., Charlet L. (2003) Aqueous cadmium uptake by calcite: A stirred flow-through reactor study, *Geochimica et Cosmochimica Acta* **67**, 2763-2774.

McGrail B.P., Schaef H.T., Ho A.M., Chien Y.-J., Dooley J.J., Davidson C.L. (2006) Potential for carbon dioxide sequestration in flood basalts, *Journal of Geophysical Research* **111**, B12201.

Parkhurst D.L., Appelo C.A.J. (2013) Description of Input and Examples for PHREEQC Version 3 - A Computer Program for Speciation, Batch-Reaction, One-Dimensional Transport, and Inverse Geochemical Calculations.

Qafoku N.P., Brown C.F., Wang G., Sullivan C., Lawter A.R., Harvey O.R., Bowden M. (2013) Geochemical Impacts of Leaking CO_2 from Subsurface Storage Reservoirs to Unconfined and Confined Aquifers. Report PNNL-22420.

Rempel K.U., Liebscher A., Heinrich W., Schettler G. (2011) An experimental investigation of trace element dissolution in carbon dioxide: Applications to the geological storage of CO_2, *Chemical Geology* **289**, 224-234.

Schaef H.T., McGrail B.P., Owen A.T. (2010) Carbonate mineralization of volcanic province basalts, *International Journal of Greenhouse Gas Control* **4**, 249-261.

Sverjensky D.A. (2003) Standard states for the activities of mineral surface sites and species, *Geochimica et Cosmochimica Acta* **67**, 17-28.

Tertre E., Beaucaire C., Juery A., Ly J. (2010) Methodology to obtain exchange properties of the calcite surface - Application to major and trace elements: Ca(II), HCO_3^-, and Zn(II), *Journal of Colloid and Interface Science* **347**, 120-6.

Tsakiroglou C.D., Payatakes A.C. (1990) A New Simulator of Mercury Porosimetry for the Characterization of Porous Materials, *Journal of Colloid and Interface Science* **137**, 315-339.

van Brakel J. (1981) Mercury Porosimetry: State of the Art, *Powder Technology* **29**, 1-12.

Villegas-Jiménez A., Mucci A., Pokrovsky O.S., Schott J. (2009) Defining reactive sites on hydrated mineral surfaces: Rhombohedral carbonate minerals, *Geochimica et Cosmochimica Acta* **73**, 4326-4345.

Viswanathan H., Dai Z., Lopano C., Keating E., Hakala J.A., Scheckel K.G., Zheng L., Guthrie G.D., Pawar R. (2012) Developing a robust geochemical and reactive transport model to evaluate possible sources of arsenic at the CO_2 sequestration natural analog site in Chimayo, New Mexico, *International Journal of Greenhouse Gas Control* **10**, 199-214.

Washburn E.W. (1921) Note on a method of determining the distribution of pore sizes in a porous material, in *Proceedings of the National Academy of Sciences of the United States of America* **7**, 115-116.

Wigand M., Carey J.W., Schutt H., Spangenberg E., Erzinger J. (2008) Geochemical effects of CO_2 sequestration in sandstones under simulated *in situ* conditions of deep saline aquifers, *Applied Geochemistry* **23**, 2735-2745.

World Health Statistics 2011 World Health Organization.

Xu T., Apps J., Pruess K. (2004) Numerical simulation of CO_2 disposal by mineral trapping in deep aquifers, *Applied Geochemistry* **19**, 917-936.

Xu T., Kharaka Y.K., Doughty C., Freifeld B.M., Dalye T.M. (2010) Reactive transport modeling to study changes in water chemistry induced by CO_2 injection at the Frio-I Brine Pilot, *Chemical Geology* **271**, 3-4, 153-164.

Zachara J.M., Kittrick J.A., Harsh J.B. (1988) The mechanism of Zn^{2+} adsorption on calcite, *Geochimica et Cosmochimica Acta* **52**, 2281-2291.

Zachara J.M., Cowan C.E., Resch C.T. (1991) Sorption of divalent metals on calcite, *Geochimica et Cosmochimica Acta* **55**, 1549-1562.

Zheng L., Apps J.A., Zhang Y., Xu T., Birkholzer J.T. (2009) On mobilization of lead and arsenic in groundwater in response to CO_2 leakage from deep geological storage, *Chemical Geology* **268**, 281-297.

6

Comparison of the Acidity of Heteropolyacids Encapsulated in or Impregnated on SBA-15

Teresa Pinto[1], Kai Szeto[1], Nesrine Oueslati[1], Nadine Essayem[2], Véronique Dufaud[1]* and Frédéric Lefebvre[1]*

[1] *Laboratory of Chemistry, Catalysis, Polymers and Processes (C2P2), University of Lyon 1, CNRS, CPE, UMR 5265, 43 Bd du 11 Novembre 1918, 69616 Villeurbanne - France*

[2] *Institut de Recherches sur la Catalyse et l'Environnement de Lyon (IRCELyon), University of Lyon 1, CNRS, UMR 5256, 2 Avenue Albert Einstein, 69626 Villeurbanne - France*

e-mail: teresapspinto@gmail.com - kai-chung.szeto@univ-lyon1.fr - nesrine.oueslati@cpe.fr nadine.essayem@ircelyon.univ-lyon1.fr - veronique.dufaud@univ-lyon1.fr - frederic.lefebvre@univ-lyon1.fr

* Corresponding authors

Abstract — *Heteropolyacids (HPA) immobilized onto SBA-15 silica were prepared by two different ways using either impregnation or encapsulation methodologies. Two Keggin-type HPA, $H_3PW_{12}O_{40}$ and $H_4SiW_{12}O_{40}$ were considered in this study. The resulting hybrid materials were fully characterized by N_2 adsorption-desorption isotherms, XRD, FT-IR, Raman, diffuse reflectance UV-Vis spectroscopies and ^{31}P MAS NMR. All characterization methods showed that at room temperature the catalysts contained well-dispersed and intact Keggin units throughout the solid. The catalytic activity of these solids was investigated in the isomerization of n-hexane. The impregnated and encapsulated phosphotungstic catalysts performed similarly in catalysis showing that the amount of active sites was nearly the same in both catalysts. On the contrary, the tungstosilicic encapsulated material was completely inactive while its impregnated counterpart was even more active than the phosphotungstic derived catalysts. The acidity of the solids was measured by various methods: microcalorimetry of ammonia adsorption, ammonia desorption followed by Temperature Programmed Desorption (TPD) and DRIFT/GC-MS and pyridine adsorption followed by infrared spectroscopy. Only pyridine adsorption and ammonia desorption followed by DRIFT/GC-MS agreed with the catalytic data. Ammonia adsorption followed by microcalorimetry was not able to differentiate between the four catalysts while the TPD experiments led to unreliable results, as not only the evolved ammonia but also other molecules such as water were taken into account in the measurements. The behavior difference between the encapsulated silico- and phosphotungstic acids was explained by a more pronounced encapsulation in the case of silicon.*

Résumé — Comparaison des propriétés acides d'hétéropolyacides encapsulés ou imprégnés dans une matrice de type SBA-15 — Deux types de catalyseurs à base d'hétéropolyacides (HPA) immobilisés sur une silice de type SBA-15 ont été préparés en utilisant soit une méthode par imprégnation soit une méthode par encapsulation. Deux hétéropolyacides à structure de Keggin ont été étudiés, $H_3PW_{12}O_{40}$ et $H_4SiW_{12}O_{40}$. Les solides ainsi obtenus ont été totalement caractérisés par diverses techniques physico-chimiques (isothermes d'adsorption-désorption d'azote, diffraction des Rayons X, spectroscopies infrarouge et Raman, spectroscopie UV-visible par réflectance et RMN du solide du phosphore-31). Toutes les techniques de caractérisation indiquent qu'à température ambiante les hétéropolyacides ont conservé leur structure de Keggin et ce quelle que soit la méthode de préparation. L'activité catalytique de ces matériaux a ensuite été étudiée dans l'isomérisation du *n*-hexane. Les systèmes formés à partir de $H_3PW_{12}O_{40}$ ont un comportement similaire indépendant de la méthode de préparation, en accord avec un nombre de sites actifs quasiment identique. Par contre, dans le cas de $H_4SiW_{12}O_{40}$, le catalyseur préparé par encapsulation est inactif alors que celui préparé par imprégnation présente une activité comparable aux systèmes phosphotungstiques précédents. L'acidité des solides a été mesurée par diverses méthodes : microcalorimétrie d'adsorption d'ammoniac, thermodésorption d'ammoniac suivie par TPD et spectroscopie infrarouge couplée à un système GC-MS (DRIFT/GC-MS) et adsorption de pyridine suivie par spectroscopie infrarouge. Seules l'adsorption de pyridine et les mesures de désorption d'ammoniac suivie par DRIFT donnent des résultats concordants avec les résultats de catalyse. La microcalorimétrie a donné des résultats comparables pour les quatre systèmes tandis que la désorption d'ammoniac suivie par TPD ne conduit pas à des résultats interprétables, les valeurs obtenues prenant en compte non seulement l'ammoniac désorbée mais aussi d'autres molécules, notamment de l'eau. La différence entre les deux matériaux encapsulés a été interprétée comme provenant d'une encapsulation plus prononcée de l'hétéropolyacide dans le cas de $H_4SiW_{12}O_{40}$.

INTRODUCTION

Heteropolyoxometalates (POM) have been widely applied in homogeneous or heterogeneous catalysis due to their unique physico-chemical, acid-base and redox properties [1]. Different types of polyoxometalates exist, such as the Keggin $H_nXM_{12}O_{40}$ or the Dawson $H_nX_2M_{18}O_{62}$ structures. However, the Keggin compounds have been extensively studied as they are the most stable and acidic when the cation of compensation is the proton. Their very strong Brønsted acidity allows them to be potentially good candidates for acid catalysis. Unfortunately, their low specific surface area in the solid state (1-10 m^2/g) and their high solubility in polar solvents are obstacles to their development in heterogeneous catalysis [2-4]. These problems can be overcome by two ways: either by immobilizing them onto high surface area supports to improve the dispersion of the active phase or through the direct synthesis of acidic porous salts. The immobilization onto oxide supports, such as silica, zirconia or γ-alumina, can also induce some changes in the redox properties and acid strength or, in some cases, may lead to some decomposition of the HPA [5, 6]. The best support seems to be silica as it does not modify strongly the properties of the polyacid. Among all types of silica, mesoporous silicas are good candidates as supports as they have very high surface areas (up to 1 000 $m^2.g^{-1}$) with large and uniform pore sizes (~ 6 nm diameter for SBA-15). These characteristics provide ample room for reactant and product diffusion, and the thick walls offer greater hydrothermal and mechanical stability [7-9]. The most known routes to successfully prepare HPA supported onto SBA-15 (or MCM-41) are the classical impregnation method and the direct synthesis. These methods, when optimized, can yield hybrid materials with structured and large pore sizes, showing a stable and uniform distribution of the active phase over the solid and intact Keggin structures [10].

Over the last decades, HPA have been used in numerous acid-catalyzed reactions requiring highly acidic sites such as the skeletal isomerization of *n*-alkanes [11-15]. This reaction is becoming an important way to produce clean fuels, as it satisfies most of the environmental specification requirements for fuel production as well as provides an efficient way to increase the octane number of motor gasoline [16].

Currently, bifunctional catalysts are used in the industrial isomerization processes [17]. One function is an acid catalyst while the other is a hydrogenation-dehydrogenation catalytic system (typically a noble metal). Typical examples are based on platinum catalysts supported either on chlorinated alumina [18] or zeolites such as mordenite [19]. The latter catalysts are particularly advantageous as they are less corrosive and less prone to deactivation by the contaminants (*i.e.* water, sulfur) present in the feed [20, 21]. Nevertheless, Pt-zeolites catalysts present some disadvantages such as high operating temperatures, high hydrogen pressure and lower acidity which, in the end, leads to lower selectivity to branched molecules [22-24]. As a consequence, the development of new solid acid catalysts with higher acid strength is highly required for such isomerization reactions.

In a previous paper, we have reported on the use of $H_3PW_{12}O_{40}$ (HPW) and $H_4SiW_{12}O_{40}$ (HSiW) supported on SBA-15 for the isomerization of *n*-hexane in the gas phase [25]. These catalysts were prepared by using classical impregnation methodologies. The best results were obtained with the silicotungstic acid but all systems were found to be active and highly selective towards branched isomers even though deactivation by coke formation occurred with time on stream.

In the present report, we wish to investigate an alternative method to HPA containing SBA-15 materials, developed recently by us, where the HPA entities are not held at the silica surface but directly encapsulated within the silica framework, and compare their performance with that of conventionally impregnated materials in the isomerization of *n*-hexane. Both types of catalysts were thoroughly characterized using different analytical and spectroscopic techniques to allow for the establishment of structure-activity relationships. In particular, the acidity of the resulting hybrid materials was examined in details using three classical methods of heterogeneous catalysis, namely TPD of adsorbed ammonia, microcalorimetry of ammonia adsorption and pyridine adsorption. We shall see that only pyridine adsorption and ammonia desorption followed by DRIFT/GC-MS measurements agreed in a meaningful manner with catalytic results, ammonia adsorption followed by microcalorimetry being not able to differentiate between the four catalysts.

1 EXPERIMENTAL

1.1 Synthesis of the Catalysts

The supported HPA were prepared by conventional impregnation (noted HPA/SBA-15 in the following), or sol-gel encapsulation method (noted HPA@SBA-15). The $H_3PW_{12}O_{40}$ and $H_4SiW_{12}O_{40}$ HPA were purchased from *Aldrich* and used as received. SBA-15 was prepared according to literature procedures [9]. The HPA/SBA-15 catalysts were prepared in methanol as described previously [25] and were dried at 80°C in an oven. The HPA@SBA-15 materials were prepared according to the procedure described previously for encapsulated $H_3PW_{12}O_{40}$ [26]. The samples were calcined in air saturated with water from room temperature until 220°C for 6 h and then heated up to 400°C for 20 h. All solids contained around 33 wt% of HPA.

1.2 Physical Characterization of the Catalysts

Transmission FT-IR spectra were recorded on a *Nicolet* 5700 spectrometer in the 400-4 000 cm^{-1} range from KBr pellets (resolution 4 cm^{-1}, 32 scans). The ^{31}P MAS NMR spectra were recorded on a *Bruker* Avance-500 spectrometer, operating at 202.40 MHz with a classical 4 mm probe-head allowing spinning rates up to 10 kHz (recycle delay 15 s, 3 000 scans). The X-ray diffraction patterns of samples were recorded on a *Bruker* D5005 diffractometer by using the Cu $K\alpha_1$ monochromatic radiation (λ = 1.54184 Å). N_2 adsorption-desorption was performed on a *Micromeritics* ASAP 2020 system. The samples were evacuated at 350°C for 12 h before the experiment. The surface area and the *C* constant were obtained from the BET equation and the pore size distribution was calculated by the BJH method applied to the desorption branch of the nitrogen adsorption/desorption isotherm. Elemental analyses were performed by ICP-AES with a ICP spectroflamme-D from a solution obtained by treatment of the solid catalysts with a mixture of HF, HNO_3 and H_2SO_4 in a Teflon reactor at 150°C. Diffuse reflectance UV-Visible spectra were acquired using a *Perkin Elmer* Lambda 1050 UV/VIS/NIR spectrometer equipped with a praying *Mantis* equipment (*Harrick^{TM}*) for solid analysis. The acquisition domain was 200-800 nm with an acquisition step of 2 nm. Raman studies were made on a DXR Raman Microscope (*Thermo Scientific*) spectrometer equipped with an automated confocal microscope, SmartLock filters and a Charge-Coupled Device (CCD) detector. The excitation was provided at 780 nm and focused with a 50× long working distance objective. The spectra were obtained with a maximum power of 7 mW on the sample and a 5 cm^{-1} resolution. The TGA/DTA analysis was performed on a *Mettler Toledo* TGA DSC 1 apparatus. The samples were heated under air from 25°C to 1 000°C at a rate of 10°C.min^{-1}.

1.3 Measurement of Catalysts Acidity

The acidity of the catalysts was studied by microcalorimetry of ammonia adsorption, TPD of ammonia desorption and infrared study of pyridine adsorption. For the calorimetry experiments, the study was carried out on a *Tian-Calvet* calorimeter coupled to volumetric equipment. The samples (~ 50 mg) were first pretreated at 200°C for 1 h under secondary vacuum (10^{-5} Torr). Then, the solid was cooled and submitted at 80°C to successive small doses of gas up to equilibrium. The differential enthalpy of each adsorption was recorded together with the amount of NH_3 adsorbed.

For the NH_3-TPD experiments, the solids were first treated under helium (P = 1 bar) at 300°C for 1 h. After cooling the sample to 100°C, a permutation of helium to a gas mixture of 15% NH_3 in argon was passed through it during 30 minutes. Ammonia was then desorbed upon heating at 750°C under helium while the evolved gases were analyzed by TCD.

Desorption of ammonia was also monitored by combined DRIFT and GC-MS. The experiments were carried out in an integrated system comprising mass flow controllers (*Brooks*), high temperature reaction chamber (*Harrick^{TM}*) with ZnSe windows, a *Nicolet* FT-IR 6700 spectrophotometer equipped with MCT detector and online GC-MS (*Agilent* 6850/5975C). The sample was individually placed on the high temperature reaction chamber in an Ar-filled glove box. This cell was fitted into a Praying Mantis optical unit (*Harrick^{TM}*) and connected to the gas lines. The samples were first treated under vacuum (P = 10^{-5} Torr) at 275°C for 20 h and then exposed to ammonia (≥ 99.95%, GHC *Gerling Holz + Co.*) in excess at room temperature. During the FT-IR acquisition, 5 mL. min^{-1} of purified (molecular sieve/deoxo catalyst) argon (99.999%, *Air Liquide*) was constantly flowing through the sample which was heated from 20 to 400°C at 2°C. min^{-1}. Each spectrum was acquired by 64 scans with a resolution of 4 cm^{-1}. The output gas was separated with a 25 m PoraBOND Q column and analyzed by mass spectrometer in SIM mode (m/z = 15-17).

The adsorption of pyridine was studied on a *Nicolet* 5700 FT-IR spectrometer equipped with a KBr beam splitter and a DTGS detector, allowing recording the spectra of the solids *in situ*. The FT-IR spectra were recorded at a 4 cm^{-1} resolution. The catalyst samples were first pressed in a self-supporting pellet. They were then activated overnight at 200°C under vacuum at 10^{-5} Torr, in a quartz IR cell equipped with CaF_2 windows. Pyridine adsorption was carried out by equilibrating the catalyst pellet for 10 min with a large amount of base at room temperature, and then the spectrum was recorded. The desorption of the pyridine was successively monitored by evacuating the sample for 1 hour at each temperature (room temperature, 50, 100, 150 and 200°C). The samples were cooled to room temperature between each step and the FT-IR spectrum was recorded.

1.4 Catalytic Experiments

Isomerization of *n*-hexane was carried out in a fixed bed dynamic flow reactor at atmospheric pressure. The catalyst was weighted and charged into a stainless steel ½' cylindrical reactor. The reactor was then placed in an oven. Before allowing the *n*-hexane feed to pass through the reactor, the catalyst was activated to remove the remaining water molecules. This was made by treatment under argon (flow rate 5 mL.min^{-1}) during 2 h at 200°C. After activation of the catalyst, the reaction mixture (5 mL.min^{-1} of argon flow in contact with liquid hexane (> 99% provided by *Sigma-Aldrich*), at 25°C, partial pressure of *n*-hexane = 151 Torr, ratio hexane/argon = 0.25) was allowed to pass through the reactor at the desired temperature. The experiment was conducted during 12 h. The reaction products (hexane isomers and cracking products) were analyzed online by Gas Chromatography (GC) equipped with a Flame Ionization Detector (FID) and a non-polar capillary KCl/Al_2O_3 column (Varian capillary column; 50 m × 530 μm × 15 μm). The conversion was calculated from the surface of the peak of *n*-hexane before and after reaction. The selectivities were calculated from the peaks of the corresponding products. Selectivity towards isomerization was defined as the amount of isomerization products (excluding *n*-hexane) *versus* the total amount of products.

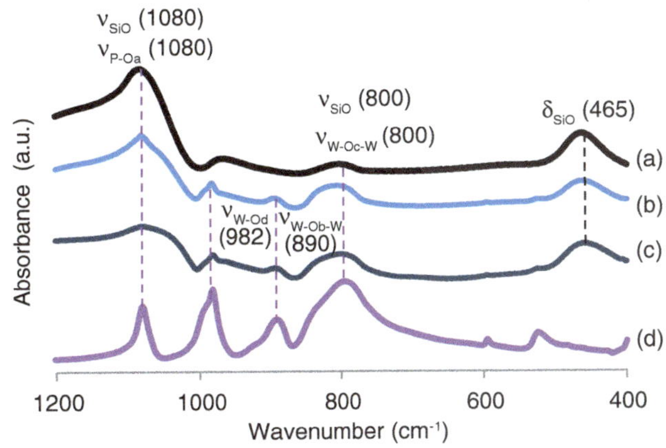

Figure 1

FT-IR spectra of supported HPW catalysts: a) pure SBA-15; b) HPW/SBA-15; c) HPW@SBA-15 and d) pure HPW.

2 RESULTS AND DISCUSSION

2.1 Structural State of the HPA in the Catalysts

Various molecular characterization techniques were used in order to determine if the Keggin structure had been preserved after the preparation of the catalysts: DR UV-Vis, Raman and FT-IR spectroscopies and ^{31}P MAS NMR.

The FT-IR of all hybrid materials are shown in Figure 1 (for the HPW family) and Figure 2 (for the HSiW family), together with those of the bulk HPA and SBA-15. The spectra of phosphotungstic acid derived catalysts, HPW@SBA-15 and HPW/SBA-15, clearly showed the characteristic bands of the $[PW_{12}O_{40}]^{3-}$ Keggin anion at 1 080 (ν_{as}(P-O$_a$)), 982 (ν_{as}(W-O$_d$)), 890 (ν_{as}(W-O$_b$-W)), and 800 cm^{-1} (ν_{as}(W-O$_c$-W)) [9]. Similarly, the FTIR spectra of the HSiW@SBA-15 and HSiW/SBA-15 materials displayed bands at 978 (ν_{as}(W-O$_d$)), 928 (ν_{as}(Si-O$_a$)), 876 (ν_{as}(W-O$_b$-W)), and 775 cm^{-1} (ν_{as}(W-O$_c$-W)) [9] suggesting that at least part of the Keggin structures was preserved during the preparation of the samples whether they were impregnated over the support or encapsulated during sol-gel synthesis.

The HPW@SBA-15 and HPW/SBA-15 catalysts were also characterized by ^{31}P MAS NMR. The spectra confirmed that the Keggin structure had been retained in the two cases, as they showed a sharp peak at −15 ppm (Supporting Information, Fig. S1) which could be attributed to the hydrated $[PW_{12}O_{40}]^{3-}$ anion [9]. A small peak at −12 ppm was also observed in the case of the encapsulated material and attributed to the presence of dehydrated polyanion [27, 28].

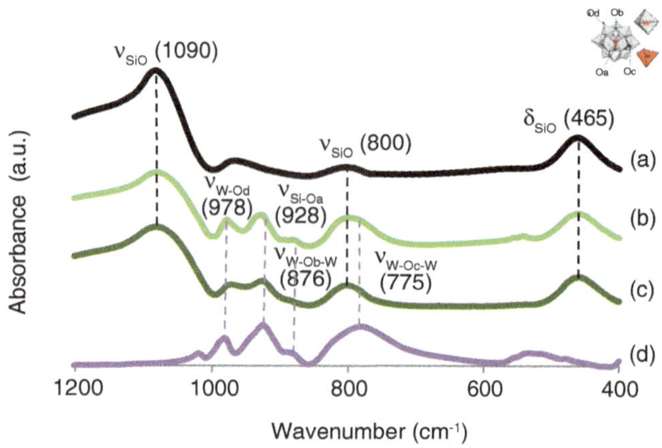

Figure 2

FT-IR spectra of supported HSiW catalysts: a) pure SBA-15; b) HSiW/SBA-15; c) HSiW@SBA-15 and d) pure HSiW.

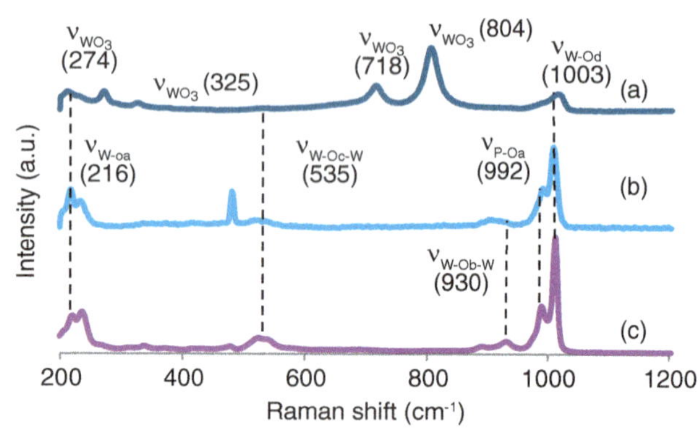

Figure 3

Raman spectra of the HPW catalysts: a) HPW@SBA-15; b) HPW/SBA-15 and c) pure HPW.

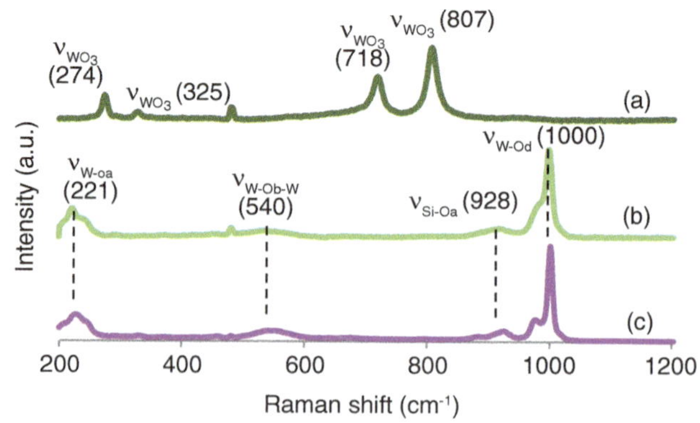

Figure 4

Raman spectra of the HSiW catalysts: a) HSiW@SBA-15; b) HSiW/SBA-15 and c) pure HSiW.

The Raman spectra of the four catalysts and of the pure HPA are shown in Figures 3 and 4. Clearly, there is a considerable difference between the spectra of the impregnated and encapsulated polyacids. The solids prepared by impregnation led to spectra quite similar to those of the pure HPA with bands at 1 003 (ν_s(W=O_d)), 995-980 (ν_s(P-O_a), ν_{as}(W=O_d)), 930 (ν_{as}(W-O_b-W)), 535 (ν_s(W-O_c-W)) and 216 cm^{-1} (ν_s(W-O_a)) [29, 30] in the case of phosphorus-based materials and bands at 1 000 (ν_s(W=O_d)), 928 (ν(Si-O_a)), 540 (ν_s(W-O_c-W)), and 221 cm^{-1} (ν(W-O_a)) [30, 31] for the silicotungstic derivatives.

The Raman spectra of the two encapsulated catalysts are quite comparable and showed mainly two bands at 807 and 718 cm^{-1} and small bands at ca. 325 and 274 cm^{-1}. All these bands can be attributed to tungsten trioxide WO_3 [29, 32]. In the case of the HPW@SBA-15 catalyst, the bands of the HPA at ca. 1 000 cm^{-1} (ν_s(W=O_d)) can also be seen but they are broadened and of low intensity. These results indicate that there is a partial (in the case of HPW@SBA-15) or complete (in the case of HSiW@SBA-15) degradation of the polyoxometalate ions which contradicts the FT-IR data for HSiW@SBA-15. One plausible explanation would be that, in the case of the encapsulated materials, the polyoxometalate species have been decomposed under the laser beam. We tried to decrease the power of the laser beam in order to eliminate this effect but due to a high fluorescence of the samples it was not possible to decrease it below 3 mW. As the main effect of the laser beam is to heat locally the sample, this suggests that the encapsulated materials are thermally less stable than the impregnated ones, leading to the following order of thermal stability:

$$\text{HSiW@SBA-15} < \text{HPW@SBA-15} < \text{HSiW/SBA-15} \approx \text{HPW/SBA-15}$$

A study by Differential Thermal Analysis (DTA) confirmed this conclusion (Fig. S2 and S3) even though the exothermic peaks associated with the decomposition of the polyoxometalates (or the crystallization of the simple oxides resulting from this decomposition) were not very intense. Decomposition temperatures, determined by the inflexion point of the curves, were found at ca.:

- 580°C for HPW/SBA-15,
- 520°C for HPW@SBA-15,
- 530°C for HSiW/SBA-15,
- 390°C for HSiW@SBA-15.

These values can be only approximated due to the small variations of the curves but they are in quite good agreement with those deduced from the Raman study, the encapsulated materials being less stable than those obtained by impregnation.

TABLE 1

Textural and physical properties of HPW and HSiW hybrid materials

Catalysts	wt% W	d_{100} [(a)] (Å)	a_0 [(b)] (Å)	S_{BET} (m^2/g)	Wall thickness [(c)] (Å)	V_p [(d)] (cm^3/g)	V_μ [(e)] (cm^3/g)	D_p [(f)] (Å)	C_{BET} [(g)]
SBA-15	-	98	113	834	37	1.10	0.05	76	161
HPW/SBA-15	24.6	97	112	452	34	0.60	0.04	77	241
HPW@SBA-15	27.7	95	109	368	34	0.48	0.04	75	333
HSiW/SBA-15	24.7	97	112	503	35	0.65	0.04	77	266
HSiW@SBA-15	22.6	101	116	392	41	0.70	0.02	75	138

[a] $d(100)$ spacing (estimated error 3 Å).

[b] $a_0 = 2d(100)/\sqrt{3}$, hexagonal lattice parameter calculated from XRD, estimated error 4 Å.

[c] Calculated by a_0 – pore size.

[d] Total pore volume at $P/P_0 = 0.980$.

[e] Micropore volume determined using the t-plot method.

[f] Pore size from desorption branch applying the BJH pore analysis.

[g] Derived from the BET equation of the nitrogen adsorption isotherms.

Finally, the diffuse reflectance UV-visible spectra of all catalysts were recorded (Fig. S4 and S5). The spectra of the pure HPA showed two bands at ca. 250 and 360 nm. attributed to octahedral (O_h) tungsten species in accord with literature data [33, 34]. Both encapsulated and impregnated catalysts displayed quite the same patterns with two bands at ca. 200 and 265 nm which can be ascribed to the charge transfer between oxygen and tungsten. In particular, the well-defined maximum at 265 nm could be attributed to an $O^{2-} \rightarrow W^{6+}$ intraframework ($W_{12}O_{36}$) charge transfer in the highly dispersed Keggin units. All HPA supported catalysts showed shifted signals with regards to pure HPA indicating a change in the symmetry, the interactions with the support decreasing the tetrahedral symmetry of the Keggin units [33, 34].

2.2 Study of the Texture of the Catalysts

The structure of all materials was analyzed by XRD measurements. Small-angle XRD patterns allowed for the identification of the mesostructure and long-range ordering of the silica framework, while wide-angle XRD patterns gave information on the presence or absence of HPA crystallites in the solid and thus on the dispersion of the active phase. The low-angle XRD patterns (Fig. S6 and S7) showed the three well-resolved peaks, indexed as (100), (110) and (200) reflections, characteristic of the hexagonal ordered mesostructure of SBA-15 [35]. These peaks are present for all hybrid materials even for the encapsulated ones, which confirm that adequate SBA-15 mesostructuration was obtained through the sol-gel encapsulation method. The wide-angle XRD patterns (also displayed in Fig. S6 and S7) did not show any peak in the 10-70° 2θ region, suggesting that there are no HPA crystallites in the materials whatever the preparation method used.

The N_2 adsorption-desorption isotherms of all materials are shown in Figures S8 and S9 and, as expected, are characteristic of mesoporous materials (type IV isotherms). The physical and textural parameters deduced from these isotherms and from the X-ray diffraction patterns are listed in Table 1. All samples have quite the same a_0 parameter and the same wall thickness if the measurement errors are taken into account. The main differences stem from the BET surface area and the porous volume. As the tungsten content in the samples is ca. 25 wt%, the HPA loading can be estimated as ca. 33 wt%. The value signifies that one gram of catalyst contains only 0.67 g of SBA-15 resulting in a surface area of $0.67 \times 834 = 560$ m^2 and a pore volume of $1.10 \times 0.67 = 0.74$ cm^3. The experimental values for both encapsulated and impregnated materials are always lower indicating that these solids are slightly less mesostructured than the starting SBA-15 and/or that some pore blocking might occur. Another parameter which can give informative indications on the the adsorbant-surface interaction and consequently on the surface philicity properties is the C constant in the BET equation. As shown in Table 1, the two HPW derived catalysts and the HSiW/SBA-15 solid exhibit higher C_{BET} values than pure SBA-15 while HSiW@SBA-15 has quite the same surface behavior than plain SBA-15. An increase of C_{BET} can be related to an increase of the strength of the interaction between sorbed nitrogen molecules and the solid and therefore to an increase of the surface polarity. For pure SBA-15 and HSiW@SBA-15, nitrogen interacts principally with the silica surface while for the three other catalysts it also interacts with HPA species suggesting that HPA accessibility in HSiW@SBA-15 material is much lower than for the other hybrids.

Taken together, the results suggest that, whether impregnated or encapsulated, the HPA have mostly retained their primary Keggin structure in the final hybrid materials as evidenced by FT-IR, DR UV-Vis and ^{31}P NMR (for HPW derived materials). The encapsulated catalysts appeared to be less thermally stable than their impregnated counterparts. This is perhaps due to the perturbation caused by the presence of a very significant amount (33 wt%) of HPA as part of the silica framework, which is relatively unperturbed in the case of impregnated materials. In addition, the accessibility to active sites in both encapsulated hybrid materials appears to be quite different, with the HPA being probably held more deeply in the interior of the walls in the case of HSiW@SBA-15.

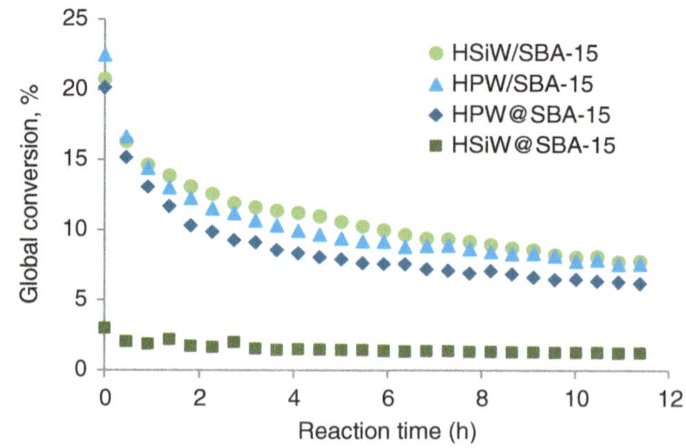

Figure 5

n-hexane conversion as a function of time for the four catalysts.

2.3 Catalytic Experiments

n-Hexane isomerization was chosen as a model reaction as it requires highly acidic sites to proceed and several studies had shown the good catalytic performance of supported HPA in this reaction [36-40]. All experiments were performed with a dynamic system during 12 h in order to study the deactivation of the catalyst. The four catalysts, which had quite the same tungsten loading, (Tab. 1), were studied in the following conditions: 1 g of catalyst, reaction temperature 200°C, atmospheric pressure, flow rate 5 mL/min argon and *n*-hexane vapor pressure at 25°C.

Figure 5 shows the variation of the conversion as a function of time for the four catalysts. It can be clearly seen that the HSiW@SBA-15 catalyst has a behavior completely different from that of the three other systems achieving a steady conversion of ca. 3% while an initial conversion of 20-25% was reached with the other catalysts. The HPW/SBA-15 and HSiW/SBA-15 catalysts give quite the same results (same initial activity and same deactivation rate) while the HPW@SBA-15 system displays a slightly higher deactivation rate. A blank test with SBA-15 did not show any conversion (Tab. 2), the small amounts of 3-methylpentane (3MP) and 2,2-dimethylbutane (2.2DMB) corresponding to impurities in the *n*-hexane used for the study.

For all catalysts, the selectivity towards isomerization increases with time on stream (Tab. 2) while the conversion decreases. This phenomenon can be explained by mechanistic considerations. At the first stages of the reaction, the number of active sites is high, allowing for the formation of an alkene via an initial carbocation formation and its subsequent reaction with another carbocation. This bimolecular mechanism is responsible for cracking and thus of the loss of selectivity. When the reaction time increases, the catalyst deactivates leading to a decrease of the amount of active sites, and so of carbocations, resulting in a lower probability for the bimolecular reaction to occur. Plotting the selectivity to isomerization *versus* the conversion clearly shows that the HPW and HSiW systems have a different behavior, the silicotungstic acid leading to more selective catalysts at the same conversion (Fig. 6). These results are quite similar to those reported previously [25, 41].

The variation of the amounts of the four branched C_6 isomers can also be rationalized by plotting their relative proportion (and not the selectivity as depicted in Tab. 2) as a function of the conversion (Fig. S12). Quite the same results are obtained for the three systems (HSiW@SBA-15 is not taken into account as the global conversion is too low). The monobranched products correspond to more than 80% of the branched isomers (2-methylpentane (2MP) being the major product), 2,3-dimethylbutane (2.3DMB) corresponds to *ca.* 15% and 2.2DMB is minor (*ca.* 5%). These data are far away from the thermodynamic equilibrium (which corresponds to a *n*-hexane conversion of 86% with the following relative proportions of the four branched isomers: 2MP 36%, 3MP 23%, 2.2DMB 30% and 2.3DMB 11% [42]).

2.4 Study of the Acidity of the Catalysts by Adsorption-Desorption of Ammonia

Ammonia is often used as a probe molecule for the determination of the acidity of the solids, for example by studying its adsorption by microcalorimetry or its desorption from a catalyst previously saturated with it (TPD) [43-45]. We used these two methods in order to compare the acidity of the impregnated and encapsulated catalysts as the catalytic reactivity had shown a difference in the behavior of the HSiW@SBA-15 system compared to the three other ones.

Blank experiments of NH_3-TPD were first carried out on bulk $H_3PW_{12}O_{40}$ and pure SBA-15. The HPA gave a single sharp peak at ca. 650°C (Fig. S13) which corresponds to the decomposition of the ammonium salt of the polyoxometalate,

TABLE 2
Conversions and selectivities at 0 and 12 h for *n*-hexane isomerization

Catalysts			SBA-15	HPW/SBA-15	HPW@SBA-15	HSiW/SBA-15	HSiW@SBA-15
Parameters		Time (h)					
Global Conv. (%)		0	0.83	22.5	20.1	20.7	3.0
		12	0.84	7.8	6.2	7.5	1.2
Selectivity (%)	3MP	0	23.0	20.6	22.5	24.3	24.1
		12	22.2	24.5	23.5	26.7	21.5
	2MP	0	0	33.9	38.0	41.3	38.8
		12	0	44.3	43.8	48.4	24.1
	2.3DMB	0	0	9.5	10.3	10.2	11.1
		12	0	12.4	11.3	12.7	7.6
	2.2DMB	0	77.0	0.3	0.4	2.4	20.3
		12	77.8	3.0	5.1	1.9	43.5
	C_5	0	0	10.6	9.1	7.3	0
		12	0	5.1	5.4	4.6	0
	C_4	0	0	17.4	13.7	10.7	3.7
		12	0	7.1	7.8	3.9	2.4
	C_3	0	0	7.7	6.0	3.8	2.0
		12	0	3.6	3.1	1.8	0.9
Selectivity C_6 (%)		0	100	64.3	71.2	78.2	94.3
		12	100	84.2	83.7	89.7	96.7

Reaction conditions: 1 bar, 200°C, 5 mL.min^{-1} argon flow, ratio hexane/argon = 0.25, 25 wt% W, 1 g of catalyst.

$(NH_4)_3PW_{12}O_{40}$. As expected, SBA-15 led to a very small amount of desorbed molecules in agreement with its weak-acidic character with respect to HPA (Fig. S14). The four supported and encapsulated HPA led to similar profiles (Fig. S15). The HPW catalysts presented two relatively broad peaks at 200 and 450°C and a relatively sharp one, at quite the same position than that observed for bulk $H_3PW_{12}O_{40}$ (650°C) and attributed to the decomposition of its ammonium salt. This attribution was confirmed by recording the NH_3-TPD of $H_3PW_{12}O_{40}/SiO_2$ (Fig. S16) with quite the same HPA loading. In that case, due to the lower surface area of *Aerosil* silica compared to SBA-15, the amount of bulk HPA was higher resulting in a stronger peak at ca. 650°C. As the two broad signals were not observed in the case of the bulk HPA and of pure SBA-15, they are likely related to HPW species in interaction with the silica support. For HSiW catalysts, only two very broad peaks were noticed around 200 and 500° C. Unlike to HPW catalysts, the sharp signal associated with the ammonium salt of bulk $H_4SiW_{12}O_{40}$, expected around 530°C (Fig. S13 and reference [46]), was not observed, only a small very broad signal being present at ca. 700°C for HSiW@SBA-15. All these results are in agreement with the data reported in the literature [43, 44, 47].

A more quantitative comparison of the data gives 1.5 mmol NH_3 desorbed per g of $H_3PW_{12}O_{40}$ for the bulk HPA, 0.3 mmol NH_3 desorbed per g of SBA-15 and around 2 mmol NH_3 desorbed per g for the SBA-15 supported and encapsulated HPA (Tab. S1). As TCD is not a specific detector, all kind of molecules can be detected and, in the present case, not only NH_3 but also water issued from the condensation of two surface hydroxyls and N_2 formed by decomposition of ammonia and reduction of the polyoxometalate have been taken into account in the measurements [48]. For the pure HPA, the amount evolved corresponds to 4.3 molecules per polyoxometalate, in agreement with 3 NH_3 and 1.5 H_2O for the loss of constitution water. For pure SBA-15,

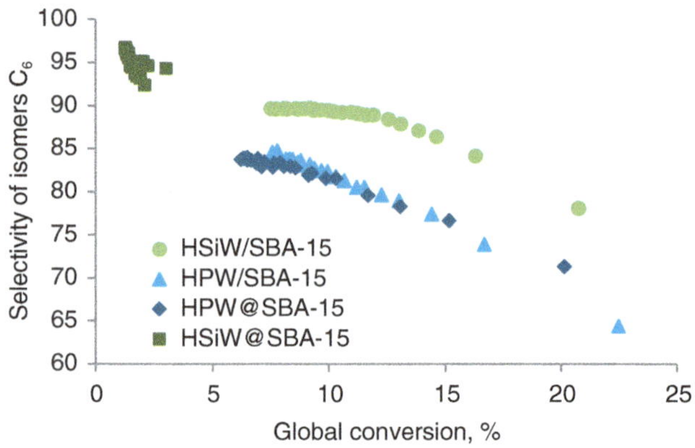

Figure 6

Variation of the selectivity for the C_6 branched isomers as a function of the conversion for all catalysts.

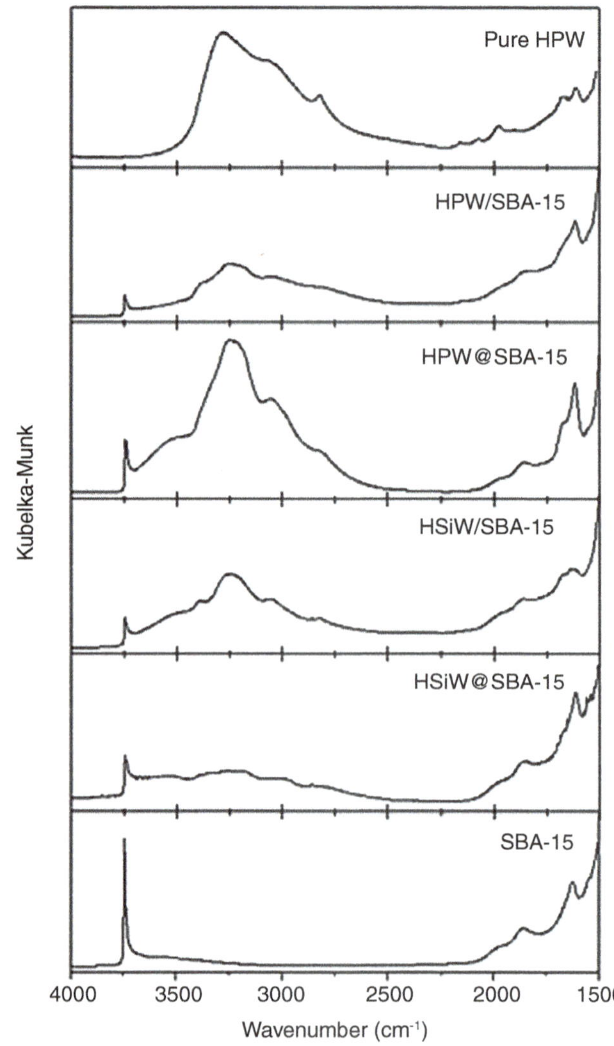

Figure 7

FT-IR spectra of catalysts after NH_3 adsorption and desorption at 20°C.

the evolved gas is very low (0.3 mmol.g^{-1}) and corresponds probably to the loss of some water by condensation of surface silanol groups, as Si-OH are known to be weakly acidic. As a consequence, we can suppose that for the encapsulated and impregnated materials the contribution of the SBA-15 support to the amount evolved is ca. 0.2 mmol (as these samples contain ca. 33 wt% of polyacid). The remaining 1.8 mmol are due to the polyoxometalate and correspond to the evolution of ca. 1.6 molecules (which are not necessary NH_3) per Keggin unit. This value is very low compared to that achieved with the pure polyacid and is probably related to the formation of covalent bonds between the polyoxometalate and the support with the concomitant evolution of water during the samples pretreatment at 300°C [49]. This method seems therefore inadequate to meaningfully compare the acidity properties of the different hybrids with their catalytic performance. The ammonia TPD was also studied by using a DRIFT/GC-MS apparatus (Fig. S17 to S20). This study confirmed the evolution of water. In addition, the IR spectra recorded after desorption of ammonia at 20°C showed a clear difference between the HSiW@SBA-15 solid and the three other catalysts as indicated by the very low intensity of the ν(N-H) band at ca. 3 300 cm^{-1} (Fig. 7). Since the intensity of this band is related to the amount of sorbed ammonia on the solid and thus to its acidity, ammonia desorption appears, here, to correlate quite well with the catalytic activity trends and could be used for the study of such materials but a special care must be taken with the detection methods.

The adsorption of ammonia by microcalorimetry was also investigated. In that case, only small pulses of ammonia were injected on the sample to prevent any decomposition of the polyoxometalate. The catalysts were pretreated under vacuum at 200°C (such treatment was previously shown to remove the crystallization water [50]) as the NH_3-TPD study had shown that a treatment at 300°C resulted in a high amount of covalent bonds between the support and the Keggin unit. The ammonia adsorption was performed at 80°C, as for classical experiments. Figure 8 shows the differential heats of ammonia as a function of the coverage for the four catalysts. As expected, SBA-15 showed only a very small amount of adsorbed ammonia with differential heats around 80 kJ.mol^{-1} (Fig. S21).

Figure S22 shows that for the four solids the adsorption curve can be decomposed in two parts, a rapid one at low pressure and a slow one at higher pressure. A comparison with the curve obtained for pure SBA-15 (Fig. S23) allows us to propose that the slow part is likely due to ammonia

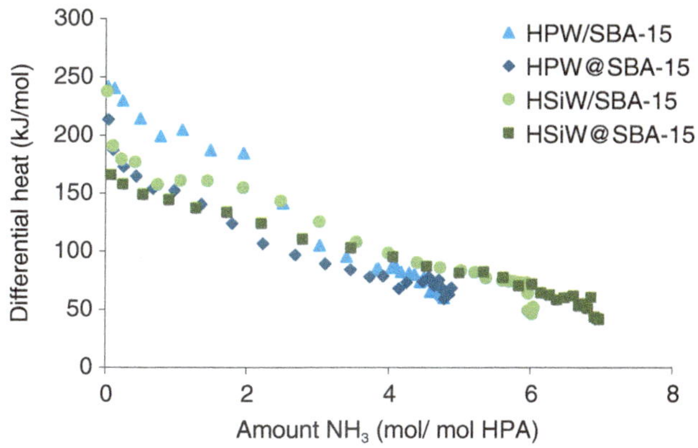

Figure 8

Differential heat of ammonia adsorption as a function of the ammonia coverage.

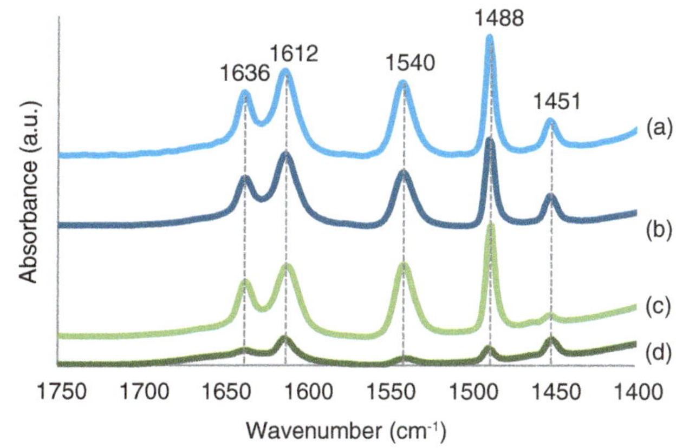

Figure 9

FT-IR spectra of the supported HPA after adsorption of pyridine at room temperature and desorption at 200°C under vacuum a) HPW/SBA-15; b) HPW@SBA-15; c) HSiW/SBA-15 and d) HSiW@SBA-15.

physisorption on the support. Extrapolation of the first part of the curves gives an adsorption of ca. 3 NH_3 per Keggin unit for the HPW-based materials and ca. 4 NH_3 for the HSiW-based ones, in agreement with the molecular formula of the corresponding HPA and the absence of covalent bonds with the SBA-15 support.

Figure 8 shows that the four solids display also quite the same behavior in terms of differential heat of ammonia adsorption with a continuous decrease as a function of the amount adsorbed. This result is in agreement with previous data for $H_3PW_{12}O_{40}$ supported on Aerosil silica [51]. In contrast, studies of the adsorption of ammonia on pure HPA had shown the presence of a plateau with a constant differential heat of adsorption [51-53]. These data were interpreted as a measure of the heat of formation of the ammonium salt of the polyoxometalate which is stable until ca. 600°C. In the present case, Figure 8 shows that HPW/SBA-15 displays higher differential heats of adsorption than the other catalysts with the presence of a small plateau at 200 kJ.mol^{-1}. The three other solids show quite the same curves even if the HSiW@SBA-15 material has a smaller initial differential heat of adsorption. However, this difference cannot account for the complete inactivity of HSiW@SBA in the isomerization of *n*-hexane. Here also, the results obtained by microcalorimetry do not fit with the catalytic experiments.

2.5 Study of the Acidity by Adsorption of Pyridine

Pyridine adsorption is often used to study the acidity of catalysts as it allows differentiating between Lewis and Brønsted acid sites. As for the above studies, the solids were pretreated under vacuum at 200°C before adsorption of pyridine at room temperature and desorption at increasing temperatures. Figure 9 shows the FT-IR spectra obtained after desorption at 200°C, which can be considered as representative of the acid sites at the catalysis temperature. The spectra after desorption at lower temperatures are given in the supplementary material, Figures S24 to S27. These spectra were scaled to each other by giving the same height to the harmonic vibrational bands of the support in the 2 000-1 800 cm^{-1} range. As a consequence, they can be directly compared. Among all the bands present on the spectra, the most important ones are those at 1 540 and ca. 1 450 cm^{-1} which are characteristic of Brønsted and Lewis acid sites respectively [54, 55]. One can notice that the HSiW@SBA-15 catalyst contains only a very small amount of Brønsted acid sites compared to the three other solids which have quite the same amount of acid sites. In all cases, the amount of Lewis acid sites is very low as demonstrated by the weak intensity of the band at 1 450 cm^{-1}. These results are in good agreement with the catalytic data and those obtained by infrared spectroscopy after adsorption/desorption of ammonia.

The difference between microcalorimetry and the other methods (TPD by taking into account only sorbed ammonia and pyridine adsorption) cannot be only due to activation problems as pyridine was adsorbed after treatment at 200°C, as for the calorimetric study. The main difference is that for TPD and pyridine adsorption the samples were desorbed at room temperature. It must then be supposed that in the case of the HSiW@SBA-15 catalyst the sorbed ammonia molecules can be more easily desorbed at low temperature. This could be related to a more pronounced encapsulation of the HPA within the walls of the SBA-15 framework which is further supported by a C_{BET} value very close to that of bare SBA-15.

CONCLUSION

Four HPA-based catalysts were prepared by encapsulation or impregnation of $H_3PW_{12}O_{40}$ and $H_4SiW_{12}O_{40}$ onto SBA-15 silica. These solids were characterized by various physicochemical methods which showed that, after the catalyst preparation, the Keggin structure of the polyoxometalate had been retained in all cases. DTA and Raman studies confirm that the encapsulated materials were thermally less stable than the impregnated ones.

The HPW/SBA-15, HPW@SBA-15 and HSiW/SBA-15 catalysts performed similarly in the isomerization of *n*-hexane with a good selectivity (> 80%). On the other hand, HSiW@SBA-15 was slightly active with a conversion of only 3% although more selective. To get deeper insight into the discrepancy in catalytic behavior observed with HSiW@SBA-15, the acid properties of the four solids were studied by adsorption and/or desorption of ammonia (using microcalorimetry, TPD and DRIFT/GC-MS desorption) and pyridine adsorption/desorption monitored by FT-IR. Only pyridine adsorption/desorption and DRIFT/GC-MS of ammonia desorption led to meaningful results in accord with catalysis data. A possible explanation to account for the inactivity of HSiW@SBA-15 is that the silicotungstate anions are located more deeply within the silica framework resulting in less acidic materials. The polyoxometalate could then be accessible to small molecules such as ammonia but it could not be reached by larger molecules such as pyridine or *n*-hexane.

ACKNOWLEDGMENTS

The authors wish to acknowledge Dr. Jean-Marc Millet from IRCE Lyon for performing the ammonia TPD experiments.

REFERENCES

1 Pope M.T. (1983) *Heteropoly and Isopoly Oxometalates*, Springer Verlag.

2 Kozhevnikov I.V. (1998) Catalysis by Heteropoly Acids and Multicomponent Polyoxometalates in Liquid-Phase Reactions, *Chem. Rev.* **98**, 171-198.

3 Mizuno N., Misono M. (1998) Heterogeneous Catalysis, *Chem. Rev.* **98**, 199-217.

4 Timofeeva M.N. (2003) Acid Catalysis by Heteropolyacids, *Appl. Catal. A: Gen.* **256**, 19-35.

5 Newman A.D., Lee A.F., Wilson K., Young N.A. (2005) On the Active Site in $H_3PW_{12}O_{40}/SiO_2$ Catalysts for Fine Chemical Synthesis, *Catal. Lett.* **102**, 45-50.

6 Okuhara T., Mizuno N., Misono M. (2001) Catalysis by Heteropoly Compounds - Recent Developments, *Appl. Catal. A: Gen.* **222**, 63-77.

7 Zhao D., Feng J., Huo Q., Melosh N., Fredrickson G.H., Chmelka B.F., Stucky G.D. (1998) Triblock Copolymer Syntheses of Mesoporous Silica with Periodic 50 to 300 Angstrom Pores, *Science* **279**, 548-552.

8 Zhao D., Huo Q., Feng J., Chmelka B.F., Stucky G.D. (1998) Nonionic Triblock and Star Diblock Copolymer and Oligomeric Surfactant Syntheses of Highly Ordered, Hydrothermally Stable, Mesoporous Silica Structures, *J. Am. Chem. Soc.* **120**, 6024-6036.

9 Zhao D., Sun J., Li Q., Stucky G.D. (2000) Morphological Control of Highly Ordered Mesoporous Silica SBA-15, *Chem. Mater.* **12**, 275-279.

10 Dufaud V., Lefebvre F. (2010) Inorganic Hybrid Materials with Encapsulated Polyoxometalates, *Materials* **3**, 682-703.

11 Falco M.G., Canavese S.A., Comelli R.A., Fígoli N.S. (2000) Influence of Pt Concentration on Tungsten Oxide-Promoted Zirconia during *n*-Hexane Isomerization, *Appl. Catal. A: Gen.* **201**, 37-43.

12 Gagea B.C., Lorgouilloux Y., Altintas Y., Jacobs P.A., Martens J.A. (2009) Bifunctional Conversion of *n*-Decane over HPW Heteropoly Acid Incorporated into SBA-15 during Synthesis, *J. Catal.* **265**, 99-108.

13 Kamiya Y., Okuhara T., Misono M., Miyaji A., Tsuji K., Nakajo T. (2008) Catalytic Chemistry of Supported Heteropolyacids and their Applications as Solid Acids to Industrial Processes, *Catal. Surv. Asia* **12**, 101-113.

14 Yang W., Billy J., Ben Taârit Y., Védrine J.C., Essayem N. (2002) $H_3PW_{12}O_{40}$ Supported on Cs Modified Mesoporous Silica: Catalytic Activity in *n*-Butane Isomerization and *in situ* FTIR Study Comparison with Microporous $Cs_xH_{3-x}PW_{12}O_{40}$, *Catal.Today* **73**, 153-165.

15 Essayem N., Ben Taârit Y., Feche C., Gayraud P.Y., Sapaly G., Naccache C. (2003) Comparative Study of *n*-Pentane Isomerization over Solid Acid Catalysts, Heteropolyacid, Sulfated Zirconia, and Mordenite: Dependence on Hydrogen and Platinum Addition, *J. Catal.* **219**, 97-106.

16 Sousa B.V., Brito K.D., Alves J.J.N., Rodrigues M.G.F., Yoshioka C.M.N., Cardoso D. (2011) *n*-Hexane Isomerization on Pt/HMOR: Effect of Platinum Content, *React. Kinet. Mech. Catal.* **102**, 473-485.

17 Ono Y. (2003) A Survey of the Mechanism in Catalytic Isomerization of Alkanes, *Catal. Today* **81**, 3-16.

18 Bernard P.M., Primet M. (1990) Influence of Hydrogen Chloride Addition on the Catalytic Isomerization Activity of Chlorinated Alumina and Chlorinated Platinum-Alumina Solids, *J. Chem. Soc., Faraday Trans.* **86**, 567-570.

19 Otten M.M., Clayton M.J., Lamb H.H. (1994) Platinum-Mordenite Catalysts for *n*-Hexane Isomerization: Characterization by X-Ray Absorption Spectroscopy and Chemical Probes, *J. Catal.* **149**, 211-222.

20 Miyaji A., Echizen T., Li L., Suzuki T., Yoshinaga Y., Okuhara T. (2002) Selectivity and Mechanism for Skeletal Isomerization of Alkanes over Typical Solid Acids and their Pt-Promoted Catalysts, *Catal. Today* **74**, 291-297.

21 Modhera B.K., Chakraborty M., Bajaj H.C., Parikh P.A. (2011) Influences of Mesoporosity Generation in ZSM-5 and Zeolite Beta on Catalytic Performance During *n*-Hexane Isomerization, *Catal. Lett.* **141**, 1182-1190.

22 Aboul-Gheit A.K., Abdel-Hamid S.M., El-Desouki D.S. (2010) Hydroisomerization of *n*-Hexane using Unloaded and low-Platinum-Loaded H-ZSM-5 and H-MOR Catalysts, *Petroleum Sci. Technol.* **28**, 582-593.

23 Guisnet M., Costa L., Ribeiro F.R. (2009) Prevention of Zeolite Deactivation by Coking, *J. Mol. Catal. A: Chem.* **305**, 69-83.

24 Weitkampf J. (1982) Isomerization of Long-Chain *n*-Alkanes on a Pt/CaY Zeolite Catalyst, *Ind. Eng. Chem. Prod. Res. Dev.* **21**, 550-558.

25 Pinto T., Dufaud V., Lefebvre F. (2014) Isomerization of *n*-Hexane on Heteropolyacids Supported on SBA-15. 1. Monofunctional Impregnated Catalysts, *Appl. Catal. A: Gen.* **483**, 103-109.

26 Dufaud V., Lefebvre F., Niccolai G.P., Aouine M. (2009) New Insights into the Encapsulation and Stabilization of Heteropolyacids inside the Pore Walls of Mesostructured Silica Materials, *J. Mater. Chem.* **19**, 1142-1150.

27 Ghanbari-Siahkali A., Philippou A., Dwyer J., Anderson M.W. (2000) The Acidity and Catalytic Activity of Heteropoly Acid on MCM-41 Investigated by MAS NMR, FTIR and Catalytic Tests, *Appl. Catal. A: Gen.* **192**, 57-69.

28 Okuhara T., Nishimura T., Watanabe H., Misono M. (1992) Catalysis by Heteropoly Compounds. XXI. Insoluble Heteropoly Compounds as Highly Active Catalysts for Liquid-Phase Reactions, *J. Mol. Catal.* **74**, 247-256.

29 Carriazo D., Domingo C., Martín C., Rives V. (2008) PMo or PW Heteropoly Acids Supported on MCM-41 Silica Nanoparticles: Characterisation and FT-IR Study of the Adsorption of 2-Butanol, *J. Solid State Chem.* **181**, 2046-2057.

30 Legagneux N., Basset J.M., Thomas A., Lefebvre F., Goguet A., Sa J., Hardacre C. (2009) Characterization of Silica-Supported Dodecatungstic Heteropolyacids as a Function of their Dehydroxylation Temperature, *Dalton Trans.*, 2235-2240.

31 Rocchiccioli-Deltcheff C., Thouvenot R., Franck R. (1976) Ir and Raman Spectra of Heteropolyanions with Keggin-Type Structure, α-$XM_{12}O_{40}^{n-}$ [X = B(III), Si(IV), Ge(IV), P(V), As(V); M = W (VI), Mo(VI)], *Spectrochim. Acta Part A* **32**, 587-597.

32 Lu D.Y., Chen J., Chen H.J., Gong L., Deng S.Z., Xu N.S., Liu Y.L. (2007) Raman Study of Thermochromic Phase Transition in Tungsten Trioxide Nanowires, *Appl. Phys. Lett.* **90**, 041919/1-041919/3.

33 Thielemann J.P., Ressler T., Walter A., Tzolova-Müller G., Hess C. (2011) Structure of Molybdenum Oxide Supported on Silica SBA-15 Studied by Raman, UV-Vis and X-Ray Absorption Spectroscopy, *Appl. Catal. A: Gen.* **399**, 28-34.

34 Lizama L., Klimova T. (2008) Highly Active Deep HDS Catalysts Prepared Using Mo and W Heteropolyacids Supported on SBA-15, *Appl. Catal. B: Env.* **82**, 139-150.

35 Lefebvre F. (1992) Phosphorus-31 MAS NMR Study of $H_3PW_{12}O_{40}$ Supported on Silica: Formation of $(\equiv SiOH_2^+)(H_2PW_{12}O_{40}^-)$, *J. Chem. Soc., Chem. Commun.*, 756-757.

36 Essayem N., Ben Taârit Y., Gayraud P.Y., Sapaly G., Naccache C. (2001) Heteropolyacid Catalysts for Hydroisomerization of n-Hexane: Effects of Alkali Salt Modification, *J. Catal.* **204**, 157-162.

37 Gherib A., Aouissi A., Rives A., Fournier M., Hubaut R. (2007) Isomerization of *n*-Hexane over Silica-Supported Hetero Polyoxometallates Promoted by Pt-Ce Oxides, *Chin. J. Catal.* **28**, 1041-1046.

38 Hubaut R., Ben Tayeb B.O., Rives A., Fournier M. (2004) Mechanical Mixtures of PdCeO and HPA/SiO_2: Structure and Reactivity for *n*-Hexane Isomerization, *Solid State Ionics* **172**, 85-87.

39 Ivanov A.V., Vasina T.V., Nissenbaum V.D., Kustov L.M., Timofeeva M.N., Houzvicka J.I. (2004) Isomerization of *n*-Hexane on the Pt-promoted Keggin and Dawson Tungstophosphoric Heteropoly Acids Supported on Zirconia, *Appl. Catal. A: Gen.* **259**, 65-72.

40 Macht J., Carr R.T., Iglesia E. (2009) Consequences of Acid Strength for Isomerization and Elimination Catalysis on Solid Acids, *J. Am. Chem. Soc.* **131**, 6554-6565.

41 Grinenval E., Garron A., Lefebvre F. (2013) *n*-Butane Isomerization over Silica-Supported Heteropolyacids: Study of Some Parameters, *J. Catal.* **2013**, ID 828962, 8 pages.

42 Alberty R.A., Gehrig C.A. (1984) Standard Chemical Thermodynamic Properties of Alkane Isomer Groups, *J. Phys. Chem. Ref. Data* **13**, 1173-1197.

43 Kamiya Y., Okuhara T., Misono M., Miyaji A., Tsuji K., Nakajo T. (2008) Catalytic Chemistry of Supported Heteropolyacids and Their Applications as Solid Acids to Industrial Processes, *Catal. Surveys Asia* **12**, 101-113.

44 Zhang J., Kanno M., Wang Y., Nishii H., Miura Y.K., Kamiya Y. (2011) Changes in Surface Acidity of Silica-Supported Dodecatungstosilicic Acid in Relation to the Loading Amount, *J. Phys. Chem. C* **115**, 14762-14769.

45 Zhang J., Kanno M., Zhang J., Ohnishi R., Toriyabe K., Matsuhashi H., Kamiya Y. (2010) Preferential Oligomerization of Isobutene in a Mixture of Isobutene and 1-Butene over Sodium-Modified 12-Tungstosilicic Acid Supported on Silica, *J. Mol. Catal. A: Chem.* **326**, 107-112.

46 Miyaji A., Echizen T., Nagata K., Yoshinaga Y., Okuhara T. (2003) Selective Hydroisomerization of *n*-Pentane to Isopentane over Highly Dispersed Pd-$H_4SiW_{12}O_{40}/SiO_2$, *J. Mol. Catal. A: Chem.* **201**, 145-153.

47 Xu Y., Zhang X., Li H., Qi Y., Lu G., Li S. (2009) Promotion Effect of Lanthanum Addition on the Catalytic Activity of Zirconia Supported Platinum and Tungstophosphoric Acid Catalyst for *n*-Pentane Isomerization, *Appl. Surf. Sci.* **255**, 6504-6507.

48 Essayem N., Frety R., Coudurier G., Védrine J.C. (1997) Ammonia Adsorption-Desorption over the Strong Solid Acid Catalyst $H_3PW_{12}O_{40}$ and its Cs^+ and NH_4^+ Salts. Comparison with Sulfated Zirconia, *J. Chem. Soc., Faraday Trans.* **93**, 3243-3248.

49 Grinenval E., Rozanska X., Baudouin A., Berrier E., Delbecq F., Sautet P., Basset J.M., Lefebvre F. (2010) Controlled Interactions between Anhydrous Keggin-Type Heteropolyacids and Silica Support: Preparation and Characterization of Well-Defined Silica-Supported Polyoxometalate Species, *J. Phys. Chem. C* **114**, 19024-19034.

50 Essayem N., Tong Y.Y., Jobic H., Védrine J.C. (2000) Characterization of Protonic Sites in $H_3PW_{12}O_{40}$ and $Cs_{1.9}H_{1.1}PW_{12}O_{40}$: a Solid-State ^1H, ^2H, ^{31}P MAS-NMR and Inelastic Neutron Scattering Study on Samples Prepared under Standard Reaction Conditions, *Appl. Catal. A: Gen.* **194-195**, 109-122.

51 Hamad B., Lopes de Souza R.O., Sapaly G., Carneiro Rocha M.G., Pries de Oliveira P.G., Gonzalez W.A., Andrade Sales E., Essayem N. (2008) Transesterification of Rapeseed Oil with Ethanol over Heterogeneous Heteropolyacids, *Catal. Commun.* **10**, 92-97.

52 Jozefowicz L.C., Karge H.G., Vasilyeva E., Moffat J.B. (1993) A Microcalorimetric Investigation of Heteropolyacids, *Microp. Mater.* **1**, 313-322.

53 Lefebvre F., Liu-Cai F.X., Auroux A. (1994) Microcalorimetric Study of the Acidity of Tungstic Heteropolyanions, *J. Mater. Chem.* **4**, 125-131.

54 Zaki M.I., Hasan M.A., Al-Sagheer F.A., Pasupulety L. (2001) *In situ* FTIR Spectra of Pyridine Adsorbed on SiO_2-Al_2O_3, TiO_2, ZrO_2 and CeO_2: General Considerations for the Identification of Acid Sites on Surfaces of Finely Divided Metal Oxides, *Colloids Surf. A* **190**, 261-274.

55 Zaki M.I., Mekhemer G.A.H., Fouad N.E., Rabee A.I.M. (2014) Structure-Acidity Correlation of Supported Tungsten (VI) -Oxo-Species: FT-IR and TPD Studies of Adsorbed Pyridine and Catalytic Decomposition of 2-Propanol, *Appl. Surf. Sci.* **308**, 380-387.

7

Methane Hydrate Formation and Dissociation in the Presence of Silica Sand and Bentonite Clay

V. Kumar Saw[1], G. Udayabhanu[2], A. Mandal[1]* and S. Laik[1]

[1] *Gas Hydrate Laboratory, Department of Petroleum Engineering, Indian School of Mines, Dhanbad, 826004 - India*
[2] *Department of Applied Chemistry, Indian School of Mines, Dhanbad, 826004 - India*
e-mail: mandal_ajay@hotmail.com

* Corresponding author

Abstract — *The formation and dissociation of methane hydrates in a porous media containing silica sand of different sizes and bentonite clay were studied in the presence of synthetic seawater with 3.55 wt% salinity. The phase equilibrium of methane hydrate under different experimental conditions was investigated. The effects of the particle size of silica sand as well as a mixture of bentonite clay and silica sand on methane hydrate formation and its dissociation were studied. The kinetics of hydrate formation was studied under different subcooling conditions to observe its effects on the induction time of hydrate formation. The amount of methane gas encapsulated in hydrate was computed using a real gas equation. The Clausius-Clapeyron equation is used to estimate the enthalpy of hydrate dissociation with measured phase equilibrium data.*

Résumé — Formation et dissociation d'hydrates de méthane en présence de sable de silice et d'argile de bentonite — La formation et la dissociation d'hydrates de méthane dans un milieu poreux contenant du sable de silice de différentes tailles et de l'argile de bentonite ont été étudiées en présence d'eau de mer synthétique de 3,55 % poids de salinité. L'équilibre de phase des hydrates de méthane sous différentes conditions expérimentales a été étudié. Les effets de la taille des particules de sable de silice ainsi que ceux d'un mélange d'argile de bentonite et de sable de silice sur la formation d'hydrates de méthane et leur dissociation ont été étudiés. La cinétique de la formation d'hydrates a été étudiée sous différentes conditions de sous-refroidissement pour observer ses effets sur le temps de latence de la formation d'hydrates. La quantité de méthane encapsulée dans des hydrates a été calculée en utilisant une équation de gaz réel. L'équation de Clausius-Clapeyron est utilisée pour déterminer l'enthalpie de la dissociation des hydrates avec des données mesurées d'équilibre de phase.

INTRODUCTION

Hydrates are thermodynamically stable if the pressure is greater than the equilibrium pressure at a given sediment temperature [1-3]. In addition to favorable temperature and pressure conditions, water salinity changes the stability conditions of gas hydrate in marine sediments [4-7]. Understanding the role of the sediment matrix in gas hydrate formation is crucial to elucidate the formation mechanisms and accumulation of methane hydrate in natural environments. Studies show that the type of medium strongly affects the formation and dissociation processes of natural gas hydrate [8-10]. Experiments have been conducted to understand the distribution of natural gas hydrate in sediments. Handa and Stupin [8] showed that the dissociation pressure of methane and propane hydrates in porous media with small pores was higher than that of those in bulk. Uchida *et al.* [9] studied the effects of different media such as sand, sandstone, glass beads and clays on methane hydrate formation. They indicated that the equilibrium conditions of gas hydrates within these media were affected mainly by pore sizes and, to a lesser degree, by surface textures and mineral components. Li *et al.* [11] and Guanli *et al.* [12] reported that hydrate formation media significantly affect the formation of gas hydrates. They also reported that the smaller the particle size of the media, the harder the methane hydrate formation within it. Oceanic gas hydrates are commonly found in sand- or clay-type sediments. The capillary inhibition of hydrate stability in narrow pores, such as fine-grained silts, muds and clays, has been considered to explain differences between predicted and actual hydrate stability zones in ocean sediments, while the equilibrium pressure for hydrate dissociation and the thermodynamic properties in large pores of coarse-grained sand sediment are nearly identical to those in the pure bulk water phase [13]. Gas hydrate stability in the ocean sediment is affected by electrolytes dissolved in seawater. Therefore, it is essential to consider the combined effects of both pores and electrolytes to predict an accurate gas hydrate stability zone in the ocean sediment. Kang and Lee [14] investigated the phase equilibria and formation kinetics of synthetic natural gas in porous silica gel. It is found that the hydrate-phase equilibria are shifted to the inhibition zone more than those in the bulk phase. The inhibition effect is more pronounced in the meso-sized pores than the macro-sized ones. Subsequently, Li and Zhang [15] also studied the dissociation kinetics of methane hydrate in silica gels as porous media. The results reveal that the rate of methane released from the hydrate dissociation increases with an increase in the initial formation pressure, decrease in the environmental temperature and increase in pore size. In subsea environments, the sediments are normally saturated with seawater. Natural marine sediment is saturated with aqueous solution of salinity 3.55 wt% TDS [16, 17]. The equilibrium conditions for the formation and dissociation of methane hydrates can be influenced by the interlayer structure and surface chemistry of the host sediments. The effects of sediment surface chemistry are largely unknown.

The objective of the present study is to observe the effects of silica sand and bentonite clay suspended in synthetic seawater of salinity 3.55% on the phase equilibrium of methane hydrate, to predict the formation and dissociation characteristics of methane hydrate in natural environments. This work is focused on the effects of silica sand and its size as well as a mixture of silica sand and bentonite clay on methane hydrate formation and dissociation. The kinetics of hydrate formation was also studied to observe the induction time of hydrate formation, which is a very important parameter of hydrate formation.

1 EXPERIMENTAL SET-UP, MATERIALS AND PROCEDURE

1.1 Apparatus

The schematic diagram of the gas hydrate autoclave apparatus is shown in Figure 1. The high-pressure hydrate autoclave procured from *Vinci Technology*, France, was used to study gas hydrate formation and dissociation. The apparatus measures the induction time of hydrate formation and pressure as a function of time during hydrate formation, and takes video graphs during the experiments. The system consists of a constant volume hydrate cell with 250 cm^3 capacity. The cell temperature is controlled by a thermostatic bath. A thermocouple measures the temperature of the cell with an

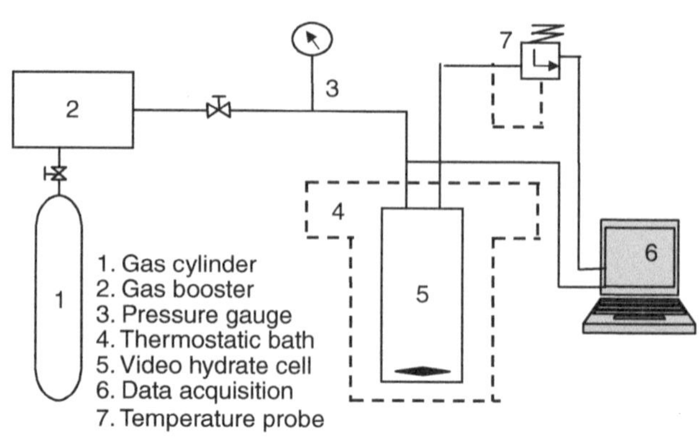

Figure 1

Schematic of gas hydrate autoclave apparatus.

accuracy of 0.1°C. The cell pressure is monitored by a pressure transducer. A magnetic stirrer is used to agitate the test fluid. A computer is connected to the hydrate cell for acquisition of temperature and pressure *versus* time data.

1.2 Materials

All the experiments were carried out with 99.99% pure methane (procured from *Chemtron Science Laboratory*, Navi Mumbai, India) and reverse osmosis water from the Millipore water system (*Millipore SA*, Molshein, France). Silica sand was purchased from *Sigma Aldrich*, Germany. The different salts (purity 98-99.9%) used for preparing synthetic seawater were procured from *Merck Specialities Pvt. Ltd.*, Mumbai, India. Three particle size ranges of sand, as given in Table 1, were obtained by sieving.

The bentonite clay was supplied by *Central Drug House*, New Delhi, India. The Particle Size Distribution (PSD) of the bentonite clay suspension was studied using a Nano-S90 Zetasizer Ver. 6.34 from *Malvern Instruments Ltd.* Three peaks (*Fig. 2*) are observed of different intensity and volume with mean sizes of 505 nm (42.2%), 62.93 nm (25.9%), and 2 590 nm (9.4%). Bentonite clay particles distributed in water ranges in size from 30 nm to 3 200 nm with an average size of 279.8 nm. The results of the PSD analysis are given in Table 2.

1.3 Experimental Procedure

The cell was filled with a test sample (130 cc) with a fixed composition of bentonite clay and silica sand prepared in synthetic seawater (3.55% TDS) and immersed in a temperature-controlled bath to maintain the constant temperature of the cell. The liquid in the bath is a mixture of water and ethylene glycol (25%). Before charging with methane gas, the cell was evacuated by a vacuum pump. The cell was then pressurized with methane gas up to the desired pressure. The cell was cooled step-wise in the programmable bath and sufficient time (1-2 hours) was given to attain the equilibrium condition at each temperature. The hydrate formation was observed with a sudden sharp pressure drop under constant cell temperature and an online video picture. After hydrate formation, the entire cell was heated slowly (1 K/h) so that the equilibrium of the system was not disturbed. Dissociation of the hydrate was observed with substantial increase in pressure. The process of dissociation is assumed to be complete at the temperature where the formation-dissociation loop is closed. The temperature and

TABLE 1

Grain Size (diameter) Range (GSR) of silica sand

Mesh size	GSR (μm)
30-50	500-300
52-60	300-250
60-100	250-150

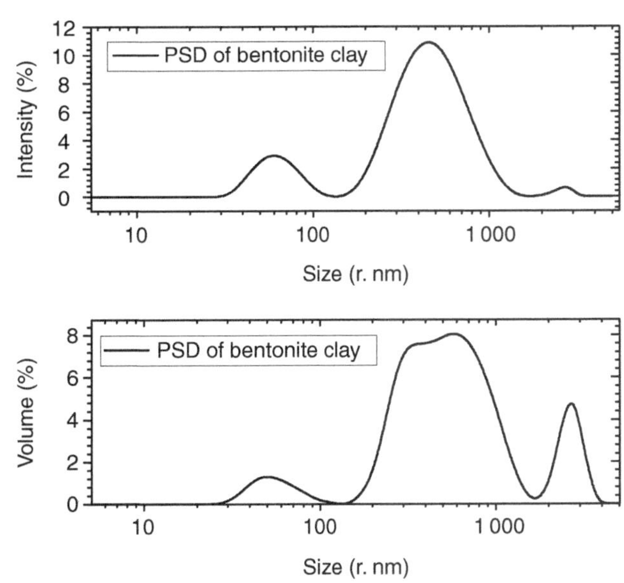

Figure 2

PSD of bentonite clay in aqueous suspension.

TABLE 2

PSD results of bentonite clay in water

Peak (s)	Size (r.nm)	% Intensity	σ	%Pd
Peak 1	62.93	14.4	16.33	42.2
Peak 2	505.0	84.2	213.3	25.9
Peak 3	2 590.0	1.4	243.1	9.4

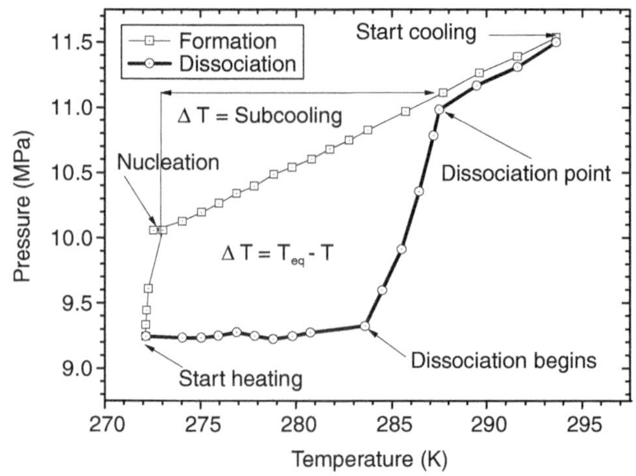

Figure 3

Pressure temperature profile of methane hydrate formation and dissociation in 10% silica sand (mesh size 30-52) in Synthetic SeaWater (SSW).

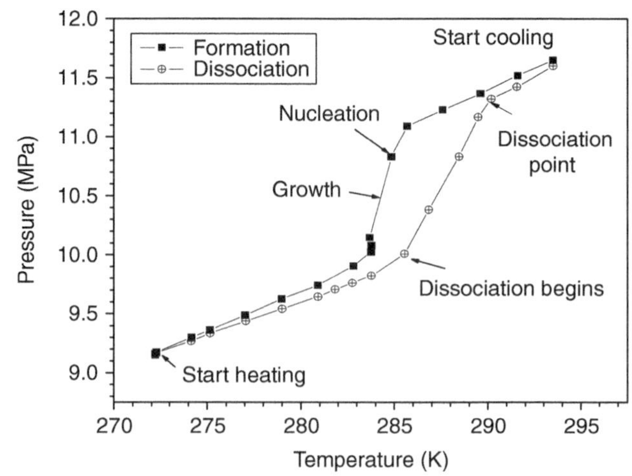

Figure 4

Pressure temperature profile of methane hydrate formation and dissociation in the presence of 10% bentonite clay.

TABLE 3

Nucleation temperature, initial cell pressure, growth temperature and pressure drop during hydrate formation in the presence of silica sand. SSW: Synthetic SeaWater

Test sample	Initial cell pressure (MPa)	Nucleation temperature (K)	Growth temperature (K)	Pressure drop ΔP (MPa)
10% bentonite in SSW	11.54	284.85	283.70	1.13
10% silica sand (mesh size 32-52) in SSW	11.54	272.27	272.19	0.49
10% silica sand (mesh size 52-60) in SSW	11.54	272.20	272.0	0.58
10% silica sand (mesh size 60-100) in SSW	11.54	271.99	271.99	0.59

pressure data during the experiments were stored in the computer and also displayed as online graphs. In the present study, methane hydrate formation was observed in the presence of silica sand in synthetic seawater (3.55% TDS) and also with a process sample with a mixture of both bentonite clay and silica sand. Variation in the size of silica sand was found to have an effect on hydrate formation as well as on dissociation. During the experiments, the reaction medium was agitated using a magnetic stirrer at a constant rate (1 000 RPM). An experiment was also performed with 10% bentonite clay to compare the results obtained for hydrate formation and dissociation in the presence of silica sand, and of both bentonite clay and silica sand.

2 RESULTS AND DISCUSSION

2.1 Formation of Hydrate in the Presence of Silica Sand

Figure 3 shows the conversion of water and methane into hydrate in the presence of silica sand and its reverse process. It can be noted that hydrate formation is initiated at 272.27 K with a sharp drop in pressure. This temperature is maintained until no pressure drop is observed, indicating complete formation of hydrate. Under similar conditions of temperature (293.5 K) and pressure (11.65 MPa), an experiment was carried out with bentonite clay. In this case, hydrate formation begins at 284.85 K, which is 12.58 K higher (*Fig. 4*, *Tab. 3*).

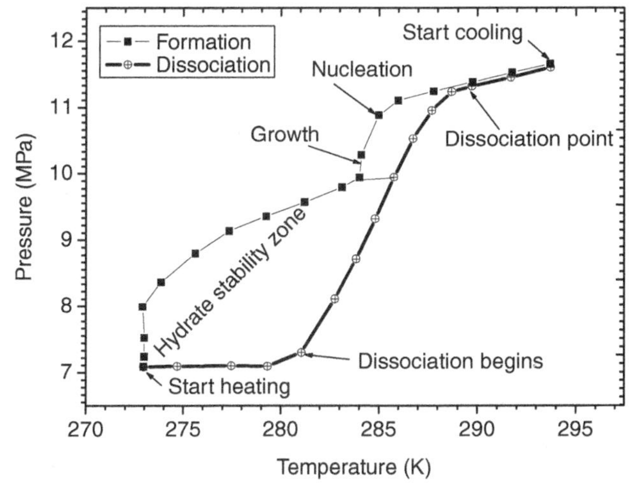

Figure 5

Pressure temperature profile of methane hydrate formation and dissociation in the presence of 10% bentonite clay and 10% silica sand (mesh size 30-52).

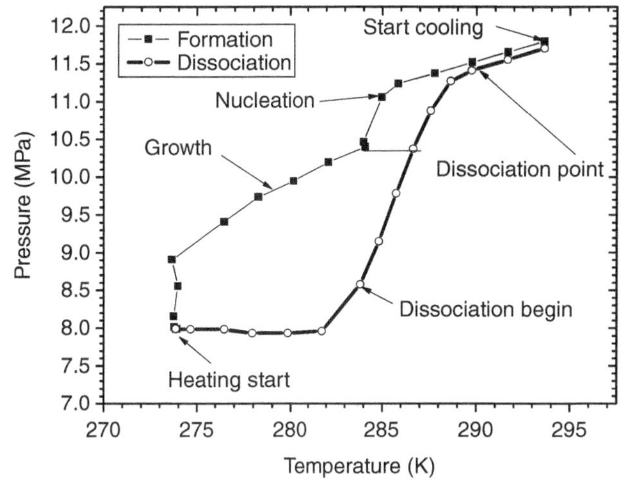

Figure 6

Pressure temperature profile of methane hydrate formation and dissociation in the presence of 10% bentonite clay and 10% silica sand (mesh size 52-60) in SSW.

Experiments were also performed with silica sand with variation in grain size to observe its effect on methane hydrate formation and dissociation. It is observed that the hydrate formation temperature is not much affected upon variation in the size of silica sand (*Tab. 3*). Higher subcooling (14.44 K) is required to initiate hydrate formation in the presence of silica sand than that of hydrate formation in the presence of bentonite clay suspension (*Fig. 3, 4*). These results suggest that hydrate formation is greatly inhibited by silica sand, while the presence of bentonite clay thermodynamically promotes hydrate formation. The amount of gas consumed during hydrate formation is found to be reduced in the presence of silica (*Tab. 3*). Earlier studies also showed that silica particles exhibit thermodynamic inhibition during methane and natural gas hydrate formation [18, 19].

2.2 Formation of Hydrate in the Presence of a Mixture of Silica Sand and Bentonite Clay

Hydrate formation was also studied in the presence of a mixture of silica sand and bentonite clay under the same experimental conditions of temperature and pressure, with variation in the grain size of the sand, to observe the effect of particle size on methane hydrate formation and dissociation. It can be seen from Figure 5 that in addition to pressure drop at 284.85 K, a fall in pressure is again observed at different temperatures depending on the grain size of the sand. This result indicates that there is an obvious effect of the grain size of silica sand on methane hydrate formation.

Figure 6 shows the variation in the temperature and pressure of methane hydrate formation and dissociation in the presence of 10% bentonite clay and 10% silica sand of mesh size 52-60. It can be seen that the first pressure drop is observed at 284.85 K, which is due to bentonite clay as it was observed in the earlier case, and again a sharp drop in pressure at 273.71 K is observed (*Fig. 6, Tab. 4*). Figure 7 illustrates the effect of silica sand of mesh size 60-100 present in the test sample with 10% bentonite clay and 10% silica sand. Pressure drop at two different temperatures is observed. Drop in pressure occurs at 284.85 K and 274.55 K on account of bentonite clay and silica sand, respectively. It can be seen that in all the four different sets of experimental conditions, pressure drop at 284.85 K is found in all cases, while the formation temperature shifts to higher temperatures when the mesh size of the silica sand is varied from smaller to larger (*Fig. 8*). Also, it can be noted that the highest drop in pressure occurred in a test sample with silica sand of mesh size 60-100 compared with that of test samples with silica sand of smaller mesh sizes (*Tab. 4*). These results indicate that a process sample containing a mixture of silica sand and bentonite clay is relatively more effective for encapsulation of methane in hydrate compared with samples containing only bentonite clay or silica sand (*Tab. 4, Fig. 8*).

It can be seen that as the size of the silica sand is decreased, the growth temperature of methane hydrate

TABLE 4

Nucleation temperature, initial cell pressure, growth temperature and pressure drop during hydrate formation in the presence of bentonite clay and silica sand

Test sample	Initial cell pressure (MPa)	Nucleation temperature (K)	Growth temperature (K)	Pressure drop ΔP (MPa)
10% bentonite in SSW	11.54	284.85	283.70	1.13
10% bentonite + 10% silica sand (mesh size 30-52) in SSW	11.54	284.85	283.70 272.91	3.80
10% bentonite + 10% silica sand (mesh size 52-60) in SSW	11.54	284.85	283.70 273.65	3.25
10% bentonite + 10% silica sand (mesh size 60-100) in SSW	11.54	284.85	283.70 274.55	4.24

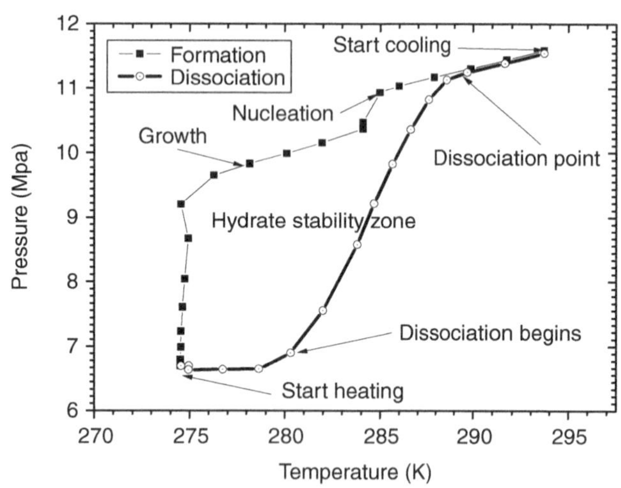

Figure 7

Pressure temperature profile of methane hydrate formation and dissociation in the presence of 10% bentonite clay and 10% silica sand (mesh size 60-100) in SSW.

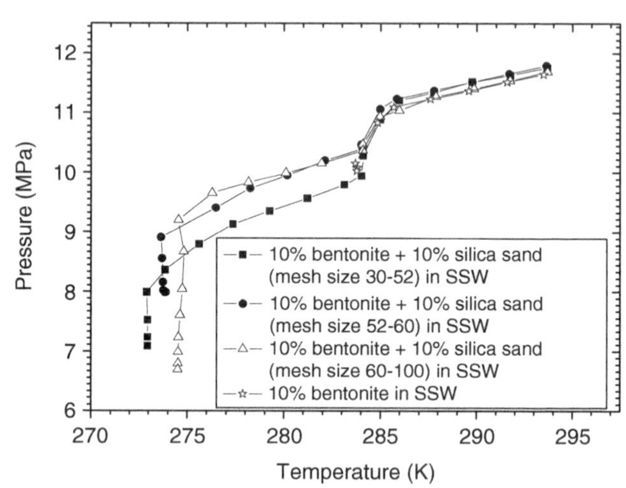

Figure 8

Comparative pressure temperature profile of methane hydrate formation in the presence of 10% bentonite clay and 10% silica sand.

is shifted to a higher temperature, and simultaneously enhances the encapsulation of methane gas inside the hydrate. It can also be seen that hydrate formation occurred in multiple stages (*Fig. 8*). Similar phenomena were observed by other authors [20, 21]. Silica sands with smaller mesh size have a larger surface area since the surface area is inversely related to particle size [22]. As a result, water with silica sand with a large mesh size will be subjected to a greater influence of the silica surface than that with a smaller mesh size of the silica sand. This implies that the proportion of water available for hydrate formation may vary with the size of the sand. The properties of bulk water are different from those of water with electrolytes, sands, silts and clays, effectively affecting hydrate formation. The particle size plays a very crucial role in determining hydrate saturation in sediments. Bagherzadeh *et al.* [23] studied methane hydrate formation in an unconsolidated bed of silica sand of different particle sizes, employing the magnetic imaging technique. It is observed that hydrate formation is not uniform in such a porous medium, but the rate of hydrate formation is found to be faster in a bed of low water content and sand of smaller particle size.

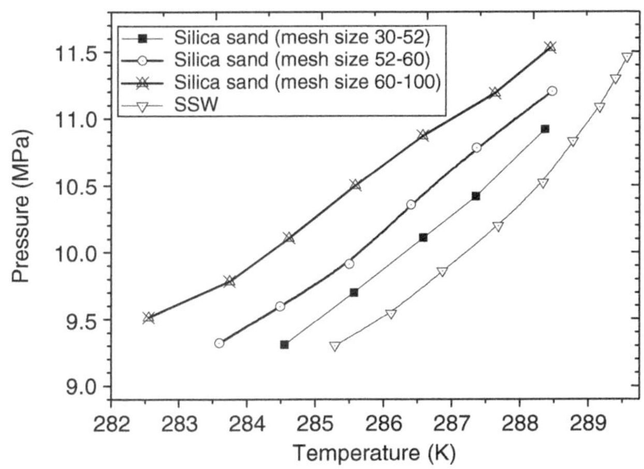

Figure 9

Methane hydrate phase equilibrium data in the presence of silica sand in synthetic seawater of 3.55% TDS.

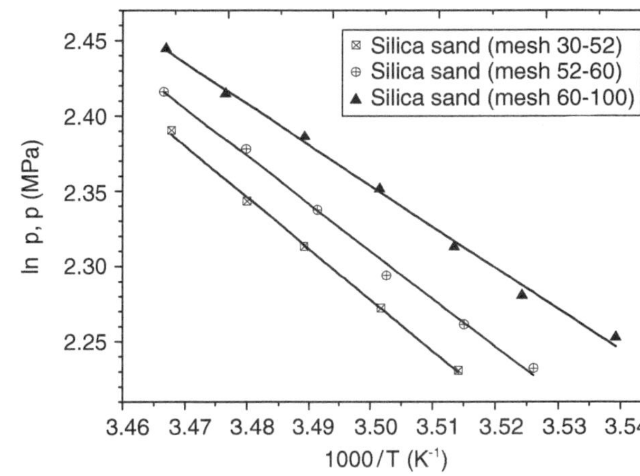

Figure 10

Semilogarithmic plot of phase equilibrium data of methane hydrate in the presence of silica sand of different mesh sizes.

2.3 Dissociation of Hydrate in the Presence of Silica Sand

The phase equilibrium of methane hydrate is significantly influenced by the presence of silica sand of different mesh sizes, because the liquid to solid transition is depressed to a lower temperature at any given pressure. The phase equilibrium curves are shifted to a relatively high-pressure and low-temperature zone as the particle size of the sand is decreased. These curves suggest that methane hydrate formed in the presence of sand with a smaller particle size would be more stable in the higher pressure zone only than hydrate formed in the presence of sand with a larger particle size (*Fig. 9*). There is a linear relationship between temperature and pressure. As the temperature increases, the pressure also increases during dissociation of the hydrate (*Fig. 10*). Such behavior of the hydrate equilibrium curves in the presence of silica sand of different mesh sizes may be induced by the high curvature of the solid-liquid interface [24].

Based on phase equilibrium data, it is possible to determine the enthalpy of hydrate dissociation using the Clausius-Clapeyron equation. The calculated enthalpy of hydrate dissociation of methane hydrate at a corresponding equilibrium temperature and pressure is presented in Table 5. The enthalpy of dissociation of methane hydrate is affected by the presence of silica sand. The dissociation enthalpy was found to vary with the grain size of the sand, since the phase equilibrium temperature and pressure of the hydrate are affected. This results in a change in the value of the compressibility factor (Z, calculated using the Peng-Robinson equation of state) of the gas.

2.4 Dissociation of Hydrate in the Presence of a Mixture of Silica Sand and Bentonite Clay

The phase equilibria of methane hydrate in the presence of bentonite clay and a mixture of silica sand and bentonite clay are shown in Figure 11. The phase equilibrium curve of methane hydrate in the presence of bentonite clay is shifted to a high-pressure region compared with the equilibrium curves of hydrate in the presence of a mixture of bentonite clay and silica sand.

It is clear that methane hydrate phase equilibrium is shifted to a low-pressure and high-temperature region, indicating higher stability in the presence of both bentonite clay and silica sand (*Fig. 11*). It can also be noted that there is no significant effect of the particle size of silica sand on the phase equilibrium of methane hydrate in a mixture of sand and clay. The phase equilibrium data of temperature and pressure obtained in the presence of bentonite clay and silica sand are presented in Table 6 and plotted as $\ln(p)$ *versus* $1/T$ (*Fig. 12*). It is also important to note that on varying the size of sand in a mixture of sand and bentonite clay the phase equilibrium curves are not significantly affected, while methane gas from the hydrate is released slowly on heating in a wide range of temperatures compared with methane gas from the hydrate

TABLE 5

Calculated dissociation enthalpy at different equilibrium points of pressure (P_{diss}) and temperature (T_{diss}) in the presence of silica sand

Test sample	P_{diss} (Mpa)	T_{diss} (K)	Z	ΔH (kJ/mol)
SSW (3.55% TDS)	9.30	285.29	0.8085	162.0258
	9.54	286.12	0.8064	162.6113
	9.85	286.87	0.8055	163.3407
	10.19	287.69	0.801	163.4212
10% silica sand (mesh 30-52)	9.69	285.57	0.8032	161.3015
	10.10	286.59	0.8009	162.0695
	10.42	287.36	0.7994	162.6956
	10.92	288.37	0.7969	163.4062
10% silica sand (mesh 52-60)	9.91	285.5	0.8003	160.6349
	10.35	286.41	0.7976	161.1853
	10.78	287.37	0.7956	161.9343
	11.20	288.47	0.7846	161.0031
10% silica sand (mesh 60-100)	9.51	282.55	0.7975	156.5563
	9.78	283.75	0.7972	157.9229
	10.10	284.62	0.7956	158.6412
	10.49	285.59	0.7938	159.4376

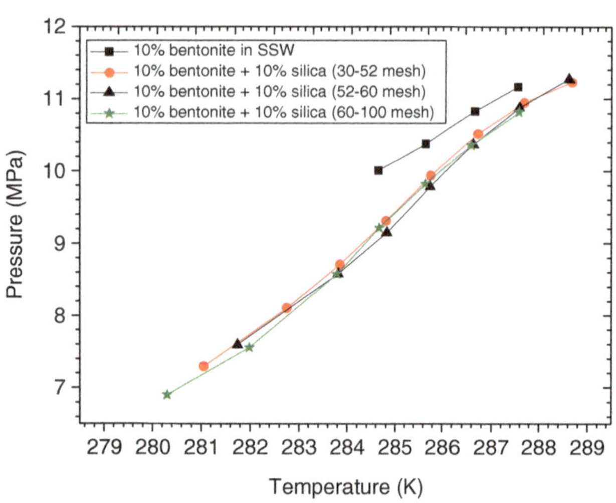

Figure 11

Methane hydrate phase equilibrium data in the presence of bentonite and silica sand.

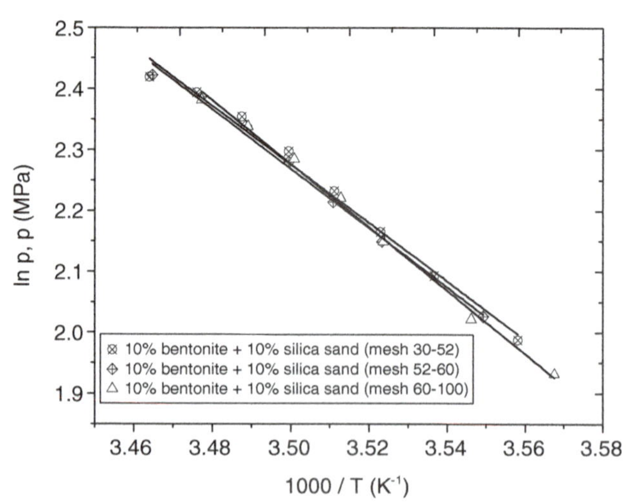

Figure 12

Semilogarithmic plot of phase equilibrium data of methane hydrate in the presence of a mixture of bentonite clay and silica sand.

with bentonite clay. There is not much effect of the particle size of the sand on dissociation enthalpy; however, the process sample with sand affects the enthalpy of dissociation appreciably more than the test sample with clay alone. Relatively more heat is required to dissociate the hydrate formed in the presence of both

TABLE 6

Calculated dissociation enthalpy at different equilibrium points of pressure (P_{diss}) and temperature (T_{diss}) in the presence of silica sand and bentonite clay

Test sample	P_{diss} (Mpa)	T_{diss} (K)	Z	ΔH (kJ/mol)
10% bentonite in SSW	10.38	285.65	0.8568	21.15
	10.83	286.68	0.8431	22.19
	11.60	287.56	0.8336	21.34
10% bentonite + 10% silica sand (mesh 30-52) in SSW	7.30	281.05	0.8621	167.32
	8.10	282.75	0.8367	164.50
	9.32	284.81	0.8202	163.78
	10.52	286.74	0.8021	162.49
	11.23	288.69	0.7921	162.80
10% bentonite + 10% silica sand (mesh 52-60) in SSW	7.59	281.75	0.8533	166.49
	9.14	284.83	0.8338	166.52
	9.78	285.75	0.8236	165.62
	10.36	286.65	0.8126	164.51
	11.27	288.63	0.7982	163.98
10% bentonite + 10% silica sand (mesh 60-100) in SSW	6.90	280.3	0.8751	168.87
	7.55	281.99	0.8554	167.21
	8.57	283.79	0.8223	162.94
	9.21	284.67	0.8134	162.25
	10.36	286.61	0.8009	162.09

sand and clay compared with hydrate formed in the presence of clay alone (*Tab. 6*).

3 KINETICS OF HYDRATE FORMATION

3.1 Measurement of Induction Time

To measure the induction time of hydrate formation, the process sample is adjusted to the temperature and pressure where hydrate is completely dissociated, *i.e.* the dissociation point or equilibrium point of the hydrate (*Fig. 3*) and allows the process sample to achieve that temperature where hydrate begins to form, *i.e.* the nucleation temperature. The time taken during this process is the induction time or nucleation time. The temperature-pressure response during the measurement of the induction time is shown in Figure 13.

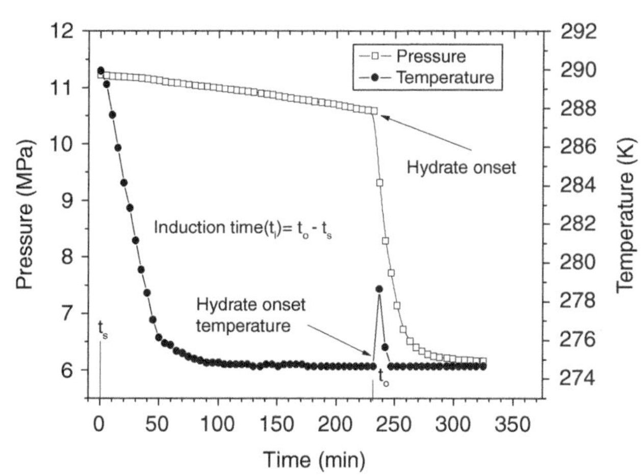

Figure 13

Response of temperature and pressure during hydrate formation of the test sample with 10% bentonite clay and 10% silica sand of mesh size 60-100.

The induction time measured is the difference in the hydrate onset time (t_o) and the start time (t_s) of the experiment where hydrate is completely dissociated, *i.e.* the dissociation point of the hydrate. The hydrate onset time is taken as t_o, where the pressure suddenly falls, followed by an increase in the cell temperature due to the latent heat of crystallization, while t_s is the time of the start of experiment from the dissociation point. The spike in temperature is observed in all the experiments. This is due to the release of the latent heat of crystallization when water and methane molecules rearrange into the hydrate state. The heat released during the process of hydrate formation is not the same in all cases. It depends on the type of medium in which the hydrate formers (water and methane) are allowed to crystallize. The induction time is measured under different subcooling conditions.

The induction time measured during hydrate formation in the presence of silica sand is presented in Table 7. It is found that the silica sand of smaller mesh size has a stronger impact on the induction time of hydrate formation, as it lengthens the induction time of hydrate nucleation, than that of silica sand with larger mesh size. The induction time is also found to be dependent on the subcooling of hydrate formation (*Tab. 7*).

Higher subcooling shortens the induction time, while lower subcooling lengthens it. The induction time of hydrate formation is shorter for the test sample with a mixture of bentonite clay and silica sand but higher than that of the sample with bentonite clay alone. The presence of bentonite clay decreases the induction time of hydrate formation (*Tab. 8*). The induction time of hydrate formation also depends on the rate of cooling: the slower the cooling rate, the longer the nucleation time. A slow cooling rate leads to lower subcooling and a longer nucleation time [12].

3.2 Gas Consumption

The number of moles of gas consumed during hydrate formation are computed using a real gas equation:

$$\Delta n = n_i - n_f = V R P_i Z_i T_i - P_f Z_f T_f \quad (1)$$

where Δn is the amount of gas consumed when hydrates form, V is the gas volume, P_i, T_i and P_f, T_f are the pressure and temperature in initial and final conditions, and R is the universal gas constant. Compressibility factors (Z_i, Z_f) were measured using the Peng-Robinson equation of state at the respective pressure and temperature.

Silica sand with smaller particle size is found to increase gas consumption. This consumption of gas is more pronounced for the test sample with the smallest particle size of silica sand. It is observed that hydrate formation takes place rapidly in a short duration after nucleation, and also the slope of the curves seems to be almost the same, indicating a constant formation rate (*Fig. 14, 15*). The addition of bentonite clay to the silica sand sample enhances the rate of gas consumption.

The conversion rate of methane into hydrate is also dominated by the cooling process. A higher cooling rate decreases the methane conversion rate [12].

TABLE 7

Measured formation temperature (T), dissociation temperature (T_{diss}), subcooling ($\Delta T = T_{diss/eq} - T$) and induction time of hydrate formation in silica sand

Test sample	T (K)	T_{diss} (K)	ΔT (K)	Induction time (min)
10% silica sand (mesh size 30-52) in	272.27	285.57	13.3	418
	272.27	286.59	14.32	389
	272.27	287.36	15.09	348
	272.27	288.37	16.1	307
10% silica sand (mesh size 52-60) in SSW10%	270.2	283.6	13.4	405
	270.2	284.49	14.29	354
	270.2	285.5	15.3	334
Silica sand (mesh size 60-100) in SSW	271.99	285.59	13.6	278
	271.99	286.58	14.69	251
	271.99	287.64	15.65	234

TABLE 8

Measured formation temperature (T), dissociation temperature (T_{diss}), subcooling ($\Delta T = T_{diss} - T$) and induction time of hydrate formation in the presence of both bentonite clay and silica sand

Test sample	T (K)	T_{diss} (K)	ΔT (K)	Induction time (min)
10% bentonite in SSW	284.85	286.68	1.83	15
	283.85	285.65	1.80	20
	283.85	284.65	0.80	25
10% bentonite + 10% silica sand (mesh size 30-52) in SSW	272.91	283.85	10.94	384
	272.91	284.81	11.9	375
	272.91	285.76	12.85	356
	272.91	286.74	13.83	334
	272.91	287.69	14.78	323
10% bentonite + 10% silica sand (mesh size 52-60) in SSW	273.65	283.82	10.17	343
	273.65	284.83	11.18	324
	273.65	285.75	12.1	310
	273.65	286.65	13.0	292
10% bentonite + 10% silica sand (mesh size 60-100) in SSW	274.55	281.99	7.44	289
	274.55	284.67	10.12	268
	274.55	286.61	12.06	241
	274.55	287.58	13.03	232

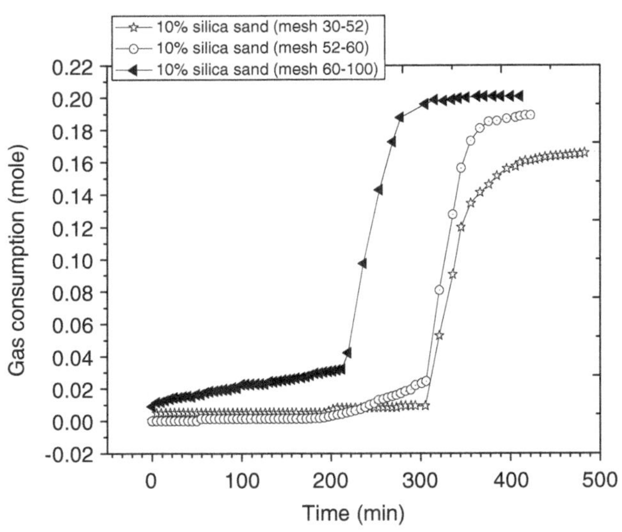

Figure 14

The rate of gas consumption during hydrate formation in the presence of silica sand in SSW.

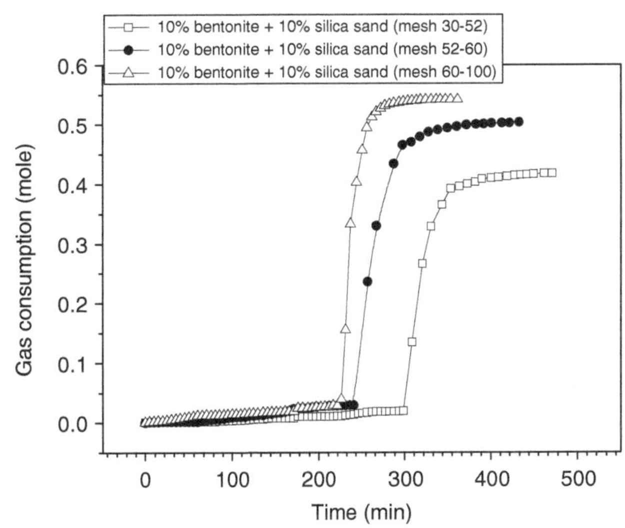

Figure 15

The rate of gas consumption during hydrate formation in the presence of both silica sand and bentonite clay in SSW.

Several authors just estimated the gas hydrate distribution and saturation level in sediments because natural gas hydrate samples have rarely been recovered intact. They revealed that particle size does play an important role in affecting the saturation level of hydrate in sediments. The conversion of methane into hydrate depends on water saturation, and the particle size of sand and clay concentration in sediments [14, 25-26].

CONCLUSIONS

The effects of bentonite clay, silica sand and the grain size of silica sand were studied on methane hydrate formation and dissociation. The addition of bentonite clay to the test sample with silica sand greatly affects the hydrate formation and dissociation temperature. Hydrate formation is shifted to a lower temperature in the presence of silica sand compared with test samples with bentonite clay and a mixture of silica sand and bentonite clay. The phase equilibrium curve is not appreciably affected upon variation in the size of the sand of test samples with a mixture of sand and clay; however, it is significantly affected by variation in the size of silica sand in the test sample with silica sand alone. There is no significant effect of the particle size of the sand on enthalpy; however, the process sample with sand affects the enthalpy of dissociation appreciably more than the test samples with clay alone. Relatively more heat is required to dissociate hydrate in the presence of sand compared with the hydrate in the presence of clay. Silica sand of smaller mesh size delayed the nucleation time of hydrate formation more than that of sand with larger mesh size. Addition of bentonite clay to the sample with silica sand significantly reduces the induction time.

ACKNOWLEDGMENTS

We gratefully acknowledge the financial assistance provided by the University Grants Commission, New Delhi, India, under the Special Assistance Program (SAP) to the Department of Petroleum Engineering, Indian School of Mines, Dhanbad, India.

REFERENCES

1 Kvenvolden K.A. (1995) A review of the geochemistry of methane in natural gas hydrates, *Org. Geochem.* **23**, 997-1008.

2 Sloan E.D. (1998) *Clathrate Hydrate of Natural Gases*, 2nd ed., Marcel Deckker, New York.

3 Sun R., Duan Z. (2007) An accurate model to predict the thermodynamic stability of methane hydrate and methane solubility in marine environments, *Chem. Geol.* **244**, 248-262.

4 Handa Y.P. (1990) Effect of hydrostatic pressure and salinity on the stability of gas hydrates, *J. Phys. Chem.* **94**, 6, 2652-2657.

5 Dixon J.E., Clague D.A., Stolper E.M. (1991) Degassing history of water, sulfur, and carbon in submarine lavas from Kilauea Volcano, Hawaii, *J. Geol.* **99**, 371-394.

6 Ginsburg G.D. (1996) *First master workshop of gas hydrates: relevance of world margin Stability and Climate Change*, Gent, Belgium, pp. 123-147.

7 Kvenvolden K.A. (1996) *First Master Workshop on Gas Hydrates: Relevance to World Margin Stability and Climate Change*, Gent, Belgium, pp. 39-80.

8 Handa Y.P., Stupin D.Yu (1992) Thermodynamic properties and dissociation characteristics of methane and propane hydrates in 70-Å-radius silica gel pores, *J. Phys. Chem.* **96**, 8599-8603.

9 Uchida T., Takeya S., Chuvilin E.M., Ohmura R., Nagao J., Yakushev V.S., Istomin V.A., Minagawa H., Ebinuma T., Narita H. (2004) Decomposition of methane hydrates in sand, sandstone, clays, and glass beads, *J. Geophys. Res.: Solid Earth* **109**, B05206, 1-12. doi:10.1029/2003JB002771.

10 Turner D.J., Cherry R.S., Sloan E.D. (2005) Sensitivity of methane hydrate phase equilibria to sediment pore size, *Fluid Phase Equilib.* **228/229**, 505-510.

11 Li G., Li X.-S., Tang L.-G., Feng Z.-P., Fan S.-S., Zhang Y. (2007) Experimental Study on the Effect of Temperature Cooling Profile on Methane Hydrate Formation, *Chin. J. Process. Eng.* **7**, 4, 723-727.

12 Guanli J., Qingbai W., Jing Z. (2010) Effect of cooling rate on methane hydrate formation in media, *Fluid Phase Equilib.* **298**, 225-230.

13 Clennell M.B., Hovland M., Booth J.S., Henry P., Winters W.J. (1999) Formation of natural gas hydrates in marine sediments: 1. Conceptual model of gas hydrate growth conditioned by host sediment properties, *J. Geophys. Res.* **104**, B10, 22985-23003.

14 Kang S.-P., Lee J.-W. (2010) Formation Characteristics of Synthesized Natural Gas Hydrates in Meso- and Macroporous Silica Gels, *J. Phys. Chem. B.* **114**, 20, 6973-6978.

15 Li X.-S., Zhang Y. (2011) Study on Dissociation Behaviors of Methane Hydrate in Porous Media Based on Experiments and Fractional Dimension Shrinking-Core Model, *Ind. Eng. Chem. Res.* **50**, 8263-8271.

16 Gieskes J.M. (1975) Chemistry of Interstitial Waters of Marine Sediments, *Annu. Rev. Earth Planetary Sci.* **3**, 433-453.

17 Saw V.K., Ahmad I., Mandal A., Udayabhanu G., Laik S. (2012) Methane hydrate formation and dissociation in synthetic seawater, *J. Nat. Gas Chem.* **21**, 6, 625-632.

18 Aladko E.Y., Dyadin Y.A., Fenelonov V.B., Larionov E. G., Mel'gunov M.S., Manakov A.Y., Nesterov A.N., Zhurko F.V. (2004) Dissociation Conditions of Methane Hydrate in Mesoporous Silica Gels in Wide Ranges of Pressure and Water Content, *J. Phys. Chem. B* **108**, 42, 16540-16547.

19 Kang S.P., Seo Y. (2008) International Gas Union Research (IGRC), Paris, 1-10.

20 Linga P., Haligva C., Nam S.C., Ripmeester J.A., Englezos P. (2009) Recovery of Methane from Hydrate Formed in a Variable Volume Bed of Silica Sand Particles, *Energy Fuels* **23**, 11, 5508-5516.

21 Prasad P.S.R., Chari V.D., Sharma D.V.S.G.K., Murthy S.R. (2012) Effect of silica particles on the stability of methane hydrates, *Fluid Phase Equilib.* **318**, 110-114.

22 Bird R.B., Stewart W.E., Lightfoot E.N. (2002) *Transport Phenomena*, Wiley, New York.

23 Bagherzadeh A.S., Moudrakovski I.L., Ripmeester J.A., Englezos P. (2011) Magnetic Resonance Imaging of Gas Hydrate Formation in a Bed of Silica Sand Particles, *Energy Fuels* **25**, 7, 3083-3092.

24 Anderson R., Llamedo M., Tohidi B., Burgass R.W. (2003) Experimental Measurement of Methane and Carbon Dioxide Clathrate Hydrate Equilibria in Mesoporous Silica, *J. Phys. Chem. B* **107**, 15, 3507-3514.

25 Riestenberg D., West W., Lee S., McCallum S., Phelps T.J. (2003) Sediment surface effects on methane hydrate formation and dissociation, *Marine Geology* **198**, 1-2, 181-190.

26 Lu H., Kawasaki T., Ukita T., Moudrakovski I., Fujii T., Noguchi S., Shimada T., Nakamizu M., Ripmeester J., Ratcliffe C. (2011) Particle size effect on the saturation of methane hydrate in sediments – Constrained from experimental results, *Marine Petrol. Geol.* **28**, 10, 1801-1805.

8

Influence of Gasoline Components on Engine Efficiency and Emissions

Guilherme B. Machado[1]*, José E.M. Barros[2], Sérgio L. Braga[3], Carlos Valois M. Braga[3], Antonio H.M. da F.T. da Silva[1] and Ellen C.Z. de Azevedo[4]

[1] Petróleo Brasileiro S.A., Petrobras/Research and Development Center, Av. Horácio Macedo, 950, Cidade Universitária, Ilha do Fundão, CEP: 21941-915, Rio de Janeiro, RJ - Brazil

[2] Department of Mechanical Engineering, Universidade Federal de Minas Gerais, UFMG, Av. Antonio Carlos, 6627, Campos Universitário, Pampulha, CEP: 31270901, Belo Horizonte, MG - Brazil

[3] Department of Mechanical Engineering, Pontifícia Universidade Católica do Rio de Janeiro, PUC-Rio, rua Marquês de São Vicente, 225, Gávea, CEP: 22453-900, Rio de Janeiro, RJ - Brazil

[4] Consultant, rua Dias da Cruz, 673, Casa 10/101, Meier, CEP: 20720-011, Rio de Janeiro, RJ - Brazil

e-mail: machadogb@petrobras.com.br - emautone@terra.com.br

slbraga@puc-rio.br - valois@puc-rio.br - ahmfts@petrobras.com.br - ezalona@gmail.com

* Corresponding author

Résumé — L'influence des composants de l'essence sur l'efficacité et les émissions du moteur — Pour les décennies à venir, il est prévu que les combustibles fossiles et les biocarburants utilisés dans les moteurs à combustion interne restent la principale source de propulsion des véhicules, ce qui justifie la recherche intense dans le monde entier et les efforts de développement pour répondre aux enjeux d'une efficacité croissante et de la réduction des émissions. La modélisation des combustibles commerciaux et les processus de combustion des moteurs présentent de grands défis. Il est également impératif de mieux comprendre comment les différents composants de carburant interagissent et interviennent sur la combustion du moteur et ses paramètres de performance. Dans le travail présenté, des composés représentatifs ont été utilisés afin d'appliquer des méthodologies et d'évaluer l'influence des composants du carburant sur ses propriétés et sur la combustion du moteur, ainsi que sur les paramètres de performance. Une attention toute particulière est accordée à l'efficacité du moteur et au comportement des émissions, de même qu'à leurs corrélations par rapport aux propriétés du carburant et à d'autres paramètres de performance du moteur. Les potentiels de chaque composant et le groupe chimique y afférent ont été identifiés pour différentes définitions techniques moteurs. Les résultats combinent des informations et des méthodologies qui peuvent être utilisées pour développer des carburants destinés à différentes applications.

Abstract* — *Influence of Gasoline Components on Engine Efficiency and Emissions — *For the next few decades, it is expected that fossil fuels and bio-fuels used in internal combustion engines will remain the primary source for vehicular propulsion. This justifies the intense worldwide research and development effort to comply with the challenges of increasing efficiency and reducing internal combustion engine emissions. The modeling of commercial fuels and engine combustion processes presents great challenges. There is also the need to better understand how different fuel components interact and influence engine combustion and performance parameters. In the present work,* *surrogate fuels were used to implement methodologies to evaluate the influence of fuel components on fuel properties and multiple engine combustion and performance parameters. Special attention is given to engine efficiency and emissions behavior and their correlations to fuel properties and others performance parameters of the engine. The potentials of each component and corresponding chemical group were identified for different engine designs. The results combine information and methodologies that can be used to develop fuels for different applications.*

ABBREVIATIONS

A/F ratio	Air/Fuel ratio
AKI	Anti-Knock Index
BTDC	Before piston Top Dead Center
CA	Crank Angle
CFR	Cooperative Fuel Research or code of federal regulations
CO	Carbon monoxide
CO_2	Carbon dioxide
E	Anhydrous ethanol
ECU	Electronic Control Unit
HC	Hydrocarbons
H/C	Hydrogen/carbon molar ratio
ICE	Internal Combustion Engine
IO	Iso-octane
LHV	Lower Heating Value
MBT	Maximum Break Torque
MON	Motor Octane Number
NH	N-heptane
R^2	Correlation factor
RON	Research Octane Number
SI	Spark Ignition
T	Toluene
T50%	Temperature at 50% v/v of fuel distilled
WOT	Wide Open Throttle

INTRODUCTION

Gasoline is a complex mixture that has a quasi-continuous spectrum of hydrocarbon constituents. Surrogate fuels that decrease the chemical and/or physical complexity of gasoline can be used to enhance the understanding of fundamental processes involved in the interaction between fuels and Internal Combustion Engines (ICE). Pitz *et al.* (2007) have presented an extensive review about surrogate fuels that describes all of the possible components, their importance and the availability of physical and thermochemical data. These studies indicate that *n*-heptane, iso-octane and toluene are interesting components to model gasoline properties in surrogate fuels. Several studies in the literature examined the kinetic models of iso-octane (Curran *et al.*, 2002; Davidson *et al.*, 2005; He *et al.*, 2005), *n*-heptane (Curran *et al.*, 1998; Herzler *et al.*, 2005; Berta *et al.*, 2006; Gauthier *et al.*, 2004; Silke *et al.*, 2005) and toluene (Davidson *et al.*, 2005; Bounaceur *et al.*, 2005; Dagaut *et al.*, 2002; Emdee *et al.*, 1992; Pitz *et al.*, 2003; Sivaramakrishnan *et al.*, 2005). For example, Andrae (2008) presented a detailed kinetic model that describes the autoignition of surrogate fuels consisting of iso-octane, *n*-heptane, toluene, diisobutylene and ethanol. These studies were based on experiments conducted at different pressures and temperatures, and in different environments, including shock tubes, rapid compression machines and constant volume chambers. The experiments were conducted to reproduce various thermodynamic conditions that represent different engine operating conditions.

In the literature, the authors found few direct correlations between surrogate fuel behavior in basic controlled experiments using shock tubes and rapid compression machines and the ability of surrogate fuels to reproduce gasoline performance in commercial SI engines. The behavior of surrogate fuels in a modern commercial ICE is not well understood.

Machado (2012) and Machado *et al.* (2011) presented the ability of surrogate fuels composed of iso-octane, *n*-heptane, toluene and ethanol to reproduce the performance of full Brazilian gasoline in commercial engines. These studies determined the optimal amount of each component in a surrogate fuel for high-octane oxygenated gasolines. It is important to mention that regular Brazilian gasoline contains 18 to 25% anhydrous ethanol by volume, which is required by law and varies depending on ethanol production and international market conditions (ANP, 2011). Machado (2012) and Machado *et al.* (2012) investigated the effect of individual components of surrogate fuels on fuel properties and commercial engine performance. That study led to the introduction of methodologies for fuel development based on response surface models derived from experimental design and engine tests. The present paper presents further results, and gives special attention to the influence of fuel components and properties on engine efficiency and emissions.

1 METHODOLOGIES

The methodologies for surrogate fuels selection, engine test cell preparation and experimental procedures are described in more details in Machado (2012) and Machado *et al.* (2011, 2012). The main points are summarized in the next sections.

1.1 Fuel Components Concentrations

Based on a mixture design approach, ten gasoline surrogate fuels were prepared with different concentrations of iso-octane (IO), *n*-heptane (NH) and toluene (T). Anhydrous ethanol (E) was incorporated into all of the mixtures at a fixed volume concentration of 25%. Table 1 lists the original volume concentration of each component in the surrogate fuels. In Section 2, the surrogate fuels are represented in ternary diagrams according to the volume fraction of each component normalized with respect to the surrogate fuel in the absence of ethanol (*Fig. 1*).

In this work, no runs were done with pure iso-octane, *n*-heptane or toluene in blends with ethanol. These additional runs could make results extrapolations more reliable and is a suggestion for future works. However, the mixture of *n*-heptane in blend with 25% of ethanol could present running problems in several engine operating conditions due to the very low octane rating of the fuel.

1.2 Fuel Properties

The properties of the surrogate fuels are presented in Table 2 and are within the typical range of ethanol oxygenated gasoline blends. The stoichiometric air/fuel ratio, density and Lower Heating Value (LHV) were calculated based on the C:H:O ratio of each mixture and on the properties of each component combined in a mass base. The T50% value, temperature at 50% v/v of fuel distilled, was calculated by averaging the results of three different methodologies. The fuels were individually tested in a Cooperative Fuel Research (CFR) engine to determine the values of MON and RON in order to calculate the Anti-Knock Index (AKI), which is the average of both measurements (Machado *et al.*, 2011, 2012; Machado, 2012).

1.3 Engine Tests

1.3.1 Test Cell Preparation

The engine utilized in this study was the *Fiat* Fire 1.4L Tetrafuel. This engine can run on pure gasoline, Brazilian gasoline (18 to 25% anhydrous ethanol by volume), mixtures of Brazilian gasoline and hydrous ethanol or natural gas. The main engine specifications are presented in Table 3.

TABLE 1

Original concentrations of each component

Fuel	Volume concentration (%)			
	IO[a]	NH[b]	T[c]	E[d]
A	25	25	25	25
B	31.25	12.5	31.25	25
C	12.5	31.25	31.25	25
D	31.25	31.25	12.5	25
E	37.5	18.75	18.75	25
F	18.75	37.5	18.75	25
G	18.75	18.75	37.5	25
H	0	37.5	37.5	25
I	37.5	37.5	0	25
J	37.5	0	37.5	25

[a] IO – Iso-octane,
[b] NH – *N*-Heptane,
[c] T – Toluene,
[d] E – Ethanol.

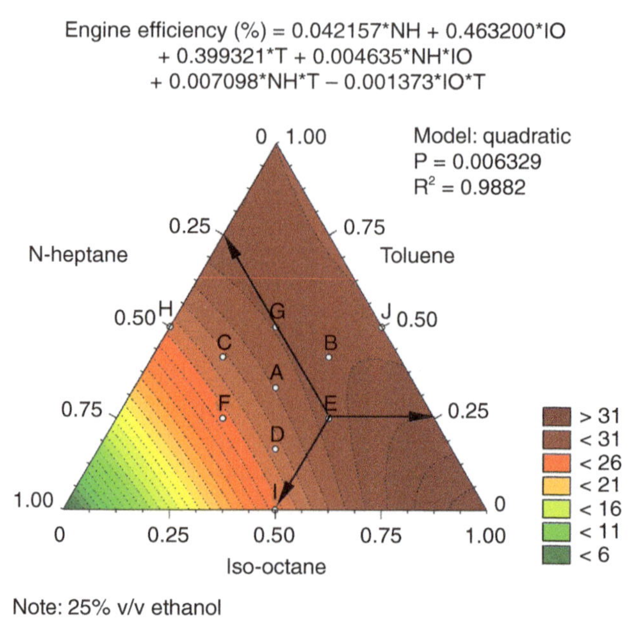

Figure 1

Engine efficiency at 3 875 rpm with WOT.

TABLE 2
Fuel properties

Fuel	Density (kg/m^3)	LHV[a] (kJ/kg)	Stoichiometric. mass A/F ratio[b]	T50%[c] (°C)	MON[d]	RON[e]	AKI[f]
A	758.3	38 712	13.04	97.0	86.0	95.2	90.6
B	770.2	38 506	12.95	97.4	90.8	103.0	96.9
C	768.7	38 531	12.95	96.6	82.2	91.2	86.7
D	735.9	39 117	13.22	96.6	82.6	89.0	85.8
E	747.8	38 898	13.12	97.4	89.4	99.0	94.2
F	746.3	38 924	13.13	96.6	77.6	84.0	80.8
G	780.6	38 331	12.87	97.4	88.2	100.2	94.2
H	779.1	38 354	12.87	96.6	78.2	87.2	82.7
I	713.6	39 546	13.41	96.6	77.4	82.0	79.7
J	782.1	38 307	12.86	97.4	96.0	108.0	102.0

[a] LHV – Lower Heating Value,
[b] A/F ratio – Air/Fuel ratio,
[c] T50% – Temperature at 50% v/v of fuel distilled,
[d] MON – Motor Octane Number (average of two tests),
[e] RON – Research Octane Number (average of two tests),
[f] AKI – Anti-Knock Index.

TABLE 3
Specifications of the *Fiat* Fire Tetrafuel engine

Swept volume	1 368 cm^3
Number of cylinders	4 in line
Cylinder diameter	72 mm
Stroke	84 mm
Piston bore	71.9 mm
Compression ratio	10.35:1
Valves per cylinder	2
Camshaft	1 (overhead)

TABLE 4
Engine operating conditions

Speed (rpm)	Load (% throttle position)	Lambda
5 500	100	0.9
3 875	100	0.9
3 875	16	1.0
2 250	100	0.9
2 250	16	1.0
1 500	16	1.0

The tests were conducted in an engine test cell according to ISO 1585 (ISO, 1992). The original Electronic Control Unit (ECU) of the engine was replaced with a programmable unit to allow for engine spark timing and air/fuel ratio calibration. Cylinder 1 was instrumented with an AVL GU 13Z-24 piezoelectric combustion pressure transducer and an IndiModul 621 data acquisition system for combustion analysis and knocking observations. The occurrence of knocking was further monitored by an audio system.

1.3.2 Experimental Test Procedures

To cover a wide range of engine operating conditions, the tests were performed at six different operating points (speed and % throttle position), as shown in Table 4.

Lambda was fixed at a value of 0.9 for full load conditions (Wide Open Throttle - WOT) and 1.0 for partial load conditions (*Tab. 4*). This approach is justified by the calibration carried out by the engine manufacturers. The calibration is conducted to optimize power at WOT and to reduce emissions during partial load conditions. The lambda value for gasoline to achieve maximum power is typically within the range of 0.88 to 0.92, while lambda to achieve maximum catalyst efficiency and minimal emissions at partial load is 1.0. To maintain a constant lambda at each engine operating point, the tests were performed in a closed loop mode.

For each fuel and engine operating condition (speed and % throttle position), the spark timing had a different calibration, adjusted to obtain the MBT (Maximum Break Torque), constrained by knocking occurrence and a maximum exhaust gas temperature of 900°C (1 173 K). This temperature limit generally preserves emissions control catalysts.

At least three measurements of torque and fuel consumption were performed for each fuel after a one minute engine stabilization period per operating condition.

1.3.3 Emissions Measurement

The emissions measurement was completed using the NAPRO Modal 2010 - AO, which was adapted for the engine test cell. This equipment has nondispersive infrared gas analyzers that perform volumetric measurements of CO, CO_2 and HC in dry base. The catalyst was removed from the engine exhaust pipe to measure the raw emissions. The measurements were performed over at least 40 seconds. At the same time, the torque and fuel consumption measurements were completed. The emission results were reported as the calculated average from the measurement period.

In addition to the analysis of the volumetric emissions, the specific emissions were also calculated in g/kWh, according to CFR, title 40 (CFR, 2009; Melo *et al.*, 2012; Machado, 2012).

1.4 Data Reduction

The influences of the different fuel components on the fuel properties and engine performance were compared by adopting the mixture design approach for typical Brazilian gasoline/ethanol 75/25% vol/vol blends. Response surfaces were statistically determined based on normalized factors (normalized concentrations of the components), and mathematical models were developed based on original factors using STATISTICA® software (Montgomery and Runger, 1994; Calado and Montgomery, 2003). Therefore, the mathematical models presented in the next sections, including the corresponding ternary plots of response surfaces, are based on the volume percent concentration of each component in the original mixture (*Tab. 1*). The mathematical models presented are more accurate in the regions near the experimental data points. The extrapolations (regions near the vertices of the domain) are also important for exploring general tendencies. Some of the figures present quadratic models that include the existence of terms that reflect interactions among different components. It is important to point out that the mathematical models were the result of a multiparametric regression analysis associated with highly sensitive statistical significance test methodologies. Each response surface presented has an associated correlation factor (R^2) and a p factor, relative to the statistical significance level of the model. To guarantee 95% model confidence, only models with $p \leq 0.05$ were considered. It should be noted that the ternary diagrams have a particular reading procedure, as illustrated in Figure 1. The concentration readings of the components on the ternary diagrams should be multiplied by 75% to obtain the original concentration used in the mathematical models presented.

The models were developed from average data. The maximum uncertainty for all of the variables with all of the fuels and engine operating conditions was 0.6%. The uncertainties were calculated combining type A and type B uncertainties, respectively related to the statistical random uncertainties present in individual measurements and instrumentation uncertainties. A coverage factor of 2 was used to guarantee 95% of confidence level (Machado, 2012).

2 RESULTS AND DISCUSSION

This section presents the results and analyses of the influence of fuel components and properties on overall engine efficiency, combustion efficiency and emissions. The trends observed for the different parameters evaluated are always similar for more than one operating condition. In each case, only one engine condition is shown to represent the general behavior of the parameter analyzed.

2.1 Influence of Fuel Components and Properties on Engine Efficiencies

The engine was operated successfully using each of the surrogate fuels tested. This section presents results for the overall engine efficiency and the combustion efficiency. It is important to note that for the engine and operating conditions evaluated in this work (*Tab. 4*), 5 500 rpm with WOT, 3 875 rpm with WOT, 2 250 rpm with WOT and 1 500 rpm with partial load were the conditions susceptible to knocking. In contrast, 3 875 rpm with partial load and 2 250 rpm with partial load did not demonstrate any susceptibility to knocking with any of the fuels tested.

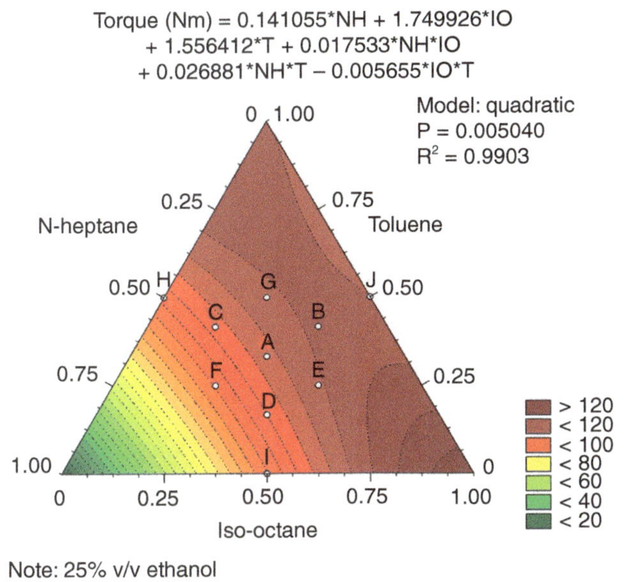

Figure 2

Engine torque at 3 875 rpm with WOT.

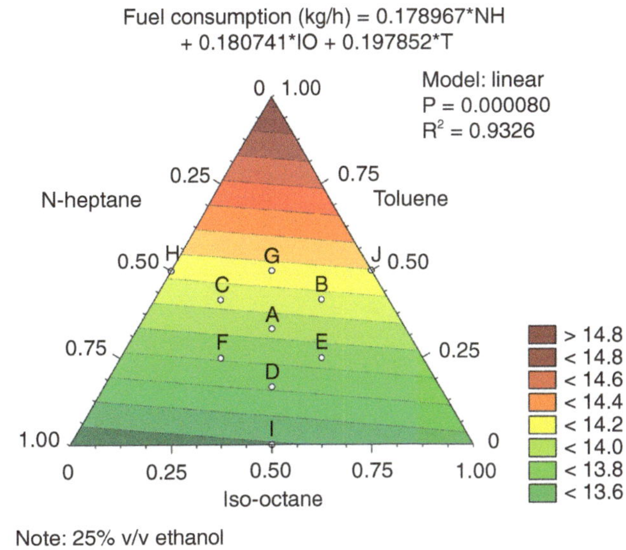

Figure 3

Fuel consumption at 3 875 rpm with WOT.

2.1.1 Overall Engine Efficiency

Several different efficiency types are studied when analyzing engine performance. The overall engine efficiency analyzed here is the ratio of the brake power to the energy flow rate delivered to the engine. The energy flow rate to the engine is the product of the fuel consumption and the LHV (fuel Lower Heating Value).

Statistically significant response surface models were obtained for all of the operating conditions with the exception of the 3 875 rpm with partial load condition. This condition had a lower efficiency, and it was difficult to distinguish differences between the fuels.

For all of the operating conditions sensitive to knocking there was good agreement between overall engine efficiency and the torque, as shown in Figures 1 and 2, which represent the operating condition of 3 875 rpm with WOT.

For surrogate fuels with higher concentrations of toluene and iso-octane, the higher the torque produced, the higher the resulting overall engine efficiency is. Iso-octane proved to be more effective than toluene in increasing the overall engine efficiency. As shown in Figure 3, toluene contributes to an increase in the fuel consumption (lower stoichiometric air/fuel ratio), while iso-octane reduces the fuel consumption (higher stoichiometric air/fuel ratio). The stoichiometric air/fuel ratio is shown in Figure 4.

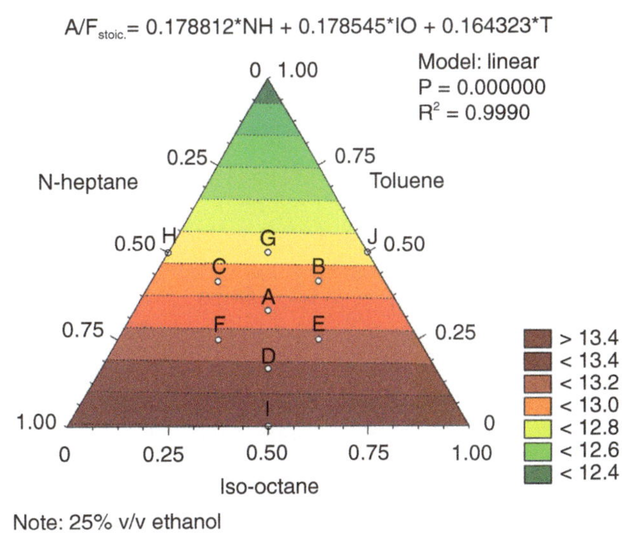

Figure 4

Stoichiometric air/fuel ratio.

The lower value of LHV per kg of stoichiometric air that is presented in Figure 5 also favors iso-octane for increasing the overall engine efficiency.

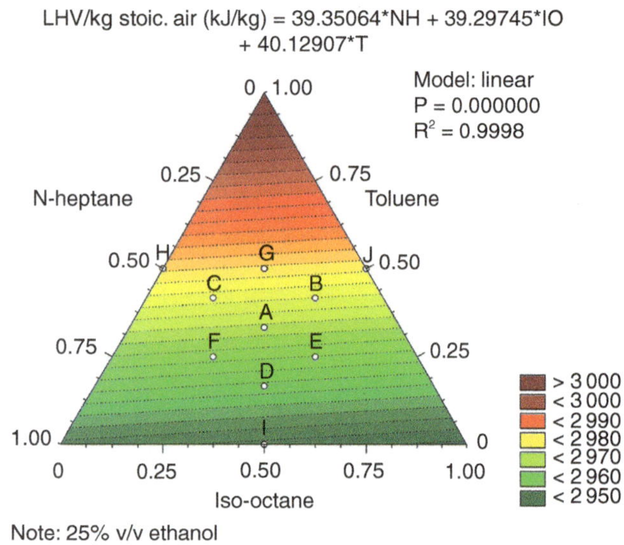

Figure 5

Fuel lower heating value per kg of stoichiometric air.

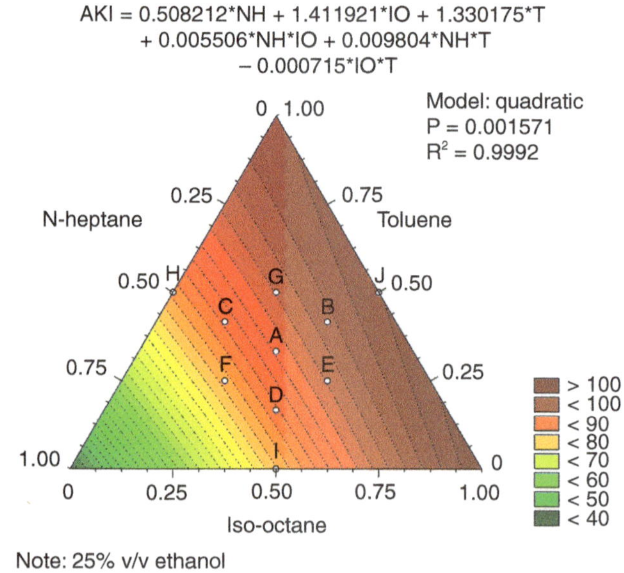

Figure 6

Fuel Anti-Knock Index (AKI).

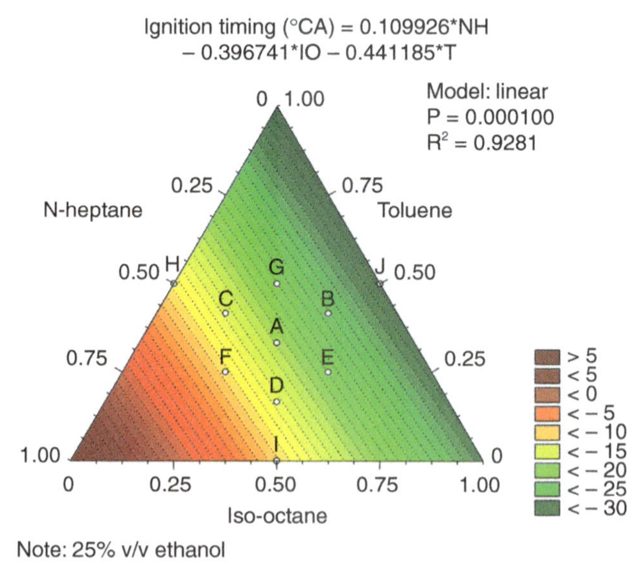

Figure 7

Ignition timing at 3 875 rpm with WOT.

For the operating conditions that are sensitive to knocking, it is generally shown that the fuel octane rating (*Fig. 6*) has a greater impact on the overall engine efficiency (*Fig. 1*) than the stoichiometric air/fuel ratio (*Fig. 4*) and the lower heating value per kg of stoichiometric air (*Fig. 5*). Figures 1 and 6 show similar trends. The higher the octane rating is, the higher the overall engine efficiency is. This phenomenon is related to more advanced ignition timing calibration, as shown in Figure 7 (negative values indicate BTDC, Before Piston Top Dead Center). Nevertheless, the stoichiometric air/fuel ratio (*Fig. 4*) and the lower heating value per kg of stoichiometric air (*Fig. 5*) contribute to the overall engine efficiency, leading to better performance characteristics with iso-octane compared to toluene.

For the operating conditions that are not sensitive to knocking, there was good agreement between the overall engine efficiency and fuel consumption. This result is clear from the comparison of Figures 8 and 9, which represent the operating condition of 2 250 rpm with partial load.

In the conditions that are not sensitive to knocking, the engine torque did not vary significantly between the fuels. Thus, the stoichiometric air/fuel ratio (*Fig. 4*) becomes a more significant fuel property than the octane rating (*Fig. 6*) for increasing the overall engine efficiency. This fact can be an important consideration for special applications, such as race fuel formulations, where higher engine speeds do not allow time for knocking and the AKI may not be as relevant.

It is possible to see that iso-octane was more effective than *n*-heptane in increasing engine efficiency for the 2 250 rpm with partial load condition, unlike the behavior observed for the fuel consumption. This occurred because there is a lower value of LHV per kg of stoichiometric air of iso-octane compared to *n*-heptane (*Fig. 5*).

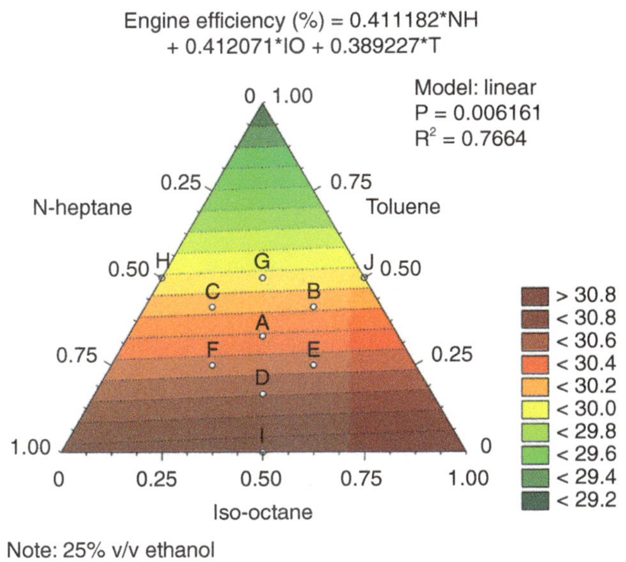

Figure 8

Engine efficiency at 2 250 rpm with partial load.

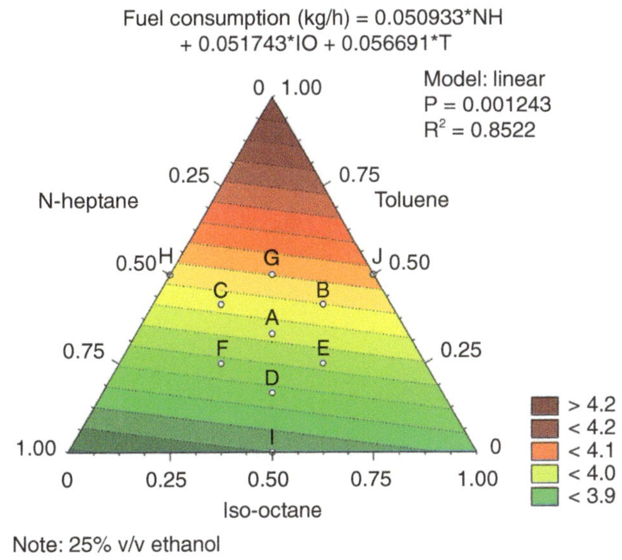

Figure 9

Fuel consumption at 2 250 rpm with partial load.

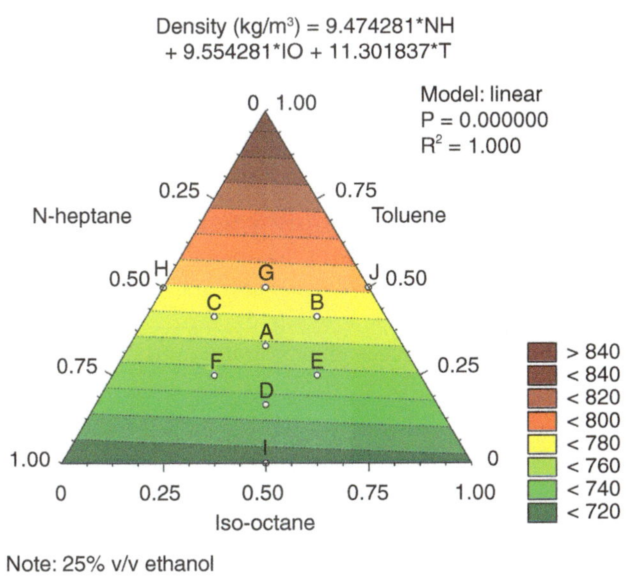

Figure 10

Fuel density.

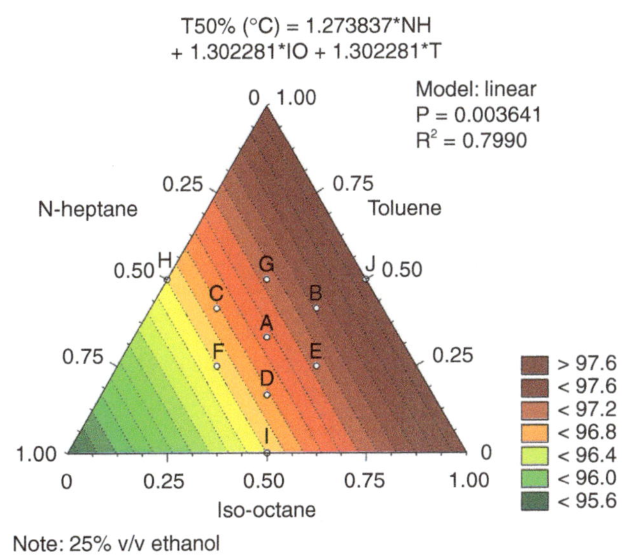

Figure 11

Fuel T50%.

As expected, the higher the stoichiometric air/fuel ratio is, the lower the fuel consumption is. The fuel consumption had the same tendency for all of the operating conditions tested, as shown in Figures 3 and 9, which represent 3 875 rpm with WOT and 2 250 rpm with partial load conditions, respectively. Despite the small differences in the stoichiometric air/fuel ratio among the fuels utilized, it was possible to observe the impact on fuel consumption. Figure 4 illustrates the slight asymmetry that can contribute to *n*-heptane being more effective than iso-octane in reducing fuel consumption. The lower density and T50% values of *n*-heptane, shown in Figures 10 and 11, respectively, help to promote improved vaporization and mixture formation. This behavior can

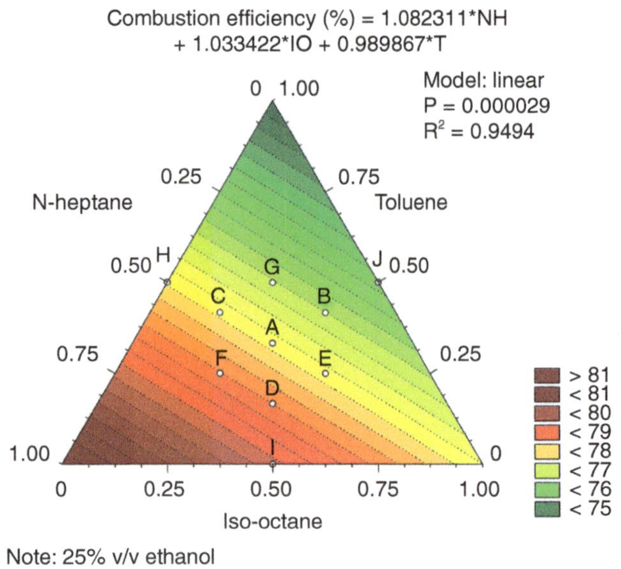

Figure 12

Combustion efficiency at 3 875 rpm with WOT.

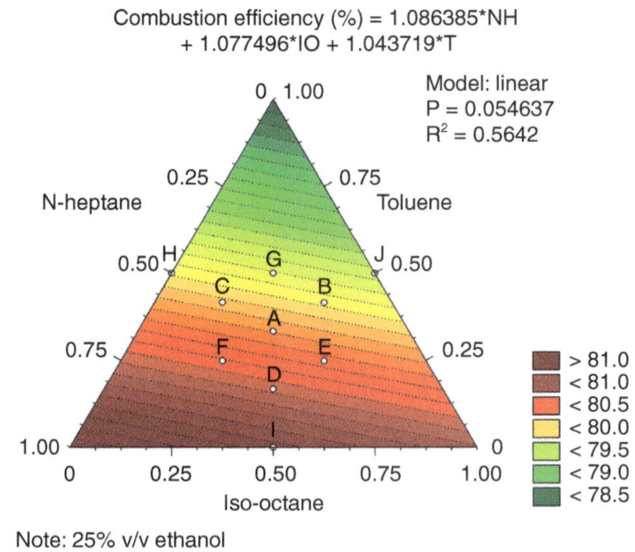

Figure 13

Combustion efficiency at 2 250 rpm with partial load.

also contribute to making *n*-heptane more effective than iso-octane in reducing fuel consumption.

2.1.2 Engine Combustion Efficiency

In this work, the combustion efficiency was defined as the ratio of the net heat release to the energy flow rate delivered to the engine in one cycle (the product of the fuel consumption and fuel lower heating value per cycle). The net heat release was calculated through combustion analysis using the average pressure curve from 300 recorded engine cycles as in Machado (2012). The net heat release excludes the amount of fuel energy transferred to the combustion chamber walls and crevice volumes and is related to the sensible energy change and work transfer to the piston (Heywood, 1988).

The engine combustion efficiency showed the same trends for all of the operating conditions. This is shown for the two examples in Figures 12 and 13, 3 875 rpm at WOT and 2 250 rpm with partial load, respectively.

The trends show higher efficiencies for the fuels with higher concentrations of *n*-heptane followed by iso-octane and toluene. This behavior can be related to the molecular structure of the components. The simpler straight chain of *n*-heptane was easier to burn followed by the branched chain of iso-octane and then the more complex aromatic chain of toluene. For the operating conditions that were not sensitive to knocking, the combustion efficiency trend (*Fig. 13*) associated with the fuel consumption behavior (*Fig. 9*) and the LHV per kg of stoichiometric air tendency (*Fig. 5*) clarifies the improved overall engine efficiency of *n*-heptane and iso-octane in relation to toluene (*Fig. 8*).

For the operating conditions that were not sensitive to knocking, *n*-heptane is an interesting component, as it simultaneously contributes to a reduction in fuel consumption and an increase in engine combustion and overall engine efficiencies. Iso-octane shows the best compromise. Iso-octane increases the octane rating and the stoichiometric air/fuel ratio contributing to increase engine efficiency for all of the operating conditions, independent of engine knocking susceptibility.

2.2 Influence of Fuel Components and Properties on Engine Emissions

This section presents the volumetric emissions results of CO_2, CO and HC. Then the results of specific emissions are presented in g/kWh to compare the different observed trends between these two emissions reporting methods.

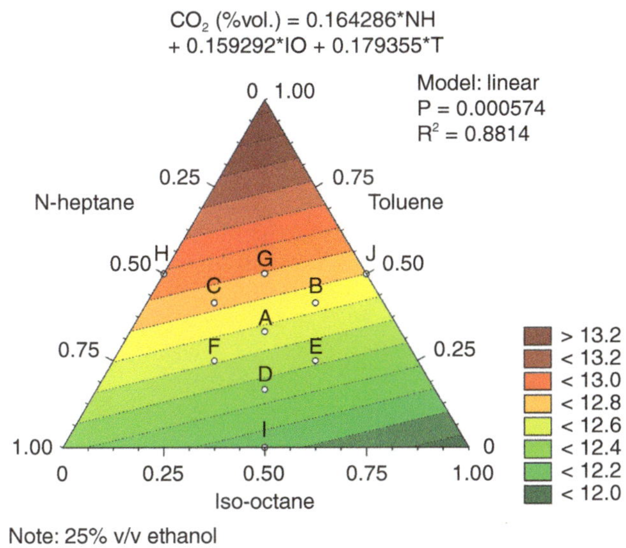

Figure 14

Volumetric emissions of CO_2 at 3 875 rpm with WOT.

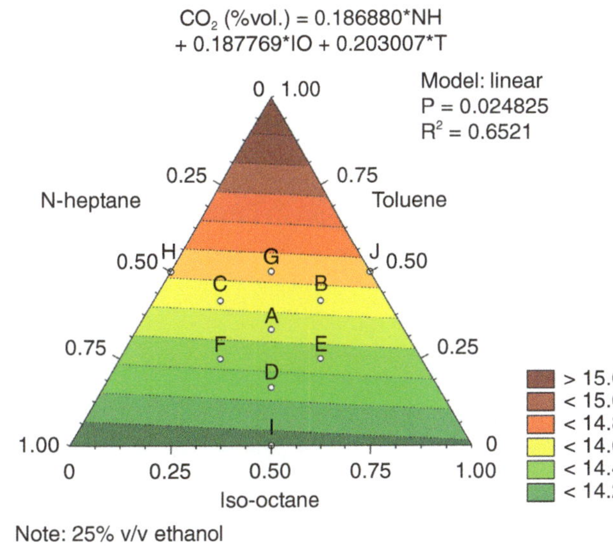

Figure 15

Volumetric emissions of CO_2 at 2 250 rpm with partial load.

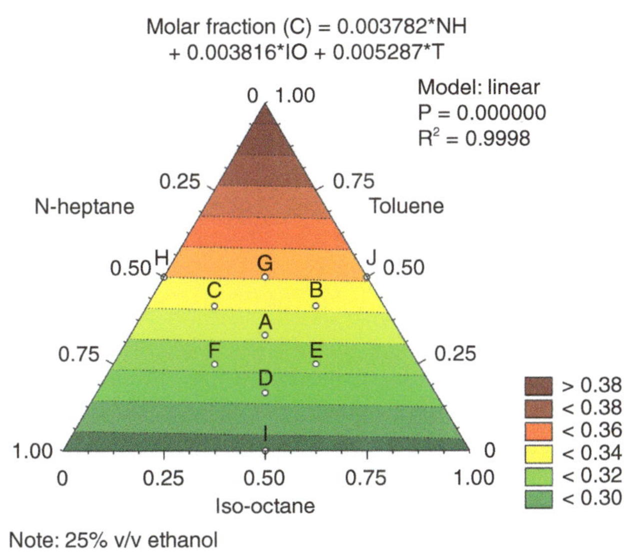

Figure 16

Fuel carbon molar fraction.

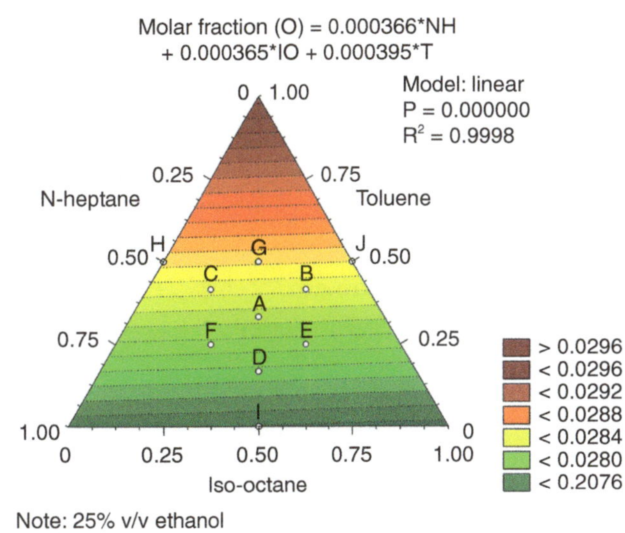

Figure 17

Fuel oxygen molar fraction.

2.2.1 Volumetric Emissions of Pollutants

Volumetric Emissions of CO_2

Regarding CO_2 emissions, response surface models with similar trends were obtained for all of the operating conditions. Two examples are shown in Figures 14 and 15, and represent 3 875 rpm with WOT and 2 250 rpm with partial load, respectively.

In general, toluene contributes to increased CO_2 emissions, while *n*-heptane and iso-octane reduce CO_2 emissions. These trends can be related to the fuel molar fractions of carbon and oxygen (*Fig. 16, 17*) and to the fuel consumption, which had the same trends for all of the operating conditions (*Fig. 3, 9*). Higher amounts of toluene contribute to an increase in the fuel carbon and oxygen contents and the fuel consumption, which increases the CO_2 emissions.

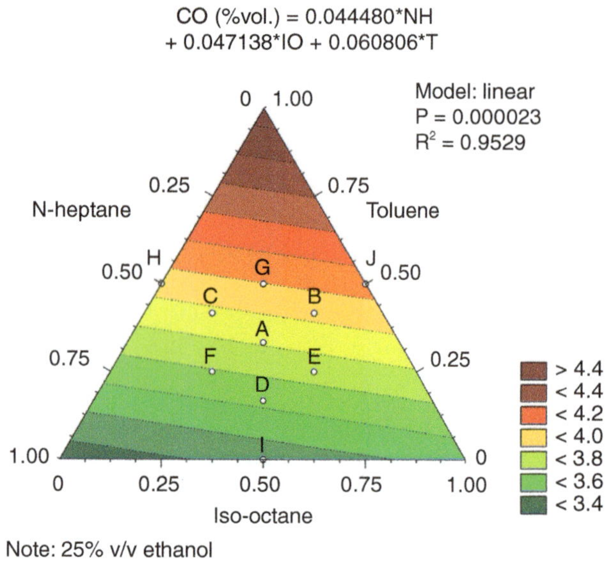

Figure 18

Volumetric emissions of CO at 3 875 rpm with WOT.

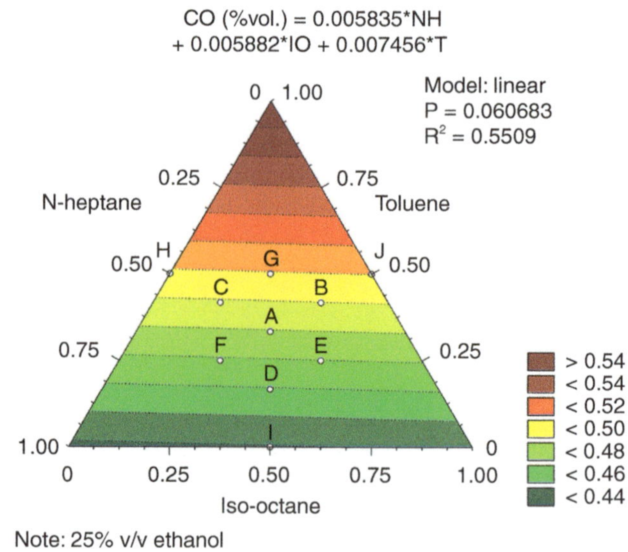

Figure 19

Volumetric emissions of CO at 2 250 rpm with partial load.

Volumetric Emissions of CO

Statistically significant response surface models of the CO emissions had similar trends for all of the operating conditions, with the exception of the 1 500 rpm with partial load condition. Two examples are shown in Figures 18 and 19 referring respectively to 3 875 rpm (WOT) and 2 250 rpm (partial load).

In general, the trends are similar to those observed for the CO_2 emissions. Higher contents of toluene presented higher levels of CO emissions. This is related to the contribution of toluene to the increase in the fuel carbon content (*Fig. 16*) and the increase in the fuel consumption (*Fig. 3, 9*) associated with its contribution to reduce combustion efficiency (*Fig. 12, 13*).

It is important to note that *n*-heptane contributed to a reduction in CO emissions more effectively than iso-octane. This is related to the higher combustion efficiency observed with the higher contents of *n*-heptane. This explains the higher CO_2 emissions for fuels with higher *n*-heptane levels compared to the fuels with higher iso-octane levels for the 3 875 rpm with WOT operating condition (*Fig. 14*). The 2 250 rpm with partial load condition presented a smaller difference between the *n*-heptane and iso-octane contributions to CO emissions (*Fig. 19*). This fact, combined with the lower fuel consumption with larger levels of *n*-heptane (*Fig. 9*), contributes to a reduction in *n*-heptane CO_2 emissions compared to iso-octane at the 2 250 rpm with partial load condition (*Fig. 15*). The carbon and oxygen contents of *n*-heptane and iso-octane are similar (*Fig. 16, 17*), which plays a minor role in explaining the differences in CO and CO_2 emissions between these components.

Volumetric Emissions of HC

Statistically significant response surface models of HC emissions were obtained for 3 875 and 5 500 rpm with WOT as well as for 1 500 and 3 875 rpm with partial load operating conditions. These trends are shown in Figures 20 and 21 for the 5 500 rpm with WOT and 1 500 rpm with partial load operating conditions, respectively. Despite the low R^2 values, the similar trends for all of the different operating conditions that presented statistically significant models support the analyses.

Fuels with higher contents of iso-octane contribute to an increase in HC emissions, while higher contents of *n*-heptane reduce HC emissions. In this case, the molar fractions of carbon, oxygen and hydrogen, observed in Figures 16, 17 and 22, respectively, do not influence this behavior, as they are similar for iso-octane and *n*-heptane.

Again, the higher combustion efficiency of the fuels with higher contents of *n*-heptane (*Fig. 12, 13*) favors the reduction of HC emissions. The differences in HC emission trends between iso-octane and toluene can be

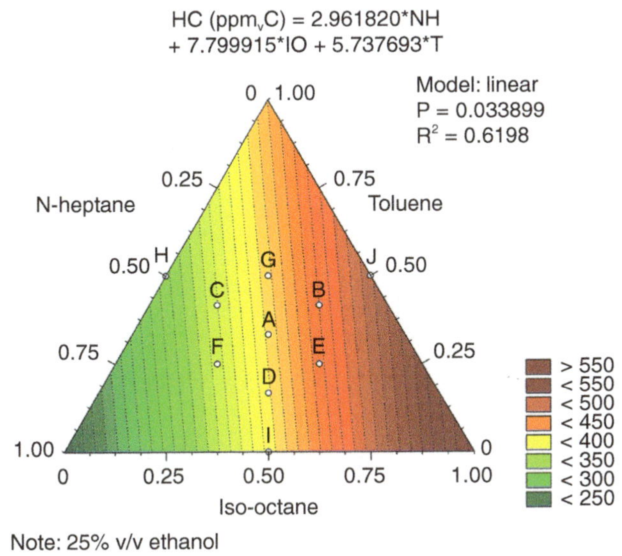

Figure 20

Volumetric emissions of HC at 5 500 rpm with WOT.

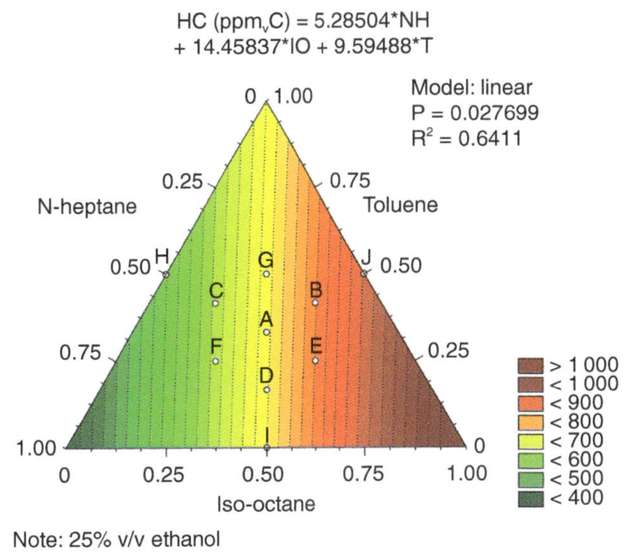

Figure 21

Volumetric emissions of HC at 1 500 rpm with partial load.

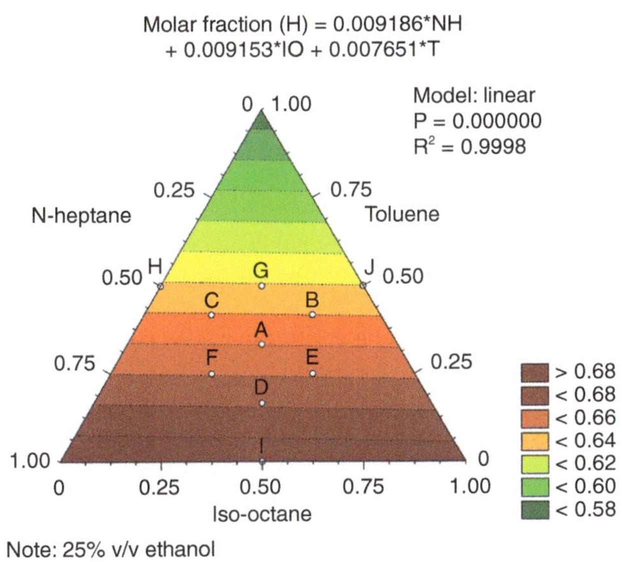

Figure 22

Fuel hydrogen molar fraction.

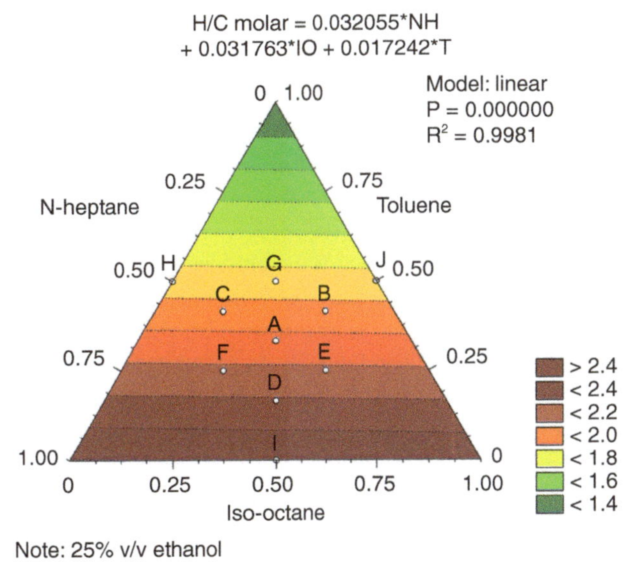

Figure 23

Fuel H/C molar ratio.

tied to the higher molar fractions of hydrogen and the higher H/C molar ratio of the fuels containing higher levels of iso-octane. This is shown in Figures 22 and 23. The availability of hydrogen atoms and the combustion efficiency directly influence the diffusion of fuel into the lubricating oil layer on the cylinder walls and the crevice volumes. These are the primary mechanisms of HC emissions in the engines (Heywood, 1988).

It is important to note that the higher molecular weight of iso-octane, compared to the other components, favors fuel absorption into the lubricating oil layer on the cylinder walls. Fuel solubility is a positive function of the molecular weight; therefore, the oil layer contributes to HC emissions depending on the solubility of different individual hydrocarbons in the lubricating oil. In fact, during compression the fuel vapor pressure increases and, according to Henry's law, absorption occurs even if the oil was saturated during intake. During combustion, the fuel vapor concentration in the burned gases goes to zero. Therefore, the absorbed fuel vapor will desorb from the liquid oil into the burned gases when the cylinder pressure decreases in the expansion stroke. By this time, complete combustion can no longer take place (AVL, 2011).

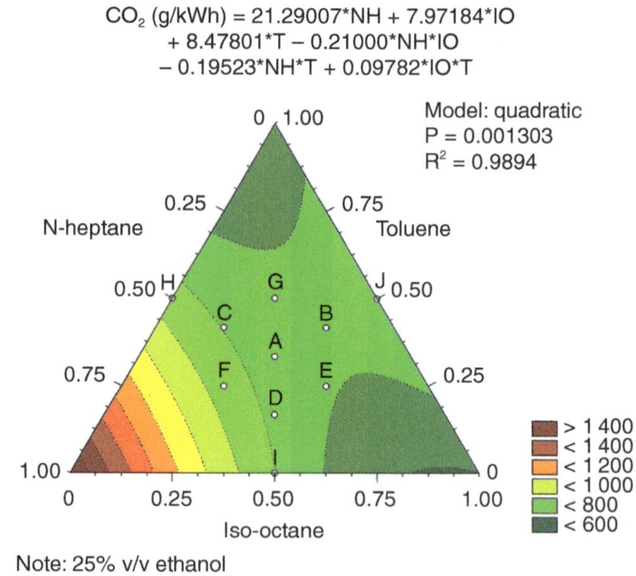

Figure 24

CO_2 specific emissions in g/kWh at 3 875 rpm with WOT.

Comments Regarding the Volumetric Emissions Results

In the present case, as the engine running points have been made in terms of % throttle position and not at a constant torque for instance, there is an interest in the volumetric emissions measurement. There are also engine applications where the volumetric emissions of pollutants are regulated. The analysis demonstrated that *n*-heptane was more effective in reducing the volumetric emissions. *N*-heptane can be an interesting component in cases where the volumetric emissions are regulated. In general, the molecular structure of the different fuels and the combustion efficiency presented a major impact on the trends observed for the volumetric emissions for all of the operating conditions. This behavior is due to the fact that all of the operating conditions, both those sensitive and those not sensitive to knocking, presented similar trends for volumetric emissions and showed similar combustion efficiency trends.

2.2.2 Specific Emissions in g/kWh

It is interesting to note that the specific emissions measured in g/kWh show different trends, due to the more significant impact of engine fuel consumption and engine power. This phenomenon will be shown in more detail in the next sections.

CO_2 Specific Emissions in g/kWh

For the operating conditions that are sensitive to knocking, the CO_2 specific emissions are exemplified with 3 875 rpm at WOT shown in Figure 24. Higher levels of *n*-heptane increased the CO_2 specific emissions. This phenomenon is explained by the reduced engine torque (and power) that occurs with increasing *n*-heptane levels (*Fig. 2*), because it was not possible to optimize spark timing at Maximum Brake Torque (MBT) in the presence of knocking (*Fig. 7*).

The tendency toward lower volumetric emissions, combined with the tendency toward higher torque using fuels with higher iso-octane levels than toluene (*Fig. 14, 2*), favors the iso-octane. This component presented lower levels of CO_2 specific emissions than toluene.

For the operating conditions that are not sensitive to knocking, the CO_2 specific emissions are exemplified with 2 250 rpm at partial load shown in Figure 25. The lower fuel consumption (*Fig. 9*) favors fuels with higher levels of *n*-heptane. This component contributed to the lowest levels of CO_2 specific emissions. For the CO_2 specific emissions results (*Fig. 25*), *n*-heptane is followed by iso-octane then toluene, maintaining the same trend observed for the volumetric emissions (*Fig. 15*) and the fuel consumption. This result occurred because for these operating conditions, there were not significant differences in the engine torque (and power) delivered with the different fuels.

CO Specific Emissions in g/kWh

The trends for CO specific emissions were similar to those observed for CO_2. For the conditions that are sensitive to knocking, the CO specific emissions are

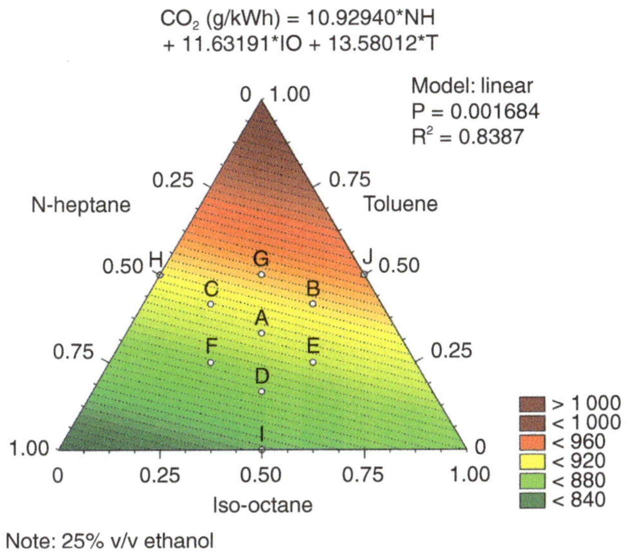

Figure 25

CO_2 specific emissions in g/kWh at 2 250 rpm with partial load.

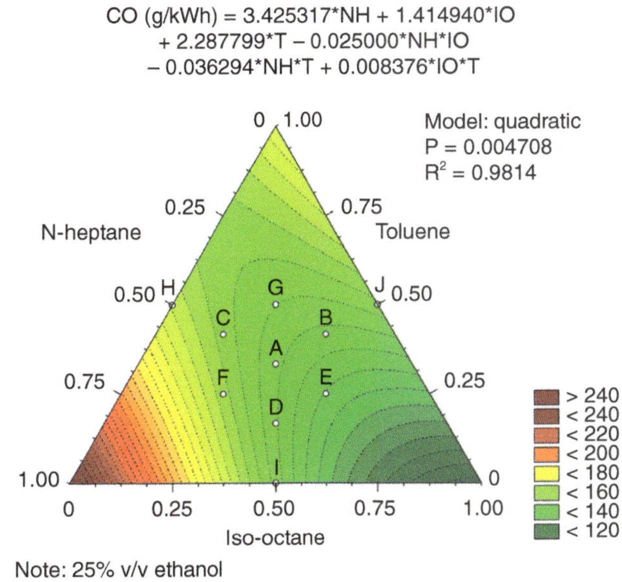

Figure 26

CO specific emissions in g/kWh at 3 875 rpm with WOT.

exemplified with 3 875 rpm at WOT shown in Figure 26. Higher *n*-heptane levels increased the emissions as a result of the lower engine torque (and power) delivered (*Fig. 2*). The trends for lower volumetric emissions (*Fig. 18*) and the trends for higher torques using fuels with higher iso-octane levels than toluene (*Fig. 2*) favor the iso-octane. This component presented lower levels of CO specific emissions than toluene.

For the operating conditions that are not sensitive to knocking, the CO specific emissions are exemplified with 2 250 rpm at partial load shown in Figure 27. The lower fuel consumption observed for higher levels of *n*-heptane (*Fig. 9*) favors the lower CO specific emissions. For the specific emissions results (*Fig. 27*), *n*-heptane is followed by iso-octane then toluene, maintaining the same trend observed for the volumetric emissions (*Fig. 19*) and the fuel consumption. This result occurred because for these operating conditions, there were not significant differences in the engine torque (and power) delivered with the different fuels.

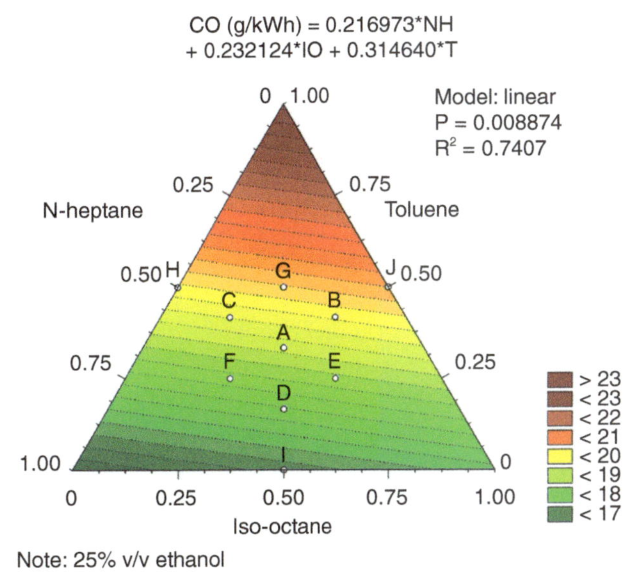

Figure 27

CO specific emissions in g/kWh at 2 250 rpm with partial load.

HC Specific Emissions in g/kWh

The trends observed for the HC volumetric emissions were maintained in g/kWh for the conditions evaluated. This is shown for the 5 500 rpm with WOT condition in Figure 28 compared to Figure 20. This behavior is explained by the larger differences observed for the volumetric emissions of HC among the different fuels tested, which had the primary impact on the specific emissions rather than the torque and fuel consumption differences. The molecular structure of the different fuels

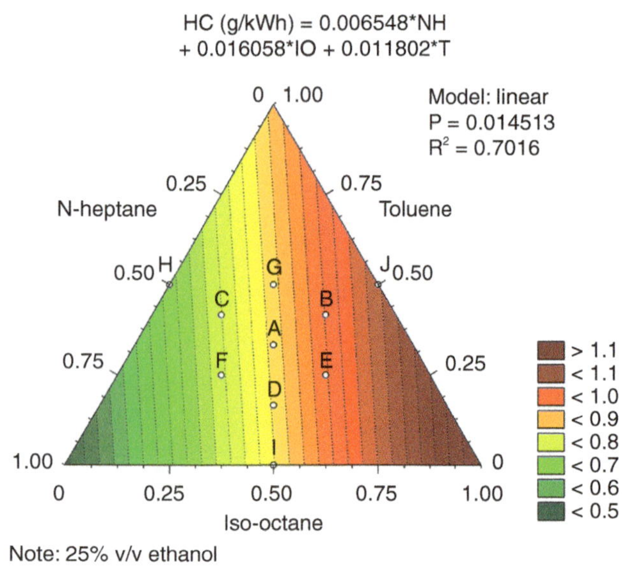

Figure 28

HC specific emissions in g/kWh at 5 500 rpm with WOT.

and the combustion efficiency had a major impact on HC emissions.

CONCLUSIONS

The analyses presented in this work show that surrogate fuels can be utilized to investigate fuel components and study their influences on fuel properties and engine combustion, performance and emissions.

Based on the results obtained, *n*-heptane can be an interesting component for fuels destined for special applications that are not sensitive to knocking. This component improves the engine combustion efficiency and overall engine efficiency, reducing CO and HC emissions without significantly increasing the levels of CO_2 emissions. The amount of this component in the fuel will depend on the engine type and its octane rating requirement. Engines with a low compression ratio and stationary engines operating in conditions that are not sensitive to knocking can be favorably affected by the use of higher amounts of *n*-heptane. Applications with high engine speeds, where there is no time available for knocking to occur, such as motorcycle and racing engines, could use higher amounts of *n*-heptane in the fuel.

For applications with a high octane rating requirement, such as high compression ratio engines, toluene and iso-octane are important components for improving engine performance and reducing CO_2 and CO specific emissions. Iso-octane can be more useful as it contributes to higher engine combustion efficiency, higher overall engine efficiency and lower levels of CO_2 and CO emissions than toluene. However, toluene presented lower HC emissions than iso-octane.

Iso-octane is an interesting and versatile component. It increases the octane rating and the stoichiometric air/fuel ratio, and helps to increase engine efficiency for all operating conditions independent of knocking susceptibility. It also contributes to the reduction of CO_2 and CO specific emissions, despite the increase observed in HC emissions. The iso-octane branched chain favors the good compromise observed for this component.

It is clear that there is a compromise among several engine combustion, performance and emission parameters. It is important to verify the relative importance of each parameter as a function of the engine type and its application to select a more adequate and efficient fuel.

In general, the trends and conclusions observed for the different components utilized could be extended for the chemical group they represent (*n*-paraffins, iso-paraffins and aromatics). This fact can help to select the fuel components and refinery processing streams to be used in fuel formulation for different applications. *N*-paraffins are common in naphthas from atmospheric distillation. Iso-paraffins are present at high concentrations in the products from the processes of alkylation and isomerization, while the aromatics are present in high concentrations in the products of catalytic reforming. This prior selection of components and processing streams can reduce the number of experimental tests, and reduce cost and time involved in the fuel development processes.

The approach introduced in this paper can also be used to investigate the influence of other fuel components, including renewable fuels, and refinery processing streams on fuel properties and engine performance. The results presented put together information and methodologies that can be used in fuel development processes for different applications.

REFERENCES

Andrae J.C.G. (2008) Development of a detailed kinetic model for gasoline surrogate fuels, *Fuel* **87**, 10-11, 2013-2022.

ANP (2011) Brazilian National Petroleum, Natural Gas and Biofuels Agency, *ANP Resolution # 57 – Regulation for Brazilian automotive gasoline specification*, ANP, Brazil.

AVL (2011) *BOOST software, version 2011 – Theory*, AVL, Graz, Austria.

Berta P., Aggarwal S.K., Puri I.K. (2006) An experimental and numerical investigation of *n*-heptane/air counterflow partially premixed flames and emission of NOx and PAH species, *Combust. Flame* **145**, 4, 740-764.

Bounaceur R., Costa I.D., Fournet R., Billaud F., Battin-Leclerc F. (2005) Experimental and modeling study of the oxidation of toluene, *Int. J. Chem. Kinet.* **37**, 1, 25-49.

Calado V.M.A., Montgomery D.C. (2003) *Design of Experiments using Statistica*, E-papers, Rio de Janeiro, Brazil.

CFR (2009) Code of Federal Regulations, *CFR Title 40 Part 91 - Control of emissions from marine spark-ignition engines, Subpart E - Gaseous exhaust test procedures*, CFR, USA.

Curran H.J., Gaffuri P., Pitz W.J., Westbrook C.K. (1998) A comprehensive modeling study of *n*-heptane oxidation, *Combust. Flame* **114**, 1-2, 149-177.

Curran H.J., Gaffuri P., Pitz W.J., Westbrook C.K. (2002) A comprehensive modeling study of iso-octane oxidation, *Combust. Flame* **129**, 253-280.

Dagaut P., Pengloan G., Ristori A. (2002) Oxidation, ignition and combustion of toluene: Experimental and detailed chemical kinetic modeling, *Phys. Chem. Chem. Phys.* **4**, 10, 1846-1854.

Davidson D.F., Gauthier B.M., Hanson R.K. (2005) Shock tube ignition measurements of iso-octane/air and toluene/air at high pressures, *Proc. Combust. Inst.* **30**, 1, 1175-1182.

Emdee J.L., Brezinsky K., Glassman I. (1992) A kinetic-model for the oxidation of toluene near 1200-K, *J. Phys. Chem.* **96**, 5, 2151-2161.

Gauthier B.M., Davidson D.F., Hanson R.K. (2004) Shock tube determination of ignition delay times in full blend and surrogate fuel mixtures, *Combust. Flame* **139**, 4, 300-311.

Herzler J., Jerig L., Roth P. (2005) Shock tube study of the ignition of lean *n*-heptane/air mixtures at intermediate temperatures and high pressures, *Proc. Combust. Inst.* **30**, 1, 1147-1153.

He X., Donovan M.T., Zigler B.T., Palmer T.R., Walton S.M., Wooldridge M.S., Atreya A. (2005) An experimental and modeling study of iso-octane ignition delay times under homogeneous charge compression ignition conditions, *Combust. Flame* **142**, 3, 266-275.

Heywood J.B. (1988) *Internal Combustion Engine Fundamentals*, McGraw Hill Inc., USA.

ISO (1992) International Organization for Standardization, *ISO 1585 - Road vehicles, engine test code, net power*, ISO, Geneva, Switzerland.

Machado G.B. (2012) Metodologias para Desenvolvimento de Combustíveis e Determinação da Velocidade de Propagação de Chama em Motores de Ignição por Centelha, *PhD Thesis*, Pontifícia Universidade Católica do Rio de Janeiro – PUC-Rio, Rio de Janeiro, Brazil.

Machado G.B., Barros J.E.M., Braga S.L., Braga C.V.M., Oliveira E.J., Silva A.H.M.F.T., Carvalho L.O. (2011) Investigations on surrogate fuels for high octane oxygenated gasolines, *Fuel* **90**, 2, 640-646.

Machado G.B., Barros J.E.M., Braga S.L., Braga C.V.M., Oliveira E.J., Silva A.H.M.F.T., Azevedo E.C.Z. (2012) Methodologies for fuel development using surrogate fuels on spark ignition engines, *SAE Technical Paper* 2012-36-0477.

Melo T.C.C., Machado G.B., Belchior C.R.P., Colaço M.J., Barros J.E.M., Oliveira E.J., Oliveira D.G. (2012) Hydrous ethanol-gasoline blends – combustion and emission investigation on a flex-fuel engine, *Fuel* **97**, 796-804.

Montgomery D.C., Runger G.C. (1994) *Applied Statistics and Probability for Engineers*, John Wiley & Sons, New York, USA.

Pitz W.J., Seiser R., Bozzelli J.W., Seshadri K., Chen C.J., Costa I.D., Fournet R., Billaud F., Battin-Leclerc F., Westbrook C.K. (2003) *Chemical Kinetic Study of Toluene Oxidation under Premixed and Nonpremixed Conditions*, UCRL-CONF-201575, Lawrence Livermore National Laboratory, California, USA.

Pitz W.J., Cernansky N.P., Dryer F.L., Egolfopoulos F.N., Farrell J.T., Friend D.G., Pitsch H. (2007) Development of an experimental database and chemical kinetic models for surrogate gasoline fuels, *SAE Technical Paper* 2007-01-0175.

Silke E.J., Curran H.J., Simmie J.M. (2005) The influence of fuel structure on combustion as demonstrated by the isomers of heptane: a rapid compression machine study, *Proc. Combust. Inst.* **30**, 2, 2639-2647.

Sivaramakrishnan R., Tranter R.S., Brezinsky K. (2005) A high pressure model for the oxidation of toluene, *Proc. Combust. Inst.* **30**, 1, 1165-1173.

9

A Novel Model of Foam Flooding Considering Multi-Factors for Enhancing Oil Recovery

J. Wang[1,2*], H. Liu[1], H. Zhang[1], G. Zhang[3], P. Liu[4] and K. Sepehrnoori[2]

[1] MOE Key Laboratory of Petroleum Engineering in China University of Petroleum, Beijing 102249 - China
[2] Department of Petroleum & Geosystems Engineering, The University of Texas at Austin, Austin 78712 - USA
[3] CNOOC Tianjin Branch Company, Tianjin 300452 - China
[4] Sinopec Shengli Oilfield Branch Company, Shandong Dongying 257000 - China
e-mail: wangjing8510@163.com

* Corresponding author

Abstract — *Foam flooding is a promising technique for achieving mobility control and diverting fluid into low-permeability strata in post-water-flooding reservoirs. However, foam flow is very complicated and is influenced by many factors which have not been studied and explored very rigorously (*i.e. *permeability, surfactant concentration, foam quality, reservoir temperature, oil saturation, water saturation and seepage velocity). Based on core flooding experiments of foam flowing and blocking rules using two kinds of foaming agents, a novel model of foam flooding considering the influences of the above factors is established and solved using a reservoir simulator which is formulated using the IMPES method in conjunction with a Runge-Kutta method. Then, the validation is performed by core flooding experiments in both the absence and presence of oil. Finally, the simulator is used to investigate the effects of the permeability max-min ratio, ratio of vertical to horizontal permeability, gas-liquid ratio, depositional sequence, foaming agent concentration, and reservoir temperature.*

Résumé — Un nouveau modèle d'injection de mousse considérant de multiples facteurs afin d'améliorer la récupération de pétrole — L'injection de mousse est une technique prometteuse pour atteindre un contrôle de la mobilité et détourner le fluide dans des couches à faible perméabilité de réservoirs après injection d'eau. Toutefois, l'écoulement de mousse est très compliqué et est influencé par de nombreux facteurs qui n'ont pas été étudiés et explorés de manière très rigoureuse (à savoir la perméabilité, la concentration en tensioactif, la qualité de la mousse, la température du réservoir, la saturation en pétrole, la saturation en eau et la vitesse d'écoulement). Sur la base d'expériences d'injection de mousse dans des carottes et de règles de blocage en utilisant deux types d'agents moussants, un nouveau modèle d'injection de mousse qui tient compte des influences des facteurs ci-dessus est établi et résolu, en utilisant un simulateur de réservoir basé sur la méthode IMPES combinée à la méthode Runge-Kutta. Ensuite, la validation est réalisée par des expériences d'injection dans des carottes à la fois en l'absence et en présence de pétrole. Enfin, le simulateur est utilisé pour étudier les effets du rapport max/min de perméabilité, du rapport de perméabilité verticale/horizontale, du rapport gaz/liquide, des séquences de dépôt, de la concentration de l'agent moussant et de la température du réservoir.

INTRODUCTION

Foam flooding is an EOR (Enhance Oil Recovery) technique taking foam as the displacing fluid. Foam not only effectively decreases the mobility ratio of gas to oil but also weakens the phenomena of gas channeling and steam override [1, 2]. Moreover, due to the particular mechanisms of foam generation and collapse, foam presents the selective blocking features of "big block and small not block" and "water block and oil not block" when it flows in porous media [3], which is very favorable for enhancing oil recovery. In the past few years, foam flooding has been carried out in many pilots in China, and most of them have shown a good performance [4].

In a study of foam triggering mechanisms, Kovscek *et al.* [5] attributed triggering mechanisms of foam to increasing the viscosity of the driving phase; Rossen *et al.* [6] attributed the triggering mechanism to declining gas phase mobility; Kovscek and Bertin [7] reported that both increasing gas phase viscosity and decreasing the gas phase relative permeability take effect simultaneously, and hence they improved the models for foam viscosity and the gas phase relative permeability. However, it is controversial that relative permeability and viscosity are considered separately [8]. As can be seen from these studies, it is recognized that the triggering mechanism of a foam system is decreasing the gas mobility, but many different approaches, *e.g.* modifying the gas phase viscosity or gas phase relative permeability or modifying both the gas phase viscosity and relative permeability are considered to achieve this mechanism. Nevertheless, foam flow in porous media is a very complicated process, since the lamellae in the flowing state will increase the viscosity of the driving phase and the lamellae adsorbed on the pore surface will reduce gas relative permeability. Unfortunately, because of the surrounding environment, especially the properties of the rock and fluid, both mechanisms are difficult to distinguish and test by experiments. Therefore, the above approaches for representing foam mechanisms are partial and difficult to quantify. In this paper, the mobility of the gas phase is considered as a whole to test and be measured by the resistance factor. This method includes increasing the viscosity of the displacing phase as well as decreasing the gas relative permeability, avoids repeated tests, and takes into account the effects of these two mechanisms. Numerical simulation is a technique to study the flow characteristics described by mathematical models [9]. It has features of low cost, high efficiency and good repeatability, and can demonstrate the entire complex flow process. Because foam is a very complex system where gas is the dispersed phase and liquid is the continuous phase, accurate prediction is difficult. There are several approaches to deal with its complexity. The first approach is called "population balance", which quantifies the relation between foam mobility and foam texture and all the mechanisms of creation and destruction of liquid films, or lamellae [10-13]. It uses a differential equation to represent foam texture *in situ*, but the basic concept is consistent with the local-equilibrium version, in which foam texture is an algebraic function of local conditions [6]. In all population balance models, the rate of creation of lamellae depends on gas velocity through pore throats where snap-off takes place [10-13]. The second approach represents foam mobility as an empirical function of many influence factors [14-17]. The third approach is the "fixed-P_c* model", which relies on the relation between capillary pressure, foam texture and foam mobility [18]. In essence, it is also a local-equilibrium version of the population balance for strong foams under conditions where capillary pressure dominates foam texture and gas mobility [6]. The first approach is more complex, and there is no way to separate the effects of lamella generation and destruction in history-matching coreflood results [6]; the second approach is simple, but it disregards the relation between foam mobility and texture; the third approach is between the other two, but it disregards the foam strength influenced by many key factors. Based on the advantages and disadvantages of the above three models, a synthesized multi-factor foam model is presented. In the novel model, the foam generation depends on the critical gas velocity, which is a function of the aqueous phase mobility fraction [12]. The foam collapse depends on limiting capillary pressure, at which foam collapses abruptly [19]. Since S_w is related to P_c through the capillary-pressure function $P_c(S_w)$, this means that foam remains at a given water saturation $S_w^* \equiv S_w(P_c^*)$. Foam strength, which is influenced by permeability, surfactant concentration, foam quality, reservoir temperature and oil saturation, affects the foam mobility, which is evaluated by the foam resistance factor. Therefore, a foam resistance factor model derived from our coreflood experiments using two kinds of foam agents is used to represent foam strength features [20]. The advantages of this approach reflect the competing goals of simplicity and completeness; however, the model is not able to describe slow generation or destruction processes thoroughly, considering the influence of flow velocity on foam mobility.

Finally, the model validation was performed by core flooding experiments; the factors which affect the EOR effects of foam flooding were investigated using the novel model: the permeability max-min ratio, the ratio of vertical to horizontal permeability, gas-water ratio, foaming agent concentration, temperature and depositional sequence.

1 GOVERNING EQUATION

1.1 Mass Conservation Equations

The assumptions imposed on the flow equations include:

- four components: oil, water, gas and surfactant;
- gas, water and surfactant generate foam which reduces the gas mobility;
- the whole process is isothermal, and the energy transfer is neglected;
- fluid flow obeys the generalized Darcy's law;
- phase equilibrium is reached instantaneously;
- foam only affects mobility (the ratio of gas relative permeability to viscosity).

Mass balance equation for the oil phase:

$$\nabla\cdot\left(\frac{\vec{k}k_{\mathrm{ro}}}{\mu_{\mathrm{o}}B_{\mathrm{o}}}\nabla\Phi_{\mathrm{o}}\right)+q_{\mathrm{o}}=\frac{\partial}{\partial t}\left(\frac{\phi S_{\mathrm{o}}}{B_{\mathrm{o}}}\right) \tag{1}$$

Mass balance equation for the aqueous phase:

$$\nabla\cdot\left(\frac{\vec{k}k_{\mathrm{rw}}}{\mu_{\mathrm{w}}B_{\mathrm{w}}}\nabla\Phi_{w}\right)+q_{\mathrm{w}}=\frac{\partial}{\partial t}\left(\frac{\phi S_{\mathrm{w}}}{B_{\mathrm{w}}}\right) \tag{2}$$

Mass balance equation for the gas phase:

$$\nabla\cdot\left(\frac{\vec{k}k_{\mathrm{rg}}}{\mu_{\mathrm{g}}B_{\mathrm{g}}R_{\mathrm{g}}}\nabla\Phi_{\mathrm{g}}\right)+q_{\mathrm{g}}=\frac{\partial}{\partial t}\left(\frac{\phi S_{\mathrm{g}}}{B_{\mathrm{g}}}\right) \tag{3}$$

Mass balance equation for the surfactant (foaming agent):

$$-\nabla\cdot(c_{\mathrm{s}}\vec{v}_{\mathrm{w}})+\nabla\cdot\left(\vec{d}_{\mathrm{s}}\phi S_{\mathrm{w}}\nabla c_{\mathrm{s}}\right)+q_{\mathrm{w}}c_{\mathrm{s}}=\frac{\partial(\phi S_{\mathrm{w}}c_{\mathrm{s}})}{\partial t}+\frac{\partial[\rho_{r}(1-\phi)\hat{c}_{\mathrm{s}}]}{\partial t} \tag{4}$$

In which k is the absolute permeability, μm^2; k_{rl} is the relative permeability of phase l; μ_{l} is the effective viscosity of phase l, mPa·s; B_{l} is the formation volume factor of phase l, m^3/m^3; ∇ is the divergence operator; q_{l} is the source and sink for phase l, m^3/(day·m^3); R_{g} is the gas resistance factor; Φ_{l} is the potential of phase l, MPa; S_{l} is the saturation of phase l; φ is porosity; c_{s} is the surfactant concentration, kg/m^3; d_{s} is the diffusion coefficient of the surfactant, m^2/s; ρ_{r} is the rock density, kg/m^3; $\hat{c}_s$ is the adsorption concentration of the surfactant, kg/kg.

1.2 Auxiliary Equations

Let S_{w}, S_{o} and S_{g} be water, oil and gas saturation, respectively. Then the saturation constraint equation is:

$$S_{\mathrm{o}}+S_{\mathrm{w}}+S_{\mathrm{g}}=1 \tag{5}$$

Also, p_{w}, p_{o} and p_{g} are water, oil and gas phase pressures, respectively; the capillary pressure equations are:

$$p_{\mathrm{cow}}=p_{\mathrm{o}}-p_{\mathrm{w}} \tag{6}$$

$$p_{\mathrm{cgo}}=p_{\mathrm{g}}-p_{\mathrm{o}} \tag{7}$$

In which p_{cow} is the capillary pressure between oil and water, Pa; and p_{cgo} is the capillary pressure between gas and oil, Pa.

1.3 Process Mechanism Models of Foam Flooding

1.3.1 Foam Resistance Factor

The effects of different factors on foam block characteristics are obtained by experiments using two kinds of surfactants. One is sodium dodecyl sulfate (#1) and another is an industrial foaming agent used in the Henan oilfield (#2). Then, a range of variograms and their correction models are applied to fit the relationship linking the foam resistance factor and other factors. The undetermined coefficients can be obtained using DATAFIT software [21] based on experimental results. Because of the consistency in physical meaning among variograms, the foam generation and collapse mechanisms, this method evades the defect from the conventional mathematical regression method that can merely be applied to a local scale but not to the extended area.

Resistance Factor *vs* Foaming Agent Concentration

A logistic model is a straightforward model used to forecast the population growth and presents an "S" tendency which can be divided into elementary, acceleration, transfer, deceleration and saturation periods. The adsorbance of active molecules on the gas/liquid interface has a similar trend, which determines the variation of foam blocking ability. As a result, the blocking capacity increases as the surfactant concentration increases, although there is a break in the curve at about 0.5% by weight. Based on experimental results (*Fig. 1*) and the Critical Micelle Concentration (CMC) introduced as a characteristic parameter, a logistic function was used to model the relationship between the resistance factor and foaming agent concentration:

$$R_{\mathrm{f}}(c_s)=\frac{R_f\big|_{c_s=CMC}}{1+a\cdot\exp(-r\cdot c_s)} \tag{8}$$

Using DATAFIT and the above model, the undetermined coefficients were obtained. For surfactant #1,

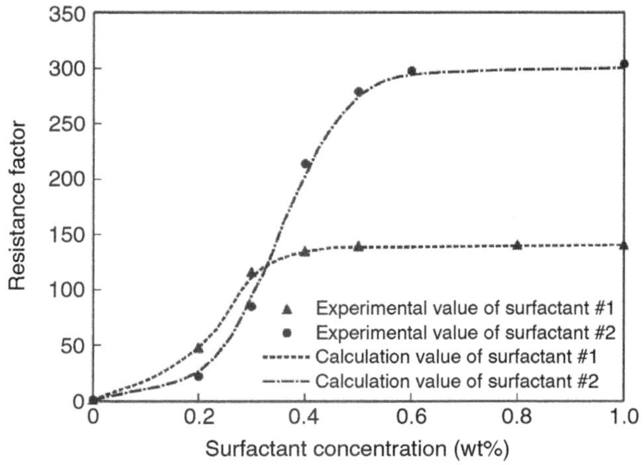

Figure 1

Comparison between experiments and the model of the resistance factor for different surfactant concentrations.

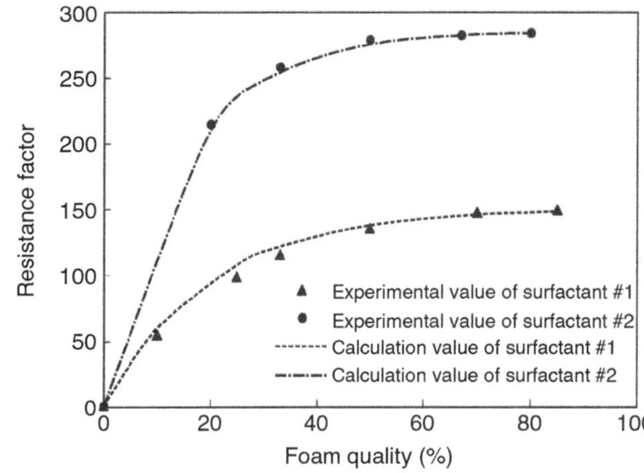

Figure 2

Comparison of experiments and the model of the resistance factor for different foam qualities.

$a_1 = 100$ and $r_1 = 19.5$; and for surfactant #2, $a_2 = 230$ and $r_2 = 16$.

Resistance Factor *vs* Foam Quality

Khatib *et al.* [19] showed that at a given gas flow rate, the foam resistance factor increases slightly with increasing foam quality ranging from 50% to 98%. De Vries and Wit [22] reported that at an imposed total flow rate, the foam resistance factor increases as foam quality increases until the break point; beyond that it decreases. Chang and Grigg [23] found that the foam resistance factor increases with increasing foam quality ranging from 33.3% to 80%. We found that the resistance factor first increases and then decreases as the foam quality increases, and the break point is at about 80%. However, it is challenging for the foam quality to reach above 80% in foam flooding pilots. Hence, the proportional region between the resistance factor and foam quality is of concern from the practical application viewpoint. Derived from the data of this region (*Fig. 2*), the variation resembles the exponential variogram model. After deliberating the endpoint feature, the relationship between the resistance factor and foam quality is:

$$R_f(\eta) = 1 + R_f|_{\eta=\eta_c} \cdot \left[1 - \exp\left(-\frac{3\eta}{b}\right)\right] \quad (9)$$

Using DATAFIT and the above model, the undetermined coefficients were obtained. For surfactant #1, $b_1 = 60$; and for surfactant #2, $b_2 = 45$.

Resistance Factor *vs* Permeability

One merit of the foam used for EOR is "big block and not small block", meaning foam has a better blocking ability in high-permeability regions or layers and the blocking ability nearly disappears in low-permeability regions or layers. This indicates shear thinning of the foam, which has been experimentally confirmed by many researchers [24-28]. In order to characterize the relation between the resistance factor and permeability (*Fig. 3*), dimensionless permeability (k_D) is used. The relation between the resistance factor and dimensionless permeability is:

$$R_f(k) = R|_{k=k_c} \cdot \left[1 - \frac{1 - k_D}{1 + m(k_D)^n}\right] \quad (10)$$

$$k_D = k/k_c \quad (11)$$

Using DATAFIT and the above model, the undetermined coefficients were obtained. For surfactant #1, $m_1 = 3$ and $n_1 = 2$; and for surfactant #2, $m_2 = 8.5$ and $n_2 = 1.05$.

Resistance Factor *vs* Temperature

Influencing mechanisms of temperature on foam resistance are diverse [29]. First of all, the dissolving capacity increases as the temperature increases, so the adsorbance of active molecules on the gas/liquid interface decreases,

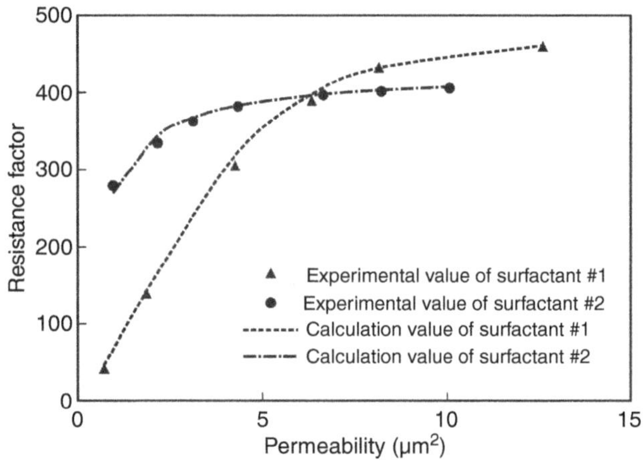

Figure 3

Comparison of experiments and the model of the resistance factor for different permeabilities.

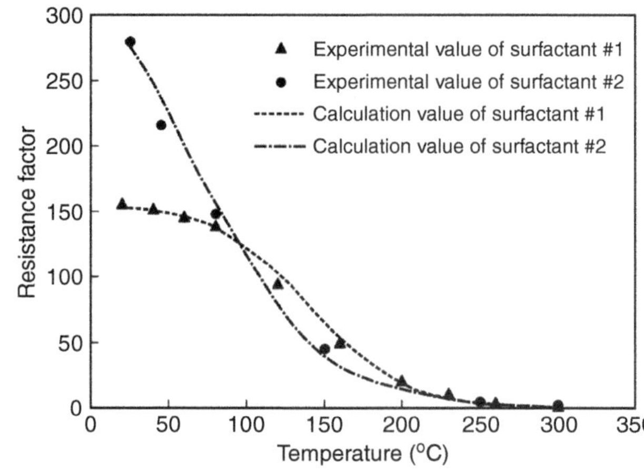

Figure 4

Comparison of experiments and the model of the resistance factor for different temperatures.

which results in a decrease in the strength of the lamellae. In addition, an increase in temperature will accelerate the drainage rate of liquid from lamellae to the liquid phase and promotes the collapse and coalescence of the bubbles. Moreover, an increase in temperature will also speed up the degradation of the foaming agent. Friedmann *et al.* [12] and Maini and Ma [30] reported that the foam resistance ability largely decreases as the temperature increases in porous media. Based on experimental results and the break point denoted as T_r (*Fig. 4*), the relation between temperature and foam quality is modeled as:

$$R_f(T) = \frac{R_f|_{T=T_r}}{1+\exp\left(\frac{T-\alpha}{\beta}\right)} \quad (12)$$

Using DATAFIT and the above model, the undetermined coefficients were obtained. For surfactant #1, $\alpha_1 = 140$ and $\beta_1 = 30$; for surfactant #2, $\alpha_2 = 70$ and $\beta_2 = 38$.

Resistance Factor *vs* Oil Saturation

The mechanisms by which oil affects the foam stability are diverse. Some studies have confirmed that the foam can be generated in the presence of oil by selecting appropriate foaming agents [31-34]. Some studies have stated that foam can be generated when the oil saturation is below a critical value [35-37], but some other studies have shown that foam is generated at relatively high oil saturation [34, 38, 39]. The characteristics of foaming

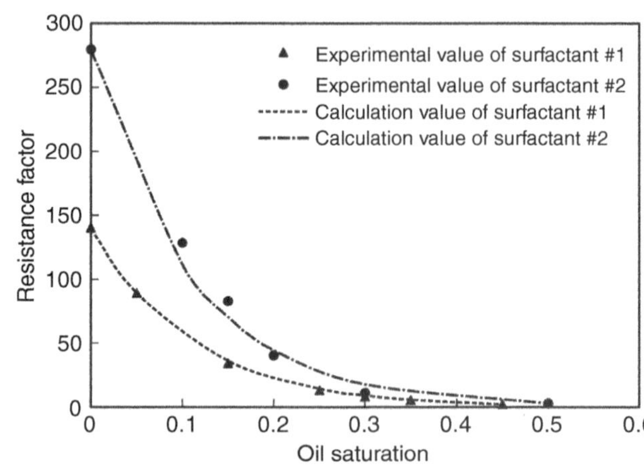

Figure 5

Comparison of experiments and the model of the resistance factor for different oil saturation ratios.

agents and oil samples used in the experiments may cause these differences. In general, it is commonly agreed that the presence of oil is detrimental to the foam stability [35, 36, 38, 40], and our experimental results also support this view. Therefore, oil saturation is more likely to be regarded as an influencing factor of the resistance factor than a critical condition of foam collapse. Based on experimental results (*Fig. 5*), the foam resistance factor exponentially decreases as the oil saturation increases, and different values of c will be achieved for different foaming agents and oil samples:

$$R_f(S_o) = R_f|_{S_o=0} \cdot \exp(-c \cdot S_o) \quad (13)$$

Using DATAFIT and the above model, the undetermined coefficients were obtained. For surfactant #1, $c_1 = 9.0$; for surfactant #2, $c_2 = 9.2$.

Seepage Velocity - The Critical Generation Condition of Foam

The generation mechanisms of foam in porous media include lamellae snap-off, lamellae lag and lamellae trapping. Many experiments indicated that the lamellae snap-off or lamellae flow will not occur until the pressure or velocity overpasses the critical value. In fact, it is a triggering mechanism. Once foam is in place, velocity can be reduced, but the foam remains. Fridmann *et al.* [12] found that the critical velocity increases as the liquid volume fraction decreases according to the following:

$$v_{gc} = \frac{1.52}{V_L^{-1.54}} \tag{14}$$

where v_{gc} is the critical velocity of foam generation, m/day, and V_L is the mobility fraction of the aqueous phase.

Water Saturation - The Critical Collapse Condition of Foam

Based on the DLVO theory [41, 42], the collapse of foam is related to critical capillary force. When the practical capillary force exceeds the critical value, foam will collapse. However, the capillary force is a function of water saturation; the higher the water saturation, the smaller the capillary force will be. Therefore, once the water saturation is lower than the critical value corresponding to the critical capillary force, the foam will collapse. Khatib *et al.* [19] found that the critical capillary force logarithmically decreases as the permeability and gas velocity increase, which also reflects the shear-thinning characteristics of the foam system. Based on the definition of the Leverett J-function and the empirical relation of the Leverett J-function [43, 44], we have:

$$J(S_w^*) = \frac{P_c^*}{\sigma_{gw}\cos\theta}\sqrt{\frac{k}{\phi}} = a_J\left(\frac{S_w^* - S_{wc}}{1 - S_{wc}}\right)^{b_J} \tag{15}$$

Khatib *et al.* [19] also found that the critical capillary pressure logarithmically decreases with increasing permeability (may reflect the effect of the pore structure) and gas velocity, which also reflects the shear-thinning characteristic of foam:

$$P_c^*(k, v_g) = -a_1 \ln(kv_g) + b_1 \tag{16}$$

Substituting Equation (16) into Equation (15), the limiting water saturation was obtained to be:

$$S_w^* = (1 - S_{wc})\left[\frac{-a_1\ln(kv_g) + b_1}{a_J\sigma_{gw}\cos\theta}\sqrt{\frac{k}{\phi}}\right]^{\frac{1}{b_J}} + S_{wc} \tag{17}$$

In Equations (15-17), $J(S_w{}^*)$ is the J function; $P_c{}^*$ is the critical capillary force, MPa; v_g is the gas velocity, cm/s; σ_{gw} is the gas-water interfacial tension, mN/m; θ is the wetting angle; a_1, b_1, a_J and b_J are undetermined coefficients; $S_w{}^*$ is the critical water saturation.

Multi-factor Gas Mobility Model

Based on a previous study [6], the resistance factor of the gas phase in porous media with the presence of foam is:

$$R_g = \begin{cases} 1 & S_W \le (S_W^* - \varepsilon) \text{ and } v < v_{gc} \\ 1 + \frac{(R_{kf}-1)(S_W - S_W^* + \varepsilon)}{2\varepsilon} & (S_W^* - \varepsilon) < S_W < (S_W^* + \varepsilon) \text{ and } v \ge v_{gc} \\ R_{kf} & S_W \ge (S_W^* + \varepsilon) \text{ and } v \ge v_{gc} \end{cases} \tag{18}$$

where ε is a special parameter; $\varepsilon = 0.001$ was used in this paper. It will not affect the simulation of the displacement effect. It is only used to avoid the jump of the calculated gas mobility due to the actual water saturation fluctuation above and below the critical value. The influencing factors of foam blocking ability in the process of foam flooding are numerous, and play a role simultaneously, hence it is very valuable to quantitatively characterize the foam resistance factor under multiple scenarios. Also, it is beneficial to the advancement of numerical simulation technology of foam flooding. Therefore, a novel foam resistance factor model that comprehensively considers the influences of the above factors is achieved:

$$R_{kf}(c_s, \eta, k, T, S_o) = A_r R_f|_{k=k_c} \frac{\left[1 - \frac{1-k_D}{1+m(k_D)^n}\right]\cdot\left[1 - \exp\left(-\frac{3\eta}{b}\right)\right]\cdot\exp(-c\cdot S_o)}{\left[1 + a\cdot\exp(-r\cdot c_s)\right]\cdot\left[1 + \exp\left(\frac{T-\alpha}{\beta}\right)\right]} \tag{19}$$

$$A_r = \frac{R_f|_{c_s=CMC}\cdot R_f|_{\eta=\eta_c}\cdot R_f|_{T=T_r}\cdot R_f|_{S_o=0}}{R_f|_{Test}{}^4} \tag{20}$$

The validation of the multi-factor gas mobility model can be verified by the experimental data, which is shown in Figure 6. The values of the undetermined coefficients are shown in Table 1.

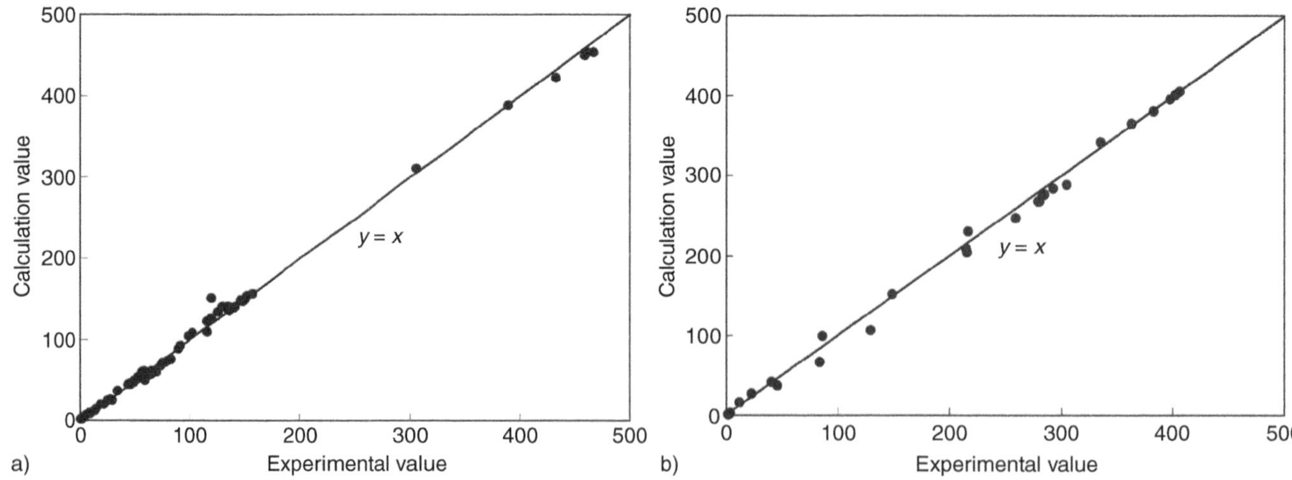

Figure 6
Validation of the multi-factor gas mobility model using experimental data. a) Surfactant #1, b) Surfactant #2.

Equation (19) not only comprehensively considers the influences of the rock properties, fluid properties and surroundings on foam block characteristics, but also avoids the discontinuous behavior of gas mobility caused by the critical foaming agent concentration and critical oil saturation, which will greatly improve the stability of the numerical simulation process.

Correction of the Resistance Factor for Weak Foam

It was reported that flow resistance is small and foam textures are coarse, consistent with the injection of unfoamed gas or weak foam in the inlet region, and there exists a strong foam piston-like front moving through the porous media [6, 44]. This phenomenon has also been verified by many experiments [45, 46]. However, most of the existing foam flow models cannot simulate this process [6]. In this model, the above resistance factor model is based on strong foam, so the following correction function is introduced to represent this effect:

$$R_f(L) = \frac{R_{kf}^0}{1 + AA \cdot \exp(-RR \cdot L)} \tag{21}$$

where R_{kf}^0 is the resistance factor of strong foam; L is the flow distance of weak foam or unfoamed gas and foaming agent solution, m; AA and RR are model coefficients.

1.3.2 Adsorption Model for the Surfactant

The adsorption of the foaming agent will occur when it flows in the porous media and obeys the typical Langmuir law:

$$\hat{c}_s = \hat{c}_{smax} \frac{b_s c_s}{1 + b_s c_s} \tag{22}$$

where b_s is the Langmuir parameter, $(kg/m^3)^{-1}$; $\hat{c}_{smax}$ is the maximum adsorbed concentration, kg/kg.

1.3.3 Capillary Desaturation and Interfacial Tension Model

The foaming agent can also reduce the oil-water interfacial tension by its surfactant characteristics and decrease the residual oil saturation [47]:

$$S_{or} = S_{or\,min} + \frac{S_{or\,max} - S_{or\,min}}{1 + A_N N_{co}} \tag{23}$$

$$N_{co} = \frac{\mu_w v_w}{\sigma_{wo}} \tag{24}$$

$$\sigma_{wo} = a_\sigma \times 10^{-b_\sigma c_s} \tag{25}$$

where N_{co} is the capillary number; σ_{ow} is the oil-water interfacial tension, mN/m; $S_{or\ min}$ is the limit of residual oil saturation under a high capillary number; $S_{or\ max}$ is the residual oil saturation under a low capillary number; a_σ, b_σ and A_N are undetermined coefficients.

TABLE 1
Values of undetermined coefficients

Surfactant #1													
a	r	b	m	n	α	β	c	$R_f\|_{Test}$	$R_f\|_{c=CMC}$	$R_f\|_{\eta=\eta 0}$	$R_f\|_{T=Tr}$	$R_f\|_{So=0}$	$R_f\|_{k=kc}$
100	19.5	60	3	2	140	30	9.0	139	139	156	150	139	430
Surfactant #2													
a	r	b	m	n	α	β	c	$R_f\|_{Test}$	$R_f\|_{c=CMC}$	$R_f\|_{\eta=\eta 0}$	$R_f\|_{T=Tr}$	$R_f\|_{So=0}$	$R_f\|_{k=kc}$
230	16	45	8.5	1.05	70	38	9.2	279	300	285	360	279	390

1.3.4 Relative Permeability Model

The Stone II model is used to calculate the three-phase relative permeability [48]:

$$k_{rw} = k_{rwro}\left[\frac{S_w - S_{wc}}{1 - S_{wc} - S_{orw}}\right]^{n_w} \tag{26}$$

$$k_{row} = k_{rocw}\left[\frac{1 - S_w - S_{orw}}{1 - S_{wc} - S_{orw}}\right]^{n_{ow}} \tag{27}$$

$$k_{rog} = k_{rocw}\left[\frac{1 - S_{wc} - S_{org} - S_g}{1 - S_{wc} - S_{org}}\right]^{n_{og}} \tag{28}$$

$$k_{rg} = k_{rgro}\left[\frac{S_g - S_{gc}}{1 - S_{wc} - S_{org} - S_{gc}}\right]^{n_g} \tag{29}$$

$$k_{ro} = f_3(S_w, S_g) = k_{rocw}\left[\left(\frac{k_{row}}{k_{rocw}} + k_{rw}\right) \times \left(\frac{k_{rog}}{k_{rocw}} + k_{rg}\right) - k_{rw} - k_{rg}\right] \tag{30}$$

where k_{rocw} is the oil phase relative permeability corresponding to connate water; k_{rw} and k_{rg} are the water and gas phase relative permeability, respectively; k_{row} and k_{rog} are the oil phase relative permeability in the oil-water system and oil-gas system, respectively; n_w, n_{ow}, n_w, n_{og} and n_g are input parameters; S_{wc}, S_{orw}, S_{gc} and S_{org} are the end-points in oil-water and oil-gas systems' relative permeability functions.

1.3.5 Physical Properties of the Gas Phase [49]

The formation volume factor is:

$$B_g = 3.458 \times 10^{-4} \times Z \times \frac{273 + t}{p_g} \tag{31}$$

The expansion factor is:

$$E_g = \frac{1}{B_g} = 2\,891.7 \times \frac{p_g}{Z(273 + t)} \tag{32}$$

where Z is the compressibility factor; t is the reservoir temperature, °C.

1.3.6 Capillary Pressure Model

Capillary pressure is a function of saturation:

$$p_{cow}(S_w) = A_1 \cdot \sqrt{\frac{\phi}{k}} \cdot \left(\frac{S_w - S_{wc}}{1 - S_{wc} - S_{or}}\right)^{B_1} \tag{33}$$

$$p_{cgo}(S_g) = A_2 \cdot \sqrt{\frac{\phi}{k}} \cdot \left(\frac{S_g - S_{gc}}{1 - S_{gc} - S_{or}}\right)^{B_2} \tag{34}$$

where A_1 and A_2 are input parameters; B_1 and B_2 are capillary pressure exponents.

2 SOLUTION AND VALIDATION OF THE MATHEMATICAL MODEL

2.1 Solution Method

In this model, gas flow velocity is used as the triggering mechanism; water saturation obtained based on DLVO theory is used for the bubble collapse mechanism. If foam exists in the porous media, the foaming agent concentration, foam quality, permeability, temperature and oil saturation come into play for calculating the resistance factor in the grid. This method avoids solving the foam population balance equation. A method for solving this mathematical model is as follows:

- the IMPES (Implicite Pressure Explicite Saturation) method is used to solve the pressure and saturation equations to obtain the pressure and saturation distributions for each phase;
- the classical four-order Runge-Kutta method is applied to solve the concentration equation of the foaming agent;
- the local gas flow velocity, critical gas flow velocity and critical water saturation based on the grid pressure and saturation are solved;

TABLE 2
Parameter values for simulations of Kovscek's experiment

a	r	b	m	n	α	β	A_r
100	19.5	60	8	1.1	140	30	2.0
$R_f\|_{k=kc}$	a_J	b_J	a_1	b_1	AA	RR	
1 200	1.3	−0.9	0.1	1.5	80	40	

- the critical generation condition and critical collapse condition are used to determine whether there is foam in the grid. If foam exists in the grid, using permeability, the foaming agent concentration, gas-liquid ratio, temperature and oil saturation are used to calculate the resistance factor; otherwise, the resistance factor is set to one;
- then the resistance factor will be substituted into the gas flow equation;
- repeat steps (1-5) until the end of the simulation time.

2.2 Validation of the Novel Model

In the Absence of Oil

The model includes all mechanisms that are:
- crucial to the process;
- known with reasonable accuracy.

With the purpose of validating the model, the foam flow data obtained from Kovscek [44] is used. Core and fluid data are based on reference [44], and other model parameters are listed in Table 2. The fitting result of the multi-factor model shown in Figure 7 is a slightly better fit than the population balance model although the curve for 0.23 PV is not a close fit due to the lack of strong foam and gas channeling.

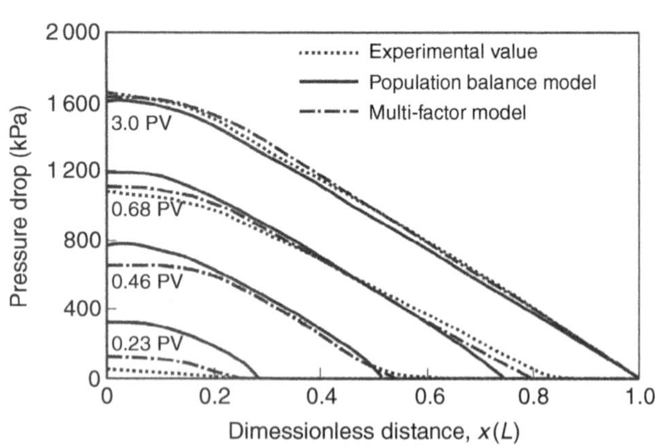

Figure 7

Comparison between the computed value using the multi-factor model, Kovscek's coreflood data, and the computed value using a population balance model

In the Presence of Oil

Because the impact of oil presence is considered, one group of foam flooding experimental data is also used. In this experiment, foaming agent solution (SDS: surfactant #1) with the concentration of 0.4 wt% was injected continuously into linear unconsolidated media of length 0.3 m at a rate of 2 mL/min, while gas was also injected continuously to give a foam quality of 80% at the sandpack entrance. The permeability is 4.5 μm^2 and porosity is 0.33. The initial oil saturation is 0.4 and oil viscosity is about 2.0 mPa·s. At 120 min, continued water flooding was performed at a rate of 5mL/min. Other model parameters are listed in Table 3. The comparison between computed and experimental data is shown in Figure 8 and the agreement is excellent.

3 RESULTS AND DISCUSSION

In order to study thoroughly the differences between the foam flooding with the gas-water flooding and the influences of many parameters on the EOR effect, a synthetic geological model is built. The reservoir scale is 10 × 10 × 2 with the dimensions of dx = dy = 10 m and dz = 5 m, as shown in Figure 9. The time step is 1 d. The reservoir temperature is 30°C and pressure is 20 MPa. The viscosity of oil is 20 mPa·s and initial oil saturation is 0.8. The porosity is 0.3, and the permeability is 1 μm^2 and 4 μm^2 for the top and bottom layers, respectively. The ratio of vertical permeability to horizontal permeability is 0.1. The production rate is 20 m^3/day, with an injection-production ratio of 1:1. The foam is injected when the water-cut reaches 90% (t = 2 000 d) and the slug size is 0.4 PV. The foaming agent concentration is 0.5 wt% and the injection gas-water ratio is 1:1 in formation conditions. The gas viscosity is 0.03 mPa·s and the water viscosity is 1.0 mPa·s. The difference between gas-water flooding and foam flooding is that the foaming agent concentration equals zero or not. Other model parameters are listed in Table 4.

TABLE 3

Parameter values for simulations of the foam flooding experiment

a	r	b	m	n	α	β	c
100	19.5	60	3	2	140	30	9.0
Ar	$R_f\|_{k=kc}$	a_J	b_J	a_l	b_l	AA	RR
2.5	430	1.3	−0.9	0.1	1.5	80	35

TABLE 4

Parameter values for simulations of foam flooding in a heterogeneous reservoir

a	r	b	m	n	α	β	c
100	19.5	60	3	2	140	30	9.0
A_r	$R_f\|_{k=kc}$	a_J	b_J	a_l	b_l	AA	RR
2.0	500	1.3	−0.9	0.1	1.5	80	35

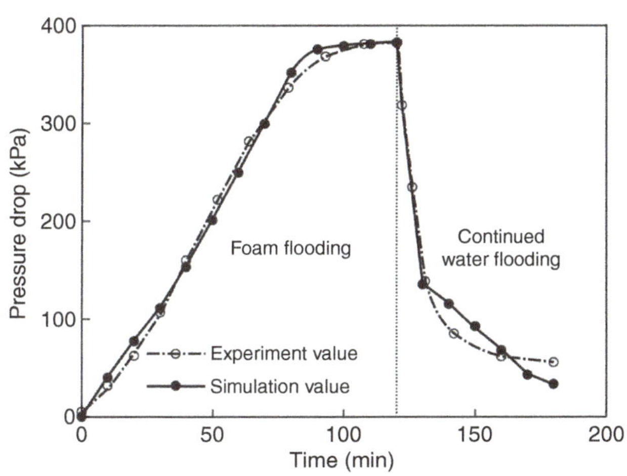

Figure 8

Comparison between computed and experimental pressure drop for foam flooding in oil-bearing conditions.

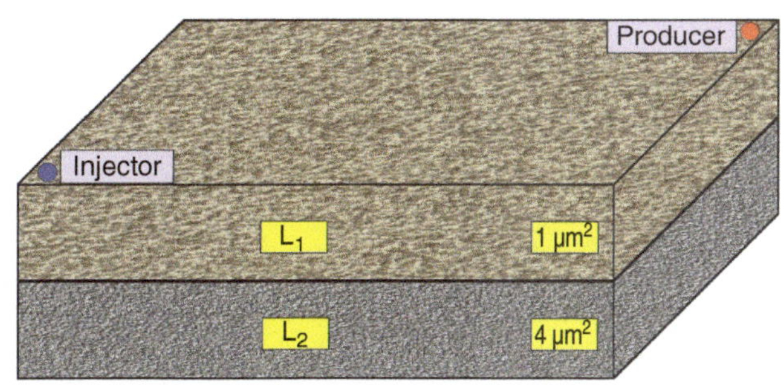

Figure 9

Geological model with vertical heterogeneity.

3.1 Comparing Foam Flooding with Water + Gas Flooding

Gas has a much smaller viscosity and lower density; therefore, if gas is used as the displacing fluid, gas channeling and override will occur. The major advantage of foam flooding is that the lamella of the foam can cut off the gas phase, effectively decrease the mobility ratio of gas to oil, and weaken the phenomena of gas channeling and override. Figure 10 shows the distribution of gas saturation at different times by foam flooding and water + gas flooding (the foam slug was injected between 2 000 d and 2 600 d). It is clear that, when t = 2 500 d, the gas phase area of L_1 is larger than L_2 for both water + gas flooding and foam flooding; the gas saturation of water + gas flooding is lower than foam flooding, but the gas phase area is larger for both L_1 and L_2; the amount of gas in L_1 is more than L_2 for gas-water flooding, which is the reverse for foam flooding. One explanation is that in transgressive depositional sequence reservoirs [50], the fractional flow of bottom strata is larger; but when the gas is injected into the reservoir, the gas is inclined to enter the top strata because of gravitational differentiation. At the same time, the gas viscosity is much lower than both water and oil; thus, the gas flows fast in both lateral and vertical directions. However, if the gas exists in the form of foam, the gas mobility will greatly decrease. Thus, the gas flows slowly and the gas saturation is larger for foam flooding. With lower permeability and

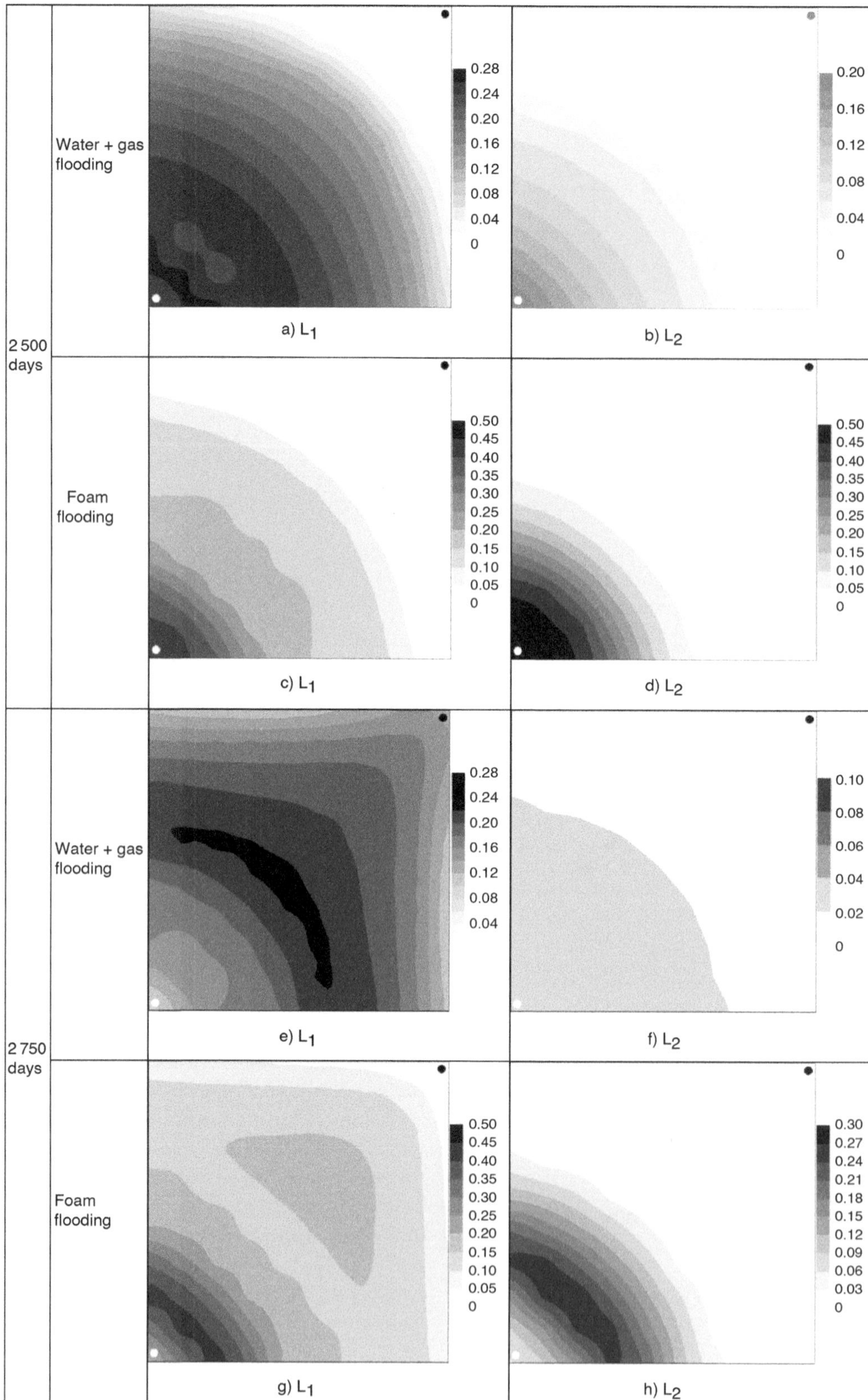

Figure 10

Distribution of gas saturation at different times by foam flooding and water + gas flooding.

higher oil saturation in L_1, the resistance factor is lower, which results in the gas phase area being larger in L_1. When t = 2 750 d, the gas has already channeled to the producer along the L_1, and the gas saturation is close to zero in the L_2 for water + gas flooding. Conversely, gas channeling does not appear and the gas saturation is still about 0.3 and exists in the form of foam in L_2 for foam flooding. The foam located in L_2 makes water divert into the lower-permeability strata (L_1) and displaces more additional remaining oil than water + gas flooding.

The development effects of different methods are shown in Figure 11. It is clear that the ultimate recovery order is foam flooding (48.8%) > water + gas flooding (46.7%) > water flooding (43.7%) and the flood-response time of water + gas flooding is earlier and shorter than foam flooding because of the densities of water, oil and gas. Due to gravitational differentiation, both top and bottom layers are well swept by gas and water, so the ultimate recovery of water + gas flooding is larger than water flooding. In the foam flooding process, some of the unfoamed gas flows in the top layer to sweep the remaining oil, and some of the gas in the form of foam flows in the bottom layer. As previously analyzed, the blocking ability of foam in the high-permeability layer is stronger than in the low-permeability layer. As a result, the gas flows in the high-permeability layer the same as in the low-permeability layer, and more water enters the lower-permeability layer and consequently displaces more oil. Hence, foam can both effectively inhibit the gas override and establish a large flow resistance to realize the fluid diversion, which is favorable for enhancing oil recovery in heterogeneous reservoirs.

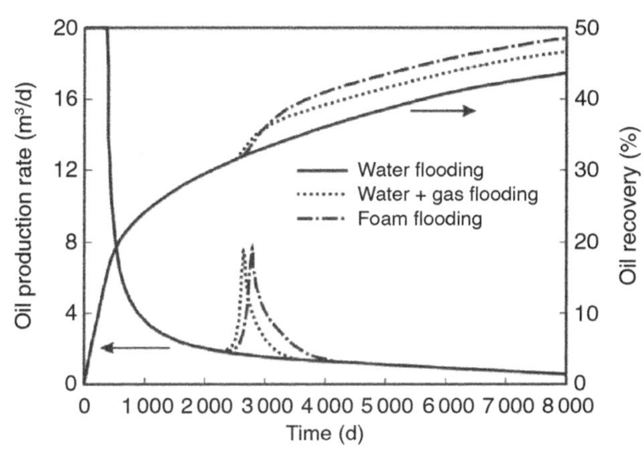

Figure 11

Development effects of different methods.

3.2 Sensitivity Analysis of Foam Flooding Enhancing Oil Recovery

In a lab study, foam flooding can enhance oil recovery by more than 20% in suitable conditions. Actually, most of the fields in China have carried out foam flooding and obtained satisfactory results [4], but the incremental oil recoveries are less than lab results due to their adverse conditions. Hence, studying the effects of reservoir and foam properties for choosing suitable reservoirs to perform foam flooding is very important. The permeability max-min ratio, ratio of vertical to horizontal permeability, depositional sequence, reservoir temperature, foam

TABLE 5

Parameters of simulated projects for sensitivity analysis

Influence factors	K_r	k_v/k_h	η (%)	Sequence	Concentration (wt%)	Temperature (°C)
K_r	2, 4, 6, 8, 10, 12, 15	0.1	50	Transgressive	0.5	30
k_v/k_h	4	0.01, 0.05, 0.1, 0.5, 1.0	50	Transgressive	0.5	30
η	4	0.1	20, 25, 33, 40, 50, 66, 75, 80, 90	Transgressive	0.5	30
Depositional sequence	4	0.1	50	Transgressive, regressive	0.5	30
Foaming agent concentration	4	0.1	50	Transgressive	0.1, 0.2, 0.3, 0.4, 0.5, 0.6, 1.0	30
Reservoir temperature	4	0.1	50	Transgressive	0.5	30, 60, 90, 120, 150, 180

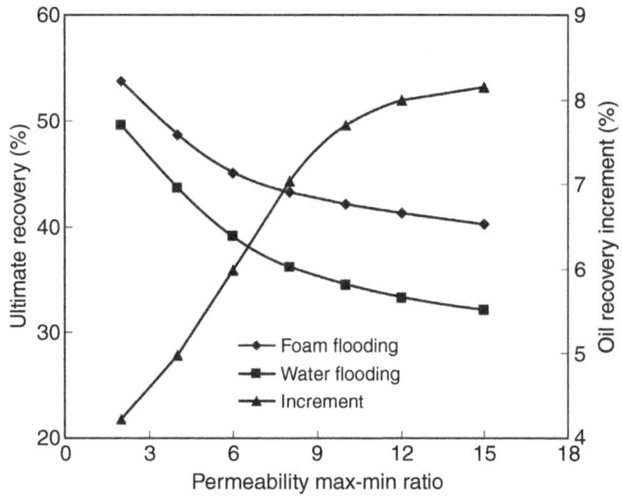

Figure 12

Ultimate recovery and increment *vs* permeability max-min ratio.

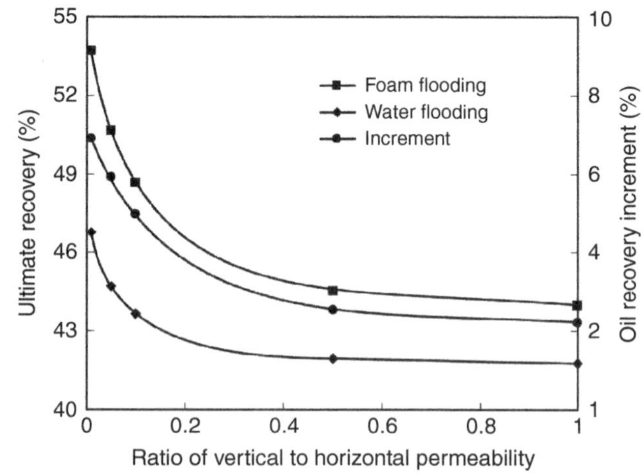

Figure 13

Ultimate recovery and the increment *vs* ratio of vertical to horizontal permeability.

quality and foaming agent concentration are very practical, common and important influencing factors in actual oilfields. Therefore, we performed a sensitivity analysis of these factors using the above heterogeneous reservoir and parameters. The parameters of the simulated projects are shown in Table 5.

3.2.1 Permeability Max-Min Ratio

The permeability max-min ratio is defined as the ratio of the highest permeability to the lowest permeability. Because foam has the feature of "big block and not small block", if the reservoir has a suitable permeability max-min ratio, the EOR effect will be superior. Figure 12 shows the ultimate and incremental oil recoveries for different permeability max-min ratios. It is shown that when the permeability max-min ratio is small, the incremental oil recovery increases sharply as the ratio increases; when it reaches 10, the oil recovery increment reaches a plateau. The reason for this is the lower the permeability max-min ratio will result in the closer the permeability of each stratum. From Equation (10), the foam resistance factor is a function of permeability. For a smaller permeability max-min ratio reservoir, after foam flooding, the water diverting into the low-permeability layers is lowered; for a larger permeability max-min ratio reservoir, the foam resistance factor is larger in the high-permeability layer and more water diverts into the low-permeability layer to sweep the remaining oil. Due to the limited blocking capacity of foam, the foam strength decreases if the permeability is too large. Therefore, the ultimate recovery is not very good for a reservoir with an extremely large permeability max-min ratio. Consequently, foam flooding is suitable for reservoirs with a moderate permeability max-min ratio.

3.2.2 Ratio of Vertical to Horizontal Permeability

Override is a problem in gas flooding. In gas flooding, the ratio of vertical to horizontal permeability should be considered. Figure 13 shows the ultimate and incremental oil recoveries for different ratios of vertical to horizontal permeability. It is shown that as the ratio of vertical to horizontal permeability increases, the incremental oil recovery sharply decreases. When the ratio is bigger than 0.2, the incremental oil recovery decreases slowly and is only about 2%. The reason is that if the ratio of vertical to horizontal permeability is large, the gas gravitational variations will be more pronounced. As a result, all the injected gas overrides into the top layer, and no foam is generated in the bottom layer. The continued water diversion would not take place. Therefore, foam flooding is suitable for reservoirs with a lower ratio of vertical to horizontal permeability.

The depositional sequence is defined as the regulation by which the sand particles deposit in sequence to form rock strata. A transgressive depositional sequence is composed of several sand layers where the sand diameter declines from the bottom to the top, *i.e.* the permeability is reduced in the vertical profile from the bottom to the top; a regressive depositional sequence is just the reverse. The depositional

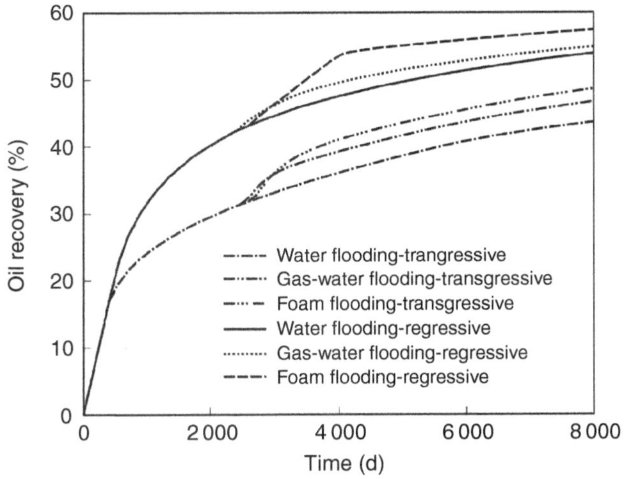

Figure 14

Comparison of the developments effects with different methods and depositional sequences.

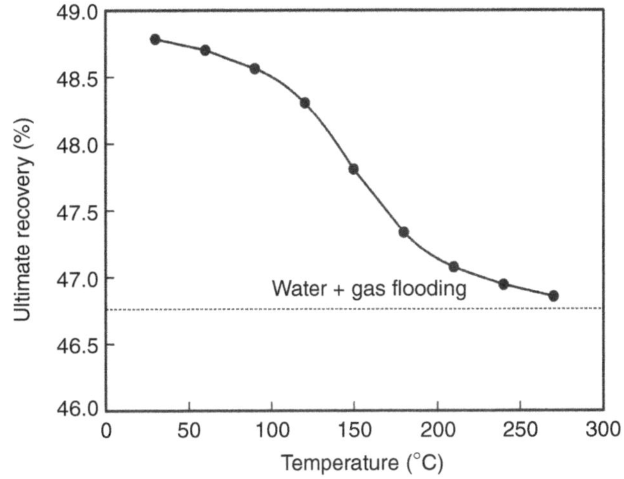

Figure 15

Ultimate recovery *vs* reservoir temperature.

sequence has a significant influence on the gas override to affect the ultimate recovery. Figure 14 shows the development effects with different displacing modes and depositional sequences. Apparently, the oil recovery in a regressive depositional sequence reservoir is higher than that in a transgressive depositional sequence reservoir for all the three displacing modes. The water + gas flooding oil recovery is higher than water flooding. However, the incremental oil recovery in regressive depositional sequence reservoirs is lower than transgressive depositional sequence reservoirs. The foam flooding oil recovery is higher than water + gas flooding. However, the incremental oil recovery in a regressive depositional sequence reservoir is higher than in a transgressive depositional sequence reservoir. For the rationale, in transgressive depositional sequence reservoirs, water is inclined to enter the high-permeability layer due to gravitational variations and lower flow resistance, which results in some remaining oil being confined within the low-permeability layer; continued gas is inclined to override into the low-permeability layer under the gravitational variations and displaces the remaining oil. In the regressive depositional sequence reservoirs, due to gravitational variations and lower flow resistance, both the high-permeability and low-permeability layers are well swept by water, and the remaining oil is lowered in the high-permeability layer (top layer); the gas is still inclined to enter the top layer due to gravitational variations and lower flow resistance, but the EOR effect is not obvious. Comparing foam flooding with water + gas flooding, in the transgressive depositional sequence reservoirs, more gas flows in the low-permeability layers, and the amount of foam in high-permeability layers is less than regressive depositional sequence reservoirs, which results in less water diverting into lower-permeability layers after foam flooding. Conversely, in the regressive depositional sequence reservoirs, since more gas flows in the high-permeability layers, consequently, this results in more flow of foam in high-permeability layers as well. As a result, more water diverts into lower-permeability layers after foam flooding.

3.2.4 Reservoir Temperature

Figure 15 shows the variation of ultimate oil recovery with reservoir temperature. The ultimate oil recovery decreases as the temperature increases. When the temperature is lower than 100°C, the ultimate oil recovery varies very little; when the temperature is between 100 and 200°C, the ultimate oil recovery decreases sharply; when the temperature exceeds 200°C, the ultimate oil recovery of foam flooding is close to that of water + gas flooding. Therefore, the reservoir temperature should not be high, or temperature-tolerant foaming agents should be used for high-temperature reservoirs.

3.2.5 Foam Quality

Foam quality is an important factor determining the blocking capacity of the foam texture. Figure 16 shows the ultimate oil recovery of different foam qualities.

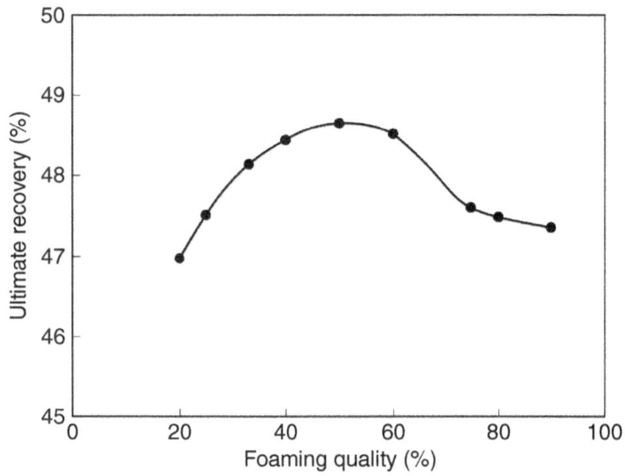

Figure 16
Ultimate recovery *vs* foam quality.

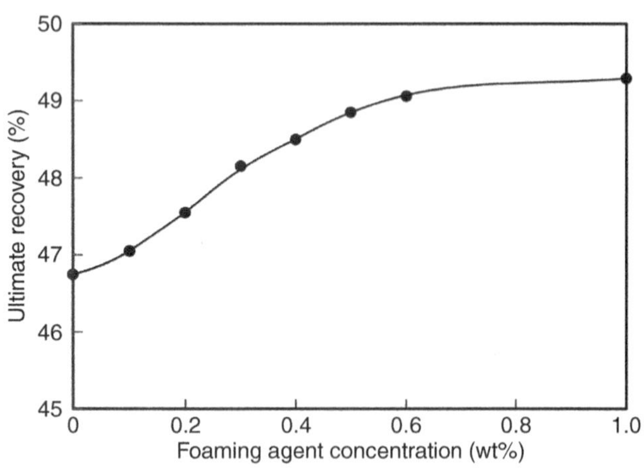

Figure 17
Ultimate recovery *vs* foaming agent concentration.

Evidently, as the foam quality increases, the ultimate oil recovery first increases and then decreases with an optimal value of 50%. This is due to the fact that a lower foam quality means a thicker lamella and a smaller bubble; under these conditions, the Jamin effect [51] is very small; conversely, a higher foam quality leads to a thinner lamella, a faster gas velocity and a bigger bubble; under these conditions, the bubble is more likely to collapse by shearing when it flows through the porous media. Therefore, a more stable bubble with a better Jamin effect can obtain a better EOR effect. Based on the relation between the foam quality and gas-water ratio, the optimal injection gas-water ratio should be chosen at the volume ratio of 1:1 under the reservoir conditions.

3.2.6 Foaming Agent Concentration

Figure 17 shows the variation of ultimate oil recovery with foaming agent concentration. Noticeably, the ultimate oil recovery reaches a plateau at 0.6 wt% after increasing along with the foaming agent concentration. Based on the parameters used in this case, the CMC is about 0.5 wt%. Optimal concentration is a little higher than the CMC since some of the foaming agent is adsorbed on the rock. So, the foaming agent concentration of foam flooding should be a little higher than the CMC of the surfactant.

CONCLUSIONS

A novel model of multi-factor foam flooding is established. The IMPES method in conjunction with a fourth-order Runge-Kutta method were used to solve the governing equation. The validation of this model was confirmed by matching the experimental data. Finally, the effects of several parameters including the permeability max-min ratio, ratio of vertical to horizontal permeability, depositional sequence, foam quality, foaming agent concentration, and reservoir temperature on oil recovery were studied. The following are the main conclusions:

1) A foam resistance factor model considering the influences of permeability, foaming agent concentration, foam quality, temperature and oil saturation is obtained. Then, taking the gas velocity as the foam generation condition and water saturation as the foam collapse condition, the gas mobility model was achieved;
2) Foam can effectively weaken the phenomena of gas override in the vertical direction due to gravitational variations and gas channeling in a horizontal direction caused by large gas mobility;
3) The mechanism of water + gas flooding enhanced oil recovery is that due to the gravitational variations, gas increases the swept area of the injected fluid; and the mechanisms of foam flooding are both the gravitational variations and the water diversion;
4) With a smaller permeability max-min ratio, the incremental oil recovery increases sharply and reaches a plateau at the ratio of 10. Foam flooding is suitable for reservoirs with a moderate permeability max-min ratio (k_r = 5-10);
5) As the ratio of vertical to horizontal permeability increases, the incremental oil recovery sharply decreases. When the ratio is bigger than 0.2, the incremental oil recovery decreases slowly, and is only about 2%. Foam flooding is appropriate for reservoirs whose ratio of vertical to horizontal permeability is lower than 0.2;

6) The oil recovery in regressive depositional sequence reservoirs is higher than that of transgressive depositional sequence reservoirs for water flooding, gas-water flooding and foam flooding. The gas-water flooding incremental oil recovery for water flooding in regressive depositional sequence reservoirs is less than that in transgressive depositional sequence reservoirs, but the foam flooding increment for water + gas flooding in regressive depositional sequence reservoirs is more than that in transgressive depositional sequence reservoirs;
7) The ultimate recovery decreases as the temperature increases. The reservoir temperature should not be high for foam flooding; otherwise, temperature-tolerant foaming agents should be used for high-temperature reservoirs;
8) As the foam quality increases, the ultimate oil recovery first increases and then decreases, and the optimal value is 50%. The optimal injection for the gas-water ratio should be chosen at the volume ratio of 1:1 under the reservoir conditions;
9) Due to adsorption of the surfactant, the foaming agent concentration during foam flooding should be a little higher than the CMC of the surfactant.

ACKNOWLEDGMENTS

This study was funded by the Chinese National Natural Science Foundation (51274212) and Important National Science & Technology Specific Projects of China (2011ZX05014-003-008HZ).

REFERENCES

1 Casteel J.F., Djabbarah N.F. (1988) Sweep Improvement in CO_2 Flooding by Use of Foaming Agents, *SPE Reserv. Eng.* **3**, 4, 1186-1192.

2 Namdar Z.M., Kam S.I., LaForce T.C., Rossen W.R. (2011) The Method of Characteristics Applied to Oil Displacement by Foam, *SPE J.* **16**, 1, 8-23.

3 Yang H., Yue X.A., Zhao R.B., Yang Z.P. (2009) Experimental Study on Validity of Foam Plugging in Porous Media, *Oilfield Chemistry* **3**, 273-275.

4 Zhu Y.Y., Hou Q.F., Weng R., Jian G.Q., Luo Y.S., Li J. G. (2013) Recent Progress and Effects Analysis of Foam Flooding Field Tests in China, *SPE paper* 165211 presented at *2013 SPE Enhanced Oil Recovery Conference*, Kuala Lumpur, Malaysia, 02-04 July.

5 Kovscek A.R., Patzek T.W., Radke C.J. (1994) Mechanistic Prediction of Foam Displacement in Multidimensions: A Population Balance Approach, *SPE paper* 27789 presented at *1994 SPE/DOE Improved Oil Recovery Symposium*, Tulsa, Oklahoma, 17-20 April.

6 Rossen W.R., Zeilinger S.C., Shi J.X., Lim M.T. (1999) Simplified Mechanistic Simulation of Foam Processes in Porous Media, *SPE J.* **4**, 3, 279-287.

7 Kovscek A.R., Bertin H.J. (2002) Estimation of Foam Mobility in Heterogeneous Porous Media, *SPE paper* 75181 presented at *2002 SPE/DOE Improved Oil Recovery Symposium*, Tulsa, Oklahoma, 13-17 April.

8 Du Q.J. (2008) Experiment and Modeling of the Seepage Characteristics in Foam Flooding. A *Dissertation* Submitted for the Degree of Doctor of Philosophy of China University of Petroleum, Shandong, China.

9 Han D.K. (1993) *Numerical Reservoir Simulation*, Petroleum Industry Press, Beijing, China.

10 Falls A.H., Musters J.J., Ratulowski J. (1989) The Apparent Viscosity of Foams in Homogeneous Bead packs, *SPE Reserv. Eng.* **4**, 2, 155-164.

11 Chang S.H., Owusu L.A., French S.B., Kovarik F.S. (1990) The Effect of Microscopic Heterogeneity on CO_2-Foam Mobility: Part 2 - Mechanistic Foam Simulation, *SPE paper* 20191 presented at *1990 Seventh Symposium on Enhanced Oil Recovery*, Tulsa, Oklahoma, 22-25 April.

12 Friedmann F., Chen W.H., Gauglitz P.A. (1991) Experimental and Simulation Study of High-Temperature Foam Displacement in Porous Media, *SPE Reserv. Eng.* **6**, 1, 37-45.

13 Kovscek A.R., Patzek T.W., Radke C.J. (1993) Simulation of Foam Transport in Porous Media, *SPE paper* 26402 presented at *1993 SPE Annual Technical Conference and Exhibition*, Houston, Texas, 3-6 Oct.

14 Low D.H.S., Yang Z.M., Stone T.W. (1992) Effect of the Presence of Oil on Foam Performance: A Field Simulation Study, *SPE Reserv. Eng.* **7**, 2, 228-236.

15 Patzek T.W., Myhill N.A. (1989) Simulation of The Bishop Steam Foam Pilot, *SPE paper* 18786 presented at *1989 SPE California Regional Meeting*, Bakersfield, 5-7 April.

16 Kular G.S., Lowe K., Coombe D. (1989) Foam Application in an Oil Sands Steam Flood Process, *SPE paper* 19690 presented at *1989 SPE Annual Tech. Conf. and Exhibition*, San Antonio, Texas, 8-11 Oct.

17 Islam M.R., Farouq-Ali S.M. (1990) Numerical Simulation of Foam Flow in Porous Media, *J. Can. Pet. Tech.* **29**, 4, 47-51.

18 Rossen W.R., Zhou Z.H., Mamun, C.K. (1991) Modeling Foam Mobility at "The Limiting Capillary Pressure", *SPE paper* 22627 presented at *1991 Annual Tech. Conf. and Exhibition*, Dallas, Texas, 6-9 Oct.

19 Khatib Z.I., Hirasaki G.J., Falls A.H. (1988) Effects of Capillary Pressure on Coalescence and Phase Nobilities in Foams Flowing through Porous Media, *SPE Reserv. Eng.* **3**, 3, 919-926.

20 Wang J., Liu H.Q., Ning Z.F., Zhang H.L. (2012) Experimental Research and Quantitative Characterization of Nitrogen Foam Blocking Characteristics, *Energy & Fuels* **26**, 8, 5152-5163.

21 http://www.oakdaleengr.com/index.html.

22 De Vries A.S., Wit K. (1990) Rheology of Gas/water Foam in The Quality Range Relevant to Steam Foam, *SPE Reserv. Eng.* **5**, 2, 185-192.

23 Chang S.H., Grigg R.B. (1996) Foam Displacement Modeling in CO_2 Flooding Processes, *SPE paper* 35401 presented at *1996 SPE/DOE Improved Oil Recovery Symposium*, Tulsa, Oklahoma, 21-24 April.

24 Isaacs E.E., McCarthy F.C., Maunder J.D. (1988) Investigation of Foam Stability in Porous Media at Elevated Temperatures, *SPE Reserv. Eng.* **3**, 2, 565-572.

25 Xu Q., Rossen W.R. (2003) Effective Viscosity of Foam in Periodically Constricted Tubes, *Colloids Surf. A.* **216**, 1-3, 175-194.

26 Cox S.J., Neethling S., Rossen W.R., Schleifenbaum W., Schmidt-Wellenburg P., Cilliers J.J. (2004) A Theory of The Effective Yield Stress of Foam in Porous Media: The Motion of A Soap Film Traversing A Three-Dimensional Pore, *Colloids Surf. A.* **245**, 1-3, 143-151.

27 Wang Z., Ganesan N. (2006) Model for Plateau Border Drainage of Power-Law Fluid with Mobile Interface and Its Application to Foam Drainage, *J. Colloid Interface Sci.* **300**, 1, 327-337.

28 Pacelli L.J., Quoc P.N., Peter K.C., Marten A.B. (2006) Coupling of Foam Drainage and Viscous Fingering in Porous Media Revealed by X-ray Computed Tomography, *Transp. Porous Med.* **64**, 3, 301-313.

29 Muijs H.M., Keijzer P.P.M., Wiersma R.J. (1988) Surfactants for Mobility Control in High-Temperature Steam-Foam Applications, *SPE paper* 17361 presented at *1988 SPE Enhanced Oil Recovery Symposium*, Tulsa, Oklahoma, 16-21, April.

30 Maini B.B., Ma V. (1986) Laboratory Evaluation of Foaming Agents for High-Temperature Applications - I. Measurements of Foam Stability at Elevated Temperatures and Pressures, *J. Can. Petrol. Tech.* **25**, 6, 65-69.

31 Nikolov A.D., Wasan D.T., Huang D.W., Edwards D.A. (1986) The Effect of Oil on Foam Stability: Mechanisms and Implications for Oil Displacement by Foam in Porous Media, *SPE paper* 15443 presented at *1986 SPE Annual Technical Conference and Exhibition*, New Orleans, Louisiana, 5-8 Oct.

32 Suffridge F.E., Raterman K.T., Russell G.C. (1989) Foam Performance under Reservoir Conditions, *SPE paper* 19691 presented at *1989 SPE Annual Technical Conference and Exhibition*, San Antonio, Texas, 8-11 Oct.

33 Emadi A., Sohrabi M., Jamiolahmady M., Ireland S., Robertson G. (2011) Mechanistic Study of Improved Heavy Oil Recovery by CO_2-Foam Injection, *SPE paper* 143013 presented at *SPE Enhanced Oil Recovery Conference*, Kuala Lumpur, Malaysia, 19-21 July.

34 Andrianov A., Farajzadeh R., Mahmoodi N.M., Talanana M., Zitha P.L.J. (2012) Immiscible Foam for Enhancing Oil Recovery: Bulk and Porous Media Experiments, *Ind. Eng. Chem. Res.* **51**, 5, 2214-2226.

35 Jensen J.A., Friedmann F. (1987) Physical and Chemical Effects of an Oil Phase on the Propagation of Foam in Porous Media, *SPE paper* 14375 presented at *1987 SPE California Regional Meeting*, Ventura, California, 8-10 April.

36 Svorstøl I., Vassenden F., Mannhardt K. (1996) Laboratory Studies for Design of a Foam Pilot in the Snorre Field, *SPE paper* 35400 presented at *1996 SPE/DOE Improved Oil Recovery Symposium*, Tulsa, Oklahoma, 21-24 April.

37 Mannhardta K., Svorstøl I. (1999) Effect of Oil Saturation on Foam Propagation in Snorre Reservoir Core, *J. Pet. Sci. Eng.* **23**, 3-4, 189-200.

38 Farajzadeh R., Andrianov A., Krastev R., Hirasaki G.J., Rossen W.R. (2012) Foam-Oil Interaction in Porous Media: Implications for Foam Assisted Enhanced Oil Recovery, *SPE paper* 154197 presented at *2012 SPE EOR Conference at Oil and Gas West Asia*, Muscat, Oman, 16-18 April.

39 Simjoo M., Dong Y.F., Andrianov A., Talanana M., Zitha P.L.J. (2012) Novel Insight into Foam Mobility Control, *SPE paper* 15338 presented at *2012 International Petroleum Technology Conference*, Bangkok, Thailand, 7-9 Feb.

40 Arnaudov L., Denkov N.D., Surcheva I., Durbut P., Broze G., Mehreteab A. (2001) Effect of Oily Additives on Foam Ability and Foam Stability. 1. Role of Interfacial Properties, *Langmuir* **17**, 22, 6999-7010.

41 Derjaguin B., Landau L. (1941) Theory of the Stability of Strongly Charged Lyophobic Sols and of the Adhesion of Strongly Charged Particles in Solutions of Electrolytes, *Acta Physico Chemica URSS* **14**, 633-662.

42 Verwey E.J.W., Overbeek J.Th.G. (1948) *Theory of the Stability of Lyophobic Colloids*, Elsevier, Amsterdam.

43 Shi J.X., Rossen W.R. (1998) Simulation and Dimensional Analysis of Foam Processes in Porous Media, *SPE Reserv. Eval. Eng.* **1**, 2, 148-154.

44 Kovscek A.R., Patzek T.W., Radke C.J. (1997) Mechanistic Foam Flow Simulation in Heterogeneous and Multidimensional Porous Media, *SPE J.* **2**, 4, 511-526.

45 Yu H.M., Ren S.R., Zuo J.L. (2012) A Mathematical Model and Numerical Simulation Method for Air-Foam, *Acta Petrolei Sinica* **33**, 4, 653-657.

46 Hou J., Du Q.J., Li Z.Q., Pan G.M., Lu X.J., Zhou K. (2013) Experiments on Foam Texture under High Pressure in Porous Media, *Flow Measurement and Instrumentation* **33**, 68-76.

47 Simjoo M., Rezaei T., Andrianov A., Zitha P.L.J. (2013) Foam Stability in the Presence of Oil: Effect of Surfactant Concentration and Oil Type, *Colloids and Surfaces A: Physicochemical and Engineering Aspects* **438**, 148-158.

48 Stone H.L. (1973) Estimation of Three-Phase Relative Permeability and Residual Oil Data, *J. Can. Pet. Tech.* **12**, 4, 53-61.

49 Qin J.S., Li A.F. (2001) *Petrophysics*, China University of Petroleum Press, Shandong, China.

50 Poston S.W., Gross S.J. (1986) Numerical Simulation of Sandstone Reservoir Models, *SPE Reservoir Engineering* **1**, 4, 423-429.

51 Wright R. (1933) Jamin Effect in Oil Production, *AAPG Bulletin* **17**, 12, 1521-1526.

10

Effect of Post-Synthesis Treatments on the Properties of ZnS Nanoparticles: An Experimental and Computational Study

Elena Balantseva[1]*, Bruno Camino[2], Anna Maria Ferrari[1] and Gloria Berlier[1]*

[1] *Università di Torino, Dipartimento di Chimica and NIS (Nanostructured Interfaces and Surfaces) Centre, Via P. Giuria 7, 10125 Torino - Italy*
[2] *Department of Chemistry, Imperial College London, South Kensington London SW7 2AZ - United Kingdom*
e-mail: ebalants@gmail.com - gloria.berlier@unito.it

* Corresponding authors

Abstract — *This work deals with the characterization of ZnS NanoParticles (NP), prepared by precipitation employing thioacetamide as sulfur source at different reaction time length. The attention is focused on the modification induced on structural, surface and electronic properties of ZnS NP by post-synthesis treatments. These were aimed at removing from the samples surface adsorbed reactants, by washing or thermal treatments, both in air or vacuum. The effect of these parameters is followed by X-Ray Diffraction (XRD), Transmission Electron Microscopy (TEM), Fourier Transform InfraRed (FTIR), gas-volumetric and ThermoGravimetric Analysis (TGA). Moreover, the effect of nanostructuration on the semiconducting material band gap is evaluated by Diffuse Reflectance UV-Vis (DR UV-Vis) spectroscopy. Density Functional Theory (DFT) calculations have been employed to clarify the role of the adsorbed reactants on the surface stability and to assess the relationship between particle size and band gap value.*

Résumé — Effet des traitements après-synthèse sur les propriétés de nanoparticules de ZnS : une étude expérimentale et computationnelle — Ce travail concerne la caractérisation de nanoparticules de ZnS préparées par précipitation en présence de ThioAcétAmide (TAA) comme source de soufre. On a étudié, en particulier, les modifications des propriétés structurelles, superficielles et électroniques en fonction des paramètres de synthèse (temps, solvant et source de soufre) et des traitements après-synthèse considérés. Ces derniers avaient pour but d'éliminer les réactifs adsorbés sur la surface par lavages ou traitements thermiques à l'air ou sous vide. L'effet des différents traitements a été suivi par diffraction des rayons X (XRD), microscopie de transmission électronique (TEM), spectroscopie infra rouge à transformée de Fourier (FTIR), analyse volumétrique des gaz et thermogravimétrie (TGA). En outre, l'effet de la nanostructure sur les propriétés de la largeur de bande du matériau semi-conducteur a été évalué par spectroscopie de reflectance UV-Vis (DR UV-Vis). Des théories de la fonctionnelle de la densité (DFT) ont été utilisées pour clarifier le rôle des réactifs adsorbés sur la stabilité de la surface et pour établir une relation entre la dimension des particules et la largeur de bande.

INTRODUCTION

Since the first discovery by Honda and Fujishima, photocatalyst hydrogen production has attracted strong consideration [1]. Many heterogeneous photocatalysts have been found, mostly working in ultraviolet region, which accounts for only 5% of the solar spectrum [2-4]. Taking this into account, it is understandable the existing wide interest in searching new visible-light-driven photocatalysts that could replace the "old" ones and extend their response to the solar spectrum. Up to now, the number of such photocatalysts is still limited [5-7]. Although CdS is reported as an effective photocatalyst with a bang gap suitable for visible light exploitation, its activity depends on expensive noble metals deposited on its surface, and environmental and safety concerns related to the well-know cadmium toxicity are serious limitations to applications in large scale [8, 9]. As an alternative ZnS can produce H_2 effectively without noble metal co-catalyst, due to its high conduction band position, but the wide band gap (3.7 eV) implies that only ultraviolet radiation can be employed to produce electron/hole couples [10]. This is the reason why band gap engineering has been proposed to modify the electronic properties of ZnS and improve its visible light harvesting efficacy [11-13].

Apart from the electronic properties (band gap, conduction and valence band position) the main requirements for an efficient photocatalysts are high crystallinity (to avoid the presence of defects, acting as electron/hole recombination traps), coupled to high surface area and small particle size (to increase the efficacy of electron/hole migration to the surface). This implies that a careful control of these parameters is crucial in the synthesis of a nanostructured photocatalyst.

Different chemical routes have been used to prepare small semiconducting NanoParticles (NP). The most common way is to use stabilizing agents which prevent growth of particles and their agglomeration by electrostatic and steric repulsion [14-19]. But at the same time they modify the structural, morphological and optical properties of nanoparticles. Growth of strongly confined nanoparticles without a capping agent is difficult to achieve and rarely reported. Among other chemical routes the use of reversed micellars solutions [20, 21], thin film deposition process, inorganic templates were reported.

For materials based on CdS and ZnS an alternative route for preparation of nanoparticles is a hydrothermal synthesis procedure [22]. The use of ThioAcetAmide (TAA) as a sulfur source allows obtaining well mono-dispersed particles, whose size depends upon the reaction conditions. The formation of small spherical CdS NP, which are then transformed into cylindrical structures while the reaction proceeds was reported [23]. The early work where TAA was applied in ZnS synthesis reported the formation of submicrometric particles [24]. Later, the effect of sulfur source (TAA in comparison with Na_2S) in the synthesis of ZnS based system (ZnS/CdS, ZnS/ZnO) was studied by different authors [25, 26]. The detailed investigation of the precipitation dynamic of ZnS from TAA decomposition in conditions close to those applied in other works was carried out with attention on product agglomeration [27]. The authors showed a crucial role of hydrodynamics in the determination of size and agglomeration, and studied the influence of parameters such as pH, electrical conductivity, concentration of sulfide ions and turbidity. The obtained results could be very helpful in control of morphology and associated properties.

Recently, we also demonstrated that this synthesis procedure works well in case of ZnS [28]. In particular, we showed that the reaction carried out at room temperature employing a methanol/ethanol mixture as a solvent was suitable for the formation of small ZnS NP with much higher yield with respect to what obtained in water. In this contribution, the attention is focused on the effect of post-synthesis treatments, aimed at removing the reactants and impurities adsorbed on the ZnS photocatalyst surface, on the material properties. This is studied on samples prepared at different reaction time length in methanol/ethanol mixture with TAA. The work reports a detailed physico-chemical characterization of the materials, concerning their morphology, structure, textural properties, adsorbed species and electronic properties. Moreover, Density Functional Theory (DFT) calculations have been employed to gain hints on the size dependence of surface and electron properties of the ZnS NP. The adsorption of formic acid (as a simple model of surface species adsorbed during the synthesis process) has been simulated. The interaction with the most relevant ZnS surfaces (110 and 111) [29] has been considered.

1 EXPERIMENTAL SECTION

1.1 Synthesis

ThioAcetAmide (CH_3CSNH_2, TAA) was purchased from *Sigma Aldrich*; analytical grade zinc acetate (Zn $(CH_3COO)_2 \cdot 2H_2O$, ZnAc) from *Baker Analyzed ACS Reagent*; methanol and ethanol from *Carlo Erba* reagents. Bulk ZnS (powder, <10 μm, 99.99%) was purchased from *Sigma Aldrich*. All the chemical and solvents were used as received without further purification.

In a typical procedure TAA and ZnAc were dissolved separately in ethanol/methanol 1/1 mixture at 313 K, to give 0.2 M solutions [28, 29]. The TAA solution was then slowly added to the ZnAc one under continuous stirring. The resulting mixture (50 mL total volume) was left under stirring for 6 h or 24 h. Afterwards the resulting white powder was separated by centrifugation (4 000 rpm), and washed first with the alcohol mixture and then trice with distilled water at Room Temperature (RT) or 353 K. The cleaned product was dried for about 3 h in oven at 353 K. The obtained samples were carefully milled before characterization. The labeling of the samples in the text is related to reaction time length (ZnS_6h or ZnS_24h) and to the water temperature in washing procedure: ZnS_*x*h_HW (x = 6 or 24) are the samples treated with hot water. The same synthetic procedure was carried out at 6 hours reaction by employing Na_2S as sulfur source instead of TAA (sample ZnS_Na_2S). This resulted in a much lower yield, so that these results are only reported for comparison.

Thermal activation in oven or in vacuum was carried out on sample ZnS_6h_HW. The resulting samples are labeled mentioning the activation temperature and conditions (ZnS_573 activated in oven at 573 K, sample ZnS_673_V activated at 673 K in vacuum).

1.2 Characterization

Powder X-Ray Diffraction (XRD) patterns were measured with PW3050/60 X'Pert PRO MPD diffractometer (Panalytical) working in Bragg–Brentano geometry using Cu $K\alpha$ radiation (λ = 1.5406 Å) and operated at 45 kV, 40 mA with a scan speed of 0.01° min^{-1} and step size of 0.0170°. The crystallite size D = 4/3 L was calculated by applying the Scherrer's equation to the (111) peak at 28.7°:

$$L = k\lambda/(\beta \cos \theta)$$

where λ is the X-ray wavelength, β the measured broadening of the diffraction peak and θ the diffraction angle [30]. The measured β value was corrected by subtracting the instrumental line broadening, estimated in 0.02°. The error on particle size estimation was calculated by including standard deviation on 5 measurements and instrumental resolution.

Transmission Electron Microscopy (HRTEM) measurements were performed on JEM 3010-UHR microscope (*JEOL Ltd.*) operating at 300 kV. For the measurements ZnS powder was dispersed on a copper grid coated with a perforated carbon film.

Gasvolumetric nitrogen adsorption/desorption isotherms were measured at Liquid Nitrogen Temperature (LNT) with ASAP 2020 physisorption analyzer (*Micromeritics*). Prior to analyses, samples were evacuated at 423 K over 2 h. The Specific Surface Area (SSA) was calculated by the Brunauer-Emmett-Teller (BET) method in the p/p_0 range from 0.05 to 0.2 [31]. The average pore size and volume were calculated on the adsorption branch of the isotherms according to the Barrett-Joyner-Halenda (BJH) method [32, 33].

ThermoGravimetric Analysis (TGA) was carried out on a TAQ600 (*TA Instruments*) heating the samples at a rate of 10 K min^{-1} from 303 to 1123 K in air flow. Before starting the measurements, samples were equilibrated at 303 K.

Fourier Transform InfraRed (FTIR) spectra were recorded with an IFS66 spectrophotometer (*Bruker*) equipped with a Deuterated TriGlycine Sulfate (DTGS) detector, working with resolution of 4 cm^{-1} over 32 scans. Samples were in the form of self-supporting pellets suitable for transmission infrared experiments and were placed in a quartz cell equipped with KBr windows, designed for *in situ* studies in vacuum and controlled atmosphere. Before infrared analysis the samples were evacuated at the desired temperature (RT or 673 K) for 1 hour.

Diffuse Reflectance (DR) UV-Vis measurements were recorded on powdered samples on a Cary 5 000 spectrophotometer equipped with a reflectance sphere. Spectra were recorded at a scan rate of 300 nm/min with a step size of 1 nm and fixed Spectral Band Width (SBW) of 2 nm. The measured intensities were converted with the Kubelka-Munk function. The fundamental absorption, which corresponds to electron excitation from the valance band to conduction band, can be used to determine the value of the optical band gap of the synthesized ZnS nanoparticles [34-37]. The relation between the incident photon energy (hv) and the absorption coefficient (α) is given by the following relation: $(\alpha hv)^{1/n} = A(hv - E_g)$, where A is constant and E_g is the band gap energy of the material and the exponent n depends on the type of transition [38, 39]. ZnS as a semiconductor is characterized by direct allowed transition where $n = 1/2$ [40], therefore the direct band gaps of the samples are calculated by plotting $(\alpha hv)^2$ versus hv and then extrapolating the straight portion of the curve on hv axis at $\alpha = 0$.

1.3 DFT Modeling

Results presented are obtained with periodic CRYSTAL09 code [41], using the hybrids B3PW (the combination of the Becke's three parameter functional [42] for the exchange and the Perdew Wang functional [43] for the correlation) and PBE0 [44]. Basis sets of triple-zed TZ quality for the valence electrons have been employed: the relativistic Steven-Krauss small core RECP with a 4211-41 [4sp2d] basis set for valence electrons for Zn atoms [45] combined to a 8-6311-1 [1s-4sp-1d] for S [45]. Within this computational set up, B3PE has been found to provide accurate estimate of the ZnS band gap (E_g = 3.71 eV) very close to the experimental value (E_g = 3.65-3.7 eV), while PBE0 was found to accurately accounts for surface properties [29, 46, 47].

In the CRYSTAL code, the level of accuracy in evaluating the Coulomb and the exchange series is controlled by five parameters [41] for which 10^{-7}, 10^{-7}, 10^{-7}, 10^{-7}, 10^{-18} values have been used for all calculations. The convergence threshold for SCF energy was set to 10^{-9} Ha. The reciprocal space was sampled according to a regular sublattice determined by the shrinking factor IS [41]. A value of IS = 8 has been adopted which corresponds to 34 independent k-points in the irreducible part of the Brillouin zone.

Nanoparticles have been modeled with bidimensional slabs of increasing thickness d (from 0.5 to about 6 nm), characterized by two infinite dimensions (x, y) and a finite thickness, and cut from the bulk so as to expose the most stable ZnS surface, the (110) [29]. Stoichiometric ZnS slabs (Zn/S ratio equal to 1/1) have been considered, in agreement with the experimental and theoretical reports [28, 48-51]. Surface Energy (E_s) is calculated as: $E_s = \frac{E_{slab} - nE_{bulk}}{2A}$ where E_{slab} is the energy of an n-layered 2D slab used to model the material that exposes the surface of interest, E_{bulk} is the energy of the bulk material and A is the area of the 2D surface cell. The factor 1/2 accounts for the existence of two limiting surfaces. E_s is then the energy per unit area required to form the surface from the bulk and it is reported in J/m^2.

The interaction with formic acid has been computed considering a coverage ($\theta = 0.5$ (simulated adopting a 2×1 supercell). Slabs of thickness of 1.2 and 1.5 nm have been selected for the (110) and (111) surfaces (corresponding to 7 and 6 ZnS layers, respectively). The Adsorption Energy (ΔE_{ads}) per molecule is calculated as: $E_{ads} = \frac{E_a - nE_m - E_{slab}}{n}$, where E_m is the energy of the optimized isolated molecule, E_{slab} is the energy of the optimized isolated slab modeling the surface and n is the number of molecules per cell. The basis set superposition error BSSE has been estimated, by employing the counterpoise method [52], to account for no more than 0.07 eV and results are corrected accordingly.

For all the systems discussed in this study, geometry has been optimized relaxing all the atoms in the cell and keeping fixed the lattice parameters at the bulk values. All the symmetry constraints have been removed. Adsorbate frequencies have been computed at the Γ point, within the harmonic approximation, by diagonalizing the mass-weighted Hessian matrix only for the adsorbate fragment, once checked that its vibrational modes are not coupled with other crystal phonons.

2 RESULTS AND DISCUSSION

2.1 Effect of Synthesis and Post-Synthesis Conditions on Crystallite Size and Textural Properties

The general properties of the prepared samples are listed in Table 1, with those of ZnS_bulk (commercial sample) for comparison. These include the crystallite size calculated by XRD (Scherrer equation applied to the 111 peak) and the textural properties, that is SSA and pore volume. The reported band gap values were calculated from DR UV-Vis spectra, as discussed in the following (Sect. 2.3).

TABLE 1
General properties of the ZnS samples depending on condition of synthesis and purification

Samples	Crystallite size (nm) XRD[a]	SSA[b] (m^2/g)	Pore volume (cm^3/g)[c]	Band gap E_g (eV)
ZnS_6h	1.8	294	0.12	4.22
ZnS_6h_HW	3.5	376	0.60	3.72
ZnS_24h	4.1	277	0.30	3.67
ZnS_24h_HW	3.4	184	0.30	3.59
ZnS_Na_2S	4.9	236	0.24	3.55
ZnS_573	3.5	108	0.28	3.44
ZnS_673_V	3.7	123	0.25	3.22
ZnS_bulk	33.2	24	0.07	3.60

[a] Calculated with the Scherrer equation, estimated error ±0.017 nm.
[b] Measured by BET equation.
[c] Measured by BJH method in the isotherm adsorption branch.

A selection of XRD patterns of the materials is reported in Figure 1. All synthesized ZnS samples have three main diffraction peaks at 28.7, 47.9 and 56.8°, corresponding to the (111), (220) and (311) lattice planes of ZnS cubic sphalerite structure (JCPDS 010772100). Minor features can be observed at 32.9, 70.3 and 77.5°, due to (200), (400) and (331) reflections. Sample ZnS_6h shows quite broad peaks, corresponding to an average crystallite size (Scherrer equation) of 1.8 nm. The peak width decreases both in relation to longer reaction time (ZnS_24h) and after washing with hot water. On the contrary, hot washing did not particularly affect the XRD patterns and related crystallite size of sample ZnS_24h.

The estimated crystallite size of all samples is listed in Table 1. The smallest crystallite size was obtained for ZnS_6h. Changing all other parameters leads to slight growing of particles. Moreover the use of Na_2S as sulfur source instead of TAA resulted in relatively big particles. As mentioned in the experimental section, this procedure gives a very low yield, and is only reported for comparison.

The changes in SSA and porosity of the samples as a function of synthesis parameters are not directly correlated to the crystallite size estimated by XRD. For instance, washing with hot water causes an increase in SSA and pore volume in sample ZnS_6h. This could be easily explained by considering that the samples porosity is due

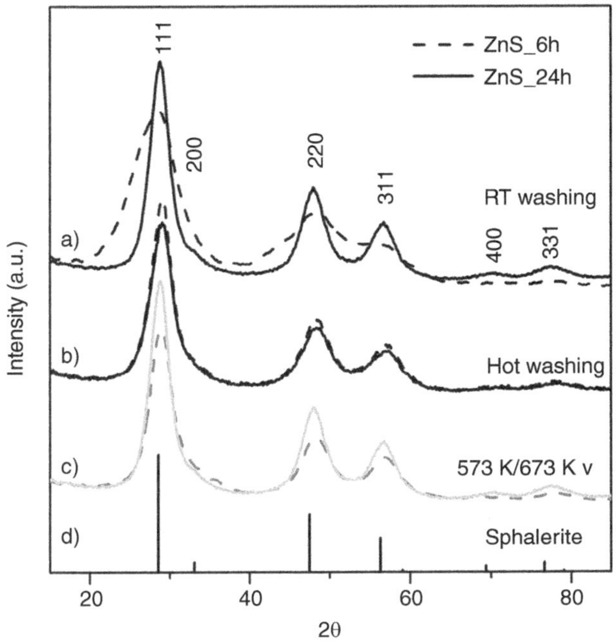

Figure 1

XRD patterns of ZnS particles as a function of different preparation conditions: a) as prepared (simply washed at RT); b) washed with hot water; c) treated in oven at 573 K or in vacuum at 673 K; d) cubic ZnS (sphalerite, JCPDS 010772100). Curves were vertically shifted for easier comparison.

to interparticle voids [29]: the removal of adsorbed products from the sample surface by hot washing increases the accessible volume and surface area of the particles. However, the same effect is not evident in sample ZnS_24h, where washing causes a slight decrease in both crystallite size and SSA. On the whole, all prepared samples show relatively high SSA, which represents an important point for application in photocatalysis.

The prepared samples were also subjected to thermal treatments, as an alternative way (often employed in catalysis) to remove adsorbed impurities and obtain a clean surface. The effect of thermal treatments in vacuum was already described in Reference [29], by employing both XRD and TEM techniques. As reported elsewhere, thermal treatments in air of ZnS causes the transformation to ZnO [53-55]. This was confirmed on our samples by both activation in static oven and in N_2/O_2 flow. In both cases, the transformation to ZnO started around 673 K (data not reported), while after activation at 573 K only the typical XRD peaks of cubic ZnS were observed (Fig. 1, grey dashed curve c). This pattern is compared in the same figure with that obtained after activation in vacuum at 673 K [29]. The crystallite size of both samples is similar to what measured before thermal activation (Tab. 1), but a decrease in SSA and pore volume is observed, suggesting further agglomeration of the primary ZnS NP.

TEM micrographs of the prepared ZnS samples revealed the presence of dense agglomerates of spheroidal particles (Fig. 2). It is however possible to observe light regions among the particles, which could be related to interparticle voids resulting in high surface area and pore volume. The high agglomeration of the material did not allow obtaining high resolution images of single particles, so that a reliable particle size analysis was not possible. The average value of *ca.* 2.6 (±0.4) nm (measured on 20-50 particles) was found for the sample ZnS_6h, in good agreement with XRD. It is important to underline the fact that the changes in crystalline domains size estimated by XRD cannot be appreciated by TEM, since the micrographs of samples ZnS_6h_HW are hardly distinguishable from those of ZnS_6h (Fig. 2, compare top panels with the left bottom one). In the case of sample ZnS_24h, the particles look more defined and crystallized with respect to the former (right bottom panel of Fig. 2). Particle size analysis was almost impossible in this case, due to very dense agglomeration of the primary particles, which however showed a primary size around 2-3 nm, which is similar to that of ZnS_6h. The different size estimated by XRD can thus be explained on the basis of higher crystallinity (less defective surfaces) and/or more dense aggregation.

Lattice fringes are visible in the TEM micrograph (bottom right panel of Fig. 2), indicating the crystalline nature of the particles. Fourier Transform (FT) analysis of the images resulted in a diffraction ring with average *d*-spacing of 3.0 Å, characteristic of the (111) peak of ZnS cubic structure.

On the whole, the results discussed in this paragraph suggest that the particles obtained at 6 h synthesis are stabilized by adsorbed reactants, so that hot washing causes a partial aggregation of the primary particles. At longer reaction time length, more crystalline domains are formed, which are densely interconnected, as shown by both TEM and surface area analysis. In this case, the role of adsorbed species in stabilizing the particles is less important.

2.2 Surface Adsorbed Species

2.2.1 Quantitative Analysis

The nature and amount of species adsorbed on the surface of the prepared ZnS sample was studied by thermogravimetric analysis. To this aim, the attention was focused on sample ZnS_6h and following modifications (ZnS_6h_HW, ZnS_573 and ZnS_673_V). These are reported in Figure 3 together the results obtained on ZnS_Na_2S for comparison. All experiments were carried out in air flow, and an analysis on ZnS_6h was repeated in nitrogen.

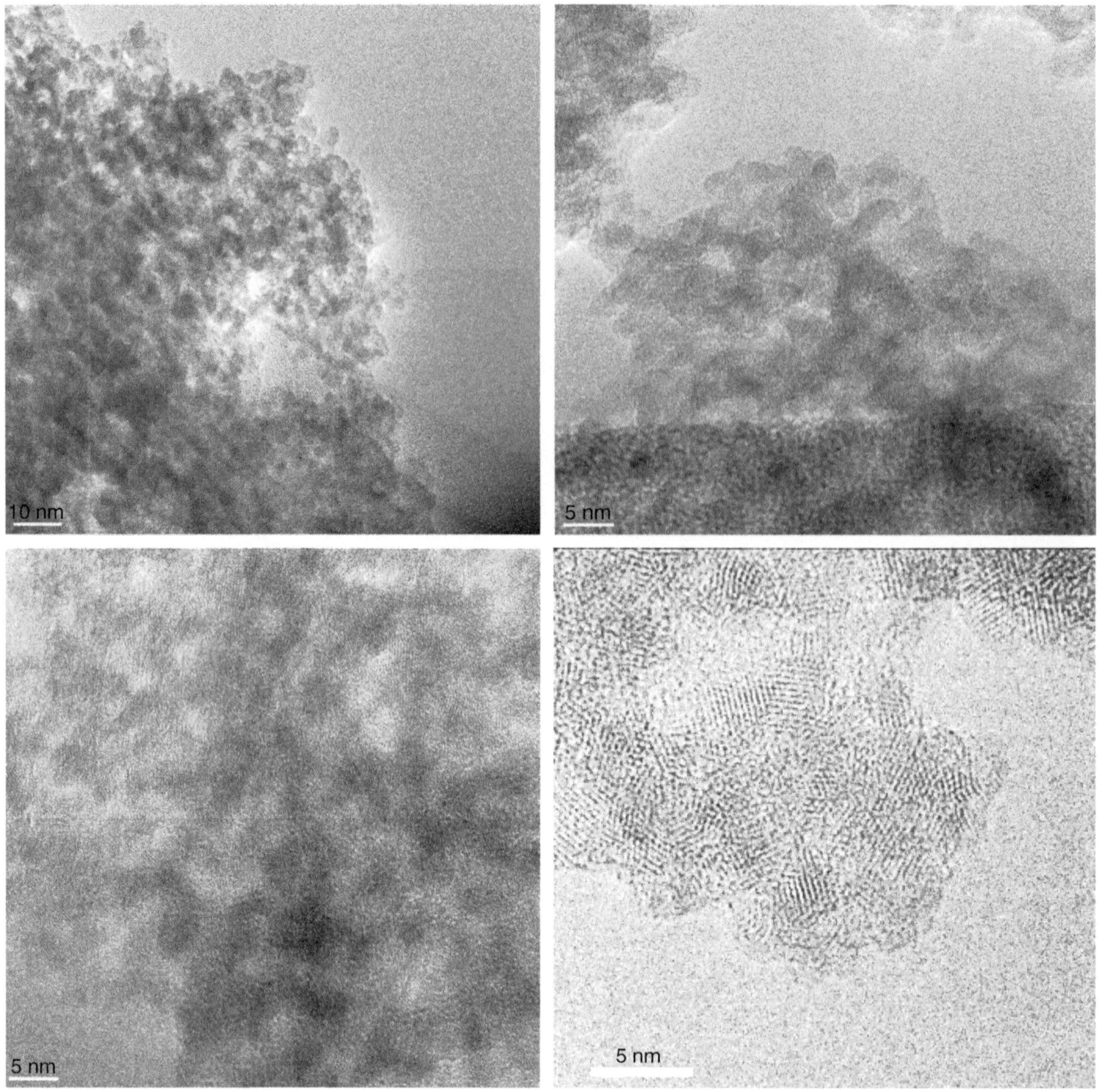

Figure 2

Representative HRTEM images of: (top) ZnS_6h (left 250000X and right 400000X; (bottom) ZnS_6h_HW and ZnS_24h (left and right respectively, both measured at 400000 X).

The TG curve of ZnS_6h is characterized by three distinct steps of weight loss accompanied by definite peaks on the Differential Thermal Analysis (DTA) curve (not shown). The first minor weight loss (4.2%) is centered around 363 K and corresponds to desorption of water molecules weakly bonded to the surface. The next two steps with larger weight losses started around 473 K (maximum of DTA at 620 K) and 773 K (DTA maximum around 830-850 K). To understand the nature of these steps it is useful to compare the profiles of the different samples and corresponding amounts listed in Table 2.

First, the amount of adsorbed water slightly changes upon synthesis and treatment conditions, in a range from 1.3 to 6.6%. This could suggest an effect of treatments or reactants on surface hydrophilicity. For instance the highest amount is observed on samples washed with hot water (ZnS_6h_HW) and prepared with Na_2S. This could be related to the fact that the surface of these samples is less covered by reactants and it is thus more available to water interaction.

This hypothesis is confirmed by comparing the TG profiles of ZnS_6h, ZnS_Na_2S and ZnS_6h_HW. The first looses 18 wt% in step II, while this is reduced to 2.1% in

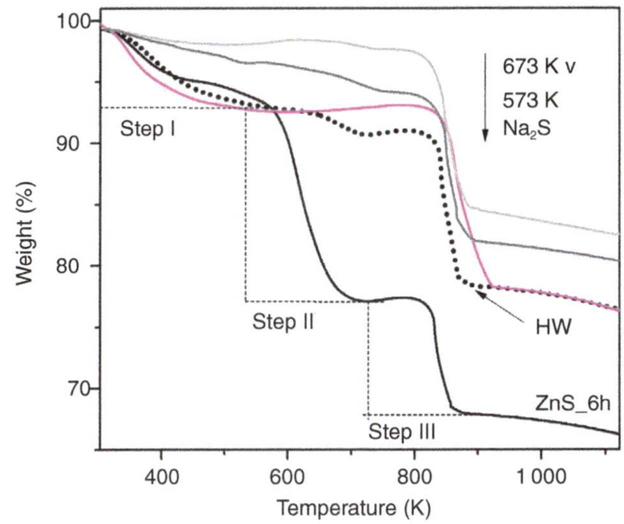

Figure 3

TGA curves of ZnS_6h, as such and after different treatments. Results about ZnS_Na$_2$S are reported for comparison.

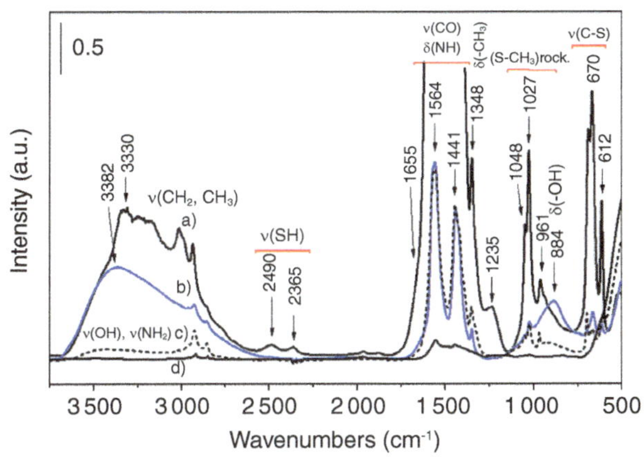

Figure 4

FTIR spectra of ZnS_6h sample as such a) and after post-synthesis treatments: b) ZnS_6h_HW; c) ZnS_573 and d) ZnS_673_V. All samples were outgassed at RT before measurements, expected ZnS_673_V which was outgassed at 673 K.

ZnS_6h_HW and to practically zero in ZnS_Na$_2$S. This clearly indicates that hot water treatment removes weakly adsorbed reactants from the synthesis, in particular TAA, which was not employed to prepare ZnS_Na$_2$S. Noticeably, a small amount of adsorbed species is still present after hot washing, thermal activation at 573 K in air or 673 K in vacuum (2.4 and 0.4 wt%, respectively). In this case, the desorption starts at higher temperature (533-593 K) with respect to untreated sample (453 K) indicating a stronger interaction of the related adsorbed species. These are explained as residual

TABLE 2

Data from thermogravimetric analysis (percentage weight loss) of ZnS_6h as such and after post-synthesis treatments. Data about ZnS_Na$_2$S are reported for comparison

Samples	Step I	Step II	Step III
ZnS_6h	4.2	18.0	9.3
ZnS_6h (N$_2$)*	4.2	20.1	0
ZnS_6h_HW	6.6	2.1	12.4
ZnS_573	2.8	2.4	12.3
ZnS_673_V	1.3	0.4	15.4
ZnS_Na$_2$S	6.6	0	14.5

* Experiment carried out in nitrogen flow instead of air.

acetate groups from the synthesis, as will be discussed in the following.

Finally, step III is similar in all samples, but is not observed when the analysis is carried out in nitrogen (curve not reported, data summarized in Tab. 2). This obviously allows assigning this phenomenon to an oxidative process, that is the transformation of ZnS to ZnO, which in these conditions starts above 800 K irrespective of the samples synthesis or treatment, in agreement with XRD findings (not reported).

2.2.2 Infrared Spectroscopy (FTIR)

The nature of adsorbed species was also studied by infrared spectroscopy, again focusing on sample ZnS_6h and its transformation upon post-synthesis treatments (Fig. 4). The as prepared sample (curve a) shows very intense bands in the whole mid infrared region, which can be easily assigned to adsorbed species, reactants or contaminants from the atmosphere. We observed a broad absorption in the OH stretching region (νOH) between 3 600 and 2 500 cm^{-1}, with superimposed bands between 3 000 and 2 800 cm^{-1} related to aliphatic νCH_2 and νCH_3 modes. At lower frequency, weak bands at 2 940 and 2 365 cm^{-1} are assigned to νS-H groups [56]. The more intense spectral features (out of scale) are found in the 1 750-1 250 cm^{-1} range, where both N-H bending (δNH, found at 1 400 cm^{-1} for TAA in KBr, spectra not reported) and νCO modes are found. Other peaks at lower frequency can be assigned to bending modes of CH_3 in CH_3-S groups, namely symmetric vibrations at 1 385 and rocking at 1 110 cm^{-1}, and to νC-S modes (611 cm^{-1}) as confirmed by comparison with KBr spectra of TAA (not reported). These bands confirm the presence of TAA molecules adsorbed on the sample surface, which could be present as CH_3-S-NH_2 in tautomeric equilibrium with CH_3-SH-NH form [57]. On the contrary, the two strong bands at 1 564 and 1 441 cm^{-1} can be ascribed to the presence of adsorbed acetate groups from the employed reactants. This hypothesis is supported by comparison with the spectra measured on samples ZnS_Na$_2$S, where acetate salts were employed in the synthesis without TAA (bands at 1 560 and 1 430 cm^{-1}, not reported). The presence of acetamide molecules resulting from the hydrolysis of TAA [58] is less evident, since it should be characterized by the typical amide I band (νCO), which is expected with high intensity around 1 690-1 694 cm^{-1} [56].

After hot water treatment (sample ZnS_6h_HW, curve b) most of the bands related to TAA (νS-H, δCH_3-S, νC-S) disappear, in agreement with TG analysis. What is observed in this case is a broad adsorption in the νOH region (3 600-2 500 cm^{-1}) suggesting the presence of OH groups involved in hydrogen bonding. This is confirmed by the observation of a broad band centered at 884 cm^{-1}, related to the corresponding δOH. This could be related to partial surface hydroxylation (Zn-OH groups) and water molecules strongly adsorbed on the surface (not removed by RT outgassing), in agreement with the relatively high affinity for water indicated by TGA. The typical δOH mode of H_2O, expected around 1 620-1 640 cm^{-1} is observable as a shoulder on the tail of the intense band at 1 564 cm^{-1}.

Apart from the bands described above, the main features of sample ZnS_6h_HW are two intense and well defined bands at 1 564 and 1 441 cm^{-1}, with weaker components at 1 348 and 1 027 cm^{-1}, which could be mainly assigned to adsorbed acetate groups (the typical amide I band of acetamide, expected around 1 690 cm^{-1} is missing). The interaction energy of these species with the ZnS NP surface is clearly high, since they are not removed after treatment at 573 K in air (curve c). In this case however, the bands related to OH groups in hydrogen bonding decreases, in agreement with TG analysis.

Finally, an almost totally cleaned surface is obtained only by thermally activating the sample in vacuum at 673 K. In this case, as already pointed out in Reference [29], only small amounts of acetate bands are observed, which could correspond to the weight loss of step II in TG curves (residual amount of 0.4 wt%, Tab. 2) and could be adsorbed on defective sites or polar surfaces of the material [29].

2.2.3 DFT Modeling

DFT modeling has been employed to understand the surface characteristics of the ZnS NP. The surface energies E_s of the lowest index ZnS surfaces, the (110), (111) and (100) are 0.6, 1.0 and 1.7 J/m^2, [29]. Dehydroxylated and "clean" surfaces were considered, which should model the situation corresponding to sample ZnS_673_V, that is of a sample activated in vacuum, with only a small fraction of reactancts adsorbed on reactive surfaces [29]. The size dependence of E_s for the (110) and (111) surfaces is reported in Figure 5

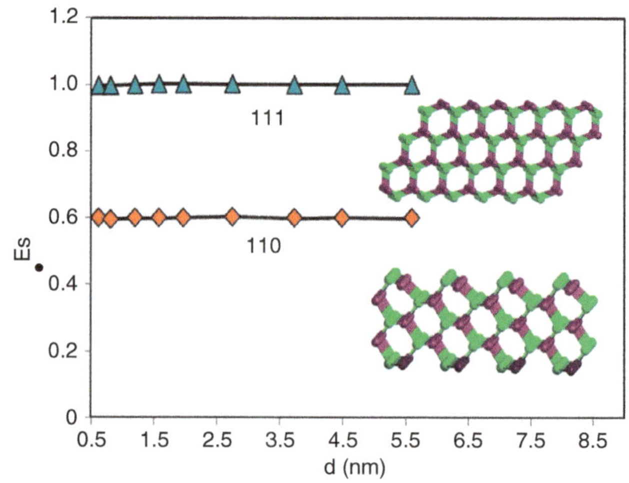

Figure 5

Surface energies E_s (in J/m^2) of the 110 and 111 ZnS surfaces as a function of the size d of the ZnS nanoparticles. Results from B3PW calculations.

(hereafter the 100 surface will be not considered anymore because of its very large E_s). Data reported in the figure, collected for d ranging from 0.5 to 6 nm, show that, in ZnS nanoparticles, the size of the particles does not affect the surface stability. The (110) surface is by far the most stable surface in nanocrystals as it is in microcrystal [55, 59]. The large differences of the surface energies of the (110) and (111) surfaces is related to their geometrical structure. According to Tasker's electrostatic model [60], in zincblende systems only the (110) surface is non polar, whereas (111) and (100) are polar surfaces of type III. Polar surfaces are characterized by two not equivalent layers of opposite charged ions that alternate along the normal to the surface; each repeat unit bears a dipole moment that increases with the number of the layers and determines the larger instability of these surfaces. Considering the large difference in E_s between the two surfaces, the (111) polar one is not likely to be exposed at the ZnS nanoparticles to a sizable extent, regardless of their size, Figure 5.

In order to shed light on the whole scenario laid out by infrared and TGA, suggesting the presence of a small fraction of strongly adsorbed species, the interaction with the (111) and (110) surfaces of formic acid has been considered. This molecule was employed as a simplified model to represent the acetate species present in solution during the synthesis, which are supposed to be the main species adsorbed on the dehydrated and cleaned ZnS surface on the basis of infrared spectroscopy. Adsorbed species are schematically depicted in Figure 6 and their main adsorption features collected in Table 3. On the (110) surface one dissociative

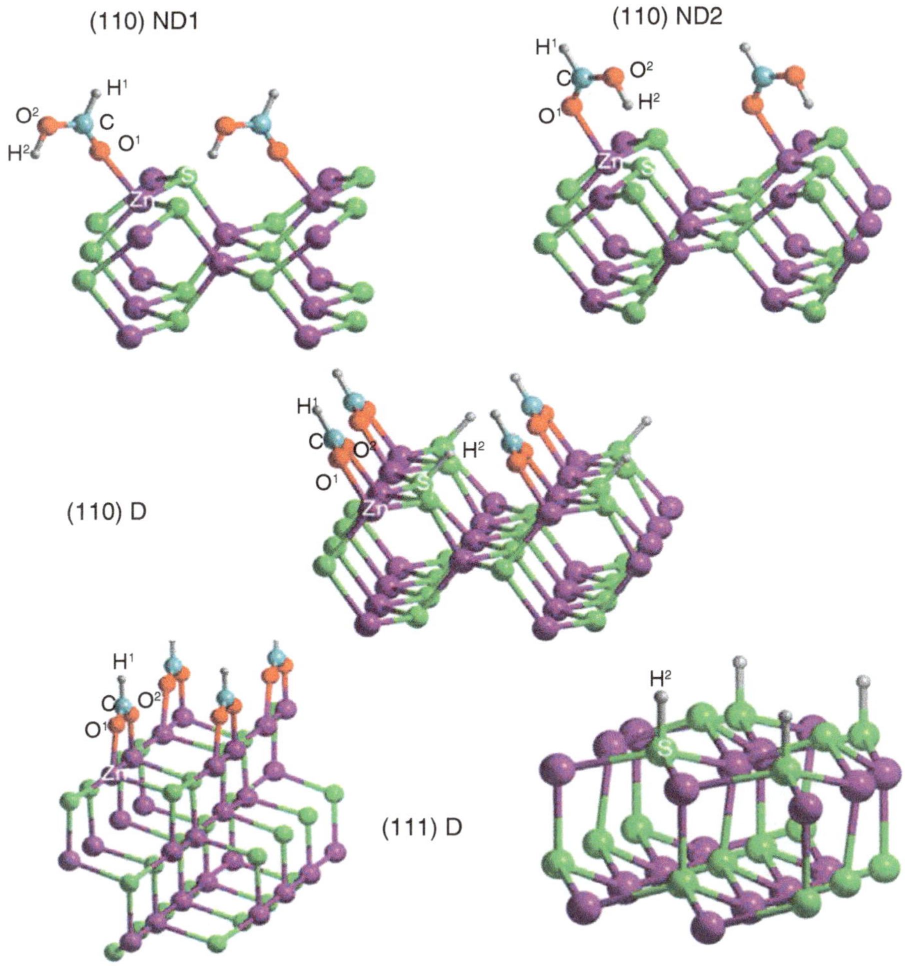

Figure 6

Schematic representation of adsorbed configuration of formic acid on the ZnS (110) and (111) surfaces. Results from PBE0 calculations.

(110-D) and two non dissociative (110-ND1 and 110-ND2) adsorption mode have been considered; only the dissociative adsorption mode can be conceived for the (111) surface (111-D): in this case, the formiate group has been adsorbed on the Zn terminated surface and the proton on the S terminated one. At the (110) surface, 110-ND1 is preferred over 110-ND2 by about 0.04 eV and over 110-D by almost 0.2 eV. The 111-D configuration is twice more stable than 110-ND1 ($E_{ads} = 2.3$ eV), Table 3. Computed vibrational frequencies provide a fingerprint for the recognition of the adsorbed species, Table 3. For the 110-ND1 carboxylic group, the asymmetric stretching ν_{as}(O-C-O) = 1 700 cm^{-1} and the bending δ(C-OH) = 1 237 cm^{-1} are found, while for the 111-D carboxylate group the asymmetric stretching ν_{as}(O-C-O) = 1 570 cm^{-1} is indeed shifted to lower wavenumbers with respect to the undissociated form. Only this latter vibration, the asymmetric stretching of dissociate formiate species, is compatible with FTIR data recorded upon the thermal treatment at 573 K (the peak at 1 564 cm^{-1}, Fig. 4c). The significant reduction of this peak only for treatment at very high T (Fig. 4d) indicates an irreversible character of the adsorption, in agreement with the strong computed adsorption energy for 111-D. This finding suggests the presence in the ZnS NP of small extension of the (111) polar surface stabilized by the adsorption of foreign species from the synthesis solution. To this extent, it is worth mentioning a recent report that combining FTIR observations and DFT calculations, has investigated the adsorption properties of CO molecules, used as a probe for acidic Lewis surface sites; this study has demonstrated that CO experienced only one type of sites belonging to the (110) surface [29].

The results discussed in the present work, confirm the finding that ZnS nanoparticles, regardless of their size, expose to a largely wider extent the most stable (110) surface;

TABLE 3

Main features of the formic acid interaction with the ZnS (110) and (111) surfaces (PBE0 results). Distances, d, in Å; adsorption energies, E_{ads} in eV, vibrational frequencies (ν is for stretching - as antisymmetric and s symmetric, δ_B bendings and δ other angle deformations) in cm^{-1}. To facilitate comparison with experiments, frequencies have been scaled by a factor 0.95521 to align the computed ν(C-H) of gas phase formic acid to the corresponding experimental frequency (2 943 cm^{-1}). PBE0 results

	110-ND1	110-ND2	110-D	111-D	Formic acid (gas phase)
d(C=O)	1.21	1.22	1.26	1.25	1.19
d(C-OH)	1.33	1.31	1.26	1.25	1.34
d(O-Zn)	2.17	2.24	2.00	2.00	–
d(C-C)	7.70	7.70	7.70	7.75	–
d(S-H)	1.98	2.13	1.35	1.35	–
ΔE_{ads}	−0.64	−0.60	−0.46	−2.28	–
ν(O-H)	2 957 (751)	–	–	–	3 612 (113)
ν(C-H)	2 978 (22.3)	–	2 898 (77.0)	2 887 (161)	2 943 (93.3)
ν(S-H)	–	–	2 503 (1.97)	2 507 (14.0)	–
ν(O-CO)	1 702 (397)		1 555 (1 878)^{as}	1 576 (1 158)	1 761 (55.3)
			1 340 (111)^{s}	1 359 (49.4)	–
ν(C-OH)	1 337 (22.5)	–	–	–	1 269
δ_B(C-H)	1 366 (70.1)	–	1 373 (87.1)	1 376 (33.8)	1 347 (42.3)
δ_B(O-H)	1 214		–	–	1 096 (34.4)
δ(C-H)	1 037	–	1 030	1 056	1 015
δ(O-H)	903	–	–	–	717
δ_B(O-C-O)	653 (86.8)	–	731 (10.1)	724 (26.9)	612

nonetheless, the presence of a minor extension of other surfaces cannot be excluded in reasons of electronic and structural reconstruction or because stabilized by adsorbed species [61].

2.3 Electronic Properties

The electronic properties of the prepared ZnS NP samples were investigated by DR UV-Vis spectroscopy (Fig. 7). This is a simple and widespread technique for the estimation of band gap energy (E_g) in semiconductors [34-37, 62, 63], by the measurement of the absorption edge position, through the Tauc plot elaboration (see Experimental for details) [38, 39]. This parameter is quite important for applications in photocatalysis, and in semiconducting materials is dependent on particle size, due to quantum size effect [34-37, 40, 63]. Notwithstanding the small size of the primary ZnS NP prepared in this work, these cannot be defined as Quantum Dots (QD), since the UV-Vis spectra of the latter are characterized by the presence of discrete bands (described by the position of their maximum) instead of a continuum absorption (defined by its edge position) [37].

The spectrum of as prepared ZnS_6h is characterized by a relatively large E_g value (4.22 eV, Tab. 1) with respect to the bulk material (3.68 eV) [40] as a result of nanostructuration [34]. On the contrary, longer reaction time and/or hot water washing result in a shift of the edge position towards higher wavelength values. Indeed, the corresponding E_g values are close to that of bulk material for most of the other ZnS samples with crystallite size (estimated by XRD peak widths) between 3.4-4.9 nm (Tab. 1, entries 2-5). Interestingly, a shift to E_g values higher with respect to bulk ZnS are observed after thermal activation, both in air (ZnS_573) or in vacuum (ZnS_673_V). This could be explained by a higher defectivity of the treated samples surface, influencing their electronic structure.

The considerable dependence of E_g on the size of the ZnS nanoparticles is also confirmed by DFT calculations. Figure 8 reports E_g as a function of the size of the nanoparticles: NP with $d \geq 4.5$-5.5 nm have an energy gap that is actually as

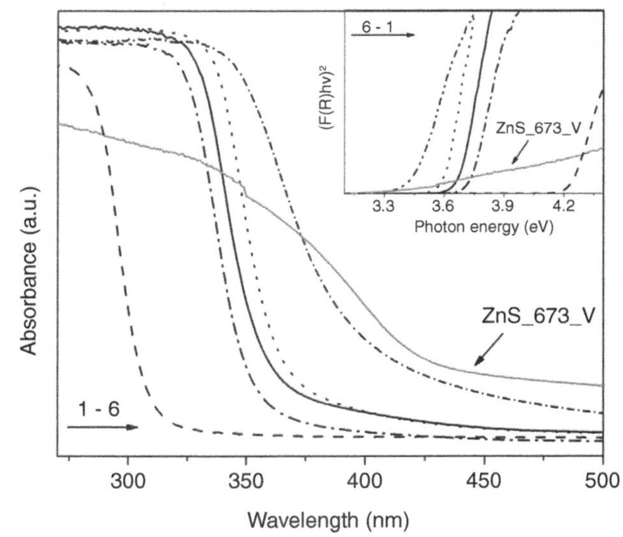

Figure 7

DR UV-Vis spectra of ZnS NP. From left to right: 1) ZnS_6h as such (dashed), 2) ZnS_6h_HW (dot-dashed), 3) ZnS_24h (full), 4) ZnS_24h_HW (dotted), 5) ZnS_573 (short dot-dashed), and 6) ZnS_673_V (full grey).

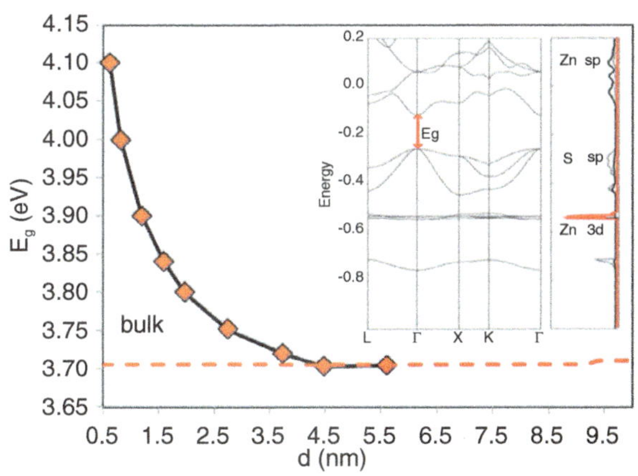

Figure 8

Energy gap, E_g, as a function of the particle size. The dashed red line is the value for bulk ZnS. In the insert band structure and projected density of states for bulk ZnS. B3PW calculations.

large as the bulk value, $E_g = 3.7$ eV, in excellent agreement with observations [40]. In smaller particles, due to quantum size confinement effects, the energy gap increases monotonically by decreasing the size of the NP. For instance $E_g = 4.1$ eV for $d = 1.5$ nm and 3.7 eV for $d = 5$ nm, in very good agreement with XRD estimation (notoriously underestimating particle size) and UV data (Tab. 1). Despite the sensitivity of E_g to the size of the NP, the electron structure is almost unaffected by *d*. As reported in the insert of Figure 8, ZnS is characterized by a direct gap in Γ (see also [40]); the top of the valence band is due to S 3sp states and the bottom of the conduction band mainly from Zn 3sp states. The Zn 3d levels are very deep in energy (about −15 eV) and do not contribute to the valence.

CONCLUSIONS

ZnS NP photocatalysts were prepared by precipitation with TAA and acetate as sulfur and zinc sources, respectively, by modifying reaction time length. The obtained materials were then treated with hot water and thermally, both in air and in vacuum, to remove the reactants adsorbed on the surface of NP during the synthesis. The effect of these procedures on the properties of the ZnS NP was studied by combining different techniques and aspects.

The results show that this synthetic methodology allows to prepare nanostructured ZnS agglomerates, in the form of white powders, with good yield and ease of preparation. Low reaction time (6 h) resulted in agglomerated crystallites with primary size around 2 nm (as determined by XRD and TEM). Particle size was found to increase after washing with hot water, as a result of removal of large amounts of adsorbed chemicals. This was also evidenced by thermogravimetric analysis and infrared spectroscopy, indicating the presence of large amounts of TAA (around 18 wt%) weakly interacting with the surface. The sample prepared at longer reaction time (24 hours) was characterized by larger (around 4 nm), more densely aggregated and crystalline NP, which were less sensitive to the treatment with hot water.

Further analysis of the data showed that a small amount of acetate species (around 2 wt%) was strongly adsorbed on the ZnS NP surface, and could be removed almost totally only after thermal activation in vacuum at 673 K. This behavior was explained on the basis of DFT calculations, showing the strong adsorption of dissociate formic acid (chosen as a model molecule to simulate the acetate salt employed in the reaction) on the (111) polar surface. This surface is unstable with respect to the apolar (110) one, irrespective to the size of the NP, (Fig. 5), but could be stabilized in a small amount by the adsorption of foreign species from the synthesis solution.

Finally, the electronic properties of the prepared materials were analyzed by DR UV-Vis spectroscopy, by calculating the band gap energy E_g with the Tauc plot method. This shows that only the sample prepared at 6 hours reaction (ZnS_6h) without any post-treatment has an E_g value sensibly higher with respect to bulk ZnS (4.2 *vs* 3.7 eV), as a result of nanostructuration. When the size of the primary crystallite moves to 3-4 nm as a result of washing with hot water, longer reaction time or employ of Na_2S, the band gap of the materials corresponds to the bulk one. This trend is rationalized by DFT calculations on ZnS NP, predicting the modification of the semiconducting electronic structure as a function of particle size.

ACKNOWLEDGMENTS

Italian Ministry for Research and University (MIUR) is gratefully acknowledged for funding project PRIN 2009BLNJC5 "Optimization of the photothermocatalytic sulfur-ammonia process for hydrogen production". A.M. F. acknowledges CINECA (Iscra project B HOME) for computational resources.

REFERENCES

1 Honda K., Fujishima A. (1972) Electrochemical Photolysis of Water at a Semiconductor Electrode, *Nature* **238**, 37-38.

2 Kato H., Asakura K., Kudo A. (2003) Highly Efficient Water Splitting into H_2 and O_2 over Lanthanum-Doped $NaTaO_3$ Photocatalysts with High Crystallinity and Surface Nanostructure, *J. Am. Chem. Soc.* **125**, 10, 3082-3089.

3 Chen D., Ye J. (2007) $SrSnO_3$ Nanostructures: Synthesis, Characterization, and Photocatalytic Properties, *Chem. Mater.* **19**, 18, 4585-4591.

4 Domen K., Kudo A., Ohnishi T. (1986) Mechanism of photocatalytic decomposition of water into H_2 and O_2 over NiO-$SrTiO_3$, *J. Catal.* **102**, 92-98.

5 Zou Z., Ye J., Sayama K., Arakawa H. (2001) Direct splitting of water under visible light irradiation with an oxide semiconductor photocatalyst, *Nature* **414**, 625-627.

6 Luo J., Maggard P.A. (2006) Hydrothermal Synthesis and Photocatalytic Activities of $SrTiO_3$-Coated Fe_2O_3 and $BiFeO_3$, *Adv. Mater.* **18**, 4, 514-517.

7 Woodhouse M., Parkinson B.A. (2008) Combinatorial Discovery and Optimization of a Complex Oxide with Water Photoelectrolysis Activity, *Chem. Mater.* **20**, 7, 2495-2502.

8 Bao N., Shen L., Takata T., Domen K. (2008) Self-Templated Synthesis of Nanoporous CdS Nanostructures for Highly Efficient Photocatalytic Hydrogen Production under Visible Light, *Chemistry Materials* **20**, 1, 110-117.

9 Jing D., Guo L. (2006) A Novel Method for the Preparation of a Highly Stable and Active CdS Photocatalyst with a Special Surface Nanostructure, *J. Phys. Chem. B* **110**, 23, 11139-11145.

10 Reber J.-F., Meier K. (1984) Photochemical production of hydrogen with zinc sulphide suspensions, *J. Phys. Chem.* **88**, 5903-5913.

11 Kudo A., Tsuji I., Kato H. (2002) AgInZn7S9 solid solution photocatalyst for H-2 evolution from aqueous solutions under visible light irradiation, *Chem. Commun.* **17**, 1958-1959.

12 Kudo A., Niishiro R., Iwase A., Kato H. (2007) Effects of doping of metal cations on morphology, activity, and visible light response of photocatalysts, *Chem. Phys.* **339**, 1-3, 104-110.

13 Kudo A., Miseki Y. (2009) Heterogeneous photocatalyst materials for water splitting, *Chem. Soc. Rev.* **38**, 1, 253-278.

14 Rockenberger J., Troger L., Kornowski A., Vossmeyer T., Eychmuller A., Feldhaus J., Weller H. (1997) EXAFS studies on the size dependence of structural and dynamic properties of CdS nanoparticles, *J. Phys. Chem. B* **101**, 14, 2691-2701.

15 Nosaka Y., Shigeno H., Ikeuchi T. (1995) Formation of polynuclear cadmium-thiolate complexes and cds clusters in aqueous-solution studied by means of stopped-flow and NMR spectroscopies, *J. Phys. Chem.* **99**, 20, 8317-8322.

16 Vogel W., Borse P.H., Deshmukh N., Kulkarni S.K. (2000) Structure and stability of monodisperse 1.4-nm ZnS particles stabilized by mercaptoethanol, *Langmuir* **16**, 4, 2032-2037.

17 Kortan A.R., Hull R., Opila R.L., Bawendi M.G., Steigerward M.L., Carroll P.J., Brus L.E. (1984) *J. Am. Chem. Soc.* **106**, 6285-6295.

18 Dance I.G., Choy A., Scudder M.L. (1984) Syntheses, properties, and molecular and crystal structures of (Me4N)4[E4M10(SPh)16] (E = sulfur or selenium; M = zinc or cadmium): molecular supertetrahedral fragments of the cubic metal chalcogenide lattice, *J. Am. Chem. Soc.* **106**, 21, 11.

19 Mokili B., Charreire Y., Cortes R., Lincot D. (1996) *Thin Solid Films* **288**, 21-28.

20 Kortan A.R., Hull R., Opila R.L., Bawendi M.G., Steigerwald M.L., Carroll P.J., Brus L.E. (1990) Nucleation and growth of CdSe on ZnS quantum crystallite seeds, and *vice versa*, in inverse micelle media, *J. Am. Chem. Soc.* **112**, 4, 1327-1332.

21 Calandra P., Longo A., Liveri V.T. (2003) Synthesis of ultra-small ZnS nanoparticles by solid-solid reaction in the confined space of AOT reversed micelles, *J. Phys. Chem. B* **107**, 1, 25-30.

22 Meneau F., Sankar G., Morgante N., Winter R., Catlow C.R.A., Greaves G.N., Thomas J.M. (2003) Following the formation of nanometer-sized clusters by time-resolved SAXS and EXAFS techniques, *Faraday Discuss.* **122**, 203-210.

23 Meneau F., Cristol S., Sankar G., Dolbnya I.P., Bras W., Catlow C.R.A., Thomas J.M., Greaves G.N. (2003) *In situ* study of the formation of CdS nanoparticles by small-angle X-ray scattering, *J. Appl. Crystallogr.* **36**, 718-721.

24 Celikkaya A., Mufit A. (1990) Preparation and mechanism of formation of spherical submicrometer zinc sulfide powders, *J. Am. Ceram. Soc.* **65**, 198329, 2360-2365.

25 Liu G., Zhao L., Ma L., Guo L. (2008) Photocatalytic H_2 evolution under visible light irradiation on a novel $Cd_xCu_yZn_{1-x-y}S$ catalyst, *Catal. Commun.* **9**, 5.

26 Karar N. (2007) Photoluminescence from doped ZnS nanostructures, *Solid State Commun.* **142**, 4.

27 Gruy F., Mekki-Berrada M.K., Cournil M. (2009) Precipitation dynamics of zinc sulfide multi-scale agglomerates, *AIChE J.* **55**, 10, 10.

28 Berlier G., Meneau F., Sankar G., Catlow C.R.A., Thomas J.M., Spliethoff B., Schueth F., Coluccia S. (2006) Synthesis and characterisation of small ZnS particles, *Res. Chem. Intermed.* **32**, 7, 683-693.

29 Balantseva E., Berlier G., Camino B., Lessio M., Ferrari A.M. (2014) Surface Properties of ZnS Nanoparticles: A Combined DFT and Experimental Study, *J. Phys. Chem. C* **118**, 41, 23853-23862.

30 Monshi A., Foroughi M.R., Monshi M.R. (2012) Modified Scherrer Equation to Estimate More Accurately World Nano-Crystallite Size Using XRD, *Journal of Nano Science and Engineering* **2**, 154-160.

31 Brunauer S., Emmet P.H., Teller E. (1938) Adsorption of Gases in Multimolecular Layers, *J. Am. Chem. Soc.* **60**, 309-319.

32 Barret E.P., Joyner L.G., Halenda P.P. (1951) The Determination of Pore Volume and Area Distributions in Porous Substances. I. Computations from Nitrogen Isotherms, *J. Am. Chem. Soc.* **73**, 1, 373-380.

33 Webb P.A., Orr C. (1997) *Analytical Methods in Fine Particle Technology*, Micromeritics Instrument Corp, Norcross, GA, USA.

34 Weller H. (1993) Colloidal Semiconductor Q-Particles - Chemistry in the Transition Region between Solid-State and Molecules, *Angew. Chem. Int. Ed.* **32**, 1, 41-53.

35 Brus L.E. (1984) Electron–electron and electron-hole interactions in small semiconductor crystallites: The size dependence of the lowest excited electronic state, *J. Chem. Phys.* **80**, 9, 4403.

36 Calandra P., Goffredi M., Liveri V.T. (1999) Study of the growth of ZnS nanoparticles in water/AOT/n-heptane microemulsions by UV-absorption spectroscopy, *Colloid Surf. A* **160**, 1, 9-13.

37 Manyar H.G., Iliade P., Bertinetti L., Coluccia S., Berlier G. (2011) Structural and spectroscopic investigation of ZnS nanoparticles grown in quaternary reverse micelles, *J. Colloid Interface Sci.* **354**, 2, 511-516.

38 Grozdanov I., Najdoski M. (1995) Optical and electrical-properties of copper sulfide films of variable composition, *J. Solid State Chem.* **1142**, 469-475.

39 Saravanan R.S.S., Pukazhselvan D., Mahadevan C.K. (2012) Studies on the synthesis of cubic ZnS quantum dots, capping and optical-electrical characteristics, *J. Alloys Compd.* **517**, 139-148.

40 Amirtharaj P.M., Seiler D.G. (2009) *Optical properties of semiconductors, in Handbook of Optics*, McGraw Hill, New York, NY.

41 Dovesi R., Saunders V.R., Roetti C., Orlando R., Zicovich-Wilson C.M., Pascale F., Civalleri B., Doll K., Harrison N.M., Bush I.J., et al. CRYSTAL09 User's Manual, University of Torino.

42 Becke A.D. (1988) Density-functional exchange-energy approximation with correct asymptotic-behavior, *Phys. Rev. A* **38**, 6, 3098-3100.

43 Perdew J.P., Wang Y. (1992) Accurate and simple analytic representation of the electron-gas correlation-energy, *Phys. Rev. B* **45**, 23, 13244-13249.

44 Adamo C., Barone V. (1999) *J. Chem. Phys.* **110**, 6158-6170.

45 Stevens W.J., Krauss M., Bash H., Jaisen P.G. (1992) *Can. J. Chem.* **70**, 612-630.

46 Mino L., Ferrari A.M., Lacivita V., Spoto G., Bordiga S., Zecchina A. (2011) CO Adsorption on Anatase Nanocrystals: A Combined Experimental and Periodic DFT Study, *J. Phys. Chem. C* **115**, 15, 7694-7700.

47 Mino L., Spoto G., Ferrari A.M. (2014) CO_2 Capture by TiO_2 Anatase Surfaces: A Combined DFT and FTIR Study, *J. Phys. Chem. C* **118**, 43, 25016-25026.

48 Hamad S., Catlow C.R.A. (2006) Computational study of the relative stabilities of ZnS clusters, for sizes between 1 and 4 nm, *J. Cryst. Growth* **294**, 1, 2-8.

49 Hamad S., Catlow C.R.A., Spano E., Matxain J.M., Ugalde J.M. (2005) Structure and properties of ZnS nanoclusters, *J. Phys. Chem. B* **109**, 7, 2703-2709.

50 Hamad S., Cristol S., Catlow C.R.A. (2005) Simulation of the embryonic stage of ZnS formation from aqueous solution, *J. Am. Chem. Soc.* **127**, 8, 2580-2590.

51 Hamad S., Woodley S.M., Catlow C.R.A. (2009) Experimental and computational studies of ZnS nanostructures, *Mol. Simul.* **35**, 12-13, 1015-1032.

52 Boys S.F., Bernardi F. (1970) The calculation of small molecular interaction by the difference of separate total energies. Some procedures with reduced errors, *Mol. Phys.* **19**, 553-566.

53 Murugadoss G. (2011) Synthesis, optical, structural and thermal characterization of Mn^{2+} doped ZnS nanoparticles using reverse micelle method, *J. Lumin.* **131**, 10, 2216-2223.

54 Wang L., Zhao C., Meng F., Huang S., Yuan X., Xu X., Yang Z., Yang H. (2010) Optical properties and simultaneous synthesis of ZnS and ZnO nanoparticles *via* one reverse micellar system, *Colloid Surf. A* **360**, 1-3, 205-209.

55 Emin S., Lisjak D., Pitcher M., Valant M. (2013) Structural and morphological transformations of textural porous zinc sulfide microspheres, *Microporous Mesoporous Mater.* **165**, 185-192.

56 Socrates G. (2006) *Infrared and Raman characteristic group frequencies*, John Wiley & Sons Ltd, Chichester, England.

57 Rosenthal D., Taylor T.I. (1957) A Study of the Mechanism and Kinetics of the Thioacetamide Hydrolysis Reaction, *J. Am. Chem. Soc.* **79**, 11, 2684-2690.

58 Gunning H.E. (1955) Thioacetamide as a Sulfide Presipitant in Qualitative and Quantitative Analysis, *J. Chem. Educ.* **32**, 5, 258-259.

59 Siriwardane R.V., Woodruff S. (1997) *In situ* Fourier transform infrared characterization of sulfur species resulting from the reaction of water vapor and oxygen with zinc sulfide, *Ind. Eng. Chem. Res.* **36**, 12, 5277-5281.

60 Tasker P.W. (1979) *J. Phys. Chem.* **12**, 4977.

61 Goniakowski J., Finocchi F., Noguera C. (2008) Polarity of oxide surfaces and nanostructures, *Rep. Prog. Phys.* **71**, 016501.

62 Alivisatos A.P. (1996) Semiconductor clusters, nanocrystals, and quantum dots, *Science* **271**, 5251, 933-937.

63 Brus L. (1986) Electronic Wave Functions in Semiconductor Clusters: Experiment and Theory, *J. Phys. Chem.* **90**, 2255-2560.

11

Solar Production of Fuels from Water and CO_2: Perspectives and Opportunities for a Sustainable Use of Renewable Energy

R. Passalacqua*, G. Centi and S. Perathoner

University of Messina and INSTM/CASPE, Department of Electronic Engineering, Industrial Chemistry and Engineering, V.le F. Stagno d'Alcontres 31, 98166 Messina - Italy

e-mail: rpassalacqua@unime.it - centi@unime.it - perathon@unime.it

* Corresponding author

Abstract — *The increasing energy demand, the depletion of fossil fuels and the concern of maintaining clean environment have become the main reasons for a worldwide attention on renewable energy production. Fuel and energy productions from sunlight represent exciting challenges in the next future thank to the recent developments in related technologies, catalysts and materials. It will be discussed the present economic data concerning energy request, the current technological issues required to face an increase of the use of renewable energy, the main drawbacks related to low conversion efficiency in energy applications, the new routes for producing renewable hydrogen and the bio-mimicking approach provided by artificial leaves. Finally, the critical role of nanoscale engineered processes for the development of efficient and cost-effective systems will be evidenced.*

Résumé — **Production solaire de carburants à partir de l'eau et de CO_2 : perspectives et opportunités pour une utilisation durable de l'énergie renouvelable** — La demande croissante d'énergie, l'épuisement des combustibles fossiles et la préoccupation de maintenir un environnement propre sont devenus les principales raisons de l'intérêt mondial pour la production d'énergie renouvelable. Les combustibles et les énergies produites par le soleil représentent des défis passionnants dans un prochain avenir du fait de récents développements dans des technologies connexes, les catalyseurs et les matériaux. Seront examinés les données économiques actuelles concernant la demande d'énergie, les enjeux technologiques nécessaires pour faire face à une augmentation de l'utilisation des énergies renouvelables, les principaux inconvénients liés à la faible efficacité de conversion dans les applications de l'énergie, les nouvelles voies de production hydrogène renouvelable et l'approche bio inspirée fournie par des feuilles artificielles. Enfin le rôle critique de l'échelle nanométrique pour les procédés d'ingénierie de développement de systèmes efficaces et rentables sera considéré.

INTRODUCTION

Energy is an essential component of the strategies to sustainability and security. To address the issues related to the increasing energy demand, the depletion of fossil fuels and the concern of maintaining a clean environment, a marked transformation of the energy system is necessary. Moving from the actual fossil-fuel-centric energy system [1, 2] to a sustainable one based on the efficiency of production and use, and on a large utilization of Renewable Energy (RE) sources [3] requires rethinking the energy system and how to transform the actual system to a carbon neutral, distributed and integrated multi-energy structure.

Solar fuels, where this term indicates fuels produced with the utilization of RE sources and having a carbon negative impact (thus in principle excluding biofuels having an almost carbon-neutral impact) [4-8], play a key role to reach this objective. However, it is necessary to produce liquid-fuel equivalent to those currently in use, in order to allow a progressive introduction in the actual energy infrastructure, without the massive investments necessary to change it, as required for an H_2-economy, for example. The use of carbon dioxide as raw material for solar fuels enables the realization of this objective, and for this reason it is a key element enabling this transformation to a sustainable energy [9] as well as to a resource-efficient, low-carbon society and economy [10-12].

The introduction of RE and raw materials in process and chemical industry is part of this general objective. While this concept has been mainly limited to the use of bio-based raw materials up to now, the combination of CO_2 and RE to produce the raw materials for process and chemical industry will have a great impact on a short-term in transforming the actual production to a new scenario for chemical and energy production based on the synergic integration between CO_2 (re)use, RE and biomass utilization [11, 12]. There are various motivations for this change, from a strategic (the need to have competitive alternative to keep low the cost of fossil fuels, as well as to diversify supplies and suppliers) to environmental and social ones (carbon emission and other environmental and political concerns) and related legislative limitations or incentives.

Although attention in literature on the conversion paths for CO_2 is rapidly increasing [13-18], the attention is mainly focused on the technical and scientific aspects of specific routes of carbon dioxide conversion, without an analysis of their sustainability and impact on the future scenario for energy and chemical production, the integration of these aspects and the opportunities opened for R&D and to meet societal challenges. This perspective contribution aims to discuss these general aspects and how they reflect on new industrial opportunities and on society. The focus is especially on the CO_2 conversion path using RE and the possibilities opened in short-medium term to enable a trading on a worldwide scale of RE equivalent to that actually present using fossil-fuels (and derived products) as energy vector and raw material source. The products of CO_2 conversion can be the equivalent energy vector and raw material sources for a sustainable future low-carbon society [19-21]. In particular, the presence of a large untapped potential of RE in remote areas will be evidenced. Its use for the conversion of CO_2 (to methanol, for example) can enable an effective transition to a larger share of RE in the energy mix, via long-distance transport of RE. It will be also shown that the potential estimated impact of this technology on the mitigation of climate change is larger than that of CCS (over 7 Gt CO_2 equivalents).

On a long-term perspective, solar fuels and chemicals from CO_2 will be at the basis for a distributed and personalized (economy 3.0) production of energy and chemicals in artificial-leave type devices [22-30]. Here, some advances in this direction will be discussed.

1 SOLAR FUELS AND CHEMICALS FOR THE NEW SUSTAINABLE SCENARIO

The use of solar energy as the backbone for future sustainable energy scenario derives from three basic characterizing elements: clean, cheap and abundant. But solar power is diffuse (ca. 170 Wm^{-2}) and intermittent. To go beyond the current limits of its utilization it is necessary to concentrate it for distribution and storage on massive scale. Solar energy can be captured and stored directly in the chemical bonds of a chemical, or "fuel", and then used when needed (*i.e.* transported and distributed). These solar fuels should be easily transported using existing distribution networks, overcoming current limits regarding long-distance distribution, including for mobile applications, intermittency of RE sources, and utilization in the current energy infrastructure.

Over the last fifteen years the drive to develop systems to produce solar fuels on large scales has been an area of increasingly intense global research activity. Significant progresses in artificial photosynthesis, natural photosynthesis and thermochemical approaches to produce solar fuels have been made [31-37]. In Europe, large interdisciplinary programs have been activated. Some examples are: NANO-structured Photoelectrodes for Energy Conversion (NANOPEC), and Nanodesigned Electrochemical Convertor of Solar Energy in Hydrogen Hosting Natural Enzymes or their Mimics (SOLHYDROMICS) projects funded as part of the European Commission's Seventh Framework Programme (FP7) for research together with SOLARH2, led by Uppsala University in Sweden. This concept of making fuels from sunlight is also now attracting commercial interest, as confirmed by joint initiative involving companies and research institutes like BioSolar Cells, founded by the Dutch government to explicitly develop an incubation structure for new industrial business, or the Collaborative Research Projects (CRP) sustained by the European Science Foundation's EuroSolarFuels programme.

In the future energy scenario a relevant role is played by CO_2 in term of reduced emissions, re-use, and conversion [38]. Carbon dioxide combined with other materials such water can be transformed into fuels and useful feedstocks using sunlight. The reduction process can either be achieved directly in a photoreactor or indirectly using solar-derived hydrogen as a reductant. The obtained products include carbon monoxide, formic acid, methanol and methane [39-53]. These carbon-based feedstocks together with hydrogen are widely used in industry to make large-volume chemicals, such as fertilizers, pharmaceuticals, plastics and synthetic fuels. Currently, these products (and the energy required to produce them) are based on fossil fuels. The replacement of the traditional (fossil-fuel-based) raw materials and energy sources would have a major breakthrough in reducing carbon footprint of our society. It is out of the scope to discuss here the wide range of possible products and options to convert CO_2, which are analysed in many papers, books and reviews [4, 10-12, 15-18, 33-56]. Here, we may shortly identify three main classes of products:

- advanced materials,
- fine chemicals,
- fuels.

Except for the area of polymers containing CO_2, in particular polyurethanes and polycarbonates, and few other materials in the area of fine chemicals, most of the investigated products are still far from the practical application.

Although there are different motivations, we believe that a general issue derives from the lack of considering the real potential of using carbon dioxide, reactions to target and synergy of the system. The value of CO_2 derives from considering its whole impact on energy and chemical production, and the synergies which can be obtained, including bio-based economy [11, 12]. It is necessary to consider the whole CO_2-economy and address large-scale productions [38], evaluate how CO_2 recycling can become a key element of sustainable carbon-resource management in chemical and energy companies [57], consider CO_2 as the strategic molecule for the progressive introduction of RE resources into the chemical and energy chain [10]. In other words, it is necessary to turn the perspective on CO_2 use, and move from specific cases to a system approach, *e.g.* a CO_2-economy, to really understand the potential of innovation. Some examples will be discussed below.

1.1 Enabling a New Scenario for Large-Scale Trading of RE through Use of CO_2

There are some key concepts which should be shortly introduced to understand this aspect. A first element is the availability on worldscale of very large amounts of unexploitable RE sources, about 10 TWh only considering hydropower. A main reason is the lack of suitable users around the possible production areas and the actual impossibility to transport on long-distance the electrical energy, the main actual output of RE technologies. The second concept is the intermittency of power output which characterizes several of the current RE technologies, such as solar cells, wind turbines, etc. The third concept is the need of an energy vector (or raw material for chemistry) which is essentially a drop-in product of those used in the actual fossil-fuel based energy (and chemical) infrastructure. The last concept is that CO_2, due to the fact that is low in the thermodynamic energy content scale, is an opportunity to store RE in the process of its transformation to more valuable (higher energy content) chemicals or fuels. Therefore, there is a unique opportunity to combine all these aspects in the challenge of enabling a new scenario for large-scale trading of RE through CO_2 use.

There are different routes to convert and utilize CO_2, the most relevant of which (to achieve the above objectives) involves the hydrogenation of CO_2 to form oxygenates and/or hydrocarbons. Possible energy vectors that derive from the hydrogenation of CO_2, either directly or through the intermediate stage of the Reverse Water Gas Shift (RWGS) reaction, are the following:

- formic acid, which may be used in formic acid fuel cells or as a vector to store and transport H_2 (the reaction of synthesis is reversible, and formic acid can be catalytically decomposed under mild conditions to form back H_2 and CO_2);
- methanol and DiMethylEther (DME);
- methane (Substituted Natural Gas, SNG);
- >C1 alcohols or hydrocarbons.

The main routes are chemical (catalytic), but electrochemical or solar thermal routes are also possible, even if the latter two routes are still not sufficiently developed. Syngas may also be produced by reaction with hydrocarbons (particularly methane) through the so-called dry reforming. However, dry reforming is an endothermic reaction occurring at high temperature (about 900-1000°C) and with formation of carbon (which deactivates the catalyst) as a side reaction.

The catalytic chemistry of the RWGS reaction, the following transformation to methanol/DME, or hydrocarbons via Fischer–Tropsch synthesis, and the subsequent production of gasoline (Methanol To Gasoline, MTG) or of Diesel via hydrocracking of the alkanes produced in the FT process (using Co-based catalysts) is well established, even if there is still need for development because of the change of feed composition starting from CO_2 rather than from syngas. Also, in terms of process development, most of the necessary knowledge is available.

Between these different routes, we consider preferable the production of methanol from CO_2, due to the efficiency and high selectivity, the formation of a liquid product highly versatile in utilization, safe and easy to transport and distribute, well integrable within the actual infrastructure. However, there are also different opinions.

There are various possible routes to produce methanol from CO_2 using RE sources. We will call here Renewable Methanol (RM) that deriving 100% from fossil-fuel-free sources. Both the H_2 utilized in the process is produced from water and RE, and the CO_2 derives from non fossil-fuel sources, for example by biomass fermentation. RM is a negative carbon footprint product, reducing of over 70% the carbon footprint (on Life-Cycle-Assessment, LCA, bases) with respect to the "conventional" methanol produced today mainly from natural gas. Also with respect to biofuels, the carbon footprint is about 30% lower (on equal energy content), allowing to move from almost carbon-neutral biofuels to really carbon-negative fuels. RM can be produced ideally from the photocatalytic or PhotoElectroCatalytic (PEC) reduction of CO_2, with the electrons and protons needed for this reaction generated by water photoelectrolysis. However, the actual production rates are still too far from those necessary for possible exploitation, even in medium term. Similarly, the electrocatalytic reduction of CO_2 to methanol is still quite far from the possible application on large scale [58, 59]. The only current route involves the catalytic conversion of CO_2 by heterogeneous catalysis [57, 60]. Several attempts, some also rather interesting, have been made to realize this process in liquid phase using homogeneous catalysts [61]. However, when the rates are analysed in terms of the industrially relevant parameters (*e.g.* productivity per reactor volume, instead that turnover frequency or number) and production costs, there is a factor of over two order of magnitude to overcome. In the short-term, also this route appears not feasible. On the other hand, methanol is currently produced in about 60 Mtons per year by heterogeneous catalysis, at reaction temperatures about 220°C and pressures of about 70-100 bar, but starting from syngas (CO/H_2 plus some CO_2 up to a maximum of about 3%). $Cu/ZnO/Al_2O_3$ are the base elements of the catalysts. These catalysts are also active in RWGS, thus in principle able to operate from CO_2/H_2 mixtures, but productivities are significantly lower. There are two main problems to overcome. First, the CO_2/H_2 mixture is more oxidizing than the $CO/CO_2(3\%)/H_2$ altering the active state of the catalyst. Second, the higher formation of water inhibits reversibly the reaction. However, by suitable doping it is possible to overcome these problems, even if a further improvement in productivity to methanol could be necessary.

The more severe actual limit regards the pressure, although often not accounted in several literature studies. To have enough productivity and conversion in methanol synthesis from CO_2, it is necessary to operate at pressures at least above 50 bar, but compression of H_2 may raise costs above an acceptable level. It is necessary to produce H_2 already at this pressure. Actual photochemical routes (water splitting, etc.) are all operating at near ambient pressure, as well as thermochemical and biological routes [62-65]. In addition to the still very low productivity (about 100 times lower than necessary), the pressure gap is a relevant issue. The only technology available in the short-middle term (except perhaps the production of H_2 from biowaste) is the water electrolysis, either at low temperature (around 70°C) on PEM-type electrolyzers or at higher temperatures (above 600°C) on Solid Oxide Electrolysis Cells (SOEC) [66, 67]. However, current technologies do not allow stable operations above about 30 bar. It is thus still necessary to improve technology and electrocatalysts in these devices, as well as to reduce overpotential, for example by a better design of the electrocatalysts [68].

Hydropower is the preferable source of RE to consider in first instance to estimate the feasibility of a new scenario for large-scale trading of RE through CO_2 use. There are various motivations. Hydropower is the largest renewable source, accounting for about 80% of the actual world renewable electricity generation. The global technically exploitable hydropower potential is estimated to be over 16 PWh per year, but nearly 80% of which not currently exploitable being in remote areas missing of the necessary local uses and too far to be transported to other areas by grid. A wide part of this hydropower potential can be utilized only if effective ways to transport this unused RE over long distances will be developed.

In addition, hydropower is the cheapest production of RE on a large scale, is a ready technology, and it provides a nearly constant energy supply, differently from most of the alternative RE sources (wind, solar, etc.) that are intermittent. The latter aspect has a relevant feature for cost estimation. In fact, the fixed costs of the technology (which account for up to 70% of the overall cost) for converting electrical energy to chemical energy (methanol), can be amortized over the entire year for a constant RE source, such as hydropower, but only for a fraction of year for the discontinuous ones (*i.e.* wind and solar).

A techno-economic assessment of the cost of production of RM by this route (CO_2 transport in remote areas, where H_2 can be produced at low cost from hydropower, conversion of CO_2 to methanol and finally transport back of RM) was made by Barbato *et al.* [69]. Figure 1a schematically illustrates the concept of remote production of methanol from CO_2, where cheap sources of electrical energy (by hydropower) exist, which can be utilized to produce H_2 and then RM. Carbon dioxide is transported there (by ship, for example) and RM transported back. The difference of energy between RM and CO_2 is nearly the imported RE. In fact, the energy loss due to transport is quite limited and there is a minimal loss of energy in the conversion of CO_2 to methanol. The main loss is associated to the production of H_2 itself by electrolysis.

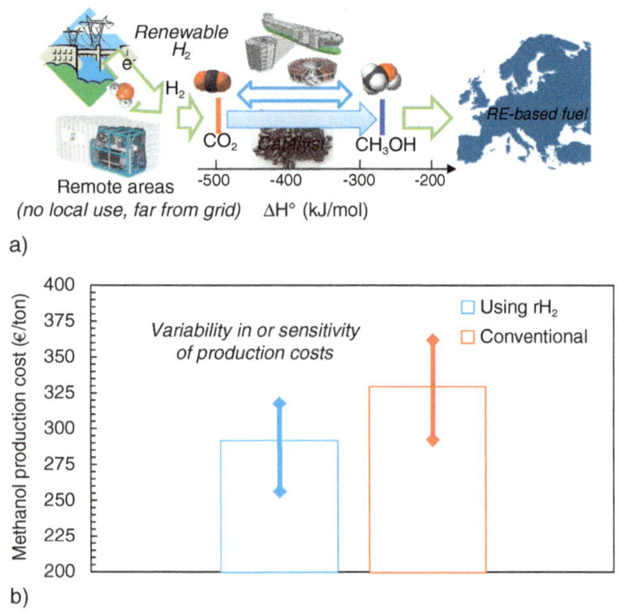

Figure 1

a) Simplified illustration of the concept of remote production of methanol from CO_2. b) Estimated cost to produce RM by this route, based on a cost of production of H_2 of 0.140 € Nm^3 H_2, in comparison with market contract price for methanol produced by "conventional" route (from natural gas). See text for comments regarding variability in or sensitivity of the production costs. Based on the data in Reference [69].

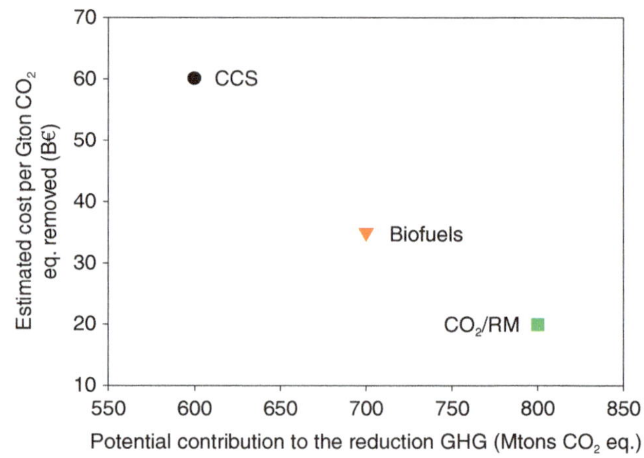

Figure 2

Estimation for Europe by year 2050 of the potential contribution on the reduction of CO_2 emissions by CCS, use of biofuels and CO_2/RE. Based from the data in Reference [69].

Figure 1b reports the estimated cost to produce RM by this route, based on a production cost of H_2 of 0.140 € Nm^3 H_2 [69]. The arrow in the bar represents the sensitivity variation of production cost with respect to a cost of carbon dioxide at site of 20-70 €/ton and of an electricity consumption for H_2 production between 3.8 and 4.8 kWh/Nm^3 H_2. The arrow in the bar for "conventional" RM cost represents the variability in market contract price (in Europe, Aug. 2013). Although these are estimations, and market price is highly variable and depends on many factors, the message in Figure 1 is that it is already possible today to produce RM and trade/import RE on world scale by using this technology, when low-cost electrical energy is available.

The potential RE from hydropower that can be potentially exploited by this technology is about 10 PWh per year of electrical energy, which correspond to a reduction of about 7 Gt of CO_2 equivalent emissions [69]. Figure 2 reports an estimation of the potential contribution to the reduction of CO_2 emissions for Europe by year 2050 for:

- Carbon Capture and Sequestration (CCS),
- the use of biofuels (average contribution, not considering land change use and other aspects such as intensified GreenHouse Gases (GHG) emissions due to the enhanced use of fertilizers consequent to monocultures),
- importing RE by conversion of CO_2 to methanol (trade RE).

It also shows the comparison of the estimated cost per Gt CO_2 removed for the three cases of CCS, use of biofuels and Trade RE via CO_2/CH_3OH [69]. This result evidences that the impact on the mitigation of climate change by this route is large, up to potentially 8 Gt CO_2 equivalents, and at least comparable with that of Carbon Capture and Storage (CCS) and biofuels.

Therefore, the proposed (CO_2/RE) technology is a cost- and environmentally effective solution for the reduction of GHG emissions ensuring:

- reduced emissions of pollutants (NO_x, SO_x, etc.) and GHG emissions, in fact, the use of unexploited RE sources decreases the needs of new power plants as well as of fossil fuel extractions;
- better diversification of energy sources (with benefits in terms of energy security) and greater flexibility in use (for example in the transportation sector);
- better flexibility in energy management to follow fluctuating energy demand, (for example, resorting to methanol-fuelled turbines).

Methanol-fuelled turbines have a rapid switch off-on time; and methanol is considered a superior turbine fuel (even if the system must be modified), with low emissions, excellent heat rate, and high power output.

The conversion of CO_2 to methanol (CO_2/RE) results in the production of energy vectors and raw materials for chemical production, making it possible to trade RE, to reduce fossil fuel consumption and contribute to Climate Change Control (CCC). RM may be considered a promising technology for commercialisation, even though a more in depth analysis of economics and some technology development are necessary.

1.2 Moving on the Technology Path for Solar Fuels

As remarked before, a systemic rather than reaction-specific approach is required to understand the real impact of CO_2 utilization, which should be analyzed not only in terms of the "traditional" techno-economic assessment. For this reason, it is relevant to discuss shortly also the technological path for the developments in this area, although a more detailed discussion is out of the scope. A first aspect regards the further developments needed to improve the preparation of solar fuels from CO_2.

We may shortly identify three main steps, although each has various sub-topics and challenges to be developed. The first one, which is currently available, is based on a 3-step approach. Electrical energy (ee) is produced using RE sources (step 1), and it is then used to produced H_2 in electrolyzers (step 2). Finally, H_2 is employed in catalytic processes, either homogeneous or heterogeneous, to convert CO_2 to different possible products, such as C1 (formic acid, methane, methanol) or >C1 chemicals (alcohols, DiMethyl-Ether, light olefins, hydrocarbons, etc.). The products could be used either as energy vectors or as base chemicals for industry. If cheap electrical energy sources are available (cost lower then about 0.05 $ per kWh), converting CO_2 by this approach may be already competitive, as demonstrated in the previous section. The clean RM process developed by the Carbon Recycling International (CRI) in Iceland is an example of industrial application, although very strongly dependent on local situation (geothermal energy). Other companies such as *Mitsui Chemicals Inc.* with their "Green House Gases-to-Chemical Resources" technology are active in this field, but with focus on hydrocarbons (via Fisher-Tropsch - FT - reaction) rather than methanol. A modification of FT is the conversion of CO_2 to olefins (Fischer-Tropsch to Olefin, FTO), which also may be interesting, when low-cost H_2 from renewable sources is available [12].

The main elements of this first approach are available, but there is still the need to develop improved catalysts (for CO_2 conversion to methanol or methane, for the reversible synthesis of formic acid to store H_2, etc.) as well related technologies (for example, compact devices using microreactors for power-to-gas or to methanol conversion). It is also necessary to improve the efficiency of electrolyzers, their scalability and stability, particularly under high-pressure operations. Advances in cell engineering and electrocatalysts design are also required, starting from the understanding on how to decrease the overpotential in H_2 generation from water electrolysis. Reasonably it can be assumed that about 5 years could be necessary to get these targets. Cost reduction of electrolyzer technology and avoid the use of noble metals in electrodes are further targets to be achieved.

The second class of routes includes electrochemical or electrocatalytic 2-step processes characterized by an intermediate production of electricity. This approach may be considered to be realized in medium-long term, *e.g.* between 5 and 10 years. A possibility is to develop inverse fuel cells, where CO_2 is directly fed on a special designed cathodic electrode, able to convert directly CO_2 using protons supplied by water oxidation from the anode and electrons coming externally from a RE source. The other one involves an electrocatalytic process, in which the adsorption and transformation of the reactant on the electro-catalyst occur according to a more complex mechanism [70-72]. This approach requires to integrate the catalysts for water oxidation (to O_2, protons and electrons) with the catalyst for H_2 generation (using the protons and electrons generated in the other step) in a PhotoVoltaic (PV) cell. There are already advanced studies in this direction (solar/H_2/chemical) but it is necessary to have a breakthrough in the productivities and cost/efficiency. Devices, in which physically separated productions of O_2 and H_2 occur, are necessary for safety and cost motivations. This aspect requires a diverse design with respect to that of devices currently investigated, and the availability of a proton-conductive membrane to separate the reactions producing O_2 and H_2.

There is the need to develop more productive and stable catalysts for water oxidation, while the catalysts for H_2 evolution are less problematic. The hydrogen produced in these devices is then used for the catalytic conversion of CO_2. It may be estimated that, within a decade, the production of H_2 from renewable sources (including from biomass, [73]) by this route at competitive costs (less than 2-3 $ per kg) should become available.

In a longer-term perspective, it is necessary to carry out a fully integrated artificial leaf-type solar cell, where a photo-anode able to photo-electrolyze water using sunlight (PEC – solar cell) is present. This is the third class of routes in which the transformation of RE to fuels or chemicals is made directly in a single (integrated) device. Photochemical processes realize on a nanoscale the separation between the processes of generation of protons/electrons from water using light and those of reduction of CO_2 using protons/electrons. The same concept was obtained on the mesoscale in PEC solar cells with the physical separation of the two processes in different compartments allowing the decrease of self-quenching effects and the separation of the products formed at the anode (O_2) from those formed at the cathode (products of CO_2 reduction), with consequent safer operation conditions. Some further aspects will be discussed in the following section.

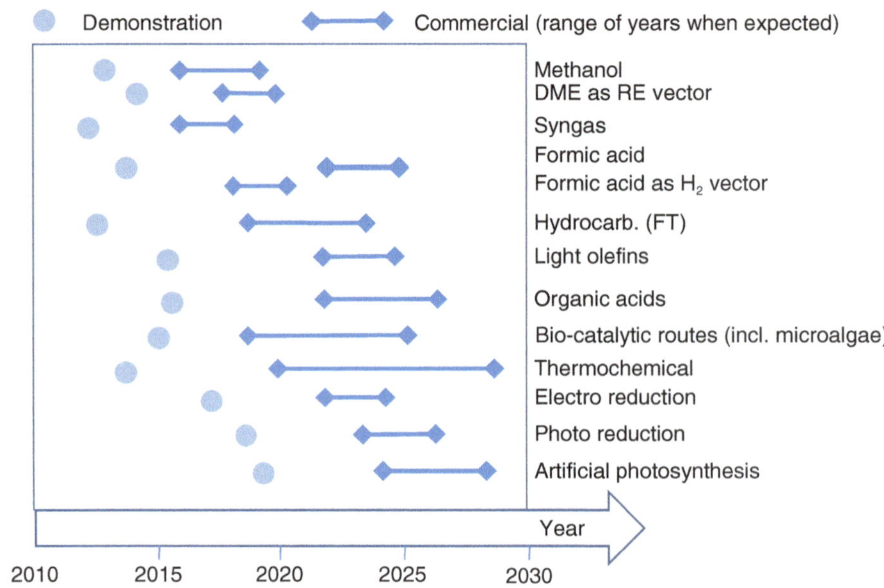

Figure 3

Estimated "technology roadmap" for commercialization of CO_2 re-use paths. Adapted from Figure 5 in Reference [9].

With respect to natural leaves, these artificial devices should allow an intensified production, be cost-competitive and be able to tune the conversion of CO_2 to the specific required products (fuels, chemicals). Their use will allow a distributed energy (or chemical) production, but there are many challenges in terms of electrocatalysts, nanodesign of elements and interfaces, charge transport and separation, etc. It is thus still a vision, whose realization should be possible only through the development of the next-generation catalysts, nanomaterials and nanosystems for energy. The realization of this vision and related knowledge on materials, nanodevices and catalysts can thus become one of the driving factors for a sustainable future of the chemical industry, because this area has the potential to become a game changer for chemistry.

There are many additional routes under development related to the broad area of solar fuels, with the limits of definition of this term indicated in the introduction. Figure 3 reports the estimated "technology roadmap" to commercialize CO_2 paths to solar fuels and chemicals. This roadmap is based on the evaluation of those presented by different institutes [74-81] corrected taking into account the results of recent workshops, meetings, reports etc. as well as considerations regarding the possible market, costs for implementation of the technology, perspectives in incentives on re-use of CO_2 in relation to the different impact of the various technologies, drop-in characteristics of the technology/ products, exploitability and technical constrains, etc. In Figure 3, the light blue circle represents the technology at pilot/demonstration scale, while the bar represents the expected time frame when commercial operation of the technology is likely. Due to different uncertainties in the conversion paths, the range of years when commercial operations will start changes from path to path.

Table 1 reports some specific examples of technologies beyond the fundamental research stage in the area of CO_2 to solar fuels and chemicals, although it should be mentioned that as for all these new areas, dominated from venture capital companies rather than large one (although all the large energy companies have active projects in the area), advertisements are higher than real demonstrations. It is thus often difficult to understand the real potential.

2 TOWARD ARTIFICIAL LEAVES

As introduced in the previous section, the design and construction of artificial-leaf type devices able to ideally utilize CO_2 (eventually captured directly from air), water and sunlight to produce tailored fuels and chemicals are the end of the transition to a new sustainable energy/chemical system. For this reason, it is a long-term (>15 years) objective, whose realization requires to start already in building the fundamental knowledge necessary to realize efficient, stable and cost-effective devices. Many activities are on-going on these aspects, as commented before and further in the reviews [82-113].

TABLE 1
Examples of technologies beyond the fundamental research stage

Organisation	Fuel product	Technology status
Carbon Sciences, US	Light hydrocarbons (via FT process)	Moving from laboratory towards commercialisation
Joule Unlimited Inc. (Joule), US	Ethanol and Diesel equivalent products (via microalgae)	Moving from laboratory towards commercialisation
Mantra Venture Group (Mantra), Canada	Formic acid (electrocatalytic)	In negotiations for first commercial demonstration
Carbon Recycling International (CRI), Island	Methanol (using geothermal energy)	First commercial plant completed at the end of 2012

Although important aspects related to the mechanism of light harvesting, charge separation and transfer, mechanism of water oxidation and nature of the active elements, etc. are discussed in the cited reviews and related works, we should observe that most of the studies on artificial photosynthesis are limited to the steps of H_2O conversion to H_2 and O_2 rather than to consider the conversion of CO_2. Although these studies contribute significantly to the advance of the knowledge, many of the proposed devices cannot operate in the presence of CO_2 but rather in water splitting. In addition, a large part of the studies on artificial photosynthesis or artificial leaves deals on the single molecular elements to realize a specific step of the whole process, with few of them investigating also the supramolecular assembling of the different functionalities, while a system approach is necessary to correctly address the topic.

Molecular assemblies bio-inspired from photosynthesis have been studied by various research groups [86]. These systems typically contain a chromophore, such as a porphyrin, which perform the first step of light harvesting. These molecules are covalently linked to one or more electron acceptors, such as fullerenes or quinones, and secondary electron donors. After chromophore excitation, the photo-induced electron transfer generates a primary charge-separation state. Electron transfer chains spatially separate the redox equivalents and reduce electronic coupling, slowing recombination of the charge-separated state to make possible their use in the redox catalytic processes. The reaction rates of the latter are typically two orders of magnitude slower, or more, than charge creation/separation processes. This aspect is the critical issue in artificial photosynthesis processes, because these charged species, if not fast consumed in the redox processes, can recombine (reducing the overall efficiency) or give rise to side reaction, including degradation of the (supra)molecular system for light-harvesting, charge separation and redox catalysis. This is one of the reasons why inorganic systems, which are more robust, are typically necessary, while organic complexes and supramolecular systems, which are better to develop and optimize as well to understand, have failed (in over twenty years of studies) to produce practically applicable artificial photosynthesis systems. Nevertheless, it has to remark the significant progresses made in many aspects, particularly on the development of:

- antenna systems, employing a variety of chromophores that absorb light throughout the whole visible spectrum, in order to increase overall efficiency,
- advanced catalytic redox centers with high turnover number.

More limited progresses have been instead made on the incorporation in the supramolecular assemblies of control and photo-protective elements borrowed from photosynthesis, although this is a key factor for the long-term stability.

The concept of artificial leaves, differently from artificial photosynthesis, is focused on the system architecture as basis for the design of the devices, rather than on the development of the single elements and their subsequent assembling into a functional device. The different perspective implies a higher consideration on the system functionality, rather than on that of the single elements, but in PEC devices, *e.g.* artificial leaf, there is a mutual influence. The behavior of the system is not the sum of the single components. In addition, a system approach emphasizes aspects such as charge transport and system reliability, which are instead less investigated in artificial photosynthesis approach. These aspects are amplified in importance in moving from water splitting to the more challenging PEC conversion of CO_2.

Another comment regards the necessity or not to have a bio-mimicking or bio-inspired approach, which has been typically highlighted in most of the cited reviews. Nature taken over billions of years to develop and optimize the highly complex machinery present in leaves that enables the use of sunlight to oxidize water and produce electrons/protons used in a different part of the cell to reduce CO_2. To remember that the products of carbon dioxide reduction are the elements (carbohydrate, lipid, other organic species) necessary for plant life and growing, while in Artificial

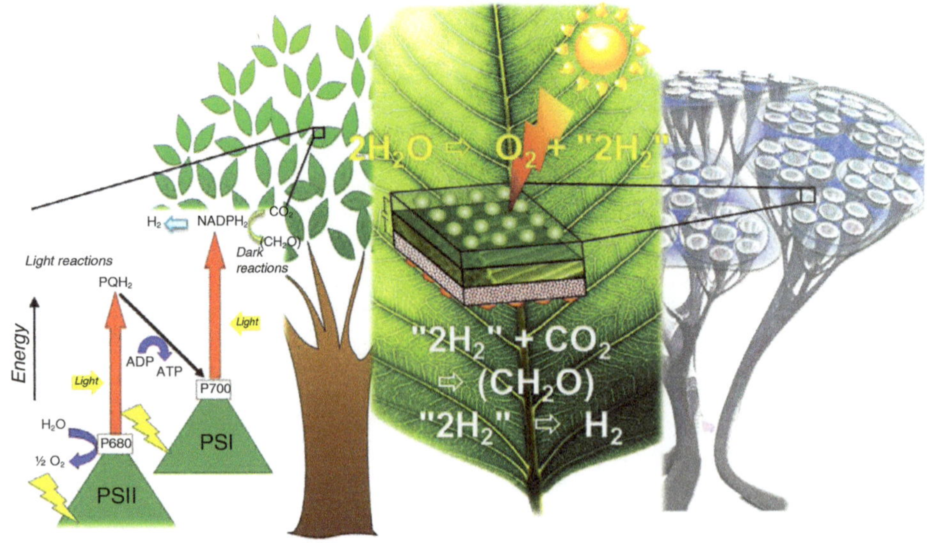

Figure 4

Going from natural to artificial leaves requires using a radically different system design taking inspiration from nature, but developing conceptually new and robust devices which overcome the limit of natural leaves. "$2H_2$" indicates the four protons and electrons produced in the water oxidation step (they are equivalent to two hydrogen molecules), while (CH_2O) indicates in general the products of CO_2 hydrogenation, which include C1 and >C1 hydrocarbons, alcohols and other organic species.

Photosynthetic Leaves (APL) the aim is to produce fuels or chemicals. In addition, plants are characterized from a series of mechanisms which allow to limit the rate of the processes (for example, light absorption) which otherwise could damage the plant itself. In artificial devices, it is necessary instead to intensify the processes, at least of one or more order of magnitude, to improve the cost-effectiveness. Therefore, the development of efficient, cost-effective artificial systems requires a new functional and robust design to realize two goals:

- intensify the process, thus allowing a higher productivity and efficiency in converting sunlight;
- use solid components which keep functionalities, but are more robust, scalable and cost-effective.

This concept is exemplified in Figure 4.

2.1 Development of PEC/Artificial Leaf-Type Solar Cells

There are various constraints, which have been typically not accounted or only in part, in cell design and characteristics derived from analysis of cell usability. These aspects have to be included already at the initial stage of cell design, because they determine the choice of the materials and their needs. An important aspect is that the products of oxidation (O_2) and reduction (H_2, or the products of CO_2 reduction) have to be produced in separate zones of the cell, to avoid safety problems, costs of separation, quenching effects. It is also necessary to have continuous operations, with easy recovery of the reaction products. In terms of usability of the cell, avoiding liquid electrolytes may be necessary to facilitate the recovery of the products, but also for stability of electrodes/cell elements and sealing. A very important element, typically not considered, is that the cell should operate with stable performances above 50-60°C, or preferably higher temperatures. It should be considered the temperature increase during exposition to sunlight, especially when a light concentrator is used to improve cost-effectiveness of the cell. The utilization of cooling systems is not possible. A final obvious aspect is that it is necessary to have a compact design, easily scalable.

These constraints determine the cell design/engineering and in turn the characteristics required for the cell elements, such as electrodes and catalytic components. However, most of the literature studies do not account for these aspects, for example, stability/performances at higher temperature than room temperature, separation of the reaction products and productivity per unit cell dimension. There are three main aspects, in particular, to take into account in the design of cells for solar fuels:

1. It is necessary to produce liquid fuels, which can be easily removed and separated; producing H_2 is not the suitable approach, but rather it is necessary to produce

protons/electrons used to selectively convert CO_2 to alcohols or hydrocarbons (or eventually to produce syngas);
2. It is mandatory to produce O_2 (from water splitting/photoelectrolysis) in a separate compartment with respect to that where the reduction occurs (formation of H_2, products of CO_2 conversion). There are many motivations, including safety, separation costs, product intensification, quenching effects, etc. PEC approach is thus mandatory.
3. The cell should have stable operations above room temperature, about 70-100°C or even higher temperatures (about 150°C or higher), when the cell operates coupled to a solar concentrator.

Practical implementation of PEC solar cells requires focusing attention on different aspects than those usually investigated, such as the following:

- productivity intensification (to improve cost-effectiveness);
- robustness even with temperatures higher than room temperature and eventually under pressure (necessary to achieve the first goal);
- chemical inertness (nascent O_2 or H_2O_2 on one side and electrons on the other side of PEC device are very reactive species);
- low cost of construction and use of non-rare materials;
- selectivity in the nature and distribution of the products obtained by CO_2 reduction, and in avoiding side reactions such as H^+/e^- recombination to give H_2 (rather than hydrogenate CO_2).

To overcome the main drawbacks related to low sunlight-to-fuel photoconversion efficiency and selectivity, a deeper understanding of mechanisms of reaction is required, in order to develop highly active photocatalysts, for what concerns adsorption of reactants, light harvesting, charge separation and transport, and CO_2 activation.

However, recent investigations, conducted by DRIFT spectroscopy and ^{13}C labelled CO_2 on photocatalytic CO_2 reduction over copper oxide promoted titania, have evidenced the participation of catalyst associated carbon residues and surface adsorbed water in the formation of CO_2 reduction products, suggesting to perform activity tests in future studies, to exclude the participation of carbon residues in the formation of CO by CO_2 photoreduction [114]. Even if clear evidences that CO is formed in the absence of carbon residues have been provided in isolated sites in Ti silicalite sieve by Ulagappan and Frei [115], in CO_2 reduction reactions performed under $^{13}CO_2$ and H_2O which showed solely ^{13}CO production, additional insights on reactive adsorption limits of CO_2, effect of surface hydration, presence and nature of species which can prevent efficient light activation of the catalyst and on the extent of water induced photocatalytic surface carbon gasification should be considered, in order to avoid overestimation of the rate of the artificial photosynthesis reactions over these materials, as well to further reveal the chemical pathways of CO_2 reduction.

The solar-driven goal in reduction of CO_2 to value-added chemical fuels is to synthesize photoactive materials able to perform the redox reactions and achieve conversion efficiency and selectivity that exceeds Nature's photosynthesis [116].

In pursuing this aim, long-time challenges have been opened in the field of catalysis intended to gain a deep understanding of surface/interface-reaction processes in order to develop novel and efficient (photo)catalysts sensitive to sunlight.

It is difficult in some cases to compare results obtained in different media and reaction conditions. Beneficial effect on CO_2 investigation derives from several studies conducted so far in heterogeneous systems. The most explored semiconductor materials for photocatalytic reduction of CO_2 include metal oxides (TiO_2, titanates, tantalate, niobate, tungstate, etc.), metal sulphides (CdS, ZnS, MnS and Cu_2ZnSnS_4), nitrides $(Zn_{1+x}Ge)(N_2O_x)$, and phosphides (*p*-InP, *p*-GaP) [31].

Recently, significant contributions in (photo)catalyst development have come from the multidisciplinary areas of material science, surface chemistry, nanotechnology, semiconductor physics and engineering, with research activities concerning:

- the electronic band-structure engineering by several techniques: quantum confinements, cation or anion doping, solid solution formation, sensitization, etc. All of them are devoted to adjust bandgap and band edge position for a wider light-harvesting and a more efficient utilization of solar energy. Also optical approaches have been applied, direct to favour light scattering to increase the optical path length, or to obtain an usable visible photon by doping with up-conversion materials able to convert two (or more) non-absorbed NIR photons;
- the possibility to arrange the materials in form of crystalline nanostructures to decrease the electron-hole recombination rate and provide direct pathway for electron transport, combined with the use of co-catalysts to lower the barrier for CO_2 activation, and to affect product selectivity and Z-scheme of heterostructures to realize the spatial separation of photoexcited e^-/h^+ pairs contributing to inhibit their recombination, and ultrathin geometries [117, 118] which promote charge carriers fast migration from the interior to the surface to participate in the photoreduction reactions;
- crystal orientation, which could influence surface energy and active sites, and modify the electronic band structure of the catalyst favouring the preferential adsorption of reactants and intermediates;
- surface oxygen vacancy engineering, which favours CO_2 adsorption and activation [116] and influences the surface

chemistry of semiconductors by changing and tuning their electronic and chemical properties; hierarchical structures, on multimodal macro-/meso-/microporous scale, characterized by easily accessed channels, shape selectivity and large surface areas which contribute to enhance reactants adsorption and to supply more surface active sites significantly enhancing gas-phase reaction efficiency. However, gas diffusion limitation is still a problem and also electrical conductivity and diffusion length of photocatalysts should be further enhanced. In some cases use of gas diffusion electrodes has been found to improve mass transport with beneficial effect on product formation rate of and selectivity [117, 119, 120].

Using electrocatalysts to reduce CO_2 dissolved in liquid, CO_2 concentration in the electrolyte and mass transport are decisive factors in determining the Faradic formation efficiencies of the different products [121].

Low temperatures are generally chosen, because performing the reaction at lower temperatures enhances CO_2 solubility as well as stability of the reduction intermediates. Whereas at higher temperatures, the selectivity to C1 products reduces and that to C2 products increases [122].

In electrochemical systems, in the attempt to overcome the problem related to the low solubility of CO_2 in aqueous solutions (33 mM at 25°C at 1 atm CO_2), organic aprotic solvents are often used as electrolytes [31]. In these nonaqueous electrolytes, the solubility of CO_2 is high but poor solution conductivity determines high ohmic losses [121]. Using nonaqueous electrolytes it is also possible to work below freezing point of water, and hence the benefit of increased solubility at low temperatures can be exploited [123]. Moreover, in nonaqueous solutions, hydrogen evolution is severely suppressed, especially at lower temperatures, and hence current faradic efficiency for CO_2 reduction is generally higher. Further, while Hydrogen Evolution Reaction (HER) and CO_2 reduction proceed in parallel competitively in aqueous electrolytes saturated with CO_2, hydrogen evolution is suppressed in nonaqueous solutions, especially at lower temperatures, and hence the current efficiency for CO_2 reduction increases.

Further useful insights may come from integrated experimental and computational research approaches to evaluate catalyst activity toward CO_2 photoreduction.

It is also beneficial pay more attention to the deactivation phenomena in the photocatalytic reduction of CO_2, attributable to:

- the adsorption or accumulation of intermediate products on the semiconductor surface which decreases the available reaction sites and also hinders the adsorption of CO_2 or H_2O, leading to semiconductor deactivation especially under the continuous irradiation;
- the desorption of the produced hydrocarbons which can affect adsorption of reactants and take part in the photocatalytic reaction;
- the surface contamination of semiconductors which can reduce light absorption.

All these aspects, up to now, have been only partially investigated and the described phenomena, at various extent, can affect the efficiency and rates of CO_2 photoreduction.

While encouraging progresses have been achieved toward CO_2 conversion, in all the described studies photo (and electro) -driven catalysts functionalities have been developed separately, independently from the cell design and/or not considering cell requirements. Although bio-mimicking approach provides inspiration, a radically different strategy, particularly in terms of system design, is required. Smart, cheap and robust devices are necessary. The use of inorganic materials would be essential, because most of the investigated molecular complexes are too costly, not robust enough and/or require special reaction conditions. Critical elements to consider, already at the initial stage, are selection and composition of artificial leaf elements (catalysts, electrodes, membranes, sensitizers) and cell engineering for which a system design approach is essential.

Therefore, the limits of natural systems should be overcome by designing conceptually different devices, but studies in this direction are still limited, notwithstanding the large number of publications around the concept of artificial photosynthesis and photosynthetic devices. Some of the critical aspects to consider are the following:

- nanoarchitecture of electrodes [124], in particular how to develop specific functional nano-architectures by mastering nano-objects and controlling charge flow [125];
- system design in approaching PEC devices to maximize cost-efficiency, productivity and charge transport;
- the role of catalysis in the design of these advanced PEC devices.

Many novel synthesis procedures for nanomaterials, including for energy applications, are available today, as reported, although effort is still necessary for a better understanding of the relationships between nanoarchitecture and functional behavior. In addition, the characteristics of these nanomaterials are often not linked to the specific needs for the design of advanced PEC solar cells. The simplified scheme of a PEC solar cell with above characteristics is shown in Figure 5.

Various attempts were made recently on the development of advanced design for artificial leaf solar cell, such as the one of the Helios Solar Energy Research Center (Helios SERC). The general idea is to develop membrane-based assemblies of oriented PEC cells with geometry similar to that of the natural photosynthetic system. The tubes are the embedded nano PV centers. A catalyst that oxidizes water

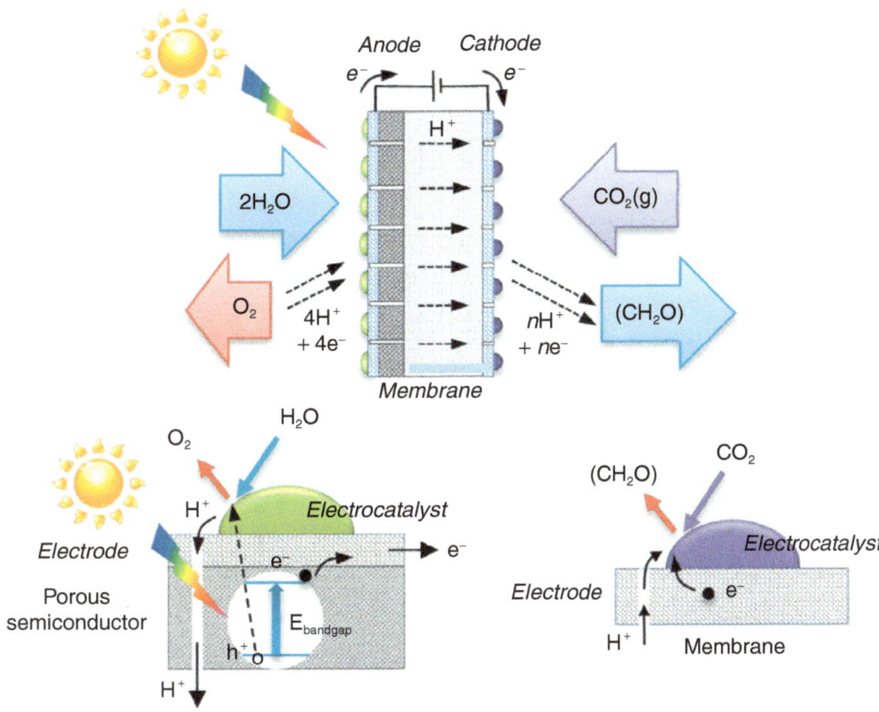

Figure 5

Simplified scheme of a PEC cell with compact design and suitable for operations at temperatures above room temperature.

is attached to the anodic end, and O_2 is formed there through the membrane and another type of catalyst at the cathodic end aids in the formation of H_2 molecules, which is typically the only product considered. Although a conceptual design was presented to utilize these devices for example in the conversion of CO_2 to methanol, the design is not really functional to this objective. The Joint Center for Artificial Photosynthesis (JCAP) is the US largest research program dedicated to the development of an artificial solar-fuel generation technology. Established in 2010 as a US Department Of Energy (DOE) Energy Innovation Hub, JCAP aims to find a cost-effective method to produce fuels using only sunlight, water, and carbon dioxide as inputs. JCAP researchers are working on different components: light absorbers, catalysts, membranes, linkers, etc. The cell is based on two semiconductors as light absorbers embedded in an electron- and ion-conducting membrane. The membrane will separate the cell into two compartments, in which either the water oxidation or reduction reactions occur. JCAP has promised to deliver a working prototype of an artificial leaf by the time its initial grant runs out in 2015.

Besides, many of the alternative approaches are being pursued elsewhere. The EU project SOLIDROMICS aims to develop hybrid artificial leaves able to split water. Competiting efforts are those of MIT, EPFL, Leiden University, University of Messina, Imperial College, CALTECH, between the others. In Santa Barbara, California, for example, a start-up firm called HyperSolar is testing a system in which coated nano- or micro-particles combining a photoabsorber and a catalyst are placed in a transparent, water-filled plastic bag. The bag will inflate as it is exposed to sunlight, and hydrogen and oxygen form inside. Such units could be deployed in sunny regions such as deserts.

There are two main classes of PEC solar cells described in literature, although with some subclasses in terms of type of electrodes and specific configurations:

- a "conventional" PEC cell operating in a liquid electrolyte (aqueous concentred electrolyte, but in principle organic or ionic liquid electrolyte could be used, when they contain enough water to drive the reaction), where water oxidation and reduction steps (to generate O_2 and H_2, respectively) occur on the two opposite electrodes in contact with the electrolyte which mediates the reactions and closes the circuit [22];
- a "modified" PEC cell based essentially on the integration of a PV cell (single or preferably multi-junction cell to obtain the necessary potential to drive water electrolysis – 1.23 eV – plus the necessary overpotential, *e.g.* around 1.8 eV in total) with catalytic elements able to use the photogenerated current/potential to oxidize water to O_2

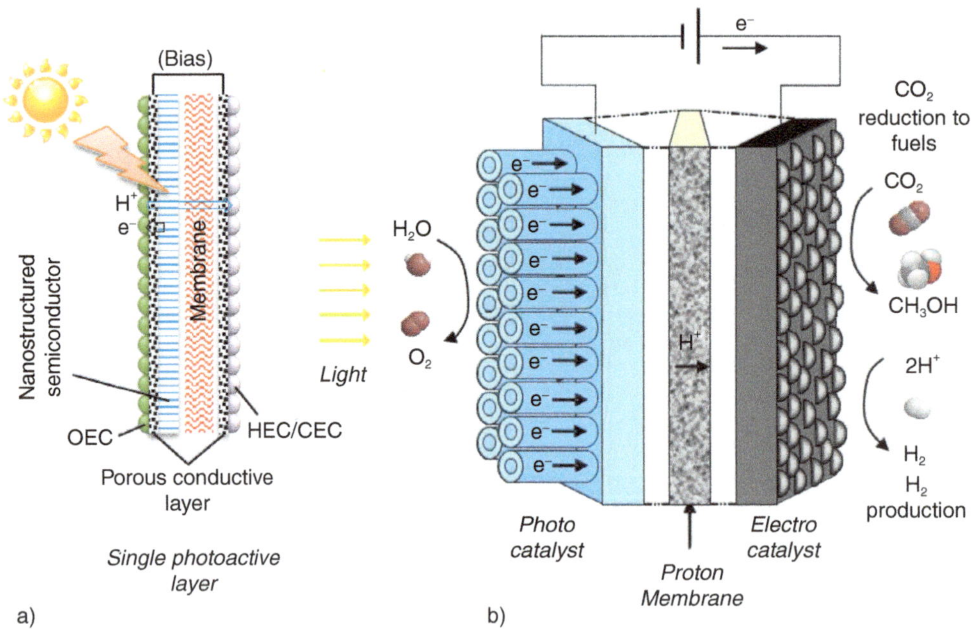

Figure 6

a) Scheme of an EL-PEC solar cell and b) realization of this type of cell at the University of Messina. Adapted from References [22, 70].

and H^+/e^- (often these elements are indicated with the acronym OEC) and reduce H^+ to H_2 on the other electrode (HEC) [30].

The temperature of operation is a critical limiting factor in both these types of cells, together with the possibility to operate efficiently in the reduction of CO_2 rather then in the evolution of H_2. Higher temperatures of operation allow the possibility to change and optimize also the charge transport across interfacial structures, a factor often limiting efficiencies. In addition, it can enhance hot-electron transfer, significantly increasing solar-to-electric power conversion efficiencies. From a practical perspective, it is necessary to consider that under full irradiation with solar light in summer, the cell may rise temperature up to about 60-70°C in the absence of cooling. The PEC cell accounts for the largest part of the costs to produce solar fuels. To reduce these costs, it may be interesting to couple the cell with low-cost solar light concentrators such as Fresnel lens. Under these conditions, the temperature of operation of the cell may easy reach temperatures above 100-150°C. The types of cell as those commented above and typically studied in literature are not suited for operations above r.t. and the largest part of the materials for PEC solar cells are also not stable for operations above r.t.

Operations without a bulk electrolyte (electrolyte-less conditions, EL) may greatly help to develop PEC solar cells for high temperature operations [126]. Interesting indications, on both the limiting steps of the process and the reaction mechanism, have been obtained comparing CO_2 reduction performed on the same electrodes in the presence (liquid-phase) or absence (gas-phase) of a bulk electrolyte. Another relevant motivation for the electrolyte-less operation is that recovery of the products of reaction is costly and significantly decreases the overall energy efficiency of the devices, which should include the energy cost to recover the product of reaction (an aspect typically not considered). In the absence of a bulk electrolyte and in a flow PEC cell, the products of reaction may be instead easily recovered from the gas phase outlet of the cell. Figure 6a illustrates the scheme of EL-PEC solar cell, and Figure 6b how the concept was applied (at University of Messina, Italy) to realize a practical cell able to convert selectively CO_2 to isopropanol, although with still low productivities due to the complexity of the conversion (a multi-electron reduction).

Thus, among the recent progresses in this field, a particular attention is devoted to the design and development of "artificial leaf" devices and to some of their critical components. This research area is very active with different concepts and ideas under investigation, although often the validity of proposed solutions needs to be proven and many constraints are not fully taken into account particularly from the perspective of system engineering. Addressing such challenging problem requires a complex multitasking approach because it is necessary not only that each artificial leaf's component works well, but also combine all of them into a complete system that works efficiently. This is even harder.

CONCLUSIONS AND OUTLOOK

In this perspective, we have presented the scenario, analyzing problems and perspectives related to the various technologies under examination for solar production of fuels from water and CO_2 and tried to discern among the various options, the current most feasible and favorable ways contributing to solve the problem of future energy supply. The promising route towards the development of artificial energy systems able to directly capture CO_2 and convert it into liquid solar fuels are also discussed, evidencing the requirements requested to the artificial leaves with respect to the natural ones, such as a higher solar-to-chemical conversion efficiency, together with robustness and cost-effectiveness.

Continued support coming from fundamental research across the range of fields regarding chemical sciences will play a key role in achieving the breakthroughs that are needed to provide practical technologies to produce solar fuels. Key research areas include catalysis, photocatalytic and electrochemical properties of organic and inorganic materials, molecular mechanisms for light harvesting and transport, nanomaterials and structure and properties of surfaces.

The challenge is exciting and cannot be postponed, the stake is the future of our energy supply system and the sustainable development of our planet.

ACKNOWLEDGMENTS

The authors would like to thank the financial contribution by *MIUR* in the frame of the Italian national project "ENERGETIC - Technologies for energy and energy efficiency - PON02_00355" and in the frame of PRIN10-11 project "Mechanisms of activation of CO_2 for the design of new materials for energy and resource efficiency".

REFERENCES

1 Annual Energy Outlook (2014) Energy Information Administration U.S. Department of Energy Washington, DC 2015, http://www.eia.gov/forecasts/aeo.

2 World Energy Outlook (WEO) 2013 and 2014, International Energy Agency, http://www.iea.org/publications.

3 National Renewable Energy Laboratory (NREL) (2013) Realizing a Clean Energy Future: Highlights of NREL Analysis, NREL/BR-6A20-60894, Dec., available at http://www.nrel.gov/docs/fy14osti/60894.pdf (24 Feb. 2015).

4 Centi G., Perathoner S. (2014) Perspectives and state of the art in producing solar fuels and chemicals from CO_2, in *Green Carbon Dioxide*, Centi G., Perathoner S. (eds), Wiley & Sons, New York, US, Chap. 1, pp. 1-24.

5 Protti S., Albini A., Serpone N. (2014) Photocatalytic generation of solar fuels from the reduction of H_2O and CO_2: a look at the patent literature, *Phys. Chem. Chem. Phys.* **16**, 19790-19827.

6 Thomas J.M. (2014) Reflections on the topic of solar fuels, *Energy Environ. Sci.* **7**, 19-20.

7 Centi G., Perathoner S. (2010) Towards Solar Fuels from Water and CO_2, *ChemSusChem* **3**, 195-208.

8 Faunce T.A. (2012) Future perspectives on solar fuels, *RSC Energy and Env. Series* **5**, 506-528.

9 Ampelli C., Perathoner S., Centi G. (2015) CO_2 utilization: an enabling element to move to a resource-and energy-efficient chemical and fuel production, *Phil. Trans. R Soc. A: Math.* **373**, 20140177.

10 Centi G., Quadrelli E.A., Perathoner S. (2013) Catalysis for CO_2 conversion: a key technology for rapid introduction of renewable energy in the value chain of chemical industries, *Energy Environ. Sci.* **6**, 1711-1731.

11 Perathoner S., Centi G. (2014) CO_2 Recycling: A Key Strategy to Introduce Green Energy in the Chemical Production Chain, *ChemSusChem* **7**, 1274-1282.

12 Lanzafame P., Centi G., Perathoner S. (2014) Catalysis for biomass and CO_2 use through solar energy: opening new scenarios for a sustainable and low-carbon chemical production, *Chem. Soc. Rev.* **43**, 7562-7580.

13 Honda M., Tamura M., Nakagawa Y., Tomishige K. (2014) Catalytic CO_2 conversion to organic carbonates with alcohols in combination with dehydration system, *Catal. Sci. Technol.* **4**, 2830-2845.

14 Das S., Wan Daud W.M.A. (2014) A review on advances in photocatalysts towards CO_2 conversion, *RSC Adv.* **4**, 20856-20893.

15 Taheri Najafabadi A. (2013) CO_2 chemical conversion to useful products: An engineering insight to the latest advances toward sustainability, *Int. J. Energy Res.* **37**, 485-499.

16 Aresta M., Dibenedetto A., Angelini A. (2013) The changing paradigm in CO_2 utilization, *J. CO_2 Utilization* **3-4**, 65-73.

17 Aresta M., Dibenedetto A., Angelini A. (2014) Catalysis for the Valorization of Exhaust Carbon: from CO_2 to Chemicals, Materials, and Fuels. Technological Use of CO_2, *Chem. Rev.* **114**, 1709-1742.

18 Markewitz P., Kuckshinrichs W., Leitner W., Linssen J., Zapp P., Bongartz R., Schreiber A., Mueller T.E. (2012) Worldwide innovations in the development of carbon capture technologies and the utilization of CO_2, *Energy Environ. Sci.* **5**, 7281-7305.

19 Centi G., Perathoner S., Passalacqua R., Ampelli C. (2011) Solar production of fuels from water and CO_2, in *Carbon-neutral Fuels and Energy Carriers: Science and Technology*, Muradov N., Veziroglu T.N. (eds), CRC Press (Taylor & Francis group), Boca Raton, FL, US, pp. 291-323.

20 Scholes G.D., Fleming G.R., Olaya-Castro A., van Grondelle R. (2011) Lessons from nature about solar light harvesting, *Nature Chem.* **3**, 763-774.

21 Haije W., Geerlings H. (2011) Efficient production of solar fuel using existing large scale production technologies, *Env. Sci. Technol.* **45**, 8609-8610.

22 Bensaid S., Centi G., Garrone E., Perathoner S., Saracco G. (2012) Towards artificial leaves for solar hydrogen and fuels from carbon dioxide, *ChemSusChem* **5**, 500-521.

23 Moore G.F., Brudvig G.W. (2011) Energy conversion in photosynthesis: a paradigm for solar fuel production, *Ann. Review Cond. Matter Phys.* **2**, 303-327.

24 Cogdell R.J., Brotosudarmo T.H.P., Gardiner A.T., Sanchez P.M., Cronin L. (2010) Artificial photosynthesis—solar fuels: current status and future prospects, *Biofuels* **1**, 861-876.

25 Bard A.J., Fox M.A. (1995) Artificial Photosynthesis: Solar Splitting of Water to Hydrogen and Oxygen, *Acc. Chem. Res.* **28**, 141-145.

26 Centi G., Perathoner S. (2013) Photoelectrochemical CO_2 activation toward artificial leaves, in *Chemical Energy Storage*, Schlögl R. (ed.), De Gruyter, Berlin, Germany, Chap. 5.1, pp. 379-400.

27 Joya K.S., de Groot H.J.M. (2014) Artificial leaf goes simpler and more efficient for solar fuel generation, *ChemSusChem* **7**, 73-76.

28 Joya K.S., Joya Y.F., Ocakoglu K., van de Krol R. (2013) Water-splitting catalysis and solar fuel devices: artificial leaves on the move, *Ang. Chem. Int. Ed.* **52**, 10426-10437.

29 Barber J., Tran P.D. (2013) From natural to artificial photosynthesis, *J. Royal Soc. Interface* **10**, 20120984/1-20120984/16.

30 Nocera D.G. (2012) The Artificial Leaf, *Acc. Chem. Res.* **45**, 767-776.

31 Roy S.C., Varghese O.K., Paulose M., Grimes C.A. (2010) Toward solar fuels: photocatalytic conversion of carbon dioxide to hydrocarbons, *ACS Nano* **4**, 3, 1259-1278.

32 Inglis J.L., MacLean B.J., Pryce M.T., Vos J.G. (2012) Electrocatalytic Pathways towards sustainable fuel production from water and CO_2, *Coord. Chem. Rev.* **256**, 2571-2600.

33 Izumi Y. (2013) Recent advances in the photocatalytic conversion of carbon dioxide to fuels with water and/or hydrogen using solar energy and beyond, *Coord. Chem. Rev.* **257**, 171-186.

34 Lei Y., Nuraje N., Yashkarova M., Kudaibergenov S. (2012) Solar fuels harvesting from carbon dioxide conversions, *Nanoscience & Nanotechn.-Asia* **2**, 110-126.

35 Sivula K. (2013) Solar-to-chemical energy conversion with photoelectrochemical tandem cells, *Chimia* **67**, 155-161.

36 Artero V., Fontecave M. (2013) Solar fuels generation and molecular systems: is it homogeneous or heterogeneous catalysis? *Chem. Soc. Rev.* **42**, 2338-2356.

37 Smestad G.P., Steinfeld A. (2012) Photochemical and Thermochemical Production of Solar Fuels from H_2O and CO_2 Using Metal Oxide Catalysts, *Ind. & Eng. Chem. Res.* **51**, 11828-11840.

38 Quadrelli E.A., Centi G., Duplan J.-L., Perathoner S. (2011) Carbon dioxide recycling: emerging large-scale technologies with industrial potential, *ChemSusChem* **4**, 9, 1194-1215.

39 Graham-Rowe D. (2008) Turning CO_2 back into hydrocarbons, *New Scientist*, March 03.

40 McKenna P. (2010) Emission control: Turning carbon trash into treasure, *New Scientist*, Sept. 29.

41 Ritter S.K. (2007) What can we do with carbon dioxide? *Chem. Eng. News* **85**, 11-17.

42 Aresta M. (ed.) (2010) *Carbon Dioxide as Chemical Feedstock*, Wiley-VCH, Weinheim, Germany.

43 Peters M., Köhler B., Kuckshinrichs W., Leitner W., Markewitz P., Müller T.E. (2011) Chemical technologies for exploiting and recycling carbon dioxide into the value chain, *ChemSusChem* **4**, 1216-1240.

44 Peters M., Müller T., Leitner W. (2009) CO_2: from waste to value, *The Chemical Engineer: TCE* **813**, 46-47.

45 Mikkelsen M., Jorgensen M., Krebs F.C. (2010) The teraton challenge. A review of fixation and transformation of carbon dioxide, *Energy Environ. Sci.* **3**, 43-81.

46 Olah G.A., Goeppert A., Surya Prakash G.K. (2009) Chemical Recycling of Carbon Dioxide to Methanol and Dimethyl Ether: From Greenhouse Gas to Renewable, Environmentally Carbon Neutral Fuels and Synthetic Hydrocarbons, *J. Org. Chem.* **74**, 2, 487-498.

47 Olah G.A., Goeppert A., Surya Prakash G.K. (2009) *Beyond Oil and Gas: The Methanol Economy*, 2nd edn., Wiley-VCH, Weinheim, Germany.

48 Sakakura T., Choi J.-C., Yasuda H. (2007) Transformation of carbon dioxide, *Chem. Rev.* **107**, 2365-2387.

49 Ma J., Sun N., Zhang X., Zhao N., Xiao F., Wei W., Sun Y. (2009) A short review of catalysis for CO_2 conversion, *Catal. Today* **148**, 221-231.

50 Song C. (2006) Global challenges and strategies for control, conversion and utilization of CO_2 for sustainable development involving energy, catalysis, adsorption and chemical processing, *Catal. Today* **115**, 2-32.

51 Whipple D.T., Kenis P.J.A. (2010) Prospects of CO_2 utilization via direct heterogeneous electrochemical reduction, *J. Phys. Chem. Lett.* **1**, 3451-3458.

52 Dorner R.W., Hardy D.R., Williams F.W., Willauer H.D. (2010) Heterogeneous catalytic CO_2 conversion to value-added hydrocarbons, *Energy Environ. Sci.* **3**, 884-890.

53 Jiang Z., Xiao T., Kuznetsov V.L., Edwards P.P. (2010) Turning CO_2 into fuel, *Phil. Trans. R. Soc. A: Math. Phys. & Eng. Sciences* **368**, 3343-3364.

54 Agarwal A.S., Zhai Y., Hill D., Sridhar N. (2011) The electrochemical reduction of carbon dioxide to formate/formic acid: engineering and economic feasibility, *ChemSusChem* **4**, 1301-1310.

55 Darensbourg D.J. (2007) Making Plastics from Carbon Dioxide: Salen Metal Complexes as Catalysts for the Production of Polycarbonates from Epoxides and CO_2, *Chem. Rev.* **107**, 6, 2388-2410.

56 Zevenhoven R., Eloneva S., Teir S. (2006) Chemical fixation of CO_2 in carbonates: Routes to valuable products and long-term storage, *Catal. Today* **115**, 1-4, 73-79.

57 De Falco M., Iaquaniello G., Centi G. (eds) (2013) *CO_2: A Valuable Source of Carbon*, Green Energy and Technology, Springer, Heidelberg, Germany.

58 Kondratenko E.V., Mul G., Baltrusaitis J., Larrazabal G.O., Perez-Ramirez J. (2013) Status and perspectives of CO_2 conversion into fuels and chemicals by catalytic, photocatalytic and electrocatalytic processes, *Energy Environ. Sci.* **6**, 3112-3135.

59 Corma A., Garcia H. (2013) Photocatalytic reduction of CO_2 for fuel production: Possibilities and challenges, *J. Catal.* **308**, 168-175.

60 Behrens M. (2014) Heterogeneous Catalysis of CO_2 Conversion to Methanol on Copper Surfaces, *Ang. Chem. Int. Ed.* **53**, 12022-12024.

61 Li Y.-N., Ma R., He L.-N., Diao Z.-F. (2014) Homogeneous hydrogenation of carbon dioxide to methanol, *Catal. Sci. Technol.* **4**, 1498-1512.

62 Babu V.J., Vempati S., Uyar T., Ramakrishna S. (2015) Review of one-dimensional and two-dimensional nanostructured materials for hydrogen generation, *Phys. Chem. Chem. Phys.* **17**, 2960-2986.

63 Ismail A.A., Bahnemann D.W. (2014) Photochemical splitting of water for hydrogen production by photocatalysis: A review, *Solar Energy Materials Solar Cells* **128**, 85-101.

64 Han Z., Eisenberg R. (2014) Fuel from Water: The Photochemical Generation of Hydrogen from Water, *Acc. Chem. Res.* **47**, 2537-2544.

65 Meier A., Steinfeld A. (2010) Solar thermochemical production of fuels, *Adv. Sci. Technol.* **74**, 303-312.

66 Bi L., Boulfrad S., Traversa E. (2014) Steam electrolysis by solid oxide electrolysis cells (SOECs) with proton-conducting oxides, *Chem. Soc. Rev.* **43**, 8255-8270.

67 Millet P., Ngameni R., Grigoriev S.A., Fateev V.N. (2011) Scientific and engineering issues related to PEM technology: Water electrolysers, fuel cells and unitized regenerative systems, *Int. J. Hydrogen Energy* **36**, 4156-4163.

68 Dinh Nguyen M.T., Ranjbari A., Catala L., Brisset F., Millet P., Aukauloo A. (2012) Implementing molecular catalysts for hydrogen production in proton exchange membrane water electrolyzers, *Coord. Chem. Rev.* **256**, 2435-244.

69 Barbato L., Centi G., Iaquaniello G., Mangiapane A., Perathoner S. (2014) Trading Renewable Energy by using CO_2: An Effective Option to Mitigate Climate Change and Increase the use of Renewable Energy Sources, *Energy Technol.* **2**, 453-461.

70 Ampelli C., Centi G., Passalacqua R., Perathoner S. (2010) Synthesis of solar fuels by a novel photoelectrocatalytic approach, *Energy Environ. Sci.* **3**, 292-301.

71 Ampelli C., Genovese C., Passalacqua R., Perathoner S., Centi G. (2012) The use of a solar photoelectrochemical reactor for sustainable production of energy, *Theor. Found. Chem. Eng.* **46**, 6, 651-657.

72 Genovese C., Ampelli C., Perathoner S., Centi G. (2013) Electrocatalytic conversion of CO_2 on carbon nanotube-based electrodes for producing solar fuels, *J. Catal.* **308**, 237-249.

73 Ampelli C., Passalacqua R., Genovese C., Perathoner S., Centi G., Montini T., Gombac V., Delgado Jaen J.J., Fornasiero P. (2013) H_2 production by selective photo-dehydrogenation of ethanol in gas and liquid phase on CuO_x/TiO_2 nanocomposites, *RSC Advances* **3**, 21776-21788.

74 Dechema (2009) *Position Paper on the Utilisation and Storage of CO_2*, Verband der Chemischen Industrie e.V, Germany.

75 Dechema/IEA/ICCA (2013) *Technology Roadmap "Energy and GHG Reductions in the Chemical Industry via Catalytic Processes"*, Dechema, Germany.

76 Frost & Sullivan Corp. (2013) Market Assessment of Energy Conservation and Recovery Technologies for CO_2 Reduction in Europe and North America, Frost & Sullivan, US.

77 Parsons Brinckerhoff/Global CCS Institute (2011) *Accelerating the uptake of CCS: industrial use of captured CO_2*, Global CCS Institute, US.

78 Thybaud N., Lebain D. (2010) *Panorama des voies de valorisation du CO_2*, Agence de l'Environnement et de la Maîtrise de l'Énergie, France.

79 SPIRE (Sustainable Process Industry through Resource and Energy Efficiency (2013) SPIRE Roadmap, http://www.spire2030.eu/uploads/Modules/Documents/spire-roadmap_broch_july2013_pbp.pdf.

80 CEFIC (The European Chemical Industry Council (2013) European chemistry for growth. Unlocking a competitive, low carbon and energy efficient future. http://www.cefic.org/Documents/PolicyCentre/Energy-Roadmap-The%20Report-European-chemistry-for-growth.pdf.

81 Centi G., van Santen R.A. (2009) *Catalysis for Renewables: From Feedstock to Energy Production*, Wiley-VCH Pub, Germany.

82 Meyer T.J. (1989) Chemical approaches to artificial photosynthesis, *Acc. Chem. Res.* **22**, 5, 163-170.

83 Liu C., Dasgupta N.P., Yang P. (2014) Semiconductor nanowires for artificial photosynthesis, *Chem. Mater.* **26**, 415-422.

84 Joya K.S., Joya Y.F., Ocakoglu K., van de Krol R. (2013) Water-splitting catalysis and solar fuel devices: artificial leaves on the move, *Angew. Chem. Int. Ed.* **52**, 10426-10437.

85 Kaerkaes M.D., Johnston E.V., Verho O., Aakermark B. (2014) Artificial Photosynthesis: From Nanosecond Electron Transfer to Catalytic Water Oxidation, *Acc. Chem. Res.* **47**, 100-111.

86 Fukuzumi S., Yamada Y. (2013) Shape- and Size-Controlled Nanomaterials for Artificial Photosynthesis, *ChemSusChem* **6**, 1834-1847.

87 Maeda K. (2013) Z-Scheme Water Splitting Using Two Different Semiconductor Photocatalysts, *ACS Catal.* **3**, 1486-1503.

88 Zhang Y.H.P., You C., Chen H., Feng R. (2012) Surpassing photosynthesis: high-efficiency and scalable CO_2 utilization through artificial photosynthesis, *ACS Symp. Series* **1097**, 275-292, (Recent Advances in Post-Combustion CO_2 Capture Chemistry).

89 Savage L. (2013) Artificial photosynthesis: saving energy for rainy day, *Optics & Photonics News* **24**, 18-25.

90 Dzhabiev T.S., Shilov A.E. (2012) Biomimetic utilization of solar energy, *Russian Chem. Rev.* **81**, 1146-1158.

91 Lee S.H., Kim J.H., Park C.B. (2013) Coupling Photocatalysis and Redox Biocatalysis toward Biocatalyzed Artificial Photosynthesis, *Chem. - A Eur. J.* **19**, 4392-4406.

92 Osterloh F.E. (2013) Inorganic nanostructures for photoelectrochemical and photocatalytic water splitting, *Chem. Soc. Rev.* **42**, 2294-2320.

93 Frischmann P.D., Mahata K., Wuerthner F. (2013) Powering the future of molecular artificial photosynthesis with light-harvesting metallosupramolecular dye assemblies, *Chem. Soc. Rev.* **42**, 1847-1870.

94 Bora D.K., Braun A., Constable E.C. (2013) "In rust we trust". Hematite - the prospective inorganic backbone for artificial photosynthesis, *Energy Env. Sci.* **6**, 407-425.

95 Concepcion J.J., House R.L., Papanikolas J.M., Meyer T.J. (2012) Chemical approaches to artificial photosynthesis, *PNAS* **109**, 15560-15564.

96 Najafpour M.M., Rahimi F., Aro E.M., Lee C.H., Allakhverdiev S.I. (2012) Nano-sized manganese oxides as biomimetic catalysts for water oxidation in artificial photosynthesis: a review, *J. Royal Soc. Interface* **9**, 2383-2395.

97 Barroso M., Arnaut L.G., Formosinho S.J. (2012) Proton-coupled electron transfer in natural and artificial photosynthesis, *RSC Catal. Series* **9**, 126-151.

98 Cogdell R.J., Gardiner A.T., Cronin L. (2012) Learning from photosynthesis: how to use solar energy to make fuels, *Phil. Trans. R. Soc. A: Math. Phys. & Eng. Sci.* **370**, 3819-3826.

99 Tachibana Y., Vayssieres L., Durrant J.R. (2012) Artificial photosynthesis for solar water-splitting, *Nature Photonics* **6**, 511-518.

100 Magnuson A., Styring S. (2012) Molecular Chemistry for Solar Fuels: From Natural to Artificial Photosynthesis, *Australian J. Chem.* **65**, 564-572.

101 Yang J., Yoon M.C., Yoo H., Kim P., Kim D. (2012) Excitation energy transfer in multiporphyrin arrays with cyclic architectures: towards artificial light-harvesting antenna complexes, *Chem. Soc. Rev.* **41**, 4808-4826.

102 Vlcek A. (2012) Solar fuels, *Coord. Chem. Rev.* **256**, 2397-2398.

103 Carraro M., Sartorel A., Toma F.M., Puntoriero F., Scandola F., Campagna S., Prato M., Bonchio M. (2011) Artificial photosynthesis challenges: water oxidation at nanostructured interfaces, *Topics in Current Chem.* **303**, 121-150.

104 Bottari G., Trukhina O., Ince M., Torres T. (2012) Towards artificial photosynthesis: Supramolecular, donor-acceptor, porphyrin- and phthalocyanine/carbon nanostructure ensembles, *Coord. Chem. Rev.* **256**, 2453-2477.

105 Rivalta I., Brudvig G.W., Batista V.S. (2012) Oxomanganese complexes for natural and artificial photosynthesis, *Current Opinion in Chem, Biol.* **16**, 11-18.

106 Fukuzumi S., Ohkubo K. (2012) Assemblies of artificial photosynthetic reaction centers, *J. Mat. Chem.* **22**, 4575-4587.

107 Larkum A.W.D. (2012) Harvesting solar energy through natural or artificial photosynthesis: scientific, social political and economic implications, *RSC Energy and Env. Series* **5**, 1-19.

108 Arifi K., Majlan E.H., Wan Daud W.R., Kassim M.B. (2012) Bimetallic complexes in artificial photosynthesis for hydrogen production: A review, *Int. J. Hydrogen Energy* **37**, 3066-3087.

109 Andreiadis E.S., Chavarot-Kerlidou M., Fontecave M., Artero V. (2011) Artificial photosynthesis: From molecular catalysts for light-driven water splitting to photoelectrochemical cells, *Photochem. Photobiol.* **87**, 946-964.

110 Hammarstrom L., Styring S. (2011) Proton-coupled electron transfer of tyrosines in photosystem II and model systems for artificial photosynthesis: the role of a redox-active link between catalyst and photosensitizer, *Energy Environ. Sci.* **4**, 2379-2388.

111 Herrero C., Quaranta A., Leibl W., Rutherford A.W., Aukauloo A. (2011) Artificial photosynthetic systems. Using light and water to provide electrons and protons for the synthesis of a fuel, *Energy Environ. Sci.* **4**, 2353-2365.

112 Song W., Chen Z., Brennaman M.K., Concepcion J.J., Patrocinio A.O.T., Iha N.Y.M., Meyer T.J. (2011) Making solar fuels by artificial photosynthesis, *Pure Applied Chem.* **83**, 749-768.

113 Hoffmann M.R., Moss J.A., Baum M.M. (2011) Artificial photosynthesis: semiconductor photocatalytic fixation of CO_2 to afford higher organic compounds, *Dalton Trans.* **40**, 5151-5158.

114 Yang C.-C., Yu Y.-H., van der Linden B., Wu J.C.S., Mul G. (2010) Artificial Photosynthesis over Crystalline TiO_2-Based Catalysts: Fact or Fiction? *J. Am. Chem. Soc.* **132**, 8398-8406.

115 Ulagappan N., Frei H. (2000) Mechanistic Study of CO_2 Photoreduction in Ti Silicalite Molecular Sieve by FT-IR Spectroscopy, *J. Phys. Chem. A* **104**, 33, 7834-7839.

116 Tu W., Zhou Y., Zou Z. (2014) Photocatalytic Conversion of CO_2 into Renewable Hydrocarbon Fuels: State-of-the-Art Accomplishment, Challenges, and Prospects, *Adv. Mater.* **26**, 4607-4626.

117 Liu Q., Zhou Y., Kou J., Chen X., Tian Z., Gao J., Yan S., Zou Z. (2010) High-Yield Synthesis of Ultralong and Ultrathin Zn_2GeO_4 Nanoribbons toward Improved Photocatalytic Reduction of CO_2 into Renewable Hydrocarbon Fuel, *J. Am. Chem. Soc.* **132**, 14385-14387.

118 Liu Q., Wu D., Zhou Y., Su H., Wang R., Zhang C., Yan S., Xiao M., Zou Z. (2014) Single-Crystalline, Ultrathin $ZnGa_2O_4$ Nanosheet Scaffolds To Promote Photocatalytic Activity in CO_2 Reduction into Methane, *ACS Appl. Mater. Interfaces* **6**, 2356-2361.

119 Kaneco S., Iiba K., Hiei N.-H., Ohta K., Mizuno T., Suzuki T. (1999) Electrochemical Reduction of Carbon Dioxide to Ethylene with Faradaic Efficiency at a Cu Electrode in CsOH/Methanol, *Electrochim. Acta* **44**, 4701-4706.

120 Hara K., Sakata T. (1997) Large Current Density CO_2 Reduction under High Pressure Using Gas Diffusion Electrodes, *Bull. Chem. Soc. Jpn* **70**, 571-576.

121 Sànchez-Sànchez C.M., Montiel V., Tryk D.A., Aldaz A., Fujishima A. (2001) Electrochemical Approaches to Alleviation of the Problem of Carbon Dioxide Accumulation, *Pure Appl. Chem.* **73**, 1917-1927.

122 Salimon J., Kalaji M. (2003) Electrochemical Reduction of CO_2 at Polycrystalline Copper in Aqueous Phosphate Buffered Solution: pH and Temperature Dependence, *Malaysian J. Chem.* **5**, 23-29.

123 Kaneco S., Iiba K., Ohta K., Mizuno T. (1999) Electrochemical Reduction of Carbon Dioxide in Methanol with Various Potassium Supporting Electrolytes at Low Temperature, *J. Solid State Electrochem.* **3**, 424-428.

124 Centi G., Perathoner S. (2011) Nanostructured Electrodes and Devices for Converting Carbon Dioxide Back to Fuels: Advances and Perspectives, in *Energy Efficiency and Renewable Energy Through Nanotechnology*, Springer, Chap. 16, pp. 561-583.

125 Centi G., Perathoner S. (2011) Creating and mastering nano-objects to design advanced catalytic materials, *Coord. Chem. Rev.* **255**, 1480-1498.

126 Centi G., Ampelli C., Genovese C., Marepally B.C., Papanikolaou G., Perathoner S. (2015) Electrocatalytic conversion of CO_2 to produce solar fuels in electrolyte or electrolyte-less configurations of PEC cells, *Faraday Discuss.*, Accepted Manuscript DOI: 10.1039/C5FD00069F (First published online 11 Jun 2015).

12

Synthesis and Reactivity of a Pentacoordinated Thiolate-Based Imido-Alkylidene W(VI) Complexes

Florian Allouche, Victor Mougel, Wolfram Grüning and Christophe Copéret*

Department of Chemistry and Applied Biosciences, ETH Zürich, Vladimir Prelog Weg 1-5, 8093 Zürich - Switzerland
e-mail: ccoperet@ethz.ch

* Corresponding author

Dedicated to the memory of Yves Chauvin

Abstract — *Attempt to synthesize a bis-thiolate tungsten alkylidene catalysts with bulky arenethiolates unexpectedly yielded upon activation of DME a pentacoordinated complex of general formula [W(NAr)(CHCMe$_3$)(SAr')(κ^2-O-CH$_2$-CH$_2$-OMe)], which is active towards self-metathesis of cis-4-nonene compared to the parent bis-benzenethiolate.*

Résumé — Synthèse et réactivité de complexes imido-alkylidène pentacoordinés du tungstène(VI) avec des ligands thiolates — Des essais dans le but de synthétiser un catalyseur bis-thiolate tungsten alkylidène comprenant des arènethiolates encombrés ont permis, via l'activation du DME, la synthèse inattendue d'un complexe de formule générale [W(NAr)(CHCMe$_3$)(SAr')(κ^2-O-CH$_2$-CH$_2$-OMe)], qui s'avère être active vis-à-vis de l'autométhatèse du cis-4-nonène en comparaison avec l'homologue bis-benzènethiolate.

INTRODUCTION

Alkene metathesis plays a key role in petrochemical and polymer industries [1, 2], and has also become widely used for organic synthesis. For these reasons, the development of homogeneous and heterogeneous metathesis catalysts still attracts major attention [3, 4]. Well-defined alkylidenes in high oxidation states, with electron-withdrawing ligands were identified as highly efficient metathesis catalysts [5-7], following the proposal by Chauvin that they were key intermediates [8, 9]. Recent progress in the understanding of the structure-activity relationship, involving surface chemistry [10-13] combined with computational studies [14-17], has highlighted trends for a more rational design of alkene metathesis catalysts and shown in particular that dissymmetry at the metal center can bring about high activity. In particular, d^0-metal alkylidene sites with two ligands of very different electron donation abilities would favor low energy barriers for coordination of the olefin while also destabilizing the metallacyclobutane intermediates, leading to overall more active catalysts [18]. MAP (MonoAlkoxide Pyrrolide) complexes illustrate perfectly this concept [2, 19-23].

Introducing a strong σ-donor ligand like thiolate [24], in combination with the weak σ-donation of surface silanolates [25], showed very good activity [26]. Here, we explore the formation of tungsten imido alkylidene systems with a bulky arenethiolate ligand. While the aim was to obtain Z-stereoselective catalysts, we show that bulky thiolate ligands lead to the unexpected formation of the pentacoordinated complex [W(NAr)(CHCMe$_3$)(SAr')(κ^2-O-CH$_2$-CH$_2$-OMe)] [27] (Scheme 1) and discuss its reactivity towards alkene.

Starting from commercial 2,6-dibromoaniline a Sandmeyer reaction affords 1,3-dibromo-2-iodobenzene. Adding a solution of three equivalents of MesMgBr in THF

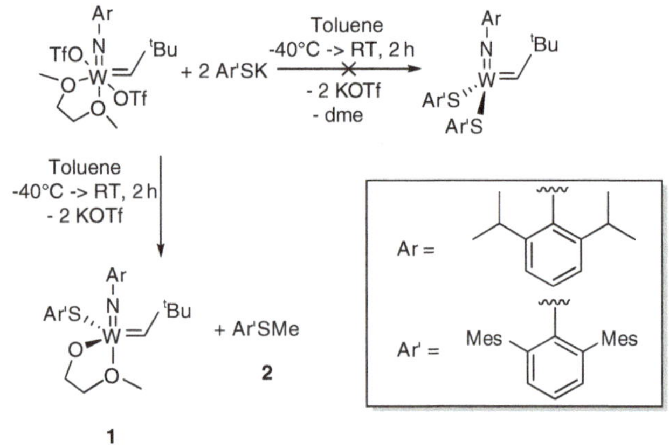

Scheme 1

Synthesis of W(NAr)(CHCMe$_3$)(SAr')(κ^2-O-CH$_2$-CH$_2$-OMe), 1.

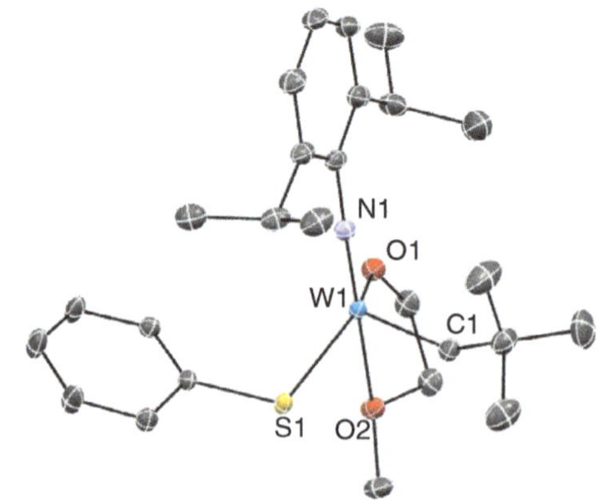

Figure 1

Thermal ellipsoid (50% probability) of the structure of **1**. Hydrogen atoms and mesityl groups on thiolate have been omitted for clarity. Selected bond distance (Å) and angles (deg): W1-N1 = 1.7486(17), W1-O1 = 1.9212(14), W1-O2 = 2.4404(15), W1-S1 = 2.3930(5), W1-C1 = 1.915(2); O1-W1-O2 = 68.90(6), O1-W1-S1 = 138.88(5), O1-W1-N1 = 103.18(7), O1-W1-C1 = 111.83(8), O2-W1-N1 = 171.80(7), O2-W1-C1 = 83.94(7), O2-W1-S1 = 81.11(4), S1-W1-N1 = 104.73(6), S1-W1-C1 = 91.43(7), C1-W1-N1 = 101.47(9).

generates (2,2″,4,4″,6,6″-hexamethyl-[1,1′:3′,1″-terphenyl]-2′-yl)magnesium bromide, before being first quenched with an excess of sulfur and then reduced with $LiAlH_4$ to give the thiol with a global 44% yield. Deprotonation by KH give the desired potassium thiolate. The reaction of two equivalents of the large thiolate ligand KSAr' by salt metathesis synthetic route [28] with [W(NAr)(CHCMe$_3$)(DME)(OTf)$_2$] lead to the clean formation of a *ca.* one-to-one mixture of [W(NAr)(CHCMe$_3$)(SAr')(κ^2-O-CH$_2$-CH$_2$-OMe)] (**1**) and MeSAr' (**2**), in place of the expected dithiolate complex (Scheme 1).

The removal of one of the methyl group of the DME ligand most likely arises from the reaction of Ar'SK with the coordinated DME on the Lewis acidic W center, similarly to the demethylation of methylethers by thiolate in the presence of Lewis acids [29].

The two reaction products **1** and **2** were obtained as crystalline materials in 62% and 72% yield, respectively. They are easily separated as pure material as a result of their difference of solubility and their tendency to form crystalline materials. Both products were fully characterized by XRD, ^{1}H and ^{13}C NMR (Nuclear Magnetic Resonance) spectroscopy. The structure of **2** is given in supplementary information (The Figure S5 is in the supplementary information. CCDC 1409694 contains the supplementary crystallographic data). The single crystal XRD study of **1** reveals a pentacoordinated tungsten complex of formula W(NAr)(CHCMe$_3$)(SAr')(κ^2-O-CH$_2$-CH$_2$-OMe) (Fig. 1).

The tungsten atom lies at 0.48 Å above the plane defined by C$_{ene}$, O1 and S, thus resulting in a slightly distorted Trigonal BiPyramidal (TBP) molecular geometry, where the imido ligand and the O2-Me group from the activated DME occupy the axial positions. The tungsten center is more sterically hindered than in W(NAr)(CHCMe$_3$)(SAr')$_2$ [26], giving a slightly longer W1-S1 bond distance (2.3930(5) Å *versus* 2.349(5) Å). The W-N bond length (1.7486(17) Å) is in the range of usual W-N bond, and so is the W-C$_{ene}$ one (1.915(2) Å). The much shorter W1-O1 bond distance (1.9212(14) Å, compared to the W1-O2 at 2.4404(15) Å), is consistent with the demethylation of DME, since the alkoxy group is a much stronger σ-donor ligand than the parent ether ligand. It is noteworthy in that such a reaction was not observed when less bulky thiolate ligands such as the 2,6-diisopropylthiophenolate or 2,4,6-triisopropylthiophenolate were used [26, 28], suggesting that the kinetics of bis-thiolate complex formation is largely reduced when the steric hindrance increases, giving a notable advantage to the demethylation pathway.

The molecular complex **1** (1 mol%) is slightly active in the self-metathesis of cis-4-nonene to yield cis/trans-mixtures of 4-octenes, 4-nonenes and 5-decenes (The Figure S3 is in the supplementary information.) with an initial TOF of 3.10^{-3} min^{-1} at 30°C. However, the catalyst deactivates, the conversion only reaching ca. 14% after 78 h. It is noteworthy that this catalyst is Z-selective (81% after ca. 1% conversion/250 minutes), yielding in fine 67% Z-selectivity after 78 h.

CONCLUSION

With the aim of synthesizing a bis-thiolate imido alkylidene tungsten complexes, the unexpected pentacoordinated complex W(NAr)($CHCMe_3$)(SAr')(κ^2-O-CH_2-CH_2-OMe) was obtained through the activation of DME by potassium thiolate, preventing the access of W(NAr)($CHCMe_3$)(SAr')$_2$ complex so far. This pentacoordinated W complex is slightly active and displays a good stereoselectivity in cis-4-nonene metathesis.

EXPERIMENTAL

General

All experiments were carried out under dry and oxygen free argon atmosphere using either standard Schlenk or glove-box techniques for organometallic synthesis. For the syntheses, reactions were carried out using high vacuum lines (10^{-5} mbar) and glove-box techniques. Pentane, toluene and diethyl ether were purified using double *MBraun* SPS (Solvent Purification System) alumina column, and were degassed using three freeze-pump-thaw cycles before being used. DME and THF were distilled from Na/Benzophenone. Et_3N, Me_3SiCl, 2,5-dimethyl-1H-pyrrole and 2,6-diisopropylamine were dried on CaH_2, distilled under reduced pressure and degassed using two freeze-pump-thaw cycles. Triflic acid was purified stirring it one day over triflic anhydride, followed by distillation under reduced pressure. All InfraRed (IR) spectra were recorded using a *Bruker* spectrometer placed in the glovebox, equipped with *OPUS* software. A typical experiment consisted in the measurement of transmission in 32 scans in the region from 4 000 to 400 cm^{-1}. The ^{1}H and ^{13}C-NMR spectra were obtained on *Bruker* DRX 200, DRX 250 or DRX 500 spectrometers. The solution spectra were recorded in C_6D_6 at room temperature. The ^{1}H and ^{13}C chemical shifts are referenced relative to the residual solvent peak. W(NAr)($CHCMe_3$)(DME)(OTf)$_2$ [28] and 1,3-dibromo-2-iodobenzene [30] were synthesized according to literature procedures.

2,2″,4,4″,6,6″-Hexamethyl-[1,1′:3′,1″-Terphenyl]-2′-Thiol (Ar′SH)

This synthesis was shortened compared to literature [31]. A solution of MesMgBr in THF (32 mL, C = 0.978 M, 31.3 mmol, 3.0 equiv) was added dropwise over 1 h at RT to a solution of 1,3-dibromo-2-iodobenzene (3.638 g, 10.1 mmol, 1 equiv) in THF (50 mL). After 3 h of stirring, the solution was cooled to −78°C and sulfur S_8 (3.22 g, 12.5 mmol, 1.25 equiv) was slowly added portionwise, resulting in an immediate colour change to deep red. The mixture was allowed to cool to RT and stirred overnight. $LiAlH_4$ (3.833g, 101 mmol, 10 equiv) was added, the reaction mixture was stirred for 8 h, and the excess of $LiAlH_4$ was quenched with water (20 mL) and 10% aqueous HCl solution (60 mL). The organic layer was separated, and the aqueous layers were extracted with Et_2O (2 × 60 mL). Organic phases were dried over $CaCl_2$ and filtered. The solvent was removed under vacuo and recrystallization at −20°C in ethyl acetate finally afforded transparent needle crystals of Ar'SH (1.61 g, 44%). ^{1}H NMR (200 MHz, [D_8]toluene) δ 2.05 (s, 12, Me_o), 2.36 (s, 6, Me_p), 3.05 (s, 1, SH), 7.00 (s, 4, H_m(Mes)), 7.06 (d, $^4J = 9$, 2, H_m(Ph)), 7.26 ($^4J = 9$, 1, H_p).

W(NAr)($CHCMe_3$)(SAr′)(κ^2-O-CH_2-CH_2-OMe)

KH (34 mg, 0.85 mmol, 0.99 equiv) was slowly added portionwise to a solution of thiol Ar'SH (300 mg, 0.86 mmol, 1 equiv) in toluene, releasing H_2 during 2 h while a white precipitate appeared. The latter was filtrated and washed twice with cold toluene (−40°C). The filtrate was concentrated and stored at −40°C overnight to yield a second crop of Ar'SK which was filtrated and washed the same way, giving 311 mg (yield: 93%) of the potassium thiolate. Ar'SK (238 mg, 0.62 mmol, 2 equiv) was added portionwise to a cold solution (−40°C) of W(NAr)($CHCMe_3$)(DME)(OTf)$_2$ (251 mg, 0.31 mmol, 1 equiv) in toluene (10 mL). The solution was left stirring 2 h, resulting in a colour change to orange, accompanied by the formation of a white precipitate of KOTf. The latter was then filtrated through Celite. The solution was taken to dryness. First crystallization at −40°C in pentane yielded transparent crystals of Ar'SMe (81 mg, 72%). After filtration, the solution was concentrated and a second crystallization in pentane at −40°C afforded orange crystals of W(NAr)($CHCMe_3$)(SAr')(κ^2-O-CH_2-CH_2-OMe) (162 mg, 62%), which were filtrated. 'H NMR (500 MHz, C_6D_6) δ 9.67 (s, 1, $CHCMe_3$), 7.11 (d, 2, J = 7.9), 7.04-6.99 (m, 1), 6.97 (s, 2), 6.92 (s, 1), 6.90 (d, 2, J = 7.1), 6.86 (s, 2), 4.06 (sept, 1, J = 3.8), 3.93 (dt, 1, J = 11.5, J' = 4.3), 3.66 (sept, 2, J = 6.8), 3.04 (s, 3, OMe), 2.96 (sex, 1, J = 4.1), 2.84 (q, 1, J = 4.3), 2.37 (s, 6, ArMe), 2.27 (s, 6, ArMe), 2.22 (s, 6, ArMe), 1.30 (6, d, J = 6.9, CH(Me_A)$_2$), 1.13 (6, d, J = 6.9, CH(Me_B)$_2$), 1.04 (9, s, $CHMe_3$); ^{13}C NMR δ 270.03, 152.47, 145.25, 144.36, 143.96, 139.82, 136.92, 136.74, 136.57, 129.66, 128.98, 126.09, 122.91, 73.73, 72.45, 58.78, 46.79, 33.04, 27.89, 24.26, 23.89, 21.73, 21.30, 20.93. Elemental Analysis: C 64.03% (62.18% expected), H 7.06% (exp. 7.00%), S 3.50% (exp. 3.77%), N 1.45% (exp. 1.75%). CCDC 1409693 contains the supplementary crystallographic data.

REFERENCES

1 Grubbs R.H. (2006) *Angew. Chem. Int. Ed.* **45**, 3760-3765.

2 Ibrahem I., Yu M., Schrock R.R., Hoveyda A.H. (2009) *J. Am. Chem. Soc.* **131**, 3844-3845.

3 Ivin K., Mol H. (1997) in *Olefin Metathesis and Metathesis Polymerization (2)*, Ivin K.J., Mol J.C. (eds), Academic Press, London.

4 Grubbs R.H. (2008) *Handbook of Metathesis*, Wiley-VCH Verlag GmbH.

5 Schrock R.R., Hoveyda A.H. (2003) *Angew. Chem. Int. Ed.* **42**, 4555-4555.

6 Hoveyda A.H., Malcolmson S.J., Meek S.J., Zhugralin A.R. (2010) *Angew. Chem. Int. Ed.* **49**, 34-44.

7 Kress S., Blechert S. (2012) *Chem. Soc. Rev.* **41**, 4389-4408.

8 Herisson J.L., Chauvin Y. (1971) *Makromol. Chem.* **141**, 161-165.

9 Chauvin Y. (2006) *Angew. Chem. Int. Ed.* **45**, 3740-3747.

10 Chabanas M., Baudouin A., Copéret C., Basset J.-M. (2001) *J. Am. Chem. Soc.* **123**, 2062-2063.

11 Blanc F., Copéret C., Thivolle-Cazat J., Basset J.-M., Lesage A., Emsley L., Sinha A., Schrock R.R. (2006) *Angew. Chem. Int. Ed.* **45**, 1216-1220.

12 Rhers B., Quadrelli E.A., Baudouin A., Taoufik M., Copéret C., Lefebvre F., Basset J.-M., Fenet B., Sinha A., Schrock R.R. (2006) *J. Organomet. Chem.* **691**, 5448-5455.

13 Blanc F., Berthoud R., Salameh A., Basset J.-M., Copéret C., Singh R., Schrock R.R. (2007) *J. Am. Chem. Soc.* **129**, 8434-8435.

14 Solans-Monfort X., Clot E., Copéret C., Eisenstein O. (2005) *J. Am. Chem. Soc.* **127**, 14015-14025.

15 Poater A., Solans-Monfort X., Clot E., Copéret C., Eisenstein O. (2007) *J. Am. Chem. Soc.* **129**, 8207-8216.

16 Solans-Monfort X., Copéret C., Eisenstein O. (2010) *J. Am. Chem. Soc.* **132**, 7750-7757.

17 Solans-Monfort X., Copéret C., Eisenstein O. (2012) *Organometallics* **31**, 6812-6822.

18 Chabanas M., Copéret C., Basset J.-M. (2003) *Chem. Eur. J.* **9**, 971-975.

19 Hoveyda A.H., Zhugralin A.R. (2007) *Nature* **450**, 243-251.

20 Malcolmson S.J., Meek S.J., Sattely E.S., Schrock R.R., Hoveyda A.H. (2008) *Nature* **456**, 933-937.

21 Marinescu S.C., Levine D.S., Zhao Y., Schrock R.R., Hoveyda A.H. (2011) *J. Am. Chem. Soc.* **133**, 11512-11514.

22 Reithofer M.R., Dobereiner G.E., Schrock R.R., Müller P. (2013) *Organometallics* **32**, 2489-2492.

23 Gerber L.C.H., Schrock R.R., Müller P. (2013) *Organometallics* **32**, 2373-2378.

24 Jakusch T., Buglyo P., Tomaz A.I., Pessoa J.C., Kiss T. (2002) *Inorg. Chim. Acta* **339**, 119-128.

25 Mougel V., Copéret C. (2014) *Chem. Sci.* **5**, 2475-2481.

26 Allouche F., Mougel V., Copéret C. (2015) *Asian J. Org. Chem.* **4**, 528-532.

27 Couturier J.L., Paillet C., Leconte M., Basset J.-M., Weiss K. (1992) *Angew. Chem. Int. Ed.* **31**, 628-631.

28 Schrock R.R., Depue R.T., Feldman J., Yap K.B., Yang D.C., Davis W.M., Park L., Dimare M., Schofield M., Anhaus J., Walborsky E., Evitt E., Kruger C., Betz P. (1990) *Organometallics* **9**, 2262-2275.

29 Weissman S.A., Zewge D. (2005) *Tetrahedron* **61**, 7833-7863.

30 Du C.J.F., Hart H., Ng K.K.D. (1986) *J. Org. Chem.* **51**, 3162-3165.

31 Ellison J.J., Ruhlandt-Senge K., Power P.P. (1994) *Angew. Chem.* **106**, 1248-1250.

13

Alkene Metathesis Catalysis: A Key for Transformations of Unsaturated Plant Oils and Renewable Derivatives

Pierre H. Dixneuf*, Christian Bruneau and Cédric Fischmeister

Institut des Sciences Chimiques de Rennes, UMR 6226 CNRS-Université de Rennes, "Organometallics: Materials and Catalysis", Centre for Catalysis and Green Chemistry, Campus de Beaulieu, 35042 Rennes - France
e-mail: pierre.dixneuf@univ-rennes1.fr

* Corresponding author

Dedicated to the memory of Yves Chauvin,
a great and inspiring pioneer in organometallic catalysis

Abstract — *This account presents the importance of ruthenium-catalysed alkene cross-metathesis for the catalytic transformations of biomass derivatives into useful intermediates, especially those developed by the authors in the Rennes (France) catalysis team in cooperation with chemical industry. The cross-metathesis of a variety of functional alkenes arising from plant oils, with acrylonitrile and fumaronitrile and followed by catalytic tandem hydrogenation, will be shown to afford linear amino acid derivatives, the precursors of polyamides. The exploration of cross-metathesis of bio-sourced unsaturated nitriles with acrylate with further catalytic hydrogenation has led to offer an excellent route to α,ω-amino acid derivatives. That of fatty aldehydes has led to bifunctional long chain aldehydes and saturated diols. Two ways of access to functional dienes by ruthenium-catalyzed ene-yne cross-metathesis of plant oil alkene derivatives with alkynes and by cross-metathesis of bio-sourced alkenes with allylic chloride followed by catalytic dehydrohalogenation, are reported. Ricinoleate derivatives offer a direct access to chiral dihydropyrans and tetrahydropyrans via ring closing metathesis. Cross-metathesis giving value to terpenes and eugenol for the straightforward synthesis of artificial terpenes and functional eugenol derivatives without C=C bond isomerization are described.*

Résumé — Catalyse de métathèse d'oléfines : une clé pour des transformations d'huiles végétales insaturées et de substances renouvelables — Cet article présente l'importance de la réaction de métathèse croisée des oléfines par catalyse au ruthénium pour la transformation de dérivés de la biomasse en produits utiles pour l'industrie. Il constitue une revue des principaux travaux réalisés dans ce domaine par les auteurs dans leur équipe de catalyse à Rennes (France) en coopération avec l'industrie chimique. La métathèse croisée d'une variété d'oléfines fonctionnelles issues des huiles végétales avec l'acrylonitrile et le fumaronitrile, suivie d'une hydrogénation tandem conduit à des dérivés d'amino acides linéaires, précurseurs de polyamides par polycondensation. L'exploration de la métathèse croisée de nitriles insaturés bio-sourcés avec des acrylates a conduit à des dérivés d'α,ω-amino acides, tandis que les aldéhydes gras ont permis un accès rapide à des aldéhydes bifonctionnalisés à longue chaîne et à des diols saturés.
Deux voies d'accès à des diènes conjugués fonctionnels impliquant des réactions de métathèse catalysées par le ruthénium à partir d'oléfines issues de la biomasse sont décrites : l'une par métathèse croisée avec un alcyne, l'autre par métathèse croisée avec un chlorure allylique suivie

d'une réaction de déhydrohalogénation catalytique. Les dérivés de l'acide ricinoléïque ont permis un accès direct à des dihydropyrannes et tétrahydropyrannes chiraux par fermeture de cycle par métathèse. Des métathèses croisées pour valoriser des dérivés de terpènes et de l'eugénol en offrant des méthodes directes de synthèse de terpènes non naturels et de dérivés d'eugénol fonctionnels sans migration de double liaisons sont aussi présentées.

INTRODUCTION

The production of chemical industry intermediates from renewable resources is a topic of current interest to offer alternatives to petrochemicals, energy savings, and to contribute to sustainable development [1-10]. Within this trend, the use of renewable unsaturated vegetable oils is attractive [1-4]. Catalytic olefin metathesis has a strong potential for selective biomass transformations. Vegetable oils by transesterification easily release unsaturated fatty esters which offer, via cross-metathesis with alkenes, selective access to terminal alkenes or bifunctional internal olefins that are crucial intermediates for industry. Thus catalytic cross-metathesis has become a key to reveal the potential of renewable plant oils or natural terpenes [11-18].

The objective of this review is to show several applications of plant oils and terpenes discovered in Rennes with ruthenium catalysts during the last 10 years such as the ethenolysis of renewable unsaturated fatty esters to produce terminal functional alkenes, their cross-metathesis with acrylonitrile in the route to produce by sequential catalyses linear amino acid derivatives that are key intermediates for polyamide production, the cross-metathesis of renewable fatty nitriles to generate linear amino esters or diamines, and selective terpene modifications. Tandem and sequential alkene metathesis/hydrogenation catalytic reactions will be presented illustrating transformations of renewable materials with atom economy and offering new contribution to green and sustainable chemistry.

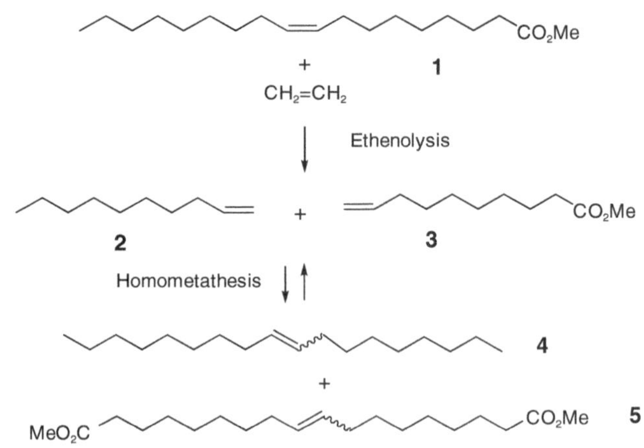

Scheme 1

Ethenolysis of methyl oleate **1**.

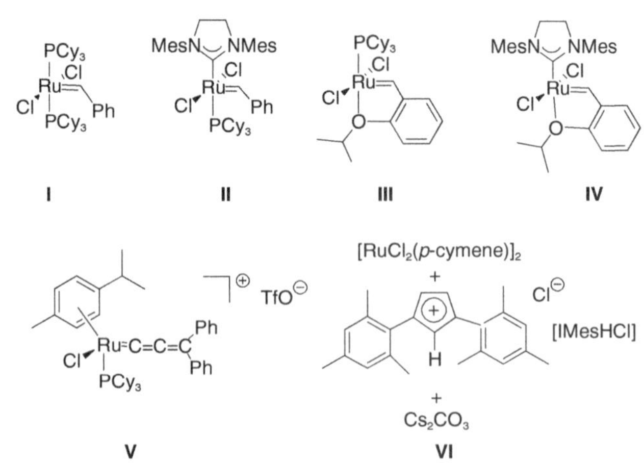

Figure 1

Ruthenium catalysts for alkene metathesis.

1 ETHENOLYSIS OF METHYL OLEATE IN IONIC LIQUID FOR TERMINAL ALKENE PRODUCTION

Catalytic ethenolysis of methyl oleate **1** has the potential to produce 1-decene **2** and methyl 9-decenoate **3**, that are two useful intermediates for the preparation of lubricants [19] or polyesters [20] (Scheme 1). Initial catalysts used for this ethenolysis were $WCl_6/SnMe_4$ [21], and the heterogeneous $Re_2O_7/Al_2O_3/SnMe_4$ [22, 23] catalysts, as well as the first well-defined ruthenium catalysts such as **I**, **II**, **IV** (Fig. 1) [24-27]. These studies revealed that double bond isomerization easily takes place, is favoured by high temperature, and produces many products. However, the formation of homometathesis products 9-octadecene **4** and symmetrical diester **5** remained low.

Ionic Liquids (IL) as solvents have shown their ability to favour several catalytic processes [28-32] and alkene metathesis can be performed in Room Temperature Ionic Liquids (RTIL) since 1995 [33-41]. We have shown the efficient ethenolysis of methyl oleate **1** under mild conditions with several ruthenium alkylidene catalysts in organic solvents and, for the first time, in imidazolium-type ionic

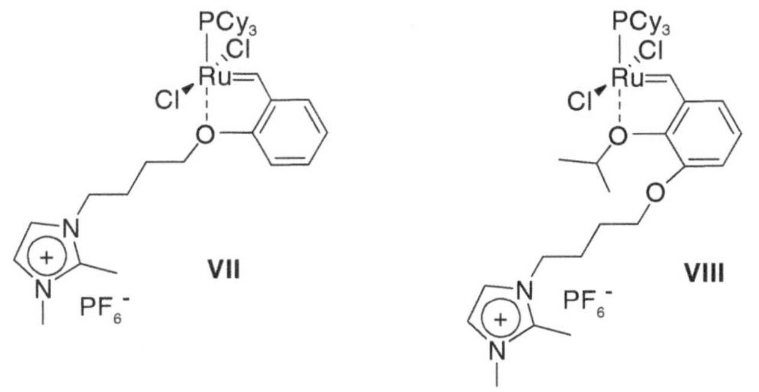

Figure 2

Ionic alkylidene-ruthenium catalysts.

liquids at room temperature without C=C bond isomerization [42].

The ethenolysis of methyl oleate was investigated with several ruthenium catalysts **I**, **III**, **IV**, and **V**, and the *in situ* prepared catalyst **VI** [43, 44] (Fig. 1) at 70°C in toluene under 1 bar of ethylene [42] (Scheme 1). These experiments demonstrated the superiority of catalyst **III**, which showed a better efficiency and selectivity. The best results were obtained under the mildest conditions at 20°C and under 1 bar pressure of ethylene leading to 93% conversion in 3.5 h, with a total selectivity to produce 1-decene and methyl 9-decenoate in 93% yield. The effect of the ethylene pressure is less pronounced but a higher pressure appears to improve the conversion of the reaction. The Hoveyda catalyst **IV** was found to allow an excellent selectivity at 20°C and at 1 bar but with lower conversion (80% instead of 93 for **III**).

Catalyst **I** was further evaluated in the ethenolysis of methyl oleate at 20°C and 10 bar of ethylene but in IL. Low conversion took place in [bmim][OTf] ([bmim] = 1-butyl-3-methylimidazolium), but in [bmim][NTf_2] and in [bdmim][NTf_2], [bdmim] = 1-butyl-2,3-dimethylimidazolium], 79% and 83% conversion were respectively obtained after only 2 hours with excellent selectivity [42].

Two Hoveyda-type catalysts **VII** and **VIII** [41], bearing an ionic tag (Fig. 2), were evaluated in the ethenolysis of methyl oleate in the ionic liquid [bdmim] [NTf_2] at 20°C and 10 bar of ethylene for 2 h. Catalyst **VIII** only displayed suitable catalytic activity leading to 89% conversion with high selectivity. However it was not possible to recycle this catalyst [42].

By contrast the neutral catalysts **I**, **III**, and **IV** (Fig. 1) in ionic liquid [bdmim] [NTf_2] at 20°C and 10 bar of ethylene for 2 h were preferable and thus subjected to consecutive runs after product extraction in hexane. We showed that catalysts **I** and **III** led to good conversion for the first run but more importantly that catalyst **III** could be recycled 3 times

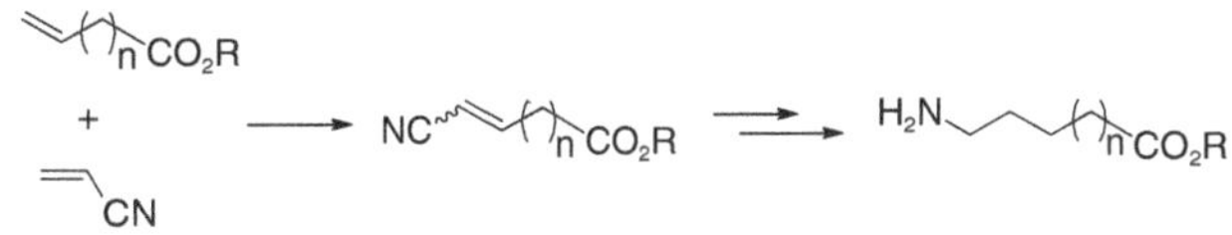

Scheme 2

From cross metathesis of fatty ester with acrylonitrile to linear aminoester.

without loss of activity and with an excellent selectivity (conversion: 1st run 95%, 2nd run 95%, 3rd run 85%) [42]. This demonstrates that it is not necessary to introduce an ionic tag in the catalyst to get excellent activity, selectivity and recyclability and that catalyst **III** in [bdmim] [NTf_2] can offer these targets for ethenolysis of methyl oleate.

2 CROSS-METATHESIS OF ACRYLONITRILE WITH UNSATURATED ACIDS OR ESTERS: A KEY STEP TOWARD LINEAR AMINO ACID DERIVATIVES

The transformation of unsaturated acids and esters arising from plant oils into linear amino acid intermediates constitutes a challenge to generate useful polyamides from renewables. Cross-metathesis of acrylonitrile and unsaturated acids or esters, associated with the further hydrogenation of both the C=C and CN bonds, constitutes a potential strategic way for the production of polyamides [14, 15] (Scheme 2).

The first example of cross-metathesis of terminal olefins with acrylonitrile was described with the Schrock catalyst $Mo(=CHCMe_2Ph)(NAr)(OCMe(CF_3)_2)$, Ar = 2,6-diisopropylphenyl, affording moderate yields [45]. Then ruthenium catalysts, especially containing a heterocyclic carbene ligand, improved the cross-metathesis of acrylonitrile with terminal alkenes [46-50], with an example of internal alkene in the presence of copper salt [51].

2.1 Cross-Metathesis of Unsaturated Esters with Acrylonitrile

We have first attempted the cross-metathesis with ruthenium catalysts of acrylonitrile with two unsaturated acids and their esters arising from renewable resources: the C11 ester **6** obtained from castor oil and its related carboxylic acid **7**, and the C18 diester **8** and diacid **9** obtained *via* biotransformation or self-metathesis of oleic acid [52] (Scheme 3).

The cross-metathesis of the unsaturated monoester **6** with 2 equivalents of acrylonitrile was performed with the

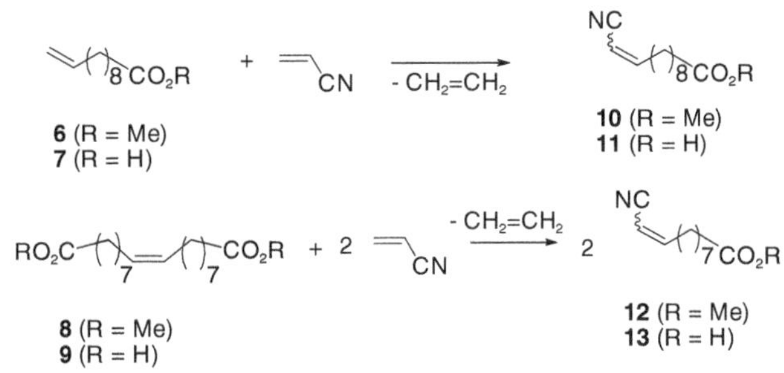

Scheme 3

Cross metathesis of fatty derivatives **6-9** with acrylonitrile.

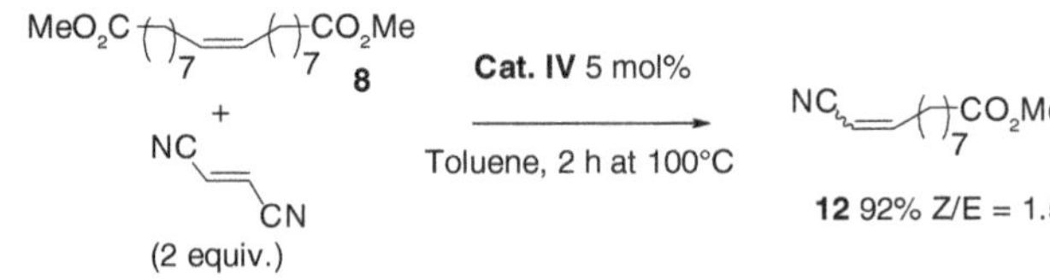

Scheme 4

Cross metathesis of diester **8** with fumaronitrile.

Grubbs **II** and Hoveyda **IV** catalysts in toluene at 100°C (Scheme 3). The Hoveyda catalyst **IV** was the most efficient. With low catalyst loading (1 mol%) it allowed better conversion (95% after 1 h) giving the C12 nitrile ester **10** as a mixture of *Z* (major) and *E* isomers (3/1). Increasing the reagent concentration resulted in lower conversion (best concentration of **6** = 0.05 M) [52] (Scheme 3).

The similar cross-metathesis reaction was performed with acid **7**. Catalyst **IV** (1 mol%, toluene) showed the best performance giving 90% conversion in 1 h at 100°C, the *Z* isomer being the major one (*Z*/*E* = 3.2/1) of nitrile acid **11** [52]. These experiments show that cross-metathesis is not significantly affected by the presence of an acidic functionality when catalyst **IV** is used.

However, it was observed that the catalyst **IV** progressively decomposed. Thus step by step addition of the catalyst allowed to reach better conversions and this effect was improved by dropwise addition. Lower catalyst loadings were then attempted at 100°C using a syringe pump equipment (*dropwise* addition) which offered a real breakthrough. Thus it was possible to decrease the catalyst **IV** loading down to 0.05 mol% still leading to a high conversion (96%) and to a TON of 1 900. The TON could be slightly increased to 2 280 using a lower catalyst loading but the conversion dropped to 55% [53].

A large excess of acrylonitrile was observed to inhibit the transformation. Simultaneous slow addition of catalyst and acrylonitrile still improved the process. It was shown that addition with two syringes, one containing a mixture of ester **6** and 1 equivalent of acrylonitrile and the other one a solution of catalyst **IV** (0.025 mol%) and the second equivalent of acrylonitrile, improved the process, which furnished 75% conversion of **6** and led to the highest TON (3 120) to produce the nitrile ester **10** [53].

Whereas the internal double bond required a higher catalyst loading of 5 mol% [52], the octadec-9-enedioic diester **8**, derived from bioconversion of sunflower oil in the presence of *Candida tropicalis* M 25 [54] or by self-metathesis of oleic acid derivatives [55-57], was evaluated for cross-metathesis with acrylonitrile. Diester **8** and acrylonitrile in the presence of catalysts **II** and **IV** lead to the C11 nitrile-ester **12** (Scheme 3). Catalyst **IV** was also the most efficient for the reaction of **8** ([**8**] = 0.1 M) with 4 equivalents of acrylonitrile in toluene at 100°C for 1 h. With 5 mol% of catalyst the nitrile ester **12** was obtained (86%) as a mixture of *Z* (major) and *E* isomers (2.9/1) with a small amount of C10 ester with a terminal double bond [52]. Thus the cross-metathesis with acrylonitrile of alkenes containing an internal double bond can be efficiently achieved, as it was also possible with acrylic esters [16, 17]. Using syringe pump addition of catalyst **IV** at 100°C for 5 h, the unsaturated nitrile ester **10** was obtained in 88% yield, with high TON of 1960 [53].

The direct cross-metathesis of the diacid **9** with acrylonitrile was performed using 5 mol% of catalyst **IV** in toluene at 100°C for 1 h to give 90% conversion of **9** and the formation of the diacid **13** (*Z*/*E* = 2.4) (Scheme 3).

2.2 Cross-Metathesis of Diester 8 with Fumaronitrile

The cross-metathesis of fumaronitrile (2 equivalents) instead of acrylonitrile with the diester **8** (Scheme 4) can be performed using ruthenium catalyst **IV** (5 mol%) and the nitrile ester **12** was obtained in 92% isolated yield [52]. These experiments show that fumaronitrile can be used for cross-metathesis, although it is more profitable to use the cheapest and readily available acrylonitrile.

2.3 Cross-Metathesis of Methyl Oleate with Acrylonitrile and Fumaronitrile

Analogously, methyl oleate **1** with an internal double bond, was reacted with an excess of acrylonitrile in toluene at 100°C for 2 h with 5 mol% of ruthenium catalyst **IV** and led to full conversion of methyl oleate and to a mixture of the expected nitrile ester **12** and unsaturated nitrile **15** in

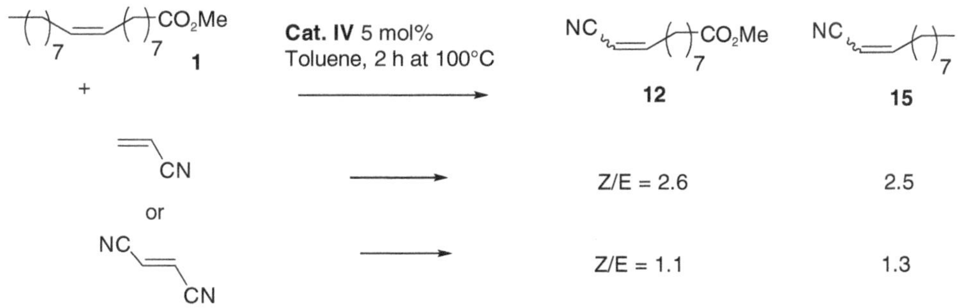

Scheme 5

Cross metathesis of methyl oleate with acrylonitrile and fumaronitrile.

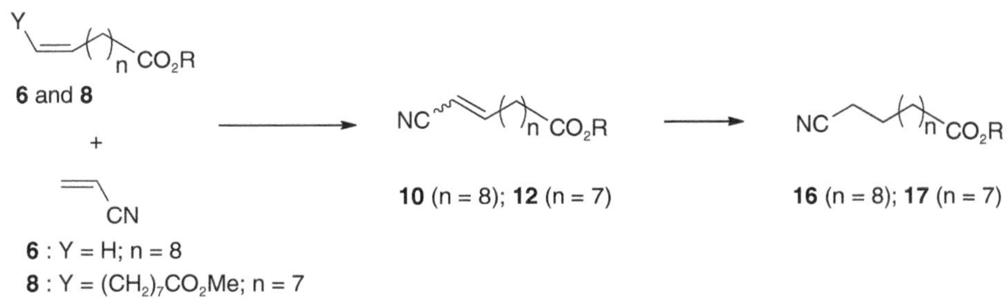

Scheme 6

Tandem catalytic cross metathesis of fatty esters with acrylonitrile and hydrogenation.

equal amounts [52] (Scheme 5). Cross-metathesis of methyl oleate with fumaronitrile in the presence of 5 mol% of catalyst **IV** in toluene at 100°C for 2 h also provided **12** and **15** as a mixture of *E* and *Z* isomers, in equal proportions [52] (Scheme 5).

3 HYDROGENATION OF NITRILE ESTERS 10 AND 12 INTO NITRILE ESTERS 16 AND 17

The possibility of synthesizing saturated nitrile-esters from renewable oils by tandem cross-metathesis/hydrogenation catalysis was investigated. After cross-metathesis of ester **6** and diester **8** with catalyst **IV**, and formation of the nitrile-esters **10** and **12**, the crude reaction mixture containing residual ruthenium catalyst was concentrated, transferred to an autoclave, and hydrogenated without additional catalyst under 10 bar of H_2 to quantitatively give the single compounds **16** and **17**, respectively (Scheme 6) [52]. This set of reactions constitutes an example of sequential alkene metathesis/hydrogenation catalytic reactions. However the hydrogenation catalyst is not powerful enough yet to reduce the nitrile group into terminal amine.

4 TANDEM ALKENE METATHESIS / HYDROGENATION CATALYSES: FROM PLANT OIL UNSATURATED ESTERS TO LINEAR AMINO ESTERS

The hydrogenation of aliphatic nitriles into primary amines requires more efficient catalysts operating in more drastic conditions. Takemoto *et al.* [58] and Li *et al.* [59] reported examples of homogeneous hydrogenation of nitriles into primary amines when alkoxide were added to ruthenium catalyst. Das *et al.* [60] and Enthaler *et al.* [61, 62] also showed that addition of ^{t}BuOK and phosphine to simple ruthenium complex led to the fast hydrogenation of various nitriles and by replacing the phosphine ligand by a *N*-heterocyclic carbene, milder conditions could be operating [63].

We have now shown that, after performance of the ruthenium alkylidene-catalyzed cross-metathesis of alkenes with acrylonitrile, when the catalytic hydrogenation of the resulting nitrile ester with the residual ruthenium catalyst was carried out in the presence of ^{t}BuOK or even KOH, the hydrogenation of the nitrile group into primary amine could be obtained, without formation of secondary amine, and that the α,ω-linear amino ester, a precursor of polyamides, could be obtained in excellent yield [64, 65] (Scheme 7).

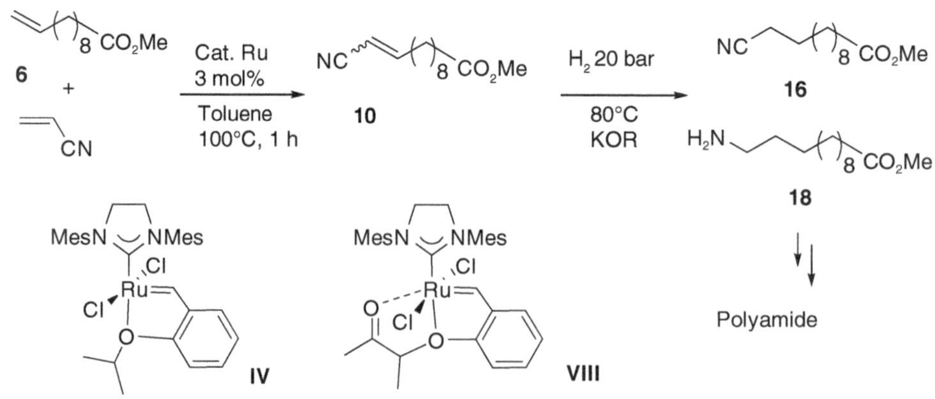

Scheme 7

Tandem catalysis cross metathesis/hydrogenation of ester **6**.

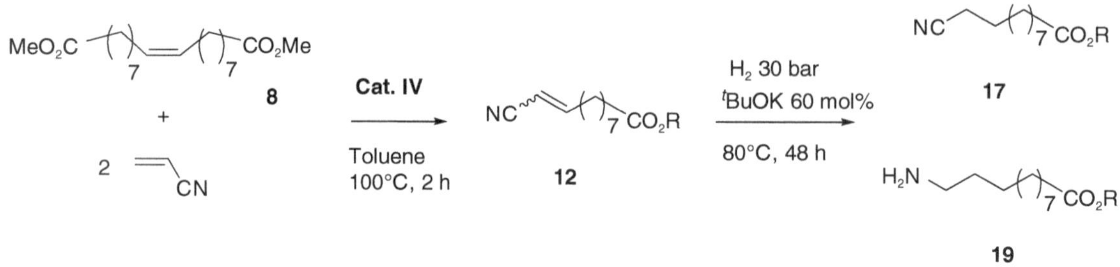

Scheme 8

Sequential catalytic synthesis of aminoester **19** from diester **8**.

Thus, after the cross-metathesis of methyl oleate with acrylonitrile in the presence of the best catalyst: the Hoveyda catalyst **IV** (3 mol%) at 100°C for 1 h in toluene, the reaction mixture containing the residual ruthenium was transferred into an autoclave. Addition of 30 mol% of ^{t}BuOK under 20 bar of hydrogen for 44 h led to the formation of 92% of methyl 12–aminododecanoate **18**. No remaining nitrile ester **16** was observed [64]. It is noteworthy that this reaction takes place without formation of secondary amine as observed in many hydrogenation of nitriles into amines.

The evaluation of other basic additives such as potassium hydroxide, sodium hydroxide, or cesium carbonate showed that KOH could be efficiently used but in 60 mol%. Using only 30 mol% of KOH led to 56% yield of **18** together with 40% yield of nitrile ester **16** [64] (Scheme 7).

We also evaluated the influence of the initial ruthenium alkylidene catalyst and we showed that the Umicore M51 catalyst **VIII** provided similar results as catalyst **IV** but with the advantage of requiring only 15 mol% of ^{t}BuOK [64] (Scheme 7).

We explored the sequential cross-metathesis with acrylonitrile and hydrogenation of the renewable fatty diester **8** [64] (Scheme 8). The reaction of **8** with 4 equivalents of acrylonitrile with 6 mol% of catalyst **IV** was first performed in toluene for 2 h at 100°C and then the reaction mixture was transferred into an autoclave charged with 60 mol% of ^{t}BuOK, 30 bar of H_2. After 48 h at 80°C, the C11 aminoester **19** was isolated in 78% yield.

5 SELECTIVE HYDROGENATION OF NITRILE ESTERS WITH RUTHENIUM ALKYLIDENE CATALYSTS

A tandem metathesis/hydrogenation process in the presence of ruthenium alkylidene catalyst has already been used to prepare saturated polymers [66-68]. Thus, we have explored whether ruthenium–benzylidene and ruthenium–indenylidene olefin metathesis catalysts could also perform hydrogenation of fatty alkyl nitriles and could be more efficient than the residual ruthenium arising from cross-metathesis.

The hydrogenation of methyl 11-cyano undec-10-enoate **10** (Scheme 9), was first performed in the presence of catalyst **IV** (3 mol%) and 30 mol% of ^{t}BuOK, under hydrogen pressure of 20 bar at 80°C for 39 h. It quantitatively

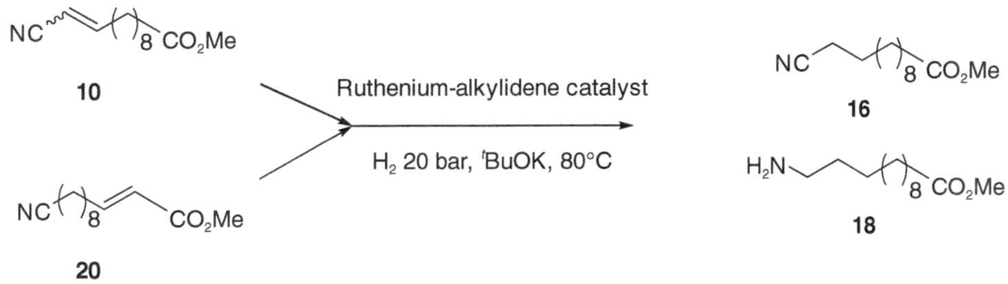

Scheme 9

Catalytic hydrogenation of nitrile esters into linear amino ester **18**.

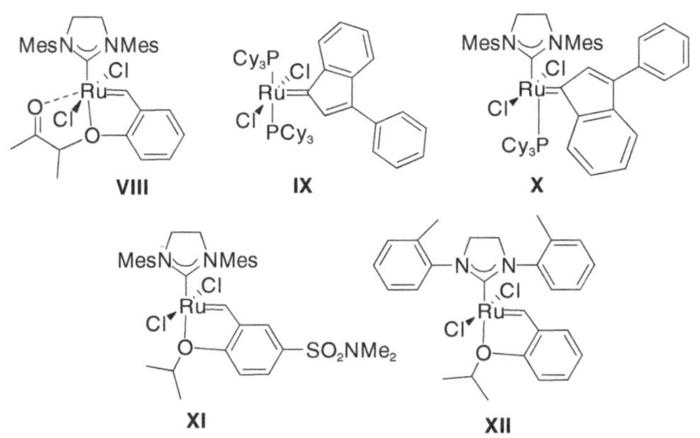

Figure 3

Alkylidene-ruthenium catalysts for alkene metathesis and hydrogenation.

provided the saturated C12 α,ω-amino ester **18**, which was isolated in 89% yield [69] (Scheme 9). The presence of ^{t}BuOK not only created the hydrogenation catalyst but was shown to inhibit the formation of secondary amines.

The catalyst activity of ruthenium-alkylidenes **I**, **II**, and **VIII-X** (Fig. 3) was evaluated on the hydrogenation of **10**, in the presence of 30 mol% of ^{t}BuOK. The best results were obtained with catalyst **X** and **II** requiring the presence of only 15 mol% of ^{t}BuOK to get quantitative formation of **18** after 16 h. More importantly these two catalysts could perform the quantitative formation of **18** by hydrogenation of nitrile ester **10** at 80°C for 16 h [69] (Scheme 9).

Another approach for the preparation of the same C12 α,ω-amino ester **18** can arise from the hydrogenation of methyl (E)-11-cyanoundec-2-enoate **20**, which is arising from cross-metathesis of 10-undecenenitrile with methyl acrylate (see Sect. 6). The catalytic hydrogenation of nitrile ester **20** using 3 mol% of catalyst **IV** under an initial hydrogen pressure of 20 bar at 80°C for 40 h in the presence of 30 mol% of ^{t}BuOK, led to complete conversion of compound **20** and 75% of compound **18** and 22% of compound **16** were obtained [69] (Scheme 9). The hydrogenation of **20** with ruthenium catalysts **I**, **II**, and **VIII–X** (3 mol%) in the presence of 30 mol% of ^{t}BuOK provided excellent yields of amino ester **18**, except for catalyst **I**. On lowering the catalyst loading to 1 mol%, and on shortening the reaction time to 16 h, ruthenium catalysts **II** and **X** offered the best efficiency for the production of **18** from nitrile ester **20** [69] (Scheme 9).

6 CROSS-METATHESIS OF RENEWABLE 10-UNDECENENITRILE AND METHYL ACRYLATE: NEW SYNTHESIS OF POLYAMIDES PRECURSORS

In some examples it was shown that acrylonitrile inhibits the ruthenium-catalysed cross-metathesis of functional alkenes, whereas methyl acrylate does not inhibit this reaction as some cross-metathesis can be performed in bulk in methyl acrylate [16, 17, 70-72]. Thus, it appeared that the linear amino ester **18** could also arise from the cross-metathesis of the 10-undecenenitrile **21** with methyl acrylate followed by hydrogenation. 10-Undecenenitrile **21** is a bio-sourced derivative arising from ammoxidation of 10-undecenoic acid [73]. 10-Undecenoic acid is produced by thermolysis of ricinoleic acid arising from castor oil followed by hydrolysis and it has already been used for the industrial production of polyamide PA11 in the Rilsan process [74].

The cross-metathesis of 10-undecenenitrile **21** with methyl acrylate (2 equivalents) in toluene was attempted with (NHC) Ru alkylidenes **II**, **IV**, **VIII**, and **XI** (0.5 mol%) (Fig. 1, 3). The catalysts **IV**, **VIII** and **XI** bearing a chelating benzylidene ligand were very efficient, at temperature of 50°C or even room temperature, to quantitatively produce nitrile ester **20**, with high percentage of the *E* isomer [75] (Scheme 10). This cross-metathesis transformation is not so easily performed than that of oleate with methyl acrylate but it occurs much more easily than the cross-metathesis of methyl oleate with acrylonitrile, thus showing the inhibition by the nitrile function especially when it is conjugated with the double bond.

To improve the TON values, catalyst **IV** was slowly injected with a syringe pump apparatus. The catalyst solution was added within 2 h 40 min and the reaction was prolonged for further 5 h. With a catalyst loading of 0.05 mol%, the temperature could be decreased down to 50°C to ensure a TON close to 2 000 with almost complete conversion.

With lower catalyst loadings, a higher temperature of 100°C was required to reach good conversion. Thus with 0.01 mol% of catalyst **IV**, 92% conversion of **21** was obtained and led to 92% of nitrile ester **20**, thus providing a high TON of 9200. The increase of the concentration of the nitrile **21** to 0.5 M improved the conversion and catalyst **IV** loading could be decreased to 0.005 mol%. At 100°C a TON of 17 200 could be reached.

Similarly to the transformation of oleate into linear amino ester (Scheme 7) we have investigated the tandem cross-metathesis/hydrogenation in the presence of ^{t}BuOK starting with cross-metathesis of nitrile **21** with methyl acrylate, followed by hydrogenation of the formed nitrile ester, taking profit of the ruthenium residue as hydrogenation catalyst. The nitrile hydrogenation could not be achieved with the very low loading of 0.05 mol% of catalyst **IV** used for the cross-metathesis reaction. Thus the metathesis was performed during the first step, with 3 mol% of catalyst loading, and hydrogenation with the corresponding residual ruthenium was performed under 20 bar of hydrogen at 100°C to give complete conversion into the saturated amino ester **18** which was obtained only in the presence of 30 mol% of ^{t}BuOK with 96% yield [75] (Scheme 10).

7 CROSS-METATHESIS OF METHYL RICINOLEATE WITH ACRYLONITRILE AND METHYL ACRYLATE

Ricinoleic derivatives are useful renewable materials that have a hydroxy functionality which offers an extra-value for further derivatization. However, the homoallylic hydroxyl group may also prevent alkene metathesis [76-79]. We thus studied the cross-metathesis of methyl ricinoleate **22** with acrylonitrile using both the single dose addition and the dropwise slow addition of catalyst **IV** (Scheme 11). This transformation led us to the formation of nitrile ester **12**

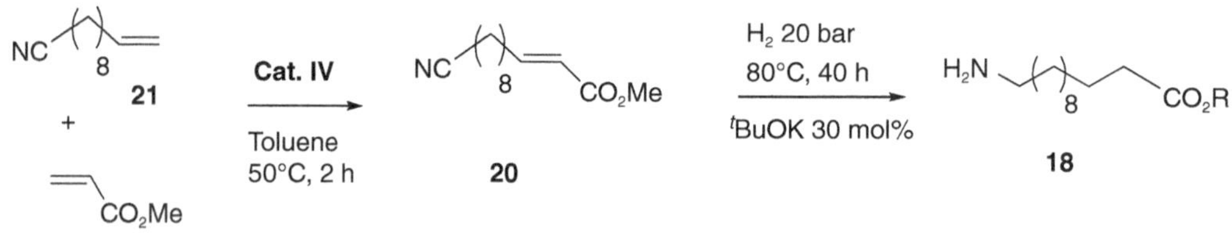

Scheme 10

Sequential catalytic synthesis of amino ester **18** from unsaturated nitrile **21**.

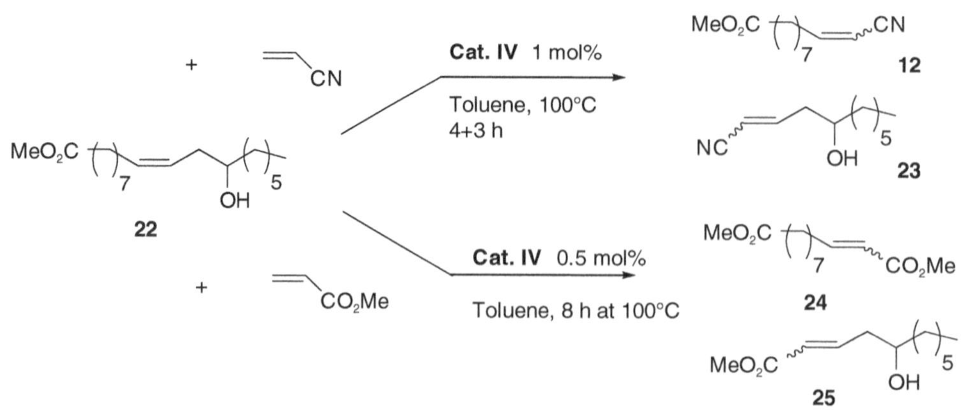

Scheme 11

Cross metathesis of methyl ricinoleate **22** with acrylonitrile and methyl acrylate.

obtained also from the cross-metathesis of diester **8** with acrylonitrile, and to the new compound **23**. Single dose addition of 1 mol% of Hoveyda catalyst **IV** at 80°C allowed full conversion of **22** providing **12** in 82% and **23** in 76% isolated yields. By using the dropwise slow addition of catalyst a higher TON (920) was reached. Using 0.1 mol% of catalyst **12** and **23** were obtained in 73% and 66% yields, respectively (Scheme 11). These first results demonstrate that the transformation of methyl ricinoleate is more difficult to perform than cross-metathesis of diester **8**, thus illustrating the inhibition presented by the homoallylic alcohol functionality with respect to a simple internal alkene [14, 53].

An alternative route for the production of the fatty diesters was explored by the cross-metathesis of methyl ricinoleate **22** with methyl acrylate, for the production of the diester **24** (89%) with concomitant formation of hydroxyl ester **25** (85%), which is a potential precursor of the fatty diol by hydrogenation (Scheme 11). The slow addition of catalyst **IV** afforded 90% conversion of **22** using 0.1 mol% of catalyst **IV** and a high TON (900) was reached [53].

8 BIFUNCTIONAL ALDEHYDE DERIVATIVES VIA SELF- AND CROSS-METATHESIS OF UNSATURATED ALDEHYDES

The aldehyde functionality is often reached in industry via hydroformylation of alkenes, and allows the access to a large variety of functional compounds by selective catalytic transformations. Alkene metathesis catalysis, especially promoted by well-defined homogeneous ruthenium alkylidenes and indenylidenes, is offering functionalization by double bond exchange and with the advantage of tolerating a large variety of functional groups including the formyl group. The previous successful results of cross-metathesis motivated our exploration of the cross-metathesis of unsaturated aldehydes arising from derivatives of plant oils, especially via cross-metathesis with acrylonitrile or via self-metathesis for the production of linear diol precursors or amino alcohols.

The 10-undecylenic aldehyde **26** is a renewable derivative obtained from thermal treatment of castor oil [76, 77]. Its self-metathesis was attempted to produce the C20 dialdehyde a key intermediate for the production of symmetrical diol or diamine and the formation of polyesters and polyamides. The self-metathesis of aldehyde **26** was attempted at 40°C in toluene first with catalysts **II** and **IV**, containing a *N*-heterocyclic carbene ligand. They showed high catalytic activity but also yielded isomerization by-products. The catalysts **I** and **III** containing the PCy_3 ligand offered better production of dialdehyde **27**. The catalyst **III** afforded 79% conversion of **26** with only 9% of self-metathesis by-product [80] (Scheme 12).

The direct synthesis of eicosane-1,20-diol **28** was then directly performed from the mono aldehyde **26**, using tandem self-metathesis/hydrogenation catalysis. The aldehyde **26** was first transformed via self-metathesis with catalyst **III** or **XII**, and the crude products containing the ruthenium residue were transferred into an autoclave and hydrogenated under 10 bar of H_2 at 50°C for 24 h. The diol **28** was isolated in 70% yield. [80] (Scheme 12).

The cross-metathesis of unsaturated aldehyde **26** with various acrylic derivatives was attempted as a route to shorter bifunctional compounds [80] (Scheme 13). The reaction of **26** with acrolein (2 equivalents) with 1 mol% of catalyst **IV** in toluene at 100°C led to the formation of the unsaturated dialdehyde **29**, which was isolated in 78% yield

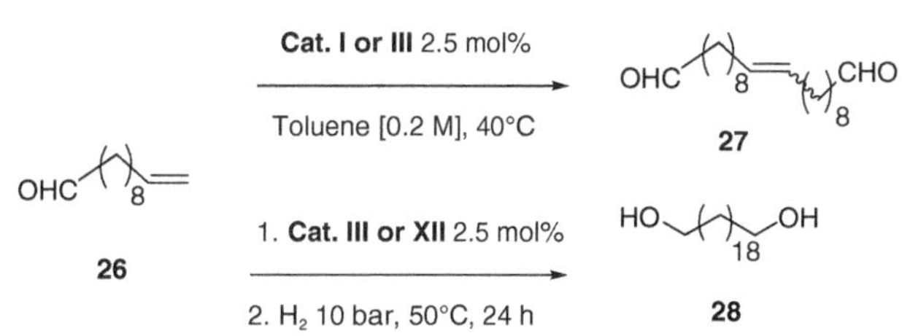

Scheme 12

Homometathesis of 10-undecylenic aldehyde **26** and hydrogenation of dialdehyde **27**.

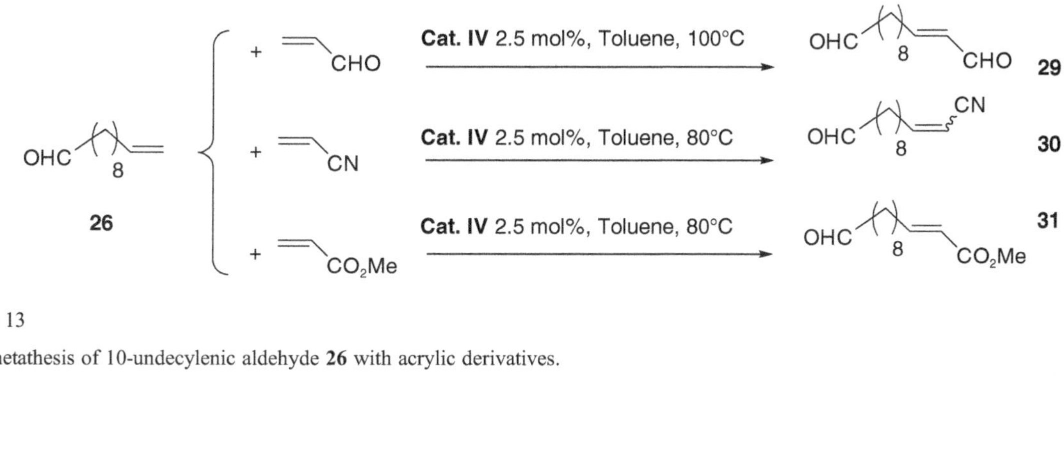

Scheme 13

Cross-metathesis of 10-undecylenic aldehyde **26** with acrylic derivatives.

as a pure *E* isomer. The corresponding C12 terminal diol was obtained directly from **26** using tandem cross-metathesis with acrolein and catalyst **IV**, followed by hydrogenation with 10 bar of H_2 at 50°C for 24 h. The saturated C12 diol was isolated in 72% yield [80].

ω-Nitrile-aldehydes have the potential to produce linear amino alcohols and amino acids as a route to polyamides and mixed polyamide-esters. The cross-metathesis of aldehyde **26** with acrylonitrile (2 equivalents) in the presence of catalyst **IV** was performed at 80°C in toluene for 5 h and led to **30** in 94% isolated yield [80] (Scheme 13).

The cross-metathesis of **26** with methyl acrylate (2 equivalents) led to the isolation of the *E* isomer of the formyl ester **31** in 92% yield. The formation of **31** was achieved with catalyst **IV** (1 mol%) in toluene at 80°C for 16 h. It is noteworthy that the cross-metathesis of **26** with acrylic acid can be performed under similar conditions and after only 14 h the corresponding formyl carboxylic acid was obtained in 84% yield [80].

9 CROSS-METATHESIS OF BIO-SOURCED FATTY NITRILES WITH ACRYLONITRILE

In order to evaluate the influence of a nitrile group at the terminal position of a fatty acid derivative on the cross-metathesis with acrylonitrile, 10-undecenenitrile **21** was subjected to cross metathesis with acrylonitrile to produce the dinitrile **32** (Scheme 14) [81]. We found that this reaction was efficiently achieved in the presence of Hoveyda **IV** and Umicore M51 **VIII** catalysts. The best results were obtained with the dropwise addition of the catalyst to the reaction mixture and a turnover number of 12 800 was reached at 80% conversion with a catalyst loading of 0.00625 mol% in diethyl carbonate at 120°C for 5 h.

These excellent results confirm that suitable experimental conditions make possible the selective cross-metathesis of acrylonitrile without isomerization side reactions in the presence of another nitrile functionality.

10 METATHESIS OF 1-DECENE WITH ALKYNES: A SYNTHESIS OF FATTY DIENES

The synthesis of conjugated dienes by ene-yne cross-metathesis [82] is another interesting pathway for the transformation of Fatty Acid Methyl Esters (FAME). In 2009, we reported the first example of such transformation applied to methyl oleate **1** and to the diester **8** [83]. However, our initial attempts to apply this transformation directly to methyl oleate failed whatever the cross-partner, *i.e.* terminal or internal alkyne. This result was likely due to the lower reactivity of vinylcarbene catalyst species generated in this reaction. To overcome this issue, we set up a sequential reaction involving an initial ethenolysis of methyl oleate followed by ene-yne cross-metathesis of the resulting terminal alkynes (Scheme 15).

This sequential reaction was performed in a one-pot fashion employing the Hoveyda first generation catalyst **III** for the initial ethenolysis step and the second generation Grubbs catalyst **II** for the ene-yne cross-metathesis step (Scheme 15). The choice of a first generation catalyst for the ethenolysis step was dictated by previous results demonstrating the better performances of this catalyst in ethenolysis of methyl oleate in particular since no C=C double-bond migration and low amounts

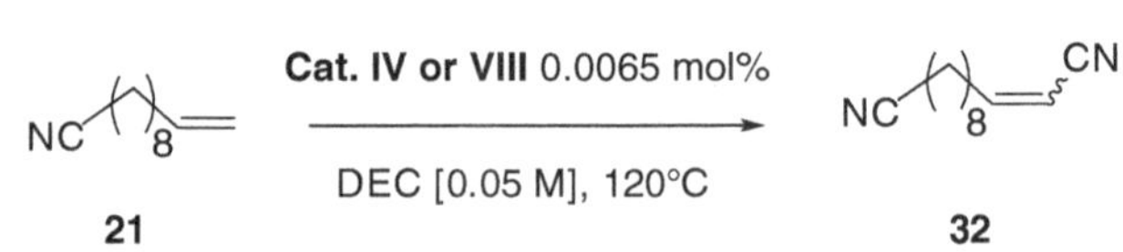

Scheme 14

Cross-metathesis of 10-undecenenitrile **21** with acrylonitrile.

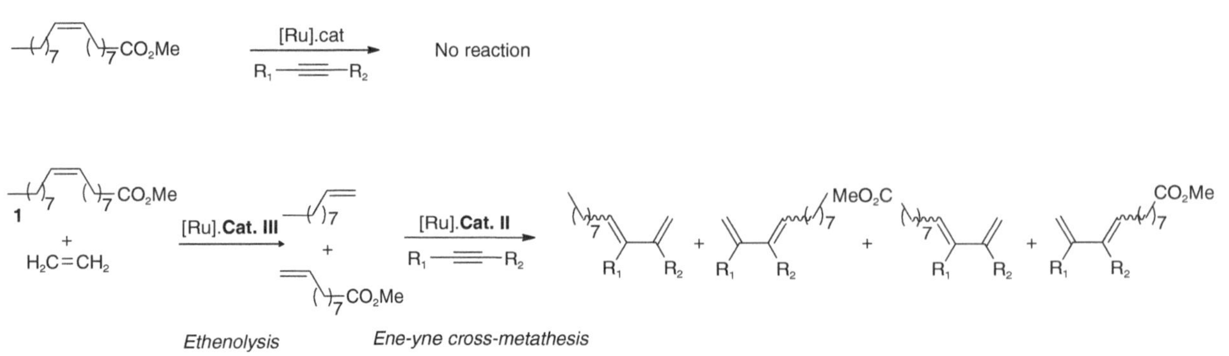

Scheme 15

Sequential ethenolysis and ene-yne cross-metathesis of methyl oleate.

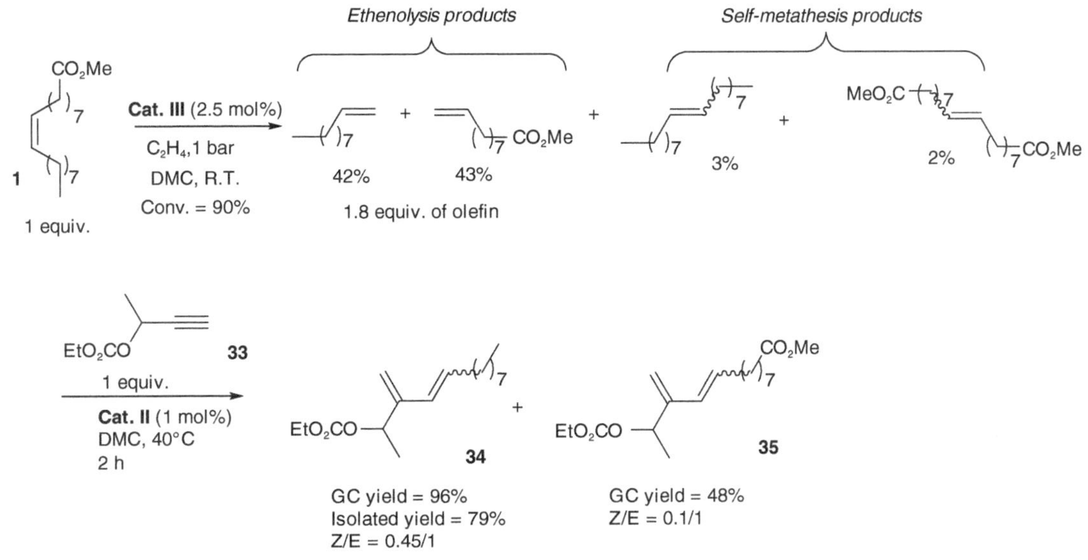

Scheme 16

One-pot ethenolysis/ene-yne cross-metathesis of methyl oleate with propargyl carbonate.

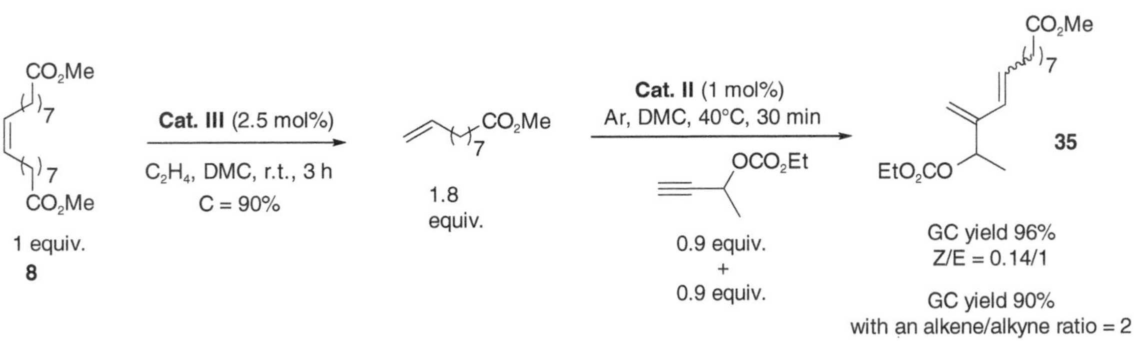

Scheme 17

One-pot ethenolysis/ene-yne cross-metathesis using continuous alkyne addition.

of self-metathesis products were observed with this catalyst [41, 42]. As depicted in Scheme 16, the initial ethenolysis of methyl oleate proceeded with high conversion furnishing the desired ethenolysis product and very low amounts of self-metathesis products. Addition of 1 equivalent of propargyl carbonate **33** and 1 mol% catalyst **II** produced the expected dienes **34, 35** in high yields as a mixture of stereoisomers. It must be underlined that these reactions were conducted in dimethyl carbonate, a solvent earlier found to be fully compatible with olefin metatheses transformations including ethenolysis [84].

Similarly to cross-metatheses of olefins, ene-yne cross-metathesis requires an excess of one reagent (usually 2 equivalents of olefinic partner) to proceed efficiently. This constitutes a major impediment regarding sustainability in particular for the transformation of a bio-sourced compound. This issue was overcome by the implementation of a slow addition of the alkyne reagent in order to maintain an excess of olefin all along the transformation [85]. As shown in Scheme 17 the diester **8** was first submitted to ethenolysis delivering 1.8 equivalent of methyl decenoate. The catalyst **II** and 0.9 equivalent of alkyne were then added and the ene-yne cross-metathesis reaction started. The second 0.9 equivalent of alkyne was then slowly added using a syringe pump apparatus. This reaction led to a higher yield in diene **35** than the previous protocol employing an excess of olefin hence improving the sustainability of this reaction sequence.

Conjugated dienes are valuable building blocks in a number of organic transformations. In particular, dienes arising

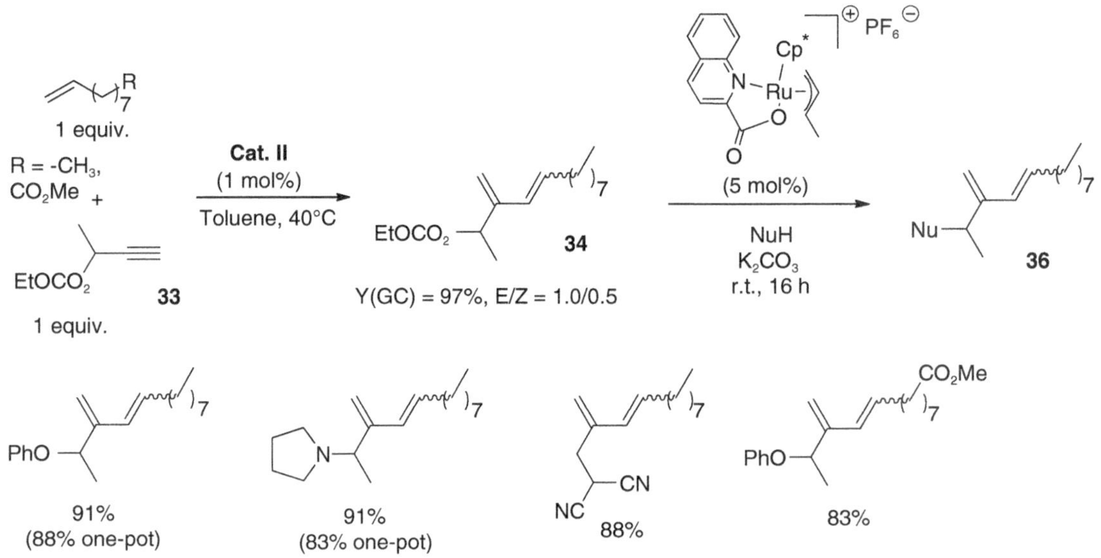

Scheme 18

Sequential ene-yne metathesis and allylic substitution from conjutated dienes.

from ene-yne metathesis have received attention in Diels-Alder transformations [86]. To extend the scope of the transformations of these conjugated dienes, we turned our attention to the nucleophilic allylic substitution reaction. The choice of the alkyne involved in the initial ene-yne cross-metathesis is crucial to allow for further allylic activation. Hence, *n*-decene and methyl decenoate were reacted with the propargylic carbonate **33** to furnish the corresponding dienes in high yields [87]. These dienes were then transformed by nucleophilic allylic substitution catalyzed by a ruthenium complex previously disclosed in our group [88]. The corresponding products **36** were obtained in high yields and regioselectivity. Of note, this transformation could be performed in high yields in a one-pot manner without isolation of the diene intermediate (Scheme 18).

11 FINE CHEMICAL FROM FATTY ACID METHYL ESTERS (FAME)

As exemplified in this article the transformation of FAME is almost exclusively dedicated to the synthesis of polymer precursors. Only rare examples deal with the transformation of FAME into fine chemicals [89]. In 2012, we reported the transformation of methyl ricinoleate into a variety of pyran and lactone derivatives of interest as fragrance ingredients. The synthesis strategy required a preliminary transformation of the hydroxyl functional group of methyl ricinoleate **37** with various olefinic reagents followed by Ring Closing Metathesis (RCM) of the resulting dienes [90]. Hence, following a Ruthenium-Catalyzed (RC) allylation of methyl ricinoleate, RCM delivered a variety of dihydropyrans **38-40** along with methyl 9-decenoate or its self-metathesis diester as co-products of the reaction (Scheme 19). The ratio of these two valuable polymer precursors could be easily tuned by a judicious selection of the RCM catalyst. An unsaturated lactone could also be prepared following the same strategy and preliminary esterification of methyl ricinoleate.

The scope of this reaction was further extended by implementation of tandem RCM/hydrogenation or domino RCM/CM transformations (Scheme 20). As depicted in Scheme 20b, the tetrahydropyran derivative **41** was obtained in good yield via RCM reaction in a single reaction, in the presence or methyl acrylate or acrylonitrile, which concomitantly delivered the bifunctional co-products, suitable as monomer precursors, upon cascade cross-metathesis without hampering the efficiency of the RCM reaction.

In a continuous effort toward the implementation of environmentally friendly processes, the potential of Thermo Morphic Solvent systems (TMS) [91] in olefin metathesis was explored with an emphasis on FAME and methyl undecenoate arising from castor oil [92].

It was demonstrated that a TMS system comprising of Propylene Carbonate (PC) as the polar solvent, Ethyl Acetate (EA) as the medium polarity solvent and CycloHexane (CH) as the apolar phase was suitable for the synthesis of a polyester precursor (Scheme 21). Interestingly, a simple phase separation and extraction of the polar phase with CH furnished the desired product with a ruthenium content

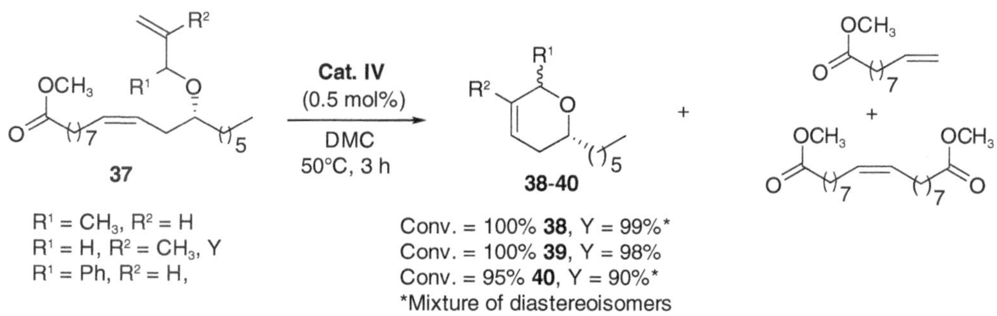

Scheme 19

Methyl ricinoleate-based dihydropyrans catalytic synthesis.

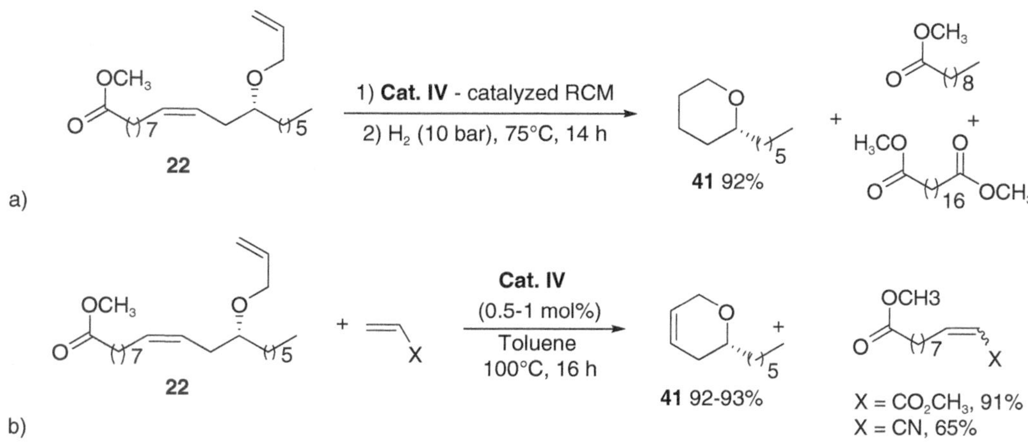

Scheme 20

Tandem and domino transformation of methyl ricinoleate derivatives.

of 62 ppm (Th ~ 2 500 ppm). A rapid filtration on a short plug of silica was sufficient to decrease the ruthenium content below 1 ppm. Further studies demonstrated that ethyl acetate could be replaced by methyl undecenoate resulting in a more efficient reaction.

12 TERPENE AND EUGENOL DERIVATIVES

12.1 Cross-Metathesis with Terpene Derivatives

Olefin metathesis was then applied to other types of unsaturated natural products including monoterpenes, and eugenol substrates derived from essential oils from cloves, cinnamon, etc. or lignin degradation. Most terpenes contain at least one carbon-carbon double bond either included in a cyclic structure or in a linear isoprenoid structure, which is prone to give olefin metathesis transformation. However, due to the presence of other double bonds or oxygenated functional groups (alcohol or aldehyde), metathesis with these substrates has not been studied extensively [93-95]. There are more examples for which terpene derivatives have been modified on purpose by introduction of an olefinic or alkynyl group prior to a metathesis step. However, direct cross-metathesis from natural terpenes represents a straightforward route to produce new functionalized derivatives, and provides more direct and greener strategies to synthesize known products previously prepared using multistep procedures.

We have investigated the cross-metathesis reaction of citronellal, citronellol and citral with methyl acrylate and methacrylate (Scheme 22) [18]. It is noteworthy that if the cross-metathesis with acrylate destroys the C10 isoprenoid structure of the initial terpenes, cross-metathesis with methacrylate gives birth to new products that belong to the terpene family.

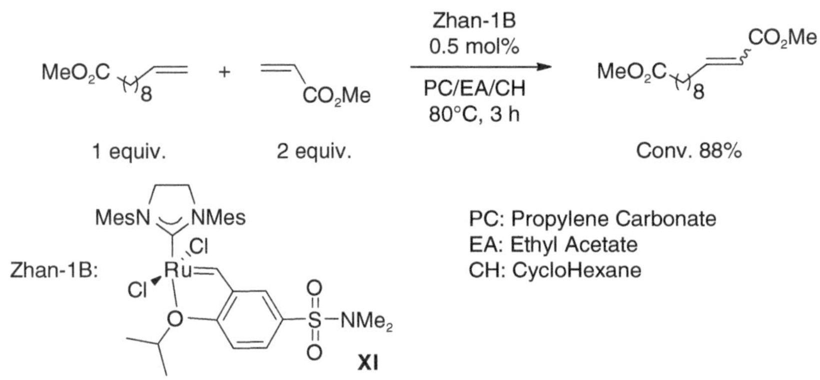

Scheme 21

Cross-metathesis of methyl undecenoate with methyl acrylate in a TMS system.

From our previous experience in cross-metathesis involving electron-deficient olefins, we selected a second generation phosphine-free ruthenium catalyst, namely the Hoveyda type catalyst **IV**.

Starting from the monoolefins citronellal and citronellol, and methyl acrylate as metathesis partner, the best conditions were obtained with 2 equivalents of acrylate in dimethyl carbonate as solvent with 2 mol% of the Hoveyda second generation catalyst **IV** at 60°C for 15 h. Under these conditions, complete conversion of the terpene was observed and the products were isolated in 68% yield (Scheme 22). GC and ^{1}H NMR analyses revealed that the sole *E*-stereoisomer was formed. In the case of the commercially available citral, which is a mixture of (*Z*)- and (*E*)-isomers (neral and geranial), full conversion also took place at 80°C in toluene or dimethyl carbonate, but the reaction was not selective and only 44% isolated yield of the desired product was obtained.

It turned out that the conditions used for cross-metathesis with methyl acrylate were not suitable for cross-metathesis with methyl methacrylate as full conversion could not be obtained even at 100°C in toluene or dimethyl carbonate. The best results were obtained when the reactions were carried out in an excess of methyl methacrylate (29 equivalents with respect to the terpene) without additional solvent in the presence of 2 mol% of catalyst **IV** at 90°C during 8 h. Selective and complete cross-metathesis reactions were observed with citronellal and citronellol in these conditions. Side polymerization of the excess of methyl methacrylate required an additional precipitation step and the desired products were isolated in 70-75% yields (Scheme 22). In accordance with the results obtained from citral with acrylate, the cross-metathesis of this dienic substrate proved to be less selective and the corresponding terpenoid was isolated in only 40% yield after reaction under neat conditions. The products arising from cross-metathesis with methyl methacrylate also presented a *E*-configuration as shown by NOE NMR experiments.

This study on cross-metathesis of terpenes with acrylates highlights that the trisubstituted isoprenoid double bond is reactive with this type of electron deficient olefins and that the reaction is *E*-stereoselective. This is not the case with acrylonitrile, which is reluctant for this type of cross-metathesis with terpenes. This strategy is much more direct and atom economic than previous multistep procedures that have been proposed to synthesize the products resulting from citronellol/methyl acrylate [96] citronellol/methyl methacrylate [97, 98] and citral/methyl methacrylate [99, 100] cross-metathesis.

12.2 Cross-Metathesis of Electron Deficient Olefins with Eugenol Derivatives

We then turned our attention to natural products featuring an allylbenzene motif. These are typically eugenol derivatives present in some essential oils or arising from hydrolytic or thermal degradation of lignocellulosic biomass. We first investigated the cross-metathesis of eugenol with methyl acrylate and then extended the study to other electron deficient olefins such as methyl methacrylate, acrylonitrile and amides.

The metathesis of eugenol with 2 equivalents of acrylate in the presence of the second generation Hoveyda catalyst **IV** (2 mol%) in dimethyl carbonate at 80°C led to complete conversion of eugenol within 3 h. However the reaction was not selective and led to 3 major products: one resulting from the expected cross-metathesis, one arising from tandem metathesis/double bond migration featuring the same number of carbon atoms, and a shorter third one resulting from tandem double bond migration/metathesis (Scheme 23) [101].

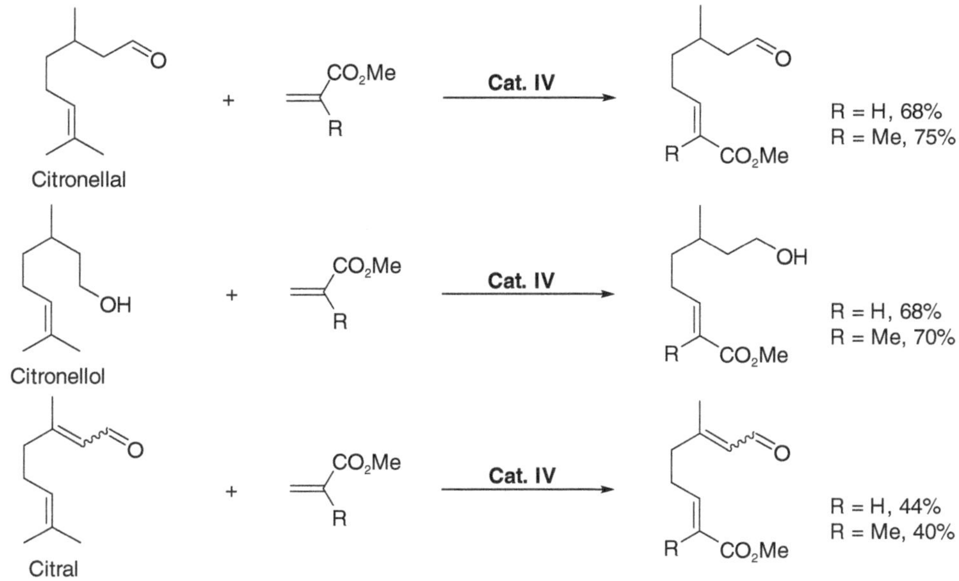

Scheme 22

Cross metathesis of 3 monoterpenoids with methyl acrylate and methacrylate.

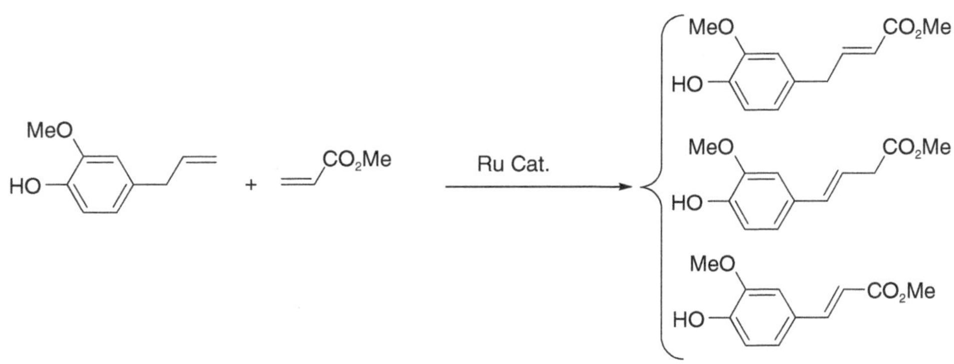

Scheme 23

Cross-metathesis of eugenol with methyl acrylate.

The possibility of isomerization via carbon-carbon double bond migration is well known in olefin metathesis with second generation ruthenium catalysts. This type of reaction is favoured in the present case because the resulting styrene products present a double bond conjugated with the phenyl ring. In order to avoid this side reaction, the use of a hydride scavenger inhibiting the formation of metal hydride species responsible for catalytic isomerization is recommended. 1,4-Benzoquinone has been extensively used to prevent this reaction, which often takes place in the presence of olefin metathesis ruthenium catalysts. Using 5 mol% of benzoquinone under the previous reaction conditions, the formation of the expected methyl (4-hydroxy-3-methoxyphenyl)-4-but-2-enoate became selective with a ratio of this product to its isomerized isomer of 91:9 leading to 78% isolated yield.

For each electron-deficient olefin, experimental conditions were found allowing full conversion of the eugenol derivative and isolation of satisfactory yields of the major products (Scheme 24) [101].

They can be summarized as follows:

- for methyl acrylate: 2 mol% of Hoveyda catalyst **IV** in dimethyl carbonate at 80°C with 5 mol% of 1,4-benzoquinone;
- for methyl methacrylate: 2 mol% of Hoveyda catalyst **IV** without solvent (30 equivalents of methacrylate) at 90°C with 5 mol% of 1,4-benzoquinone;

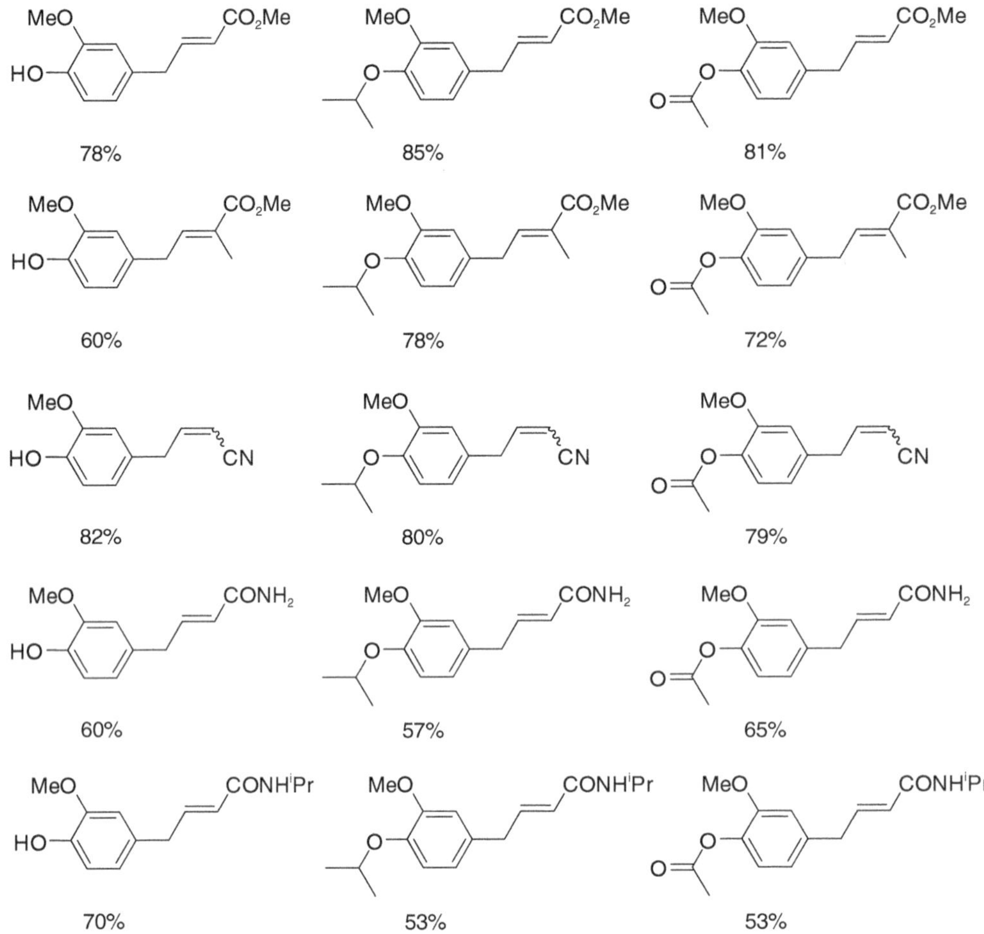

Scheme 24

Eugenol products arising from cross-metathesis with various electron-deficient olefins.

- for acrylonitrile: 2 mol% of Hoveyda catalyst **IV** in diethyl carbonate at 100°C with slow addition of the catalyst in the presence of 5 mol% of 1,4-benzoquinone;
- for amides: 2 mol% of Hoveyda catalyst **IV** in dimethyl carbonate at 80°C with slow addition of the catalyst in the presence of 5 mol% of 1,4-benzoquinone.

The new disubstituted olefins resulting from these cross-metathesis reactions exhibited the expected stereochemistry. They were all *E*-isomers except the nitrile derivatives, which contained both *Z*- and *E*-isomers with a *Z*/*E* ratio of 2:1.

13 SEQUENTIAL METATHESIS/ELIMINATION REACTION: STRAIGHTFORWARD ACCESS TO CONJUGATED DIENES

Following our works on cross-metathesis from natural products, we explored the possibility of associating two RC transformations in a sequential manner to produce conjugated bio-sourced dienes. Indeed, we have recently shown that $[Cp^{*}Ru(MeCN)_3][PF_6]$ catalyzed elimination reactions from allylic carbonates to generate dienes via an allylic activation/elimination mechanism [102]. Thus we applied a 2 step strategy involving first cross-metathesis with allylic chloride followed by HCl elimination as shown in Scheme 25.

The cross-metathesis step was achieved using 6 equivalents of allylic chloride in refluxing dichloromethane in the presence of 2 mol% of the Umicore M51 catalyst **VIII** [103]. Under these conditions, no isomerization of fatty aliphatic chain and only trace amounts of self-metathesis of the natural substrate were observed. Five natural products and two allylic chlorides were used to exemplify the generality of the transformation. 1-Decene, methyl 10-undecenoate and 10-undecenenitrile as fatty acid derivatives, citronellal as a terpene and eugenyl acetate were cross-metathesized

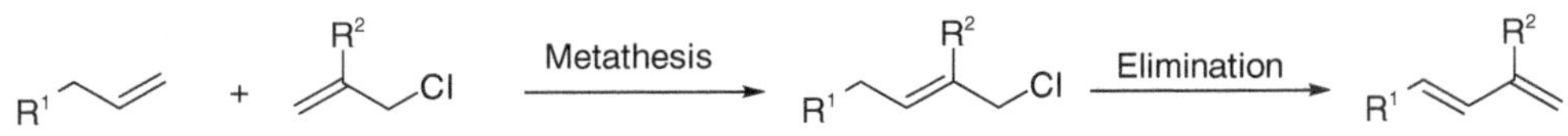

Scheme 25

Sequential procedure to produce 1,3-dienes.

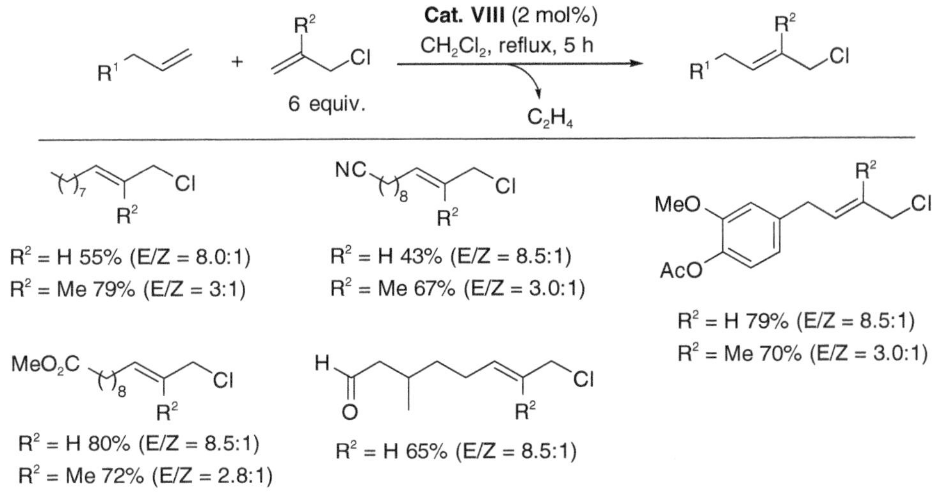

Scheme 26

Allylic products from cross-metathesis of allylic chlorides with natural products.

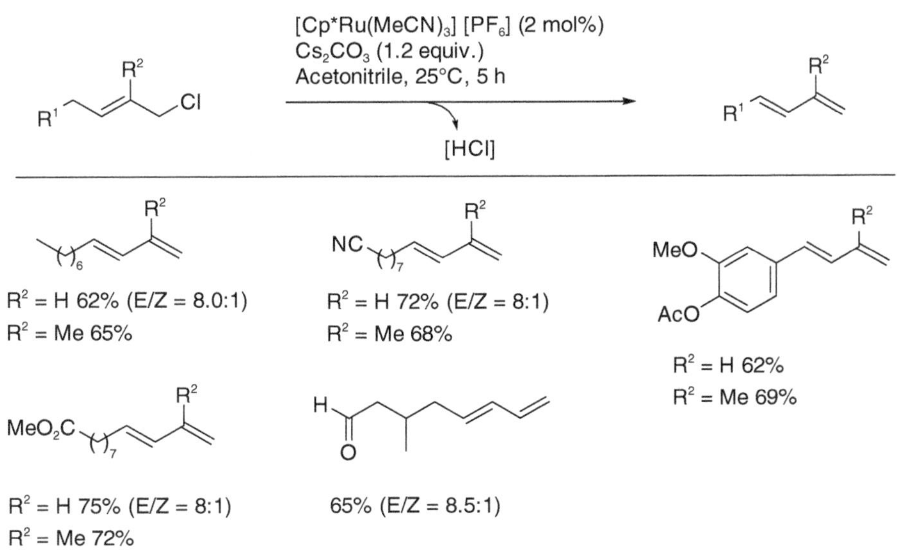

Scheme 27

Formation of terminal dienes from allylic chlorides.

with allyl and 2-methylprop-2-enyl chlorides. The corresponding internal olefins were isolated in good yields as mixtures of *Z*- and *E*-isomers with the *E*-isomer as the major product (Scheme 26).

The elimination reaction was then studied in details from methyl 12-chlorododec-10-enoate and we showed that the reaction could be performed at room temperature in acetonitrile in the presence of 5 mol% of $[Cp^*Ru(MeCN)_3][PF_6]$ as catalyst and 1.2 equivalent of Cs_2CO_3 as a base. The terminal dienes were isolated in good yields (Scheme 27). For the dienes resulting from elimination from unsubstituted allylic chlorides, the internal double bond exhibited both *E* and *Z*-configuration with *E*/*Z* ratio of 8 to 1, except in the case of the eugenol derivative where the internal double bond conjugated with the phenyl ring presented only the *E*-configuration. For the other dienes obtained from 2-methylprop-2-enyl chloride in the first metathesis step, only the *E*-isomers were obtained in a highly stereoselective manner [103].

This catalytic sequence requires the elimination of the solvent of the metathesis step which is not suitable for the elimination reaction, where the presence of the coordinating acetonitrile is required.

However, this protocol appears to be quite general as far as the substrate contains a removable hydrogen atom at the α-carbon of the initial olefin, and it can be applied as a green solution liberating only HCl trapped by the carbonate base as by product to produce conjugated terminal dienes.

CONCLUSION

The above results show that several commercially available alkylidene-ruthenium catalysts allow selective, efficient transformations of unsaturated renewables, arising from plant oils, via cross-metathesis with electron deficient olefins: acrylates, methacrylates, acrylonitrile, acrylamides. They provide access directly to bifunctional linear products without C=C bond isomerization, especially they allow the access to a variety of nitrile esters via cross-metathesis with acrylonitrile and, after hydrogenation with the ruthenium catalyst, to linear amino acid derivatives, useful as polymer precursors. The introduction of catalyst solution via syringe pump allows to reach high TON. Several routes from alkenes to functional conjugated dienes have been established via enyne metathesis of renewable alkenes with alkyne, and via cross-metathesis with simple allylic chlorides followed by ruthenium-catalyzed dehydrohalogenation. Olefin metathesis opens new opportunities allowing access to cyclic molecules with olfactory properties from natural products and giving value to natural terpenes and eugenol via cross-metathesis without isomerization.

There is no doubt that further investigation in cascade, tandem and sequential catalytic transformations of renewables involving at least one metathesis step will provide chemical industry with new atom economic and sustainable processes.

ACKNOWLEDGMENTS

The authors thank *IFP Energies nouvelles*, *Arkema* company, the European Union Seventh Framework Program (FP7/2007-2013) under grant agreement No. 241718 *Euro-BioRef*, *ADEME*, The Region Bretagne, the bilateral France-Tunisia PHC-UTIQUE No. 09G1203, *CNRS*, the French Ministry of Research, the University of Rennes 1, for support. Thanks also go to *Umicore Precious Metals Chemistry*, Hanau (Germany) for loan of metathesis catalysts.The authors wish to acknowledge all the students and researchers from academy and industry, whose names appear in the Rennes Catalysis team references for their contribution to the development of this topic.

REFERENCES

1 Biermann U., Bornscheuer U., Meier M.A.R., Metzger J.O., Schäfer H.J. (2011) Oils and fats as renewable raw materials in chemistry, *Angewandte Chemie International Edition* **50**, 3854-3871.

2 Hill K. (2000) Fats and oils as oleochemical raw materials, *Pure & Applied Chemistry* **72**, 1255-1264.

3 Winkler M., Meier M.A.R. (2014) Olefin cross-metathesis as a valuable tool for the preparation of renewable polyesters and polyamides from unsaturated fatty acid esters and carbamates, *Green Chemistry* **16**, 3335-3340.

4 Chikkali S., Mecking S. (2012) Refining of plant oils to chemicals by olefin metathesis, *Angewandte Chemie International Edition* **51**, 5802-5808.

5 Corma A., Iborra S., Velty A. (2007) Chemical routes for the transformation of biomass into chemicals, *Chemical Reviews* **107**, 2411-2502.

6 Deuss P.J., Barta K., de Vries J.G. (2014) Homogeneous catalysis for the conversion of biomass and biomass-derived platform chemicals, *Catalysis Science & Technology* **4**, 1174-1196.

7 Montero de Espinosa L., Meier M.A.R. (2012) Olefin metathesis of renewable platform chemicals, *Topics in Organometallic Chemistry* **39**, 1-44.

8 Marshall A.-L., Alaimo P.J. (2010) Useful products from complex starting materials: common chemicals from biomass feedstocks, *Chemistry: An European Journal* **16**, 4970-4980.

9 Vennestrøm P.N.R., Osmundsen C.M., Christensen C.H., Taarning E. (2011) Beyond petrochemicals: The renewable chemicals industry, *Angewandte Chemie International Edition* **50**, 10502-10509.

10 Gallezot P. (2012) Conversion of biomass to selected chemical products, *Chemical Society Reviews* **41**, 1538-1558.

11 Connon S.J., Blechert S. (2003) Recent developments in olefin cross-metathesis, *Angewandte Chemie International Edition* **42**, 1900-1923.

12 Mol J.C. (2004) Catalytic metathesis of unsaturated fatty acid esters and oils, *Topics in Catalysis* **27**, 97-104.

13 Mol J.C. (2002) Application of olefin metathesis in oleochemistry: an example of green chemistry, *green chemistry* **4**, 5-13.

14 Bruneau C., Fischmeister C., Miao X., Malacea R., Dixneuf P.H. (2010) Cross-metathesis with acrylonitrile and applications to fatty acid derivatives, *European Journal of Lipid Science and Technology* **112**, 3-9.

15 Miao X., Dixneuf P.H., Fischmeister C., Bruneau C. (2011) A green route to nitrogen-containing groups: the acrylonitrile cross-metathesis and applications to plant oils derivatives, *Green Chemistry* **13**, 2258-2271.

16 Rybak A., Meier M.A.R. (2007) Cross-metathesis of fatty acid derivatives with methyl acrylate: renewable raw materials for the chemical industry, *Green Chemistry* **9**, 1356-1361.

17 Rybak A., Meier M.A.R. (2008) Cross-metathesis of oleyl alcohol with methyl acrylate: optimization of reaction conditions and comparison of their environmental impact, *Green Chemistry* **10**, 1099-1104.

18 Bilel H., Hamdi N., Zagrouba F., Fischmeister C., Bruneau C. (2011) Cross-metathesis transformations of terpenoids in dialkyl carbonate solvents, *Green Chemistry* **13**, 1448-1452.

19 Yadav G.D., Doshi N.S. (2002) Development of a green process for poly-α-olefin based lubricants, *Green Chemistry* **4**, 528-540.

20 Warwel S., Tillack J., Demes C., Kunz M. (2001) Polyesters of ω-Unsaturated Fatty Acid Derivatives, *Macromolecular Chemistry and Physics* **202**, 1114-1121.

21 van Dam P.B., Mittlmeijer M.C., Boelhouwer C. (1972) Metathesis of unsaturated fatty acid esters by a homogeneous tungsten hexachloride–tetramethyltin catalyst, *Journal of the Chemical Society, Chemical Communications* 1221-1222.

22 Verkuijlen E., Kapteijn F., Mol J.C., Boelhouwer C. (1977) Heterogeneous metathesis of unsaturated fatty acid esters, *Journal of the Chemical Society, Chemical Communications* 198-199.

23 Bosma R.H.A., van Aardweg F., Mol J.C. (1981) Cometathesis of methyl oleate and ethylene; a direct route to methyl dec-9-enoate, *Journal of the Chemical Society, Chemical Communications* 1132-1133.

24 Newman T.H., Rand C.L., Burdett K.A., Maughon B.R., Morrison D.L., Wasserman E.P., WO02/076920.

25 Grubbs R.H., Nguyen S.-B.T, Johnson L.K., Hillmyer M.A., Fu G.C., WO96/04289.

26 Thurier C., Olivier-Bourbigou H., Dixneuf P.H., Hillion G., EP1698686, US 2006/0079704.

27 Burdett K.A., Harris L.D., Margl P., Maughon B.R., Mokhtar-Zadeh T., Saucier P.C., Wasserman E.P. (2004) Renewable Monomer Feedstocks via Olefin Metathesis: Fundamental Mechanistic Studies of Methyl Oleate Ethenolysis with the First-Generation Grubbs Catalyst, *Organometallics* **23**, 2027-2047.

28 Chauvin Y., Gilbert B., Guibard I. (1990) Catalytic dimerization of alkenes by nickel complexes in organochloroaluminate molten salts, *Journal of the Chemical Society, Chemical Communications* 1715-1716.

29 Olivier-Bourbigou H., Magna L. (2002) Ionic liquids: perspectives for organic and catalytic reactions, *Journal of Molecular Catalysis A: Chemical* **182**, 419-437.

30 Zhao D., Wu M., Kou Y., Min E. (2002) Ionic liquids: applications in catalysis, *Catalysis Today* **74**, 157-189.

31 Welton T. (2004) Ionic liquids in catalysis, *Coordination Chemistry Reviews* **248**, 2459-2477.

32 Pârvulescu V.I., Hardacre C. (2007) Catalysis in Ionic Liquids, *Chemistry Reviews* **107**, 2615-2665.

33 Chauvin Y., Olivier-Bourbigou H. (1995) Nonaqueous ionic liquids as reaction solvents, *Chemtech* **25**, 26.

34 Gürtler C., Jautelat M., EP1035093A2.

35 Buijsman R.C., van Vuuren E., Sterrenburg J.G. (2001) Ruthenium-Catalyzed Olefin Metathesis in Ionic Liquids, *Organic Letters* **3**, 3785-3787.

36 Mayo K.G., Nearhoof E.H., Kiddle J.J. (2002) Microwave-Accelerated Ruthenium-Catalyzed Olefin Metathesis, *Organic Letters* **4**, 1567-1570.

37 Ding X., Lv X., Hui B., Chen Z., Xiao M., Guo B., Tang W. (2006) Olefin self-cross-metathesis catalyzed by the second-generation Grubbs carbene complex in room temperature ionic liquids, *Tetrahedron Letters* **47**, 2921-2924.

38 Williams D.B.G., Ajam M., Ranwell A. (2006) Highly Selective Metathesis of 1-Octene in Ionic Liquids, *Organometallics* **25**, 3088-3090.

39 Sémeril D., Olivier-Bourbigou H., Bruneau C., Dixneuf P.H. (2002) Alkene metathesis catalysis in ionic liquids with ruthenium allenylidene salts, *Chemical Communications* 146-147.

40 Csihony S., Fischmeister C., Bruneau C., Horvath I.T., Dixneuf P.H. (2002) First ring-opening metathesis polymerization in an ionic liquid, Efficient recycling of a catalyst generated from a cationic ruthenium allenylidene complex, *New Journal of Chemistry* **26**, 1667-1670.

41 Thurier C., Fischmeister C., Bruneau C., Olivier-Bourbigou H., Dixneuf P.H. (2007) Ionic imidazolium containing ruthenium complexes and olefin metathesis in ionic liquids, *Journal of Molecular Catalysis A: Chemistry* **268**, 127-133.

42 Thurier C., Fischmeister C., Bruneau C., Olivier-Bourbigou H., Dixneuf P.H. (2008) Ethenolysis of Methyl Oleate in Room-Temperature Ionic Liquids, *ChemSusChem* **1**, 118-122.

43 Ackermann L., Bruneau C., Dixneuf P.H. (2001) Simple new three-component catalytic system for enyne metathesis, *Synlett* 397-399.

44 Sémeril D., Bruneau C., Dixneuf P.H. (2002) Imidazolium and Imidazolinium Salts as Carbene Precursors or Solvent for Ruthenium-Catalysed Diene and Enyne Metathesis, *Advanced Synthesis & Catalysis* **344**, 585-595.

45 Crowe W.E., Goldberg D.R. (1995) Acrylonitrile Cross-Metathesis: Coaxing Olefin Metathesis Reactivity from a Reluctant Substrate, *Journal of the American Chemical Society* **117**, 5162-5163.

46 Gessler S., Randl S., Blechert S. (2000) Synthesis and metathesis reactions of a phosphine-free dihydroimidazole carbene ruthenium complex, *Tetrahedron Letters* **41**, 9973-9976.

47 Love J.A., Morgan J.P., Trnka T.M., Grubbs R.H. (2002) A Practical and Highly Active Ruthenium-Based Catalyst that Effects the Cross-metathesis of Acrylonitrile, *Angewandte Chemie International Edition* **41**, 4035-4037.

48 Hoveyda H., Vezina M. (2005) Synthesis of Unsaturated Amino Alcohols through Unexpectedly Selective Ru-Catalyzed Cross-Metathesis Reactions, *Organic Letters* **7**, 2113-2116.

49 Bieniek M., Bujok R., Cabaj M., Lugan N., Lavigne G., Arlt D., Grela K. (2006) Advanced Fine-Tuning of Grubbs/Hoveyda Olefin Metathesis Catalysts: A Further Step toward an Optimum Balance between Antinomic Properties, *Journal of the American Chemical Society* **128**, 13562-13564.

50 Michaut A., Boddaert T., Coquerel Y., Rodriguez J. (2007) Reluctant cross-metathesis reactions: The highly beneficial effect of microwave irradiation, *Synlett* 2867-2871.

51 Rivard M., Blechert S. (2003) Effective and Inexpensive Acrylonitrile Cross-Metathesis: Utilisation of Grubbs II Precatalyst in the Presence of Copper(I) Chloride, *European Journal of Organic Chemistry* 2225-2228.

52 Malacea R., Fischmeister C., Bruneau C., Dubois J.-L., Couturier J.-L., Dixneuf P.H. (2009) Renewable materials as precursors of linear nitrile-acid derivatives *via* cross-metathesis of fatty esters and acids with acrylonitrile and fumaronitrile, *Green Chemistry* **11**, 152-155.

53 Miao X., Malacea R., Fischmeister C., Bruneau C., Dixneuf P.H. (2011) Ruthenium–alkylidene catalysed cross-metathesis of fatty acid derivatives with acrylonitrile and methyl acrylate: a key step toward long-chain bifunctional and amino acid compounds, *Green Chemistry* **13**, 2911-2919.

54 Fabritius D., Schäfer H.J., Steinbüchel A. (1998) Bioconversion of sunflower oil, rapeseed oil and ricinoleic acid by Candida tropicalis M25, *Applied Microbiology and Biotechnology* **50**, 573-578.

55 Dinger M.B., Mol J.C. (2002) High Turnover Numbers with Ruthenium-Based Metathesis Catalysts, *Advanced Synthesis & Catalysis* **344**, 671-677.

56 Ngo H.L., Jones K., Foglia T.A. (2006) *Journal of the American Oil Chemists' Society* **83**, 629-634.

57 Marvey B.B., Segakweng C.K., Vosloo M.H.C. (2008) Ruthenium Carbene Mediated Metathesis of Oleate-Type Fatty Compounds, *International Journal of Molecular Sciences* **9**, 615-625.

58 Takemoto S., Kawamura H., Yamada Y., Okada T., Ono A., Yoshikawa E., Mizobe Y., Hidai M. (2002) Ruthenium Complexes Containing Bis(diarylamido)/Thioether Ligands: Synthesis and their Catalysis for the Hydrogenation of Benzonitrile, *Organometallics* **21**, 3897-3904.

59 Li T., Bergner I., Haque F.N., Iuliis M.Z.-D., Song D., Morris R.H. (2007) Hydrogenation of Benzonitrile to Benzylamine Catalyzed by Ruthenium Hydride Complexes with P–NH–NH–P Tetradentate Ligands: Evidence for a Hydridic–Protonic Outer Sphere Mechanism, *Organometallics* **26**, 5940-5949.

60 Das S., Zhou S., Addis D., Enthaler S., Junge K., Beller M. (2010) Selective catalytic reductions of amides and nitriles to amines, *Topics in Catalysis* **53**, 979-984.

61 Enthaler S., Junge K., Addis D., Erre G., Beller M. (2008) A practical and benign synthesis of primary amines through ruthenium-catalyzed reduction of nitriles, *ChemSusChem* **1**, 1006-1010.

62 Enthaler S., Addis D., Junge K., Erre G., Beller M. (2008) A practical and benign synthesis of primary amines through ruthenium-catalyzed reduction of nitriles, Chemistry: an, *European Journal* **14**, 9491-9494.

63 Addis D., Enthaler S., Junge K., Wendt B., Beller M. (2009) Ruthenium N-heterocyclic carbene catalysts for selective reduction of nitriles to primary amines, *Tetrahedron Letters* **50**, 3654-3656.

64 Miao X., Fischmeister C., Dixneuf P.H., Bruneau C., Dubois J.-L., Couturier J.-L. (2012) Tandem Catalytic Acrylonitrile Cross-Metathesis and Hydrogenation of Nitriles with Ruthenium Catalysts: Direct Access to Linear α, ω-Aminoesters from Renewables, *ChemSusChem* **5**, 1410-1414.

65 Couturier J.-L., Dubois J.-L., Miao X., Fischmeister C., Bruneau C., Dixneuf P.H. (2011) Brevet: FR 2959742A120111111; PCT Int. Appl., WO 2011138051A1 20111110.

66 Bielawski C.W., Louie J., Grubbs R.H. (2000) Tandem Catalysis: Three Mechanistically Distinct Reactions from a Single Ruthenium Complex, *Journal of the American Chemical Society* **122**, 12872-12873.

67 Watson M.D., Wagener K.B. (2000) Tandem Homogeneous Metathesis/Heterogeneous Hydrogenation: Preparing Model Ethylene/CO2 and Ethylene/CO Copolymers, *Macromolecules* **33**, 3196-3201.

68 Drouin S.D., Zamanian F., Fogg D.E. (2001) Multiple Tandem Catalysis: Facile Cycling between Hydrogenation and Metathesis Chemistry, *Organometallics* **20**, 5495-5497.

69 Miao X., Bidange J., Dixneuf P.H., Fischmeister C., Bruneau C., Dubois J.-L., Couturier J.-L. (2012) Ruthenium-benzylidenes and -indenylidenes as efficient catalysts for the hydrogenation of aliphatic nitriles into primary amines, *ChemCatChem* **4**, 1911-1916.

70 Ho T.T.T., Jacobs T., Meier M.A.R. (2009) A Design-of-Experiments Approach for the Optimization and Understanding of the Cross-Metathesis Reaction of Methyl Ricinoleate with Methyl Acrylate, *ChemSusChem* **2**, 749-754.

71 Behr A., Perez Gomes J., Bayrak Z. (2011) Cross-metathesis of methyl 10-undecenoate with diethyl maleate: Formation of an α, ϖ-diester *via* a metathesis reaction network, *European Journal of Lipid Science and Technology* **113**, 189-196.

72 Djigoué G.B., Meier M.A.R. (2009) Improving the selectivity for the synthesis of two renewable platform chemicals *via* olefin metathesis, *Applied Catalysis A* **368**, 158-162.

73 Dubois J.-L., Gillet J.-P. (2008) Procédé de co-production de carbonates cycliques et de nitriles et/ou d'amines gras, Patent WO2008145941A9.

74 Chauvel A., Lefebvre G. (1989) Petrochemical Processes, *Editions Technip* **2**, 274.

75 Miao X., Fischmeister C., Dixneuf P.H., Bruneau C., Dubois J.-L., Couturier J.-L. (2012) Polyamide precursors from renewable 10-undecenenitrile and methyl acrylate *via* olefin cross-metathesis, *Green Chemistry* **14**, 2179-2183.

76 Das G., Trivedi R.K., Vasishtha A.K. (1989) Heptaldehyde and undecylenic acid from castor oil, *Journal of the American Oil Chemists' Society* **66**, 938-941.

77 van der Steen M., Stevens C.V., Eeckhout Y., De Buyck L., Ghelfi F., Roncaglia F. (2008) Undecylenic acid: A valuable renewable building block on route to Tyromycin A derivatives, *European Journal of Lipid Science and Technology* **110**, 846-852.

78 van der Steen M., Stevens C.V. (2009) Undecylenic Acid: A Valuable and Physiologically Active Renewable Building Block from Castor Oil, *ChemSusChem* **8**, 692-713.

79 Mutlu H., Meier M.A.R. (2010) Castor oil as a renewable resource for the chemical industry, *European Journal of Lipid Science and Technology* **112**, 10-30.

80 Miao X., Fischmeister C., Bruneau C., Dixneuf P.H. (2009) A direct route to bifunctional aldehyde derivatives *via* self- and cross-metathesis of unsaturated aldehydes, *ChemSusChem* **2**, 542-545.

81 Bidange J., Fischmeister C., Bruneau C., Dubois J.-L., Couturier J.-L. (2015) Cross-metathesis of bio-sourced fatty nitriles with acrylonitrile, *Monatshefte für Chemie* **146**, 1107-1113.

82 Fischmeister C., Bruneau C. (2011) Ene-yne cross-metathesis with ruthenium carbene catalysts, *Beilstein Journal of Organic Chemistry* **7**, 156-166.

83 Le Ravalec V., Fischmeister C., Bruneau C. (2009) First transformation of unsaturated fatty esters involving enyne cross-metathesis, *Advanced Synthesis and Catalysis* **351**, 1115-1122.

84 Miao X., Fischmeister C., Bruneau C., Dixneuf P.H. (2008) Dimethyl Carbonate: An Eco-Friendly Solvent in Ruthenium-Catalyzed Olefin Metathesis Transformations, *ChemSusChem* **1**, 813-816.

85 Le Ravalec V., Dupé A., Fischmeister C., Bruneau C. (2010) Improving sustainability in ene-yne cross-metathesis for transformation of unsaturated fatty esters, *ChemSusChem* **3**, 1291-1297.

86 Kotha S., Meshram M., Tiwari A. (2009) Advanced approach to polycyclics by a synergistic combination of enyne metathesis and Diels Alder reaction, *Chemical Society Reviews* **38**, 2065-2092.

87 Dupé A., Le Ravalec V., Fischmeister C., Bruneau C. (2013) Stepwise catalytic transformations of renewable feedstock arising from plant oils, *European Journal of Lipid Science and Technology* **115**, 490-500.

88 Zhang H.-J., Demerseman B., Toupet L., Xi Z., Bruneau C. (2008) Novel [Ru(C_5Me_4R)(2-quinolinecarboxylato)(allyl)] [PF_6] Complexes as Efficient Catalysts for Highly Regioselective Nucleophilic Substitution of Aliphatic Allylic Substrates, *Advanced Synthesis & Catalysis* **350**, 1601-1609.

89 Biermann U., Bornscheuer U., Meier M.A.R., Metzger J.O., Schäfer H.J. (2011) Oils and Fats as Renewable Raw materials in Chemistry, *Angewandte Chemie International Edition* **50**, 3854-3871.

90 Dupé A., Achard M., Fischmeister C., Bruneau C. (2012) Methyl ricinoleate as platform chemical for simultaneous production of fine chemicals and polymer precursors, *ChemSusChem* **5**, 2249-2254.

91 Behr A., Toslu N. (1999) One-and two-phase reaction engineering of the hydrosilylation reaction, *Chemie- Ingenieur Technik* **71**, 490-493.

92 Huang S., Bilel H., Zagrouba F., Hamdi N., Bruneau C., Fischmeister C. (2015) Olefin metathesis transformations in thermomorphic multicomponent solvent system, *Catalysis Communications* **63**, 31-34.

93 Yoshikai K., Hayama T., Nishimura K., Yamada K.I., Tomioka K. (2005) Thiol-catalyzed acyl radical cyclization of alkenals, *Journal of Organic Chemistry* **70**, 681-683.

94 Gutierrez S., Tlenkopatchev M.A. (2011) Metathesis of renewable products: degradation of natural rubber *via* cross-metathesis with *β*-pinene using Ru-alkylidene catalysts, *Polymer Bulletin* **66**, 1029-1038.

95 Borré E., Dinh T.H., Caijo F., Crévisy C., Mauduit M. (2011) Terpenic compounds as renewable sources of raw materials for cross-metathesis, *Synthesis* 2125-2130.

96 Hanessian S., Dhanoa D.S., Beaulieu P.L. (1987) Synthesis of carbocycles from ω-substituted α, β-unsaturated esters *via* radical-induced cyclizations, *Canadian Journal of Chemistry* **65**, 1859-1866.

97 Brown R.T., Mayalarp S.P., Watts J. (1997) Synthesis of methyl secolonitoside, *Journal of the Chemical Society, Perkin Transactions* **1**, 1633-1637.

98 Yamagushi K., Shinohara C., Kojima S., Sodeoka M., Tsuji T. (1999) (2*E*,6*R*)-8-Hydroxy-2,6-dimethyl-2-octenoic Acid, a Novel *Anti*-osteoporotic Monoterpene Isolated from *Cistanche salsa, Bioscience Biotechnology, Biochemistry* **63**, 731-735.

99 Zhang Z., Chen J., Yang Z., Tang Y. (2010) Rapid Biomimetic Total Synthesis of (±)-Rossinone B, *Organic Letters* **12**, 5554-5557.

100 Zhao Y.-J., Loh T.-P. (2008) Practical synthesis of 1,5-dimethyl substituted conjugated polyenes from geranyl acetate, *Tetrahedron* **64**, 4972-4978.

101 Bilel H., Hamdi N., Zagrouba F., Fischmeister C., Bruneau C. (2012) Eugenol as a renewable feedstock for the production of polyfunctional alkenes *via* olefin cross-metathesis, *RSC Advances* **2**, 9584-9589.

102 Beydoun K., Zhang H.-J., Sundararaju B., Demerseman B., Achard M., Xi Z., Bruneau C. (2009) Efficient ruthenium-catalyzed synthesis of [3]-dendralenes from 1,3-dienic allylic carbonates, *Chemical Communications* 6580-6582.

103 Bilel H., Hamdi N., Zagrouba Z., Fischmeister C., Bruneau C. (2014) Terminal conjugated dienes *via* a ruthenium-catalyzed cross-metathesis/elimination sequence: application to renewable resources, *Catalysis Science & Technology* **4**, 2064-2071.

14

Cationic Tungsten(VI) Penta-Methyl Complex: Synthesis, Characterization and its Application in Olefin Metathesis Reaction

Raju Dey[1], Manoja K. Samantaray[1], Emmanuel Callens[1], Ali Hamieh[1], Abdul-Hamid M. Emwas[2], Edy Abou-hamad[1], Santosh Kavitake[1] and Jean-Marie Basset[1]*

[1] *King Abdullah University of Science & Technology, KAUST Catalysis Center (KCC), Thuwal 23955-6900 - Saudi Arabia*
[2] *Imaging and Characterization Core Lab, King Abdullah University of Science and Technology (KAUST), Thuwal 23955-6900 - Saudi Arabia*
e-mail: jeanmarie.basset@kaust.edu.sa

* Corresponding author

Abstract — *Tungsten-hexa-methyl readily reacts with $B(C_6F_5)_3$ in dichloromethane and generates the corresponding well-defined cationic tungsten-penta-methyl complex which was identified precisely by ^1H NMR, ^{13}C NMR, ^1H-^{13}C NMR correlation spectroscopy. Unlike WMe_6, this cationic complex has low energy barrier to form tungsten carbene intermediate, which was further supported by the fact that WMe_6 alone has no activity in olefin metathesis reaction whereas the cationic complex shows catalytic activity for self-metathesis of 1-octene.*

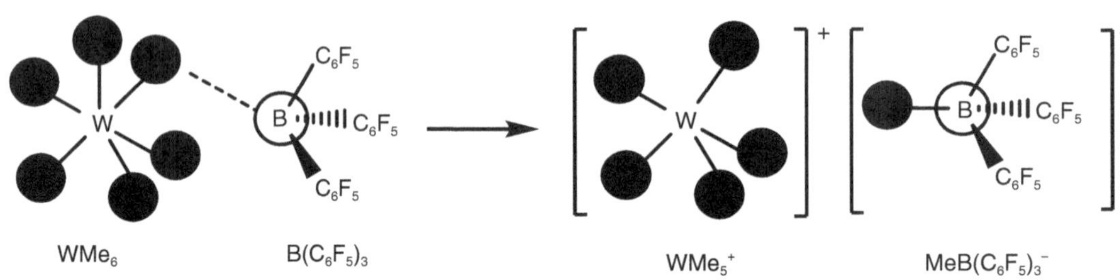

Résumé — Complexe pentaméthyle de tungstène(VI) cationique : synthèse, caractérisation et son application dans la réaction de métathèse des oléfines — L'hexaméthyle de tungstène réagit avec le $B(C_6F_5)_3$ dans du dichlorométhane et génère le complexe pentaméthyle de tungstène cationique bien défini qui a été précisément identifié par spectroscopie de corrélation ^1H RMN, ^{13}C RMN, ^1H-^{13}C RMN. À la différence du WMe_6, ce complexe cationique a une barrière énergétique faible pour former un intermédiaire de carbène tungstène, ce qui a été ultérieurement étayé par le fait que le WMe_6 seul n'a pas d'activité dans la réaction de métathèse des oléfines, tandis que le complexe cationique montre une activité catalytique pour l'autométathèse d'1-octène.

INTRODUCTION

In 1973, Shortlan and Wilkinson described the first report on Tungsten hexamethyl [1], [2-4]. After its synthesis, the structure was established in 1996 by Seppelt [5], [6, 7] and Kaupp [8]. The geometry was calculated by Demolliens *et al.* [9] and Kang *et al.* [10]. However, until 2015 and to our knowledge there was no report on the catalytic activity of this homoleptic complex. WMe_6 is indeed difficult to prepare, extremely unstable, and very sensitive to even few ppm levels of oxygen or moisture. Handling this compound is challenging due to its explosive nature as reported by Shortlan and Wilkinson [1, 2], Galyer *et al.* [3], Galyer and Wilkinson [4]. In our previous communication, we demonstrated that grafting this unstable WMe_6 to the silica surface by SOMC (Surface OrganoMetallic Chemistry) strategy and methodology at −80°C can enhance its thermal stability due to the formation of the stable grafted complex, [≡SiO-W(Me)$_5$] which proved to be a nice precursor of alkane metathesis catalysts via the formation of surface monopodal tungsten carbyne [11]. This carbyne is in equilibrium with a bis carbene upon addition of a donor ligand [12] (Scheme 1). Also, we demonstrated that WMe_6 on silica-alumina shows relatively higher reactivity than on silica because of the formation of very reactive cationic intermediate [13]. Let us recall that Kress and Osborn were the first to report the synthesis of a cationic W-carbene very active in olefin metathesis [14]. Based on the above fact we speculate that in order to prepare more reactive catalysts in terms of reactivity and selectivity a higher electrophilicity on tungsten center is essential.

However, high oxidation state cationic complexes of tungsten are very rare, especially those which do not contain a coordinating ligand [15] and these complexes are expected to be more reactive than the corresponding neutral complexes [14]. Cationic tungsten complex containing one or more stabilizing ligands *e.g.* oxygen, nitrogen, or oxygen, nitrogen, phosphorous containing ligand, cyclopentadienyl ligand are relatively stable and are described in the literature [14, 16-20]. Very recently, Schowner *et al.* have reported the synthesis of first cationic tungsten-oxo-alkylidene-N-heterocyclic-carbene complexes where the cationic metal center was stabilized by N-heterocyclic carbene [20].

To the best of our knowledge, the only example for homoleptic cationic tungsten-alkyl complex was reported in the literature by Wilkinson *et al.* Even in this paper they mentioned the reaction of WMe_6 with $AlCl_3$ which would lead to the formation of WMe_5^+ but without any proof [1-4].

1 EXPERIMENTAL

All experiments were carried out by using standard Schlenk and glovebox techniques under an inert nitrogen atmosphere. The syntheses and the treatments of the surface species were carried out using high-vacuum lines ($<10^{-5}$ mbar) and glovebox techniques. Pentane was distilled from a Na/K alloy under N_2 and dichloromethane from CaH_2. Both solvents were degassed through freeze–pump–thaw cycles. $SiO_{2\text{-}700}$ was prepared by calcination at 300°C in the presence of air followed by dehydroxylation at 700°C under high vacuum ($<10^{-5}$ mbar) for 24 h. It contained 0.5-0.7 OH per nm^2.

^1H NMR spectra were recorded at 600 MHz in CD_2Cl_2 unless otherwise stated. Chemical shifts are reported in ppm with the solvent resonance (CD_2Cl_2: 5.32 ppm). Data are reported as follows: chemical shift, integration, and coupling constants (Hz). ^{13}C NMR were recorded at 150 MHz in CD_2Cl_2 unless otherwise stated with complete proton decoupling. Chemical shifts are reported in ppm from solvent peak as the standard (CD_2Cl_2: 53.84 ppm).

1.1 1-Octene Metathesis

A mixture of catalytic species and dry 1-octene were mixed inside the glovebox. The ampoules were sealed under vacuum, kept at room temperature for 12 h. At the end of the reaction, the ampoules were frozen under liquid nitrogen. Then, the catalytic run was quenched by addition of a fixed amount of CH_2Cl_2 and after filtration the resulting solution was analyzed by GC and GC/MS.

GC measurements were performed with an *Agilent* 7890 A Series (FID detection). Method for GC analyses: column HP-5; 30 m length 0.32 mm ID X 0.25 mm film thickness; flow rate: 1 mL.min^{-1} (N2); split ratio: 50/1; inlet temperature: 250 °C; detector temperature: 250 °C; temperature program: 40 °C (3 min), 40-250 °C (12 °C.min^{-1}), 250 °C (3 min), 250-300 °C (10 °C.min^{-1}), 300 °C (3 min); *n*-decane retention time: $tR = 9.6$.

1.2 Preparation of Hexamethyltungsten, WMe_6

The molecular precursor WMe_6 was prepared from WCl_6 and $(CH_3)_2Zn$, following the literature procedure [1, 6]. To a mixture of WCl_6 (1.80 g, 4.5 mmol) in dichloromethane (25 mL) was added $(CH_3)_2Zn$ (13.6 mmol, 1.0 M in heptane) at −80 °C, and after addition, the reaction mixture was warmed to −35 °C and stirred at this temperature for another 30 min. After successive filtrations with pentane and removal of the solvent, the red solid corresponding to WMe_6 was obtained (0.16 g, 12%).

Caution! This $12e^-$ compound is highly unstable and is prone to violent decomposition [1].

^1H NMR (CD_2Cl_2, 600 MHz): δ (ppm) 1.65 (s, 18H, WCH_3). ^{13}C NMR (CD_2Cl_2, 150 MHz): δ (ppm) 82 (s, 6C, J183 W-^{13}C = 47 Hz, WCH_3). HSQC (Heteronuclear Single Quantum Coherence) confirms the correlation between the ^1H and ^{13}C NMR signals.

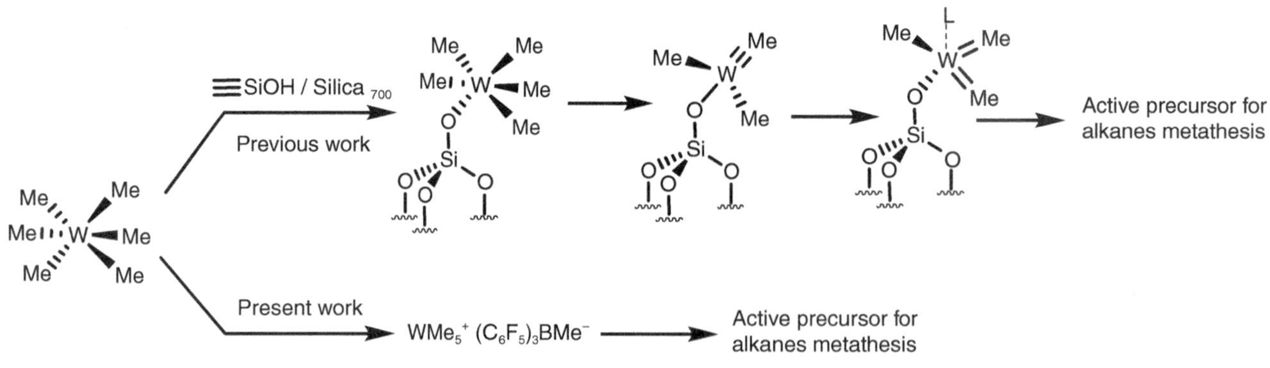

Scheme 1

Application of WMe_6 in SOMC and in solution (present work).

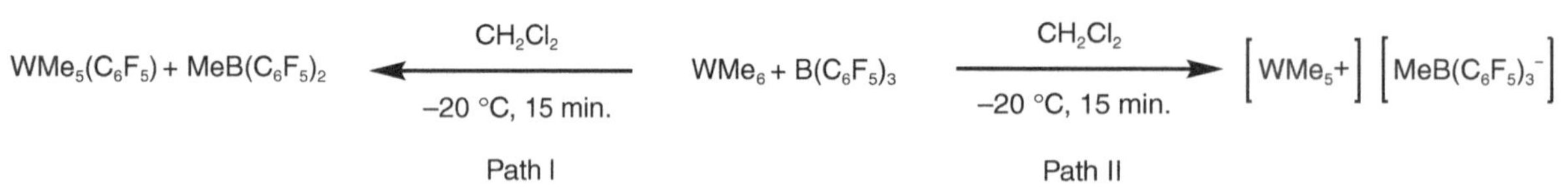

Scheme 2

Possible reaction between WMe_6 and $B(C_6F_5)_3$.

The ^{13}C-enriched $W(^{13}CH_3)_6$ was synthesised as described below: ^{13}C-enriched $(^{13}CH_3)_2Zn$ was prepared from a suspension of $^{13}CH_3Li$ and $ZnCl_2$ (2/1), with subsequent synthetic steps being analogous to those provided above.

1.3 Preparation of $[WMe_5^+\ B(C_6F_5)_3Me^-]$

A cold solution (−20 °C) of $B(C_6F_5)_3$ (100 mg) in dichloromethane was added drop wise to the cold (−20 °C) solution of hexamethyltungsten (55 mg, 1 equiv. with respect to $B(C_6F_5)_3$) in dichloromethane. The mixture was stirred for 15 min at −20 °C. Color of the solution intensified to reddish indicating the formation of ionic complex. At temperature below −60 °C, this ionic complex is precipitated out from the solution however this $10e^-$ compound is highly unstable and decompose into black tungsten power while drying.

Caution! This $10e^-$ compound is highly unstable and is prone to violent decomposition.

^{1}H-NMR(600 MHz) δ (ppm) 0.5 (s, 3H, BCH_3), 2.7 (s, 15H, WCH_3). ^{13}C NMR(150 MHz) δ (ppm) 10.8 (s, 1C, BCH_3), 103.2 (s, 5C, WCH_3).

1.4 Grafting of $[WMe_5^+\ B(C_6F_5)_3Me^-]$ on $Silica_{700}$

A solution of $[WMe_5^+\ B(C_6F_5)_3Me^-]$ (85 mg, 1.2 equiv. with respect to WMe_6 and with respect to the amount of surface-accessible silanols) in dichlomethane (15 mL) was mixed with silica ($SiO_{2\text{-}700}$; 1.0 g) at −50 °C for one hour, allowed to warm to −30 °C, and then stirred for an additional 2 h. At the end of the reaction, the resulting brown solid was washed with pentane (3×20 mL) and dried under dynamic vacuum (1 mPa, 1 h).

^{1}H solid-state NMR(400 MHz): δ (ppm) 2.0 (b, W-CH_3). ^{13}C CP/MAS solid-state NMR(100 MHz): δ (ppm) 83.0 (W-CH_3), 46.0(W-CH_3) [11].

2 RESULTS AND DISCUSSION

As a part of our continuing program to explore the novel applications of tungsten hexamethyl, we report here the synthesis of the cationic tungsten-penta-methyl complex $[WMe_5]^+[MeB(C_6F_5)_3^-]$, starting from WMe_6 and its application in olefin metathesis. The experimental procedure is very simple. Simple mixing of bulky and non-coordinating Lewis acid, $B(C_6F_5)_3$ with WMe_6 at very low temperature (−20 °C) generates quantitatively the cationic tungsten-penta-methyl complex $[WMe_5]^+[MeB(C_6F_5)_3^-]$.

In a preliminary study, we tried to synthesise WMe_5^+ according to literature reference using $AlCl_3$ as a de-methylating reagent but after several attempts we were unable to get the desired product [1-4]. Subsequently, we switched to tris(pentafluorophenyl)-boron which is known to form a

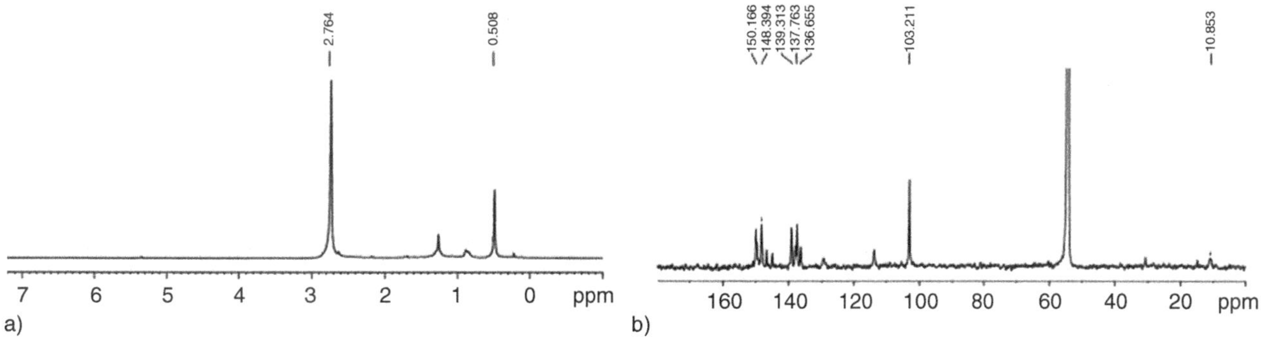

Figure 1

a) ^{1}H NMR spectra, b) ^{13}C NMR spectra of $[WMe_5]^{+}[MeB(C_6F_5)_3^{-}]$.

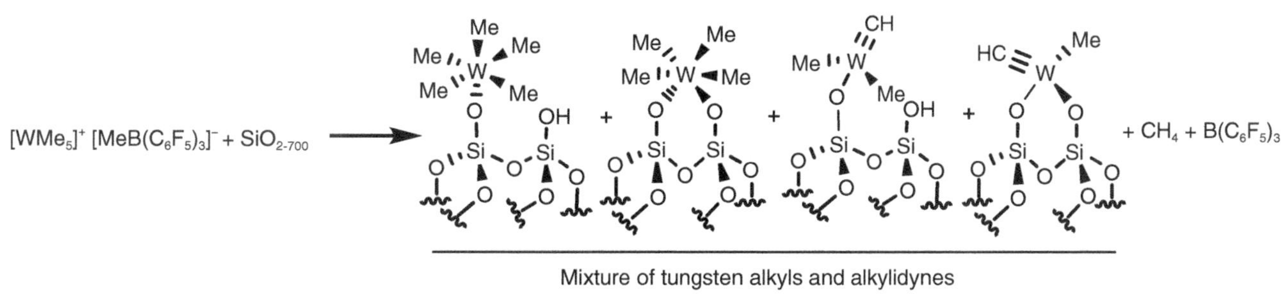

Scheme 3

Grafting of $[WMe_5]^{+}[MeB(C_6F_5)_3^{-}]$ onto the $silica_{700}$ support.

non-nucleophilic anion after de-methylation reaction. When the reaction was monitored by NMR at −20 °C, it was found that the reaction between WMe_6 and $B(C_6F_5)_3$ was very fast and the peak corresponding to WMe_6 at 1.75 ppm in ^{1}H-NMR almost completely disappeared in less than 15 minutes and two new peaks at 2.68 ppm and 0.48 ppm were formed (Fig. 1). The product generated by the above reaction has a fair stability below −40 °C in the absence of light. However, upon warming this reaction mixture to 0 °C from −40 °C, a very first decomposition of the cationic complex was observed with release of gaseous methane. Based on the above experimental facts, one can assume the two possibilities; where Path I (Scheme 2) is the formation of neutral tungsten complex by the ligand metathetical exchange and Path II (Scheme 2) is de-methylation from WMe_6, formation of cationic tungsten complex.

The ^{1}H-NMR chemical shift at +0.5 ppm corresponds to $[\mathbf{CH_3}\text{-}B(C_6F_5)_3]^{-}$ which is well documented in the literature [21, 22] and as expected the resonance for the methyl proton of WMe_6 at 1.75 ppm is shifted downfield to 2.7 ppm, and ascribed to $[WMe_5]^{+}$. Similarly the ^{13}C-NMR spectra also clearly shows that the peak at 83 ppm of WMe_6 is also completely replaced by two peaks at 10.8 ppm $[\mathbf{CH_3}\text{-}B(C_6F_5)_3]^{-}$ and 103.2 ppm corresponding to $[W(\mathbf{CH_3})_5]^{+}$ which is good agreement with the literature value reported for $[\mathbf{CH_3}\text{-}B(C_6F_5)_3]^{-}$ (Fig. 1). We also found in the ^{1}H-^{13}C correlation spectra that the peak at 0.5 ppm in ^{1}H-NMR was correlating with the peak at 10.8 ppm in ^{13}C-NMR and the peak at 2.7 ppm in ^{1}H-NMR was correlating with the peak at 103.2 ppm in ^{13}C-NMR. These spectroscopic data strongly support the formation of $[WMe_5]^{+}[MeB(C_6F_5)_3]^{-}$. In separate experiments, we synthesised $^{13}\mathbf{C}$ label $W(\mathbf{C}^{13}H_3)_6$ and upon treatment with $B(C_6F_5)_3$ we have thus identified incorporation of the labeled methyl in the final $[(\mathbf{C}^{13}H_3)B(C_6F_5)_3]^{-}$ anion. The above experiment clearly indicates that in $[MeB(C_6F_5)_3]^{-}$, 'CH_3' came from WMe_6 and strongly supports Path II (Scheme 2).

We then tried to graft cationic complex on the surface of silica-$_{700}$, in order to get a cationic tungsten catalyst onto the heterogeneous support. However, after several attempts we were unable to identify any well define heterogeneous cationic tungsten complex except a mixture of neutral tungsten-methyl complex (Scheme 3).

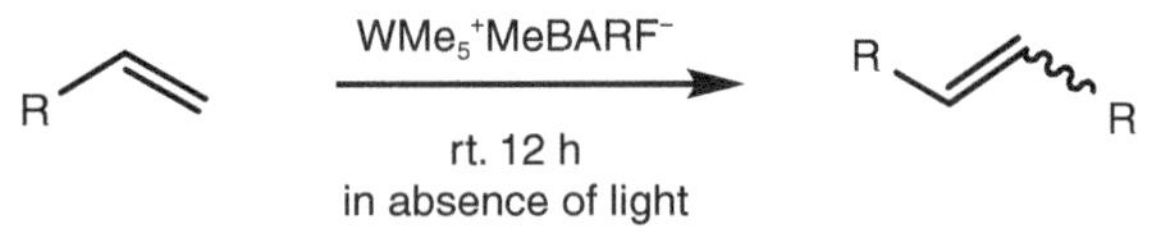

Substrate	Product	Yield
1-octene	7-tentradecene	40%(E/Z= 3/1)
2-octene	no reaction	------
Cyclooctene	Polyoctenamer	90%(E/Z= 7/3)

Scheme 4

Metathesis of olefin by cationic tungsten-methyl complex. $BARF^- = B(C_6F_5)_3$

Tungsten cationic complex was found to be very effective for the metathesis of 1-octene and ROMP (Ring Opening Metathesis Polymerization) for cyclooctene. Cyclooctene was polymerized to polyoctenamer by ROMP at even −40 °C by this cationic tungsten complex. However, this complex was inactive for metathesis of trans 2-octene (Scheme 4). The detailed mechanism and the precise nature of the active catalytic species are still unknown for this catalytic system. However from the control experiment, we found that WMe_6 does not metathetise cyclooctene nor 1-octene. From the above experiments, we can assume that the formation of a tungsten carbene intermediate is more favoured when starting from the cationic complex rather than neutral WMe_6. A very similar observation was reported by Kress and Osborn, where they showed that cationic complexes are more active towards olefin metathesis than the corresponding neutral complexes [14]. Based on this, we speculate that a tetra coordinated cationic $[WMe_3(=CH_2)]^+[-MeB(C_6F_5)_3]^-$ intermediate is formed from $[WMe_5]^+[MeB(C_6F_5)_3]^-$ by α-H transfer into neighbouring methyl groups. However, we have never seen any carbene intermediates by NMR during the decomposition of $[WMe_5]^+[MeB(C_6F_5)_3]^-$ while the experiments were carried out inside the NMR instrument for long time.

CONCLUSION

In summary, we have developed a straight forward strategy for the synthesis of extremely unstable high oxidation state cationic tungsten(VI)-methyl complex in a precise way, starting from WMe_6. We fully characterised this unstable ionic species by ^{1}H-NMR, ^{13}C-NMR, ^{1}H-^{13}C correlation, etc., and also, we showed that this cationic complex is reactive in olefin metathesis reaction. To the best of our knowledge, this is the first example of a well defined cationic tungsten complex. Further studies are in progress to assess the reactivity of this tungsten complex and its application in other catalysis.

ACKNOWLEDGMENTS

This work was supported by funds from King Abdullah University of Science and Technology. RD would like to acknowledge SABIC for his SABIC postdoctoral fellowship.

REFERENCES

1 Shortlan A., Wilkinson G. (1973) *J. Chem. Soc. Dalton* **8**, 872-876.

2 Shortlan A., Wilkinson G. (1972) *J. Chem. Soc. Chem. Comm.* **6**, 318.

3 Galyer L., Mertis K., Wilkinson G. (1975) *J. Organomet. Chem.* **85**, 37-38.

4 Galyer A.L., Wilkinson G. (1976) *J. Chem. Soc. Dalton* **21**, 2235-2238.

5 Seppelt K. (1996) *Science* **272**, 182-183.

6 Kleinhenz S., Pfennig V., Seppelt K. (1998) *Chem.-Eur. J.* **4**, 1687-1691.

7 Seppelt K. (2003) *Accounts Chem. Res.* **36**, 147-153.

8 Kaupp M. (1996) *J. Am. Chem. Soc.* **118**, 3018-3024.

9 Demolliens A., Jean Y., Eisenstein O. (1986) *Organometallics* **5**, 1457-1464.

10 Kang S.K., Albright T.A., Eisenstein O. (1989) *Inorg. Chem.* **28**, 1611-1613.

11 Samantaray M.K., Callens E., Abou-Hamad E., Rossini A.J., Widdifield C.M., Dey R., Emsley L., Basset J.M. (2014) *J. Am. Chem. Soc.* **136**, 1054-1061.

12 Callens E., Abou-Hamad E., Riache N., Basset J.M. (2014) *Chem.Commun.* **50**, 3982-3985.

13 Samantaray M.K., Dey R., Abou-Hamad E., Hamieh A., Basset J.-M. (2015) *Chem.-Eur. J.* **21**, 6100-6106.

14 Kress J., Osborn J.A. (1983) *J. Am. Chem. Soc.* **105**, 6346-6347.

15 Schrock R.R., Hoveyda A.H. (2003) *Angew. Chem. Int. Ed.* **42**, 4592-4633.

16 Ward B.D., Orde G., Clot E., Cowley A.R., Gade L.H., Mountford P. (2004) *Organometallics* **23**, 4444–4461.

17 Cremer C., Burger P. (1999) *J. Chem. Soc. Dalton Trans.* 1967-1974.

18 Nakazawa H., Kishishita M., Miyoshi K. (1999) *Phosphorus, Sulfur, and Silicon and the Related Elements* **144**, 45-48.

19 Eide E.F., Piers W.E., Parvez M., McDonald R. (2007) *Inorg. Chem.* **46**, 14-21.

20 Schowner R., Frey W., Buchmeiser M.R. (2015) *J. Am. Chem. Soc.* **137**, 6188-6191.

21 Spitzmesser S.K., Gibson V.C. (2003) *J. Organomet. Chem.* **673**, 95-101.

22 Gillis D.J., Tudoret M.-J., Baird M.C. (1993) *J. Am. Chem. Soc.* **115**, 1993, 2543.

15

Gaseous Hydrocarbon Separations Using Functionalized Ionic Liquids

Leila Moura[1,2], Catherine C. Santini[2] and Margarida F. Costa Gomes[1]*

[1] *Clermont Université, Université Blaise Pascal, Institut de Chimie de Clermont-Ferrand, UMR 6296 CNRS, Équipe Thermodynamique et Interactions Moléculaires, ICCF-TIM, BP 80026, 63171 Aubière - France*

[2] *Université de Lyon, UMR 5265 CNRS, Université de Lyon 1-ESCPE Lyon, C2P2, Équipe Chimie Organométallique de Surface, 69616 Villeurbanne - France*

e-mail: margarida.c.gomes@univ-bpclermont.fr

* Corresponding author

Abstract — *The functionalization of the side chains on the cation or the anion of an ionic liquid is a common approach to tailor its properties for different processes including the separation of gases. In this paper, we present the current state of the art concerning the usage of ionic liquids for hydrocarbon separations. We also show how the functionalization of ionic liquids or the appropriate anion/cation combinations can contribute to the increase of the performance of the ionic liquids for the separation of gaseous hydrocarbons – either by improving the capacity of the ionic liquid to absorb a given gas or by increasing the selectivity towards a particular hydrocarbon. Original results concerning the usage of olefin-complexing metal salts of lithium (I), nickel (II) and copper (II) dissolved in ionic liquids for selectively absorbing light olefins are presented. It is observed that the absorption capacity of an imidazolium-based ionic liquid is doubled by the addition of a copper (II) salt. This result is compared with the effect of the functionalization of the ionic liquid and the advantages and difficulties of the two approaches are analyzed.*

Résumé — **Liquides ioniques fonctionnalisés pour la séparation d'hydrocarbures gazeux** — La fonctionnalisation des chaînes latérales sur le cation ou l'anion d'un liquide ionique est une approche commune pour adapter ses propriétés pour les différents processus, y compris la séparation de gaz. Dans cet article, nous présentons l'état actuel des recherches concernant l'utilisation de liquides ioniques pour les séparations d'hydrocarbures gazeux. Nous montrons comment la fonctionnalisation de liquides ioniques ou des combinaisons d'anions/cations appropriées peuvent contribuer à l'augmentation de leur performance pour la séparation des hydrocarbures gazeux – soit par l'amélioration de la capacité du liquide ionique pour absorber un gaz donné ou en augmentant la sélectivité pour un hydrocarbure particulier. Des résultats originaux concernant l'utilisation de sels métalliques de lithium (I), le nickel (II) et le cuivre (II) dissous dans des liquides ioniques pour absorber sélectivement des oléfines légères sont présentés. On constate que la capacité d'absorption d'un liquide ionique à base d'imidazolium est doublée par addition d'un sel de cuivre (II). Ce résultat est comparé avec l'effet de la fonctionnalisation du liquide ionique. Les avantages et les difficultés de ces deux approches sont analysés.

INTRODUCTION

Separation techniques like distillation, crystallization, liquid-liquid extraction or absorption are commonly used and are based on the creation or on the addition of a new phase. They require either an energy transfer or the use of a mass-transfer agent. Changing from an energy-separating agent to a mass-separating agent can potentially reduce the process energy requirements, but the selected absorbent should be easy to recycle, have low volatility, and the solubilities of the species to separate should be significantly different. These constraints, together with the need to obey tougher environmental regulations, render the choice of the absorbent difficult.

Membranes make use of liquid or solid (frequently polymeric) barriers to separate species that present different permeability. Membranes are used in small, compact and clean units that require low energy to achieve separation, but are still difficult to scale-up. Adsorption is a surface-based process that frequently relies on a solid agent to achieve separation. The separation is achieved due to different interacting forces at the interface (originating physisorption or chemisorption processes) that selectively interacts with certain components in detriment of others from a liquid or gas mixture. The adsorbent material should have high surface area, good mechanical properties, fast adsorption kinetics and the ability to be regenerated without loss of its properties. Commonly used supports are alumina, ion exchanging resins, high-surface-area silica zeolites and molecular sieves. Electrophoresis and centrifugation are examples of separation techniques that exploit the differences in the responses of the constituents of a feed to an external force or gradient. They are especially useful and versatile for separating biochemical [1]. Cryogenic distillation is used for the separation of large volumes of hydrocarbon gases (*i.e.* ethane/ethylene or propane/propylene) and is viewed as an expensive, energy demanding process whose replacement by a more economic and environmentally friendly alternative performance would represent a major advance in the field.

1 IONIC LIQUIDS AS SEPARATING AGENTS

Ionic liquids have been suggested as new separating agents for gaseous hydrocarbon separations namely for olefin/paraffin gas separation. They can also act as absorbents or as solvents for the chemical complexation of olefins with silver or copper salts. Ionic liquids are composed uniquely of ions and have a melting point below 100°C. Many present unique properties such as negligible vapor pressure, high thermal, chemical and electrochemical stability, flame retardant effect, and high ionic conductivity. Ionic liquid are also called "designer solvents" due to the large variety of combination possibilities of cations and anions, leading to tunable physical chemical properties.

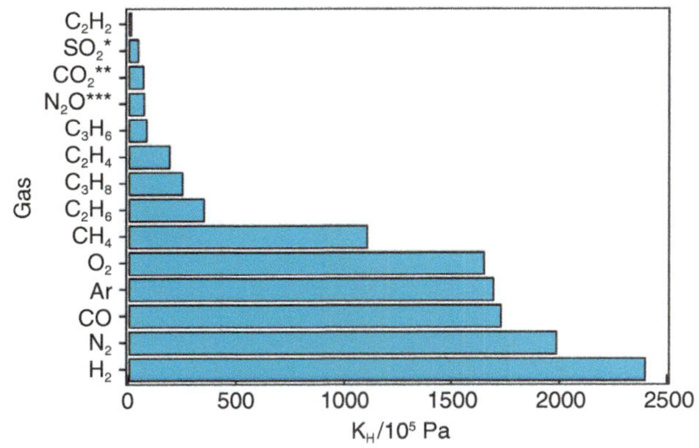

Figure 1

Henry's law constant, K_H, for several gases in ionic liquid $[C_1C_4Im][BF_4]$ at 313 K. * 293 K; ** 314 K; *** 298 K [2].

In Figure 1, are shown the Henry's law constants, K_H, of different gases in one ionic liquid – 1-butyl-3-methylimidazolium tetrafluoroborate, $[C_1C_4Im][BF_4]$ – [2]. As can be observed, the mole fraction solubility spans over more than two orders of magnitude, depending on the nature of the gas. These differences can be explored by proposing some ionic liquids as selective absorbents for gases.

Monoatomic, diatomic, non-polar or gases with low polarizability present the lowest solubilities in the ionic liquid $[C_1C_4Im][BF_4]$. Unsaturated hydrocarbons frequently present a higher solubility in ionic liquids than their saturated counterparts and so these fluids are considered promising for saturated/unsaturated hydrocarbon separation. The exact reasons for the observed trends are still poorly understood, even if they are essential for the development of new gas separation processes [3, 4].

The functionalization of the side chains or the modification of the ionic liquid anion are popular approaches to tailor its properties for particular applications such in catalysis [5], energy storage [6], capture of carbon dioxide [3] and in separation technology [7]. In this work, we have studied the possibility of using ionic liquids for the separation of light saturated and unsaturated hydrocarbons and have started by compiling the literature results on the solubility of ethane and ethylene in different ionic liquids formed by the cations and anions in Table 1. The Henry's law constants for ethane and ethylene are reported in Figure 2 [2, 8-26] and Table 2 [2, 8-26].

2 FUNCTIONALIZATION OF THE IONIC LIQUIDS

There are few examples of an experimental rational approach for the design of an ionic liquid to optimize the

TABLE 1

Abbreviation, full designation and structure of some cations of ionic liquids. $C_{n/m}$ represent alkyl chains of variable size. In the anions, R_n represents alkyl chains of variable size

Abbreviation	Full name	Structure
	Cations	
$C_nC_mIm^+$	1-C_n-3-C_mImidazolium	
$C_1(C_3H_5CH_2)Im^+$	1-(buten-3-yl)-3-methylimidazolium	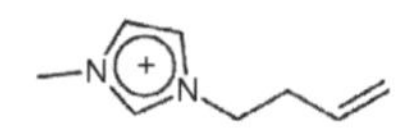
$C_1(C_2H_2CH)Im^+$	1-methyl-3-(propyn-3-yl)imidazolium	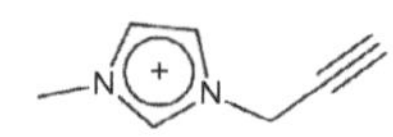
$C_1(CH_2C_6H_5)Im^+$	1-benzyl-3-methylimidazolium	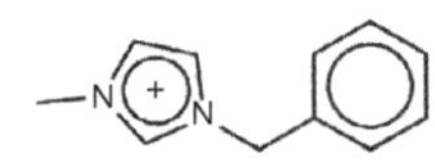
$C_1C_3CNIm^+$	1-(3-cyanopropyl)-3-methylimidazolium	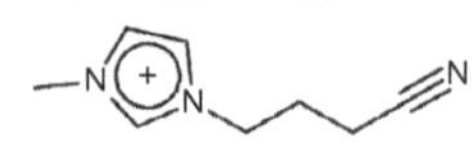
$(C_3CN)_2Im^+$	1,3-(3-cyanopropyl)Imidazolium	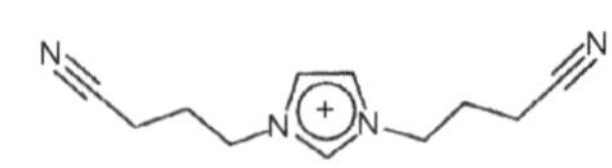
$C_1(C_8H_{16}C_2H_2C_8H_{17})Im^+$	1-(Z-octadec-9-enyl)-3-methylimidazolium	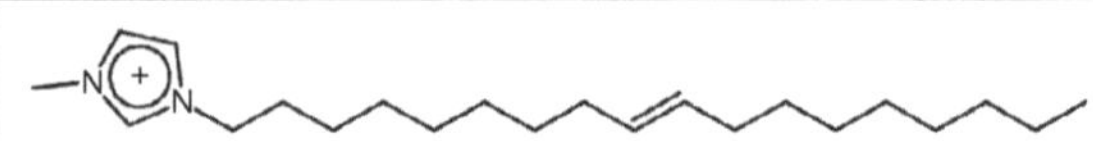
$C_nC_mPyrr^+$	C_nC_mPyrrolidinium	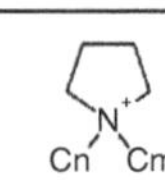
C_nPyr^+	C_nPyridinium	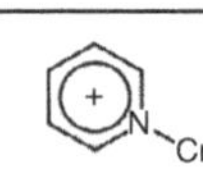
$P_{n\,m\,p\,q}^{\;+}$	$C_nC_mC_pC_q$Phosphonium	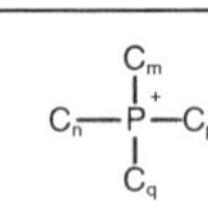
	Anions	
NTf_2^-	Bis(trifluoromethylsulfonyl)imide	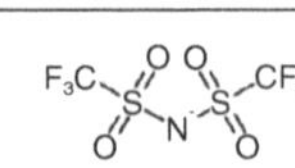
DCA^-	Dicyanamide	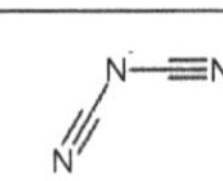
$C_nHPO_3^-$	C_nPhosphite	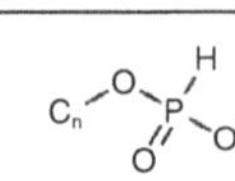
PF_6^-	Hexafluorophosphate	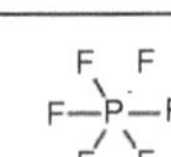
$CF_3SO_3^-$	Trifluromethanesulfonate	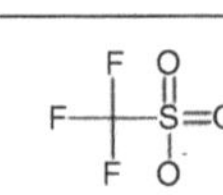
BF_4^-	Tetrafluoroborate	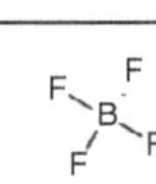

(continued)

TABLE 1 *(continued)*

Abbreviation	Full name	Structure
$TMPP^-$	Bis(2,4,4-trimethylpentyl)phosphinate	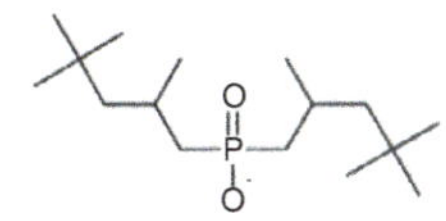
$BETI^-$	Bis(perfuoroethylsulfonyl)imide	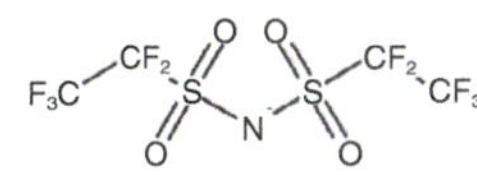
FAP^-	Tris(pentafluoroethyl)trifluorophosphate	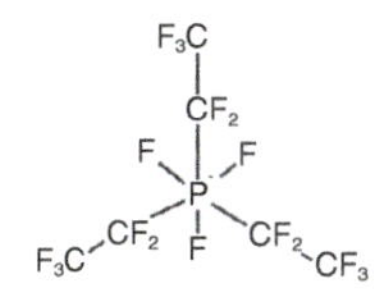
OAc^-	Acetate	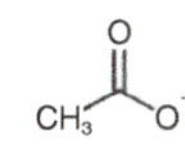
$C_nCO_2^-$	Carboxylate	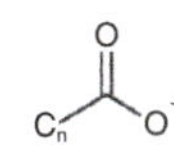
TFA^-	Trifluoroacetate	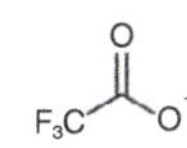
$C_nSO_4^-$	C_nSulfate	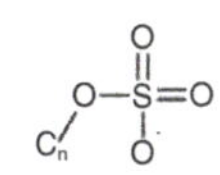
DBS^-	Dodecylbenzenesulfonate	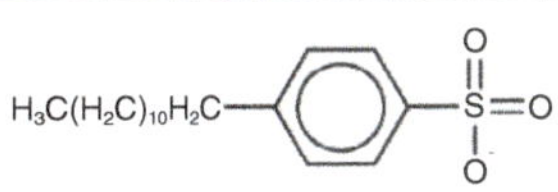

solubility of one of the gases in the separation of ethane/ethylene or propane/propene. Moura *et al.* [27, 28] have selected several functionalized ionic liquids designed to maximize the solubility and interactions of ethylene in the ionic liquid in detriment of the solubility of ethane. The functionalizations used were the increasing alkyl side-chain on the cation of the ionic liquid ($[C_1C_4Im][NTf_2]$ and $[C_1C_8Im][NTf_2]$); unsaturations on the alkyl side-chain on the cation of the ionic liquid ($[C_1(C_2H_2CH)Im][NTf_2]$, $[C_1(C_3H_5CH_2)Im][NTf_2]$ and $[C_1(CH_2C_6H_5)Im][NTf_2]$); cyano groups in the cation or anion of the ionic liquid or both ($[C_1C_3CNIm][NTf_2]$, $[C_1C_3CNIm][DCA]$ and $[C_1C_4Im][DCA]$) and a phosphite based anion in the ionic liquid ($[C_1C_4Im][C_1HPO_3]$). The authors observed that small changes in the structure of the ionic liquid can have large effects in the ethane and ethylene solubility as illustrated in Figure 2 and can significantly influence the ideal selectivity of ethylene as shown in Figure 3 [9-26, 28, 29].

Adding unsaturations in the alkyl chain of the cation of the ionic liquid, cyano groups in the alkyl chain of the cation or anion and changing the anion of the ionic liquid from an NTf_2^- to a $C_1HPO_3^-$ or DCA, lead to decreases in the

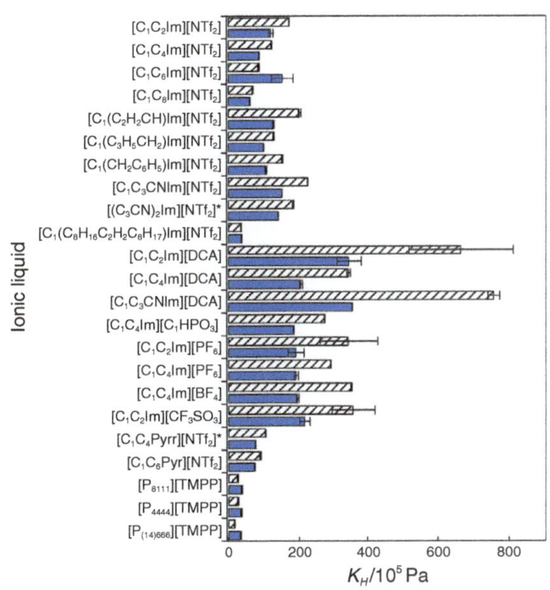

Figure 2

Henry's constant, K_H, of ethane (empty bars) and ethylene (full bars) in the ionic liquids at 313 K (* at 303 K) (Tab. 2)[2, 8-26].

TABLE 2

Ionic liquid	References
$[C_1C_2Im][NTf_2]$	[8-14]
$[C_1C_4Im][NTf_2]$	[9, 15-20]
$[C_1C_6Im][NTf_2]$*	[15, 16, 21]
$[C_1C_8Im][NTf_2]$	[10]
$[C_1(C_2H_2CH)Im][NTf_2]$	[10]
$[C_1(C_3H_5CH_2)Im][NTf_2]$	[9]
$[C_1(CH_2C_6H_5)Im][NTf_2]$	[9]
$[C_1C_3CNIm][NTf_2]$*	[10, 19, 21, 22]
$[(C_3CN)_2Im][NTf_2]$*	[19]
$[C_1(C_8H_{16}C_2H_2C_8H_{17})Im][NTf_2]$	[23]
$[C_1C_2Im][DCA]$	[12, 21]
$[C_1C_4Im][DCA]$	[10]
$[C_1C_3CNIm][DCA]$	[10]
$[C_1C_4Im][C_1HPO_3]$	[10]
$[C_1C_2Im][PF_6]$	[12]
$[C_1C_4Im][PF_6]$*	[15-17, 19-22, 24]
$[C_1C_4Im][BF_4]$	[2, 12, 21, 24]
$[C_1C_2Im][CF_3SO_3]$	[12]
$[C_1C_4Pyrr][NTf_2]$	[12, 24]
$[C_6Pyr][NTf_2]$	[24]
$[P_{8111}][TMPP]$	[25]
$[P_{4444}][TMPP]$	[25]
$[P_{(14)666}][TMPP]$	[25]

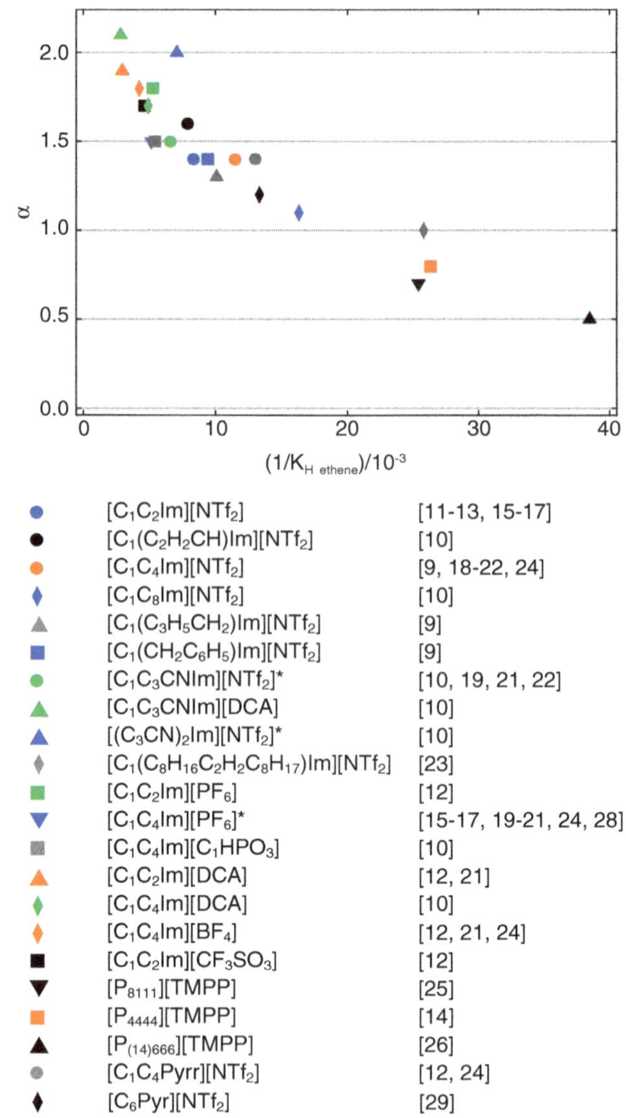

Figure 3

Ethane/ethylene ideal selectivity *versus* Henry's constant, K_H, of ethylene in the ionic liquids at 313 K (* at 303 K).

solubility of ethane and ethylene, when compared to a reference ionic liquid, $[C_1C_4Im][NTf_2]$ (Fig. 2). The only exception is for $[C_1(C_3H_5CH_2)Im][NTf_2]$, where ethane solubility is not greatly affected. These modifications in the ionic liquids also lead to increases in the ethane/ethylene separation selectivity (Fig. 3). The solubility of ethane is more affected by changes in the ionic liquid than that of ethylene. For example, increasing the alkyl side chain of the cation from $[C_1C_4Im][NTf_2]$ to $[C_1C_8Im][NTf_2]$ leads to an increase in both the ethane and the ethylene solubility but the increase is larger for the saturated gas. The only exceptions are for the ionic liquid containing a double bond in the alkyl side-chain of the cation (where the solubility of ethane is similar and the solubility of ethylene decreases 14%) and that containing a benzyl group (where the solubility of both gases decreases in a similar way). The authors also found the ionic liquid with the highest separation selectivity for the ethane/ethylene reported in the literature – $[C_1C_3CNIm][DCA]$.

Green *et al.* [29] determined the solubilities of ethane and ethylene in an imidazolium based ionic liquid, with an 18 carbon long alkyl side chain, $[C_1(C_8H_{16}C_2H_2C_8H_{17})Im][NTf_2]$, designed to have increased solubility for the nonpolar species. For this ionic liquid and for ionic liquids with large nonpolar content, the solubility of ethane or propane is slightly higher than that of their unsaturated counterparts (Fig. 2; Fig. 4, [11-22, 24, 26]), in contrast to the relative

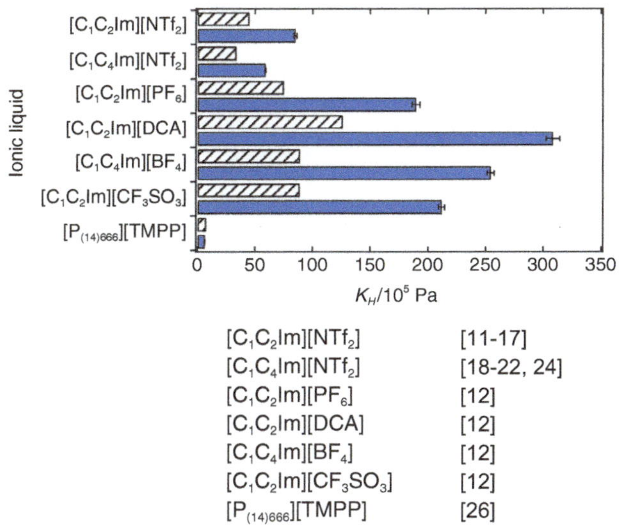

Figure 4

Henry's constant, K_H, of propane (empty bars) and propene (full bars) in the ionic liquids at 313 K (* at 303 K).

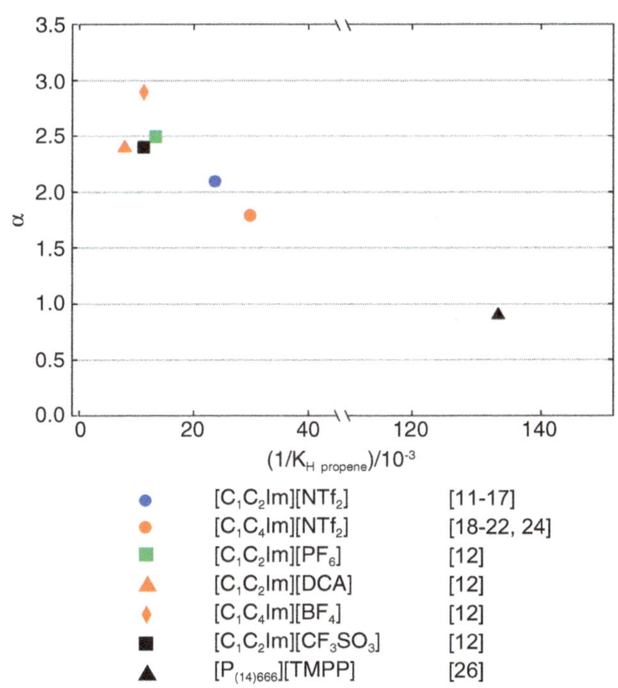

Figure 5

Propane/propene ideal selectivity *versus* Henry's constant, K_H, of propene in the ionic liquids at 313 K (* at 303 K).

solubilities in conventional, more polar ionic liquids. The same observation was made for phosphonium-based ionic liquids:

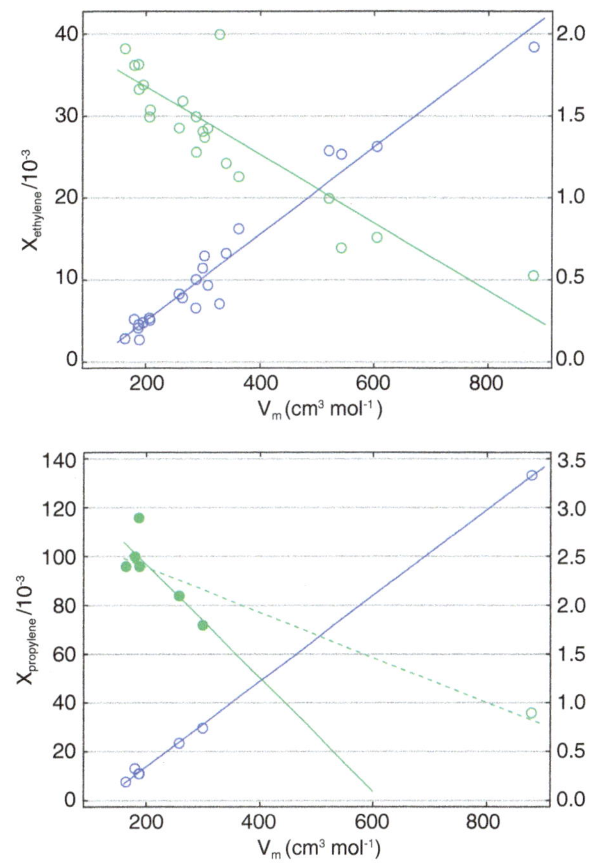

Figure 6

Upper plot: mole fraction absorption capacity for ethylene ($x_{ethylene}$) in blue and ionic liquid ideal separation selectivity for ethane/ethylene (α) in green *versus* the molar volume (V_m), in $cm^3.mol^{-1}$, of the ionic liquids. Lower plot: mole fraction absorption capacity for propylene ($x_{propylene}$) in light green and ionic liquid's ideal separation selectivity for propane/propylene (α) in blue *versus* the molar volume (V_m), in $cm^3.mol^{-1}$, of the ionic liquids. The lines represent the simple linear regression of the data. In the lower plot, the full green line corresponds to the full circles and the dotted line includes also the open circle.

– $[P_{8111}]$[TMPP] [23],
– $[P_{4444}]$[TMPP] [25],
– $[P_{(14)666}]$[TMPP] [14],

(Fig. 2) by Liu *et al.* for the separation of ethane and ethylene and for the separation of propane and propene in $[P_{(14)666}]$[TMPP] [26](Fig. 4). $[P_{(14)666}]$[TMPP] presents the lowest selectivity for the ethane/ethylene and propane/propene separation.

For both separations in the studied ionic liquids we observe that the selectivity increases with the decrease of the solubility of the unsaturated gas. (Fig. 3; Fig. 5, [11-22, 24, 26]). These results indicate that the solubility

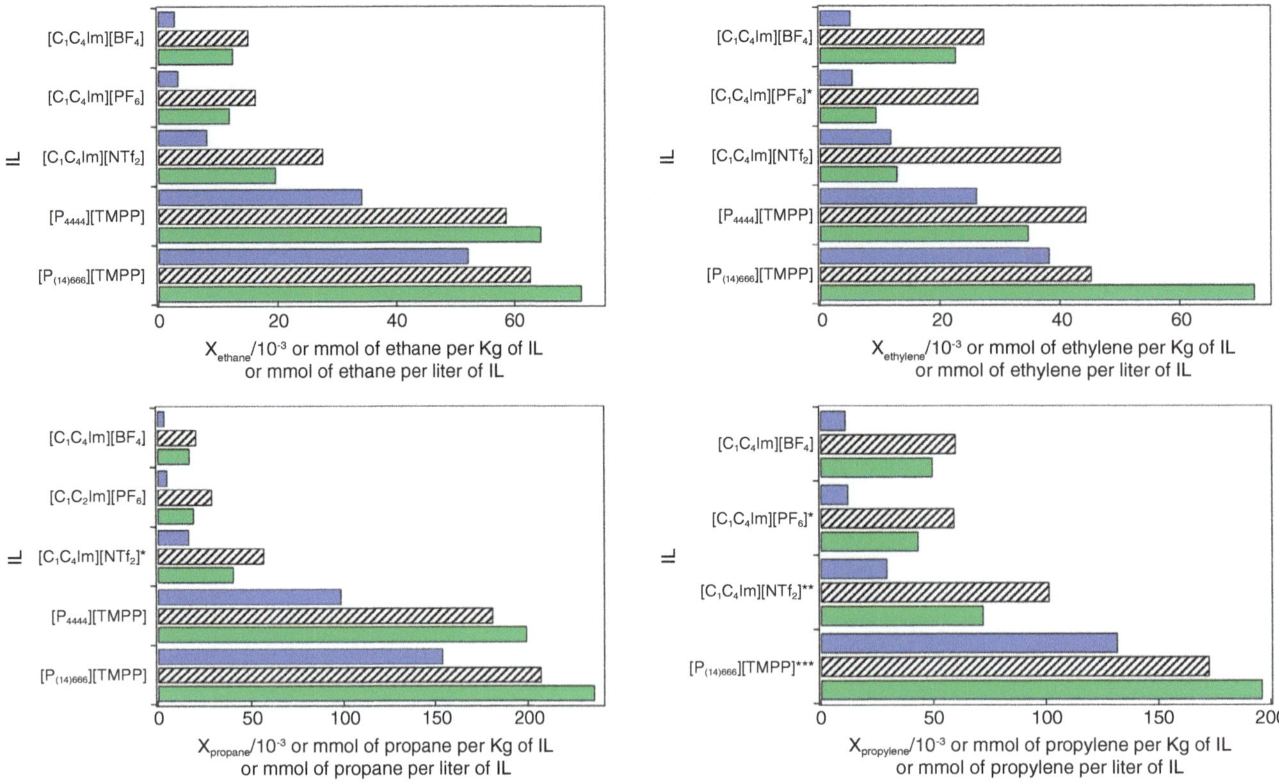

Figure 7

Upper plots: mole fraction ($\times 10^3$, blue bars), quantity of gas per mass of solvent ($\times 10^4$, patterned bars) or per volume of solvent (green bars) of ethane (on the left) and ethylene (on the right) for different ionic liquids. Lower plots: mole fraction ($\times 10^3$, blue bars), quantity of gas per mass of solvent ($\times 10^4$, patterned bars) or per volume of solvent (green bars) of propane (on the left) and propylene (on the right) for different ionic liquids. All the values refer to 1 bar partial pressure of gas and 313 K (* values at 320 K).

of ethane, ethylene, propane and propylene in the ionic liquids tested are controlled by non-specific interactions, even for ionic liquids containing side-chain functionalization designed to optimize the interactions with one of the hydrocarbons to be separated. This can be further confirmed in Figure 6 since the ionic liquids that present higher ethylene or propene absorption capacity and lower ideal selectivity are the ones with highest molar volume and *vice-versa*.

3 IONIC LIQUIDS WITH COMPLEXING AGENTS

Certain metal transition cations, such as Cu(I) or Ag(I) are able to form π-complexes with unsaturated hydrocarbons. The metal and alkene act as an electron donor and acceptor, forming a coordinative bond if the orbitals involved present the correct symmetry and if the metal possesses high electron affinity for good σ-bond accepting properties and low promotion energy for good π-backbonding [26]. This complexation reaction is used to increase the absorption capacity of the ionic liquid for unsaturated light hydrocarbons.

The influence in ethylene solubility of the presence of three different cations, lithium (I), nickel (II) and copper (II) in $[C_1(CH_2C_6H_5)Im][NTf_2]$ ionic liquid was studied by Moura [30]. The authors concluded that cations such as lithium and nickel slightly affect the solubility of ethylene. The addition of copper salts almost doubles the absorption capacity of ethylene in the ionic liquid $[C_1(CH_2C_6H_5)Im][NTf_2]$, a value compatible with the ones reported in the literature for other metallic cations [31, 32].

Sanchéz *et al.* [31] studied the absorption of ethane and ethylene in silver salt solutions with the ionic liquids:

- $[C_1C_2Im][NTf_2]$,
- $[C_1C_2Im][CF_3SO_3]$,
- $[C_1C_4Im][NO_3]$,
- $[C_0C_2Im][NO_3]$,
- $[C_1C_{2\text{-}OH}Im][NO_3]$,
- $[4\text{-butyl-}C_4Pyr][NO_3]$,
- $[N_{1112\text{-}OH}][NO_3]$,

with Ag^+ concentrations ranging from 0.45 $mol.L^{-1}$ to 4.4 $mol.L^{-1}$. The results were compared to the solubility of ethane and ethylene in $[C_1C_2Im][NTf_2]$. The authors

TABLE 3

Solubility of ethane, ethylene, propane and propylene in several ionic liquids expressed in Henry's law constant, K_H, the majority at 313 K

Ionic liquid	Gas: Ethane	Ethylene	Propane	Propylene
$[C_1C_2Im][NTf_2]$				
$[C_1(C_2H_2CH)Im][NTf_2]$*				
$[C_1C_4Im][NTf_2]$*				
$[C_1(C_3H_5CH_2)Im][NTf_2]$*				
$[C_1C_6Im][NTf_2]$				
$[C_1(CH_2C_6H_5)Im][NTf_2]$*				
$[C_1C_8Im][NTf_2]$*				
$[C_1C_{10}Im][NTf_2]$				
$[C_1C_3CNIm][NTf_2]$*				
$[(C_3CN)_2Im][NTf_2]$				
$[C_1C_1C_3CNIm][NTf_2]$				
$[C_1C_1COOC_5Im][NTf_2]$				
$[C_1C_1COOC_2OC_2Im][NTf_2]$				
$[C_1C_1COOC_2OC_2OC_4Im][NTf_2]$				
$[C_1C_4Im][BETI]$				
$[C_1C_4Im][FAP]$				
$[C_1C_6Im][FAP]$				
$[C_1C_2Im][DCA]$				
$[C_1C_4Im][DCA]$*				
$[C_1C_3CNIm][DCA]$*				
$[C_1C_1C_3CNIm][DCA]$				
$[C_1C_4Im][OAc]$				
$[C_1C_4Im][n\text{-}C_{15}H_{31}COO]$				
$[C_1C_4Im][n\text{-}C_{17}H_{35}COO]$				
$[C_1C_2Im][PF_6]$				
$[C_1C_4Im][PF_6]$				
$[C_1C_4Im][BF_4]$				
$[C_1C_8Im][BF_4]$				
$[C_1C_4Im][TFA]$				
$[C_1C_1Im][C_1HPO_3]$				
$[C_1C_2Im][C_2HPO_3]$				
$[C_1C_4Im][C_1HPO_3]$*				
$[C_1C_4Im][C_4HPO_3]$				
$[C_2C_4Im][C_2HPO_3]$				
$[C_1C_1Im][(C_1)_2PO_4]$				
$[C_1C_4Im][(C_1)_2PO_4]$				

Solubility range $K_H/10^{-5}$ Pa	Color
No data	
0-10	
11-20	
21-30	
31-40	
41-50	
51-100	
101-150	
151-200	
201-300	
301-400	
401-500	
501-600	
over 600	

observed that the ionic liquid-silver salt solution selectivities were always higher than for $[C_1C_2Im][NTf_2]$, which presents an ethylene selectivity under 2, at 303 K and 1 bar. The highest selectivity was obtained for the 1.8 $mol.L^{-1}$ silver solution in $[C_1C_2Im][CF_3SO_3^-]$, with a selectivity of 100 at 1 bar and 333 K. The authors defined selectivity as the proportion between mole of ethylene and the mole of ethane at a certain pressure.

Mortaheb *et al.* [32] studied the effect of temperature and silver salt concentration on the absorption of ethane and ethylene in ionic liquid $[C_1C_4Im][NO_3]$. A maximum selectivity of 15 was obtained for 5 $mol.L^{-1}$ silver salt at 278 K. The authors defined selectivity as the proportion between amount of ethylene and the amount of ethane.

Faiz and Li [33] reviewed the application of ionic liquid/metal salt composite membranes for olefin/paraffin separations. They concluded that even though the concept was successfully applied for separation, the long-term stability of the membranes is still a major issue.

A task specific ionic liquid complex salt, $[Ag(olefin)_x][NTf_2]$, formed as described in Scheme 1, provided an ideal selectivity of 40, a much larger selectivity when compared to the silver salt-$[C_1C_4Im][BF_4]$ solution of 16, at 298 K:

$$AgNTf_2 \xrightarrow{olefin} [Ag(olefin)_x][NTf_2]$$

Scheme 1

Synthesis of ionic liquid $[Ag(olefin)_x][NTf_2]$.

TABLE 4

Solubility of ethane, ethylene, propane and propylene in several ionic liquids expressed in Henry's law constant, K_H, the majority at 313 K

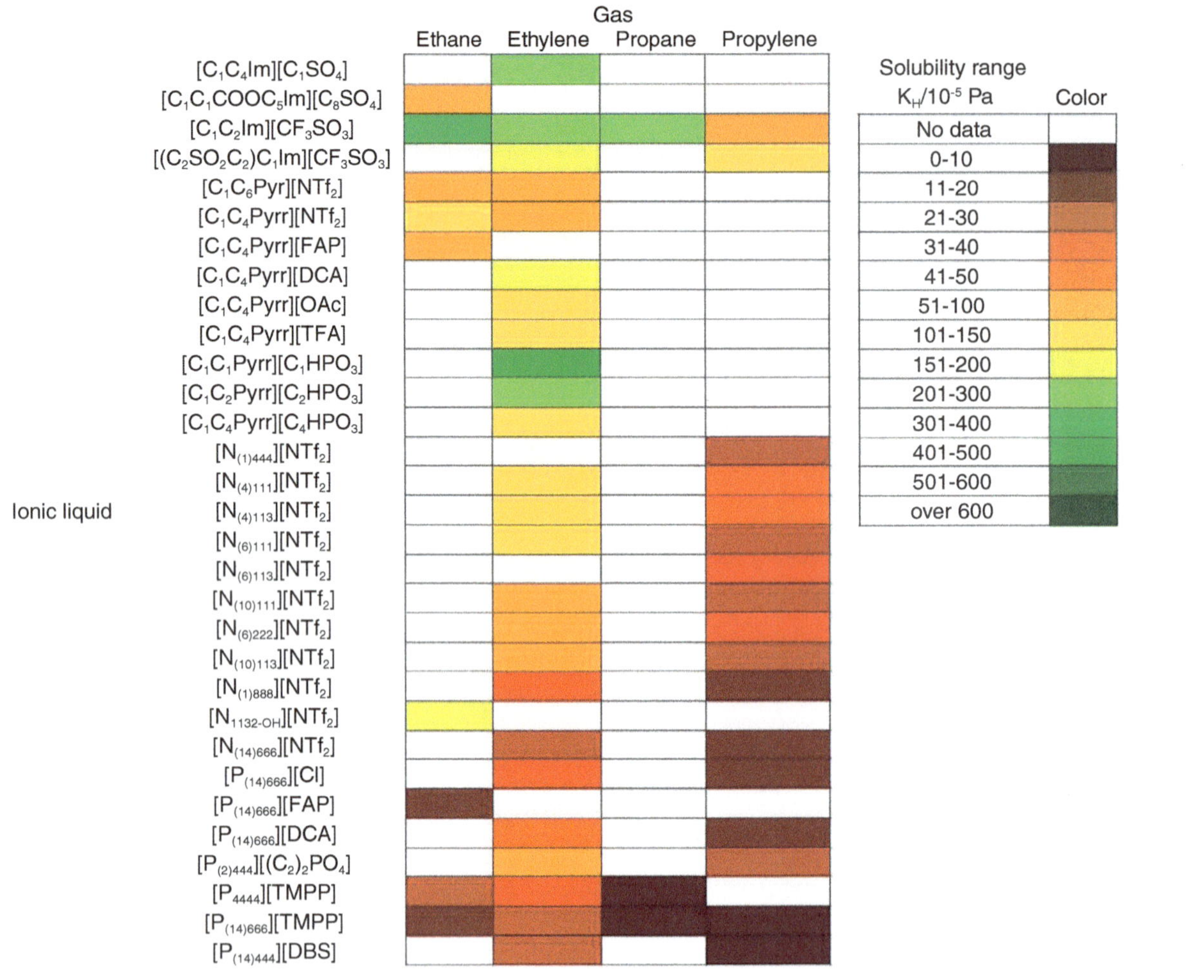

This salt was also successfully used to create an adaptive self-healing propane/propene separation membrane, although the details of its long term stability were not provided [34, 35].

Ortiz *et al.* [36] studied the selective absorption of propylene from propane/propylene mixtures in $[C_1C_4Im][BF_4]$ with, and without, silver ions, at several temperatures and pressures. They demonstrated that the propylene selectivity is less than 3 in absence of the silver salt in the ionic liquid and 8 to 16 upon addition of 0.25 $mol.L^{-1}$ silver salt in the ionic liquid at 298 K and between 1 and 3 bar. Higher pressures proved to lower the selectivity for the silver salt containing ionic liquid, probably due to the saturation of the silver ions. The propylene absorption capacity for the ionic liquids-silver salt solution was compared to aqueous solutions with the same silver salt concentration and it was found that the aqueous solutions presented lower propylene capacities. Although allowing the increase of the capacity and selectivity of the selected support for the unsaturated gas, the use of metals presents drawbacks due to their cost, contamination, degradation, safety problems, limited metal salt solubility and high viscosities of the resulting liquid absorbent [26, 35, 37, 38, 39].

A detailed engineering analysis is out of the scope of the present paper but would most certainly be necessary in order to properly compare these new technologies based on the use of ionic liquids with traditional processes such as cryogenic distillation. Although costs may considerably vary with time, the current process is so expensive energetically and presents such a negative environmental impact that the detailed analysis of possible alternatives is very pertinent.

CONCLUSIONS AND PERSPECTIVES

Contrary to what is suggested in the literature, the solubility of unsaturated gases, such as ethylene and propylene in ionic liquids is not always higher than of their saturated counterparts, ethane and propane. In fact, their solubility ranges overlap. As can be seen in Figure 7, this fact is true independently of the way the gas solubility is expressed – in mole fraction, per mass or per volume of solvent.

The tendencies observed for ethane, ethylene, propane and propylene are similar. The solubility of these gases increases with the size of the non-polar domains of the cation or anion of the ionic liquid, suggesting that the solubility of these gases in the group of ionic liquids studied is ruled by nonspecific interactions. The solubility order in terms of cations is:

$$P_{nmpq}^{+} > C_nPyr^{+} > C_nC_mPyrr^{+} > C_nC_mIm^{+}$$

In comparison, the solubility dependency in terms of the nature of the anion is much smaller, for the same gases.

It would be of great interest to develop ionic liquids containing functionalizations or anions (such as $AlCl_4^-$) that could promote a specific interaction with ethylene or propylene. Another promising possibility is to develop ionic liquids containing metal ions in their structure [40]. However, the more specific are the interactions between the gas and the ionic liquid, the harder the recycling of the absorbent will be, and a balancing effect between specificity and recovery should be taken into account on the development of an ionic liquid for separation purposes.

The published solubility data on ethane, ethylene, propane and propylene (Tab. 3, 4) provides a good start on a database in light hydrocarbon solubility in ionic liquids. This information can be used as a good basis for the development of a model that would allow the prediction of the solubility of these gases in other ionic liquids.

ACKNOWLEDGMENTS

The authors thank Dr. Y. Chauvin (Emerit Research Director 1995-2015 at C2P2, Équipe COMS, Lyon) for the insightful discussions in the early stages of this project.
L.M. was financed by a Cluster Excellence of the Rhône-Alpes region with the participation of IFPEN. The authors thank Dr Méthivier of IFPEN for fruitful discussions during this project.

REFERENCES

1 Henley E.J., Seader J.D., Roper D.K. (2011) *Separation Process Principles*, Third Edition, Jonh Wiley & Sons.

2 Jacquemin J., Costa Gomes M.F., Husson P., Majer V. (2006) Solubility of Carbon Dioxide, Ethane, Methane, Oxygen, Nitrogen, Hydrogen, Argon, and Carbon Monoxide in 1-Butyl-3-methylimidazolium Tetrafluoroborate Between Temperatures 283 K and 343 K and at Pressures Close to Atmospheric, *J. Chem. Thermodyn.* **38**, 490-502.

3 Hu Y., Liu Z., Xu C., Zhang X. (2011) The Molecular Characteristics Dominating the Solubility of Gases in Ionic Liquids, *Chem. Soc. Rev.* **40**, 3802-3823

4 Lei Z., Dai C., Chen B. (2014) Gas Solubility in Ionic Liquids, *Chem. Rev.* **114**, 1289-1326.

5 Olivier-Bourbigou H., Magna L., Morvan D. (2010) Ionic liquids and catalysis: Recent progress from knowledge to applications, *Appl. Catal. A-Gen.* **373**, 1-56.

6 Armand M., Endres F., MacFarlane D.R., Ohno H., Scrosati B. (2009) Ionic-liquid materials for the electrochemical challenges of the future, *Nat. Mater.* **8**, 621-629.

7 Han D., Row K.H. (2010) Recent Applications of Ionic Liquids in Separation Technology, *Molecules* **15**, 2405-2426.

8 Hong G., Jacquemin J., Deetlefs M., Hardacre C., Husson P., Costa Gomes M.F. (2007) Solubility of Carbon Dioxide and Ethane in Three Ionic Liquids Based on the Bis{(trifluoromethyl) sulfonyl}imide Anion, *Fluid Phase Equilibr.* **257**, 27-34.

9 Camper D., Becker C., Koval C., Noble R. (2005) Low Pressure Hydrocarbon Solubility in Room Temperature Ionic Liquids Containing Imidazolium Rings Interpreted Using Regular Solution Theory, *Ind. Eng. Chem. Res.* **44**, 1928-1933.

10 Camper D., Scovazzo P., Koval C., Noble R. (2004) Gas Solubilities in Room-Temperature Ionic Liquids, *Ind. Eng. Chem. Res.* **43**, 3049-3054.

11 Kilaru P.K., Scovazzo P. (2008) Correlations of Low-Pressure Carbon Dioxide and Hydrocarbon Solubilities in Imidazolium-, Phosphonium-, and Ammonium-Based Room-Temperature Ionic Liquids. Part 2. Using Activation Energy of Viscosity, *Ind. Eng. Chem. Res.* **47**, 910-919.

12 Kilaru P.K., Condemarin R.A., Scovazzo P. (2008) Correlations of Low-Pressure Carbon Dioxide and Hydrocarbon Solubilities in Imidazolium-, Phosphonium-, and Ammonium-Based Room-Temperature Ionic Liquids. Part 1. Using Surface Tension, *Ind. Eng. Chem. Res.* **47**, 900-909.

13 Morgan D., Ferguson L., Scovazzo P. (2005) Diffusivities of Gases in Room-Temperature Ionic Liquids: Data and Correlations Obtained Using a Lag-Time Technique, *Ind. Eng. Chem. Res.* **44**, 4815-4823.

14 Liu X., Afzal W., Yu G., He M., Prausnitz J.M. (2013) High Solubilities of Small Hydrocarbons in Trihexyl Tetradecylphosphonium Bis(2,4,4-trimethylpentyl) Phosphinate, *J. Phys. Chem. B* **117**, 10534-10539.

15 Gan Q., Zou Y., Rooney D., Nancarrow P., Thompson J., Liang L., Lewis M. (2011) Theoretical and Experimental Correlations of Gas Dissolution, Diffusion, and Thermodynamic Properties in Determination of Gas Permeability and Selectivity in Supported Ionic Liquid Membranes, *Adv. Colloid Interfac.* **164**, 45-55.

16 Xing H., Zhao X., Li R., Yang Q., Su B., Bao Z., Yang Y., Ren Q. (2013) Improved Efficiency of Ethylene/Ethane Separation Using a Symmetrical Dual Nitrile-Functionalized Ionic Liquid, *ACS Sustainable Chem. Eng.* **1**, 1357-1363.

17 Anthony J.L., Anderson J.L., Maginn E.J., Brennecke J.F. (2005) Anion Effects on Gas Solubility in Ionic Liquids, *J. Phys. Chem. B* **109**, 6366-6374.

18 Zhang J., Zhang Q., Qiao B., Deng Y. (2007) Solubilities of the Gaseous and Liquid Solutes and Their Thermodynamics of Solubilization in the Novel Room-Temperature Ionic Liquids at Infinite Dilution by Gas Chromatography, *J. Chem. Eng. Data* **52**, 2277-2283.

19 Zhang Q., Li Z., Zhang J., Zhang S., Zhu L., Yang J., Zhang X., Deng Y. (2007) Physicochemical Properties of Nitrile-Functionalized Ionic Liquids, *J. Phys. Chem. B* **111**, 2864-2872.

20 Palgunadi J., Kim H.S., Lee J.M., Jung S. (2010) Ionic Liquids for Acetylene and Ethylene Separation: Material Selection and Solubility Investigation, *Chem. Eng. Process.* **49**, 192-198.

21 Costa Gomes M.F. (2007) Low-Pressure Solubility and Thermodynamics of Solvation of Carbon Dioxide, Ethane, and Hydrogen in 1-Hexyl-3-methylimidazolium Bis(trifluoromethylsulfonyl)amide between Temperatures of 283 K and 343 K, *J. Chem. Eng. Data* **52**, 472-475.

22 Anthony J.L., Maginn E.J., Brennecke J.F. (2002) Solubilities and Thermodynamic Properties of Gases in the Ionic Liquid 1-*n*-Butyl-3-methylimidazolium Hexafluorophosphate, *J. Phys. Chem. B* **106**, 7315-7320.

23 Liu X., Ruiz E., Afzal W., Ferro V., Palomar J., Prausnitz J.M. (2014) High Solubilities for Methane, Ethane, Ethylene, and Propane in Trimethyloctylphosphonium Bis(2,4,4-trimethylpentyl) Phosphinate ([P8111][TMPP]), *Ind. Eng. Chem. Res.* **53**, 363-368.

24 Anderson J.L., Dixon J.K., Brennecke J.F. (2007) Solubility of CO_2, CH_4, C_2H_6, C_2H_4, O_2, and N_2 in 1-Hexyl-3-methylpyridinium Bis(trifluoromethylsulfonyl)imide: Comparison to Other Ionic Liquids, *Acc. Chem. Res.* **40**, 1208-1216.

25 Liu X., Afzal W., Prausnitz J.M. (2013) Solubilities of Small Hydrocarbons in Tetrabutylphosphonium Bis(2,4,4-trimethylpentyl) Phosphinate and in 1-Ethyl-3-methylimidazolium Bis (trifluoromethylsulfonyl)imide, *Ind. Eng. Chem. Res.* **52**, 14975-14978.

26 Safarik D.J., Eldridge R.B. (1998) Olefin/Paraffin Separations by Reactive Absorption: A Review, *Ind. Eng. Chem. Res.* **37**, 2571-2581.

27 Moura L., Mishra M., Bernales V., Fuentealba P., Padua A.A.H., Santini C.C., Costa Gomes M.F. (2013) Effect of Unsaturation on the Absorption of Ethane and Ethylene in Imidazolium-Based Ionic Liquids, *J. Phys. Chem. B* **117**, 7416-7425.

28 Moura L., Darwich W., Santini C.C., Costa Gomes M.F. (2015) Imidazolium-based ionic liquids with cyano groups for the selective absorption of ethane and ethylene, *Chem. Eng. J.* **280**, 755-762.

29 Green B.D., O'Brien R.A., Davis Jr. J.H., West K.N. (2015) Ethane and Ethylene Solubility in an Imidazolium-Based Lipidic Ionic Liquid, *Ind. Eng. Chem. Res.* **54**, 5165-5171.

30 Moura L. (2014) Ionic liquids for the separation of gaseous hydrocarbons, *PhD Thesis*, Université Lyon 1.

31 Sánchez L.M.G., Meindersma G.W., Haan A.B. (2009) Potential of Silver-Based Room-Temperature Ionic Liquids for Ethylene/Ethane Separation, *Ind. Eng. Chem. Res.* **48**, 10650-10656.

32 Mortaheb H.R., Mafi M., Mokhtarani B., Sharifi A., Mirzaei M., Khodapanah N., Ghaemmaghami F. (2010) Experimental Kinetic Analysis of Ethylene Absorption in Ionic Liquid [Bmim]NO_3] with Dissolved $AgNO_3$ by a Semi-Continuous Process, *Chem. Eng. J.* **158**, 384-392.

33 Faiz R., Li K. (2012) Olefin/Paraffin Separation Using Membrane Based Facilitated Transport/Chemical Absorption Techniques, *Chem. Eng. Sci.* **73**, 261-284.

34 Pitsch F., Krull F.F., Agel F., Schulz P., Wasserscheid P., Melin T., Wessling M. (2012) An Adaptive Self-Healing Ionic Liquid Nanocomposite Membrane for Olefin-Paraffin Separations, *Adv. Mater.* **24**, 4306-4310.

35 Agel F., Pitsch F., Krull F.F., Schulz P., Wessling M., Melin T., Wasserscheid P. (2011) Ionic Liquid Silver Salt Complexes for Propene/Propane Separation, *Phys. Chem. Chem. Phys.* **13**, 725-731.

36 Ortiz A., Ruiz A., Gorri D., Ortiz I. (2008) Room Temperature Ionic Liquid with Silver Salt as Efficient Reaction Media for Propene/Propane Separation: Absorption Equilibrium, *Sep. Purif. Technol.* **63**, 311-318.

37 Staudt-Bickel C., Koros W.J. (2000) Olefin/paraffin Gas Separations with 6FDA-Based Polyimide Membranes, *J. Membrane Sci.* **170**, 205-214.

38 Morisato A., He Z., Pinnau I., Merkel T.C. (2002) Transport Properties of PA12-PTMO/$AgBF_4$ Solid Polymer Electrolyte Membranes for Olefin/Paraffin Separation, *Desalination* **145**, 347-351.

39 Hsiue G., Yang J. (1993) Novel Methods in Separation of Olefin/Paraffin Mixtures by Functional Polymeric Membranes, *J. Membrane Sci.* **92**, 117-128.

40 Lin I.J.B., Vasam C.S. (2005) Metal-Containing Ionic Liquids and Ionic Liquid Crystals Based on Imidazolium Moiety, *J. Organomet. Chem.* **690**, 3498-3512.

16

Synthesis and Ring-Opening Metathesis Polymerization of Second-Generation Dendronized Poly(ether) Monomers Initiated by Ruthenium Carbenes

Pablo E. Guzmán[1]*, Victoria A. Piunova[2], Garret M. Miyake[3] and Robert H. Grubbs[1]

[1] *Arnold and Mabel Beckman Laboratories for Chemical Synthesis, Division of Chemistry and Chemical Engineering, California Institute of Technology, Pasadena, California 91125 - United States*

[2] *IBM Almaden Research Center, 650 Harry Road, San Jose California 95134 - United States*

[3] *Department of Chemistry and Biochemistry, Materials Science and Engineering Program, University of Colorado, Boulder, Colorado 80309 - United States*

e-mail: peguzman@caltech.edu - vapiunov@us.ibm.com - garret.miyake@colorado.edu - rhg@caltech.edu

* Corresponding author

Abstract — *The Ring-Opening Metathesis Polymerization (ROMP) of second-generation dendronized monomers is described. Using the highly active and fast-initiating third-generation ruthenium complex [(H_2IMes)(pyr)$_2$$Cl_2$RuCHPh], moderate to high molecular weight polymers (430-2230 kDa) are efficiently synthesized with low dispersities (Đ = 1.01-1.17). This study highlights the power of the metathesis approach toward polymer synthesis in a context where monomer structure can significantly impede polymerization.*

Résumé — La synthèse et la polymérisation par ouverture de cycle par métathèse des monomères poly(éther) de deuxième génération initié par le catalyseur de ruthénium de troisième génération — La polymérisation par ouverture de cycle par métathèse (ROMP, *Ring-Opening Metathesis Polymerization*) de monomers dendronisés de deuxième génération est décrite. En utilisant le complexe de ruthénium ayant une activité élevée et une initiation rapide de troisième génération [(H_2IMes)(pyr)$_2$$Cl_2$RuCHPh], des polymères de masses moléculaires modérées à élevées (430 à 2230 kDa) sont synthétisés de manière efficace avec de basses polydispersités (Đ = 1,01 à 1,17). Cette étude met en évidence la puissance de l'approche de métathèse vers la synthèse de polymères dans un contexte où la structure de monomère peut empêcher de manière significative la polymérisation.

INTRODUCTION

Since its discovery nearly six decades ago, olefin metathesis has evolved into a powerful method for the formation of C—C bonds, enabling the synthesis of simple to complex organic molecules [1]. Y. Chauvin proposed the commonly accepted mechanism for metathesis (Scheme 1), which illustrates the crucial involvement of a metal carbene and metallacyclobutane intermediate [2].

Ring-Opening Metathesis Polymerization (ROMP), an application of the metathesis reaction to polymer synthesis, has made a tremendous impact on synthetic polymer chemistry [3-8]. Well-defined and complex polymer architectures are readily synthesized through judicious identification of monomer and initiator. Norbornene and its functionalized derivatives have proven to be archetypical polymer precursors because of facile synthesis, affordability, and reactivity [9]. Moreover, the combination of inherent living characteristics

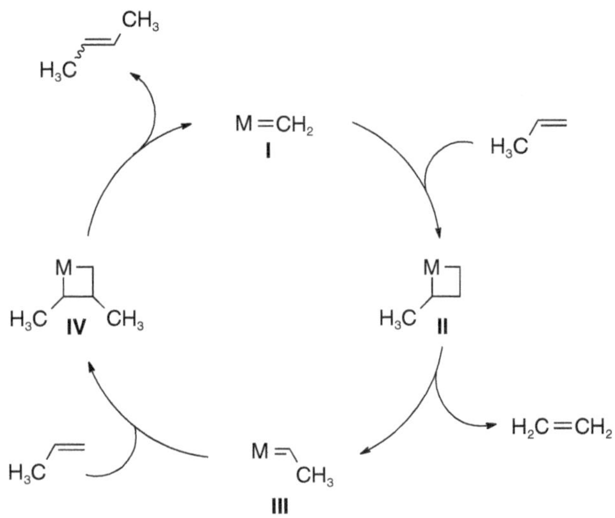

Scheme 1

The Chauvin mechanism of olefin metathesis.

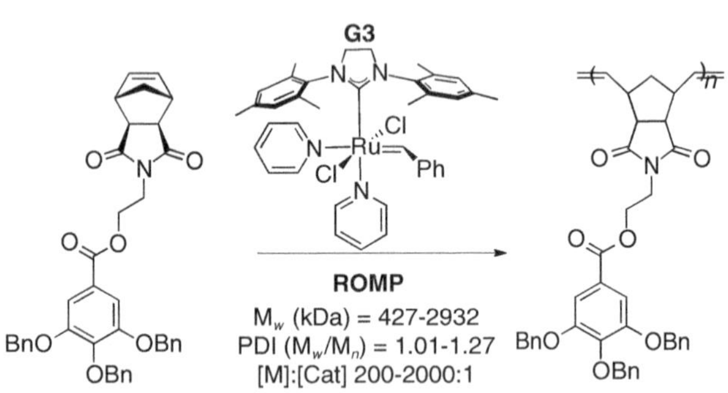

Scheme 2

Polymerization of first-generation wedge-type monomers with **G3**.

and practicality associated with ROMP has made it a powerful tool among chemists and materials scientists [10].

Recently, our interest in ROMP has been directed towards the synthesis of molecular brush and dendronized block copolymers and their self-assembly to visible light reflecting one-Dimensional Photonic Crystals (1D PC) [11-19]. Photonic crystals are periodic nanostructured materials that possess a photonic band gap that inhibits select frequencies of light from propagating through the bulk material [15-19]. The access of periodicities large enough to interact with visible light is a challenge with block copolymers because of their inherent macromolecular chain-entanglement [20]. However, though the design and synthesis of polymer architectures that reduce chain-entanglement, we have demonstrated that such block copolymers can rapidly

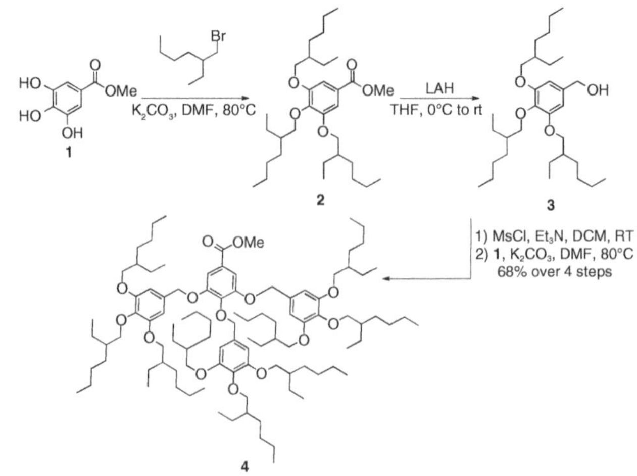

Scheme 3

Synthesis of dendronized ester **4**.

self-assemble to photonic crystals that reflect wavelengths of light across the visible spectrum and into the IR.

For instance, in the presence of the third-generation bis-pyridine initiator (**G3**), an efficient polymerization of discrete first-generation wedge-type monomers containing a functionalized *exo*-norbornene backbone tethered to a 1, 3, 4, 5-tetrasubstituted aromatic pendant/anchor group was achieved (Scheme 2) [21]. Ratios ([monomer]:[initiator]) ranging from 200-2000:1, provided a broad range of Molecular Weights (MW) (weight average MW (M_w) = 427 to 2932 kDa) along with low dispersities (*Ðs*) (PDI = 1.01-1.27). The sequential copolymerization with an appropriate first-generation wedge-type monomer provided facile entry to block copolymers capable of self-assembling to 1D PC. However, at high MW (*e.g.* M_w = 1390-1940 kDa), self-assembly proved challenging, presumably due to chain entanglement. Thus, we hypothesized that an additional generation to the existing wedge-type system would increase polymer rigidity and decrease the propensity for chain entanglement.

Notably, previous studies have shown that if the linker length between the polymerizing *exo*-norbornene group and the pendant/anchor group is too short (<10 methylene units), similar second-generation dendronized monomers could not be efficiently polymerized to high MW polymers using similar carbene initiators [22]. In this manuscript, we challenge the ROMP method and showcase the ability of the highly active **G3** initiator to ring-open and polymerize sterically hindered dendronized MacroMonomers (MM) with short linker lengths to yield high MW homopolymers under ambient conditions. We envisioned that the incorporation of a highly reactive *exo*-norbornene carbodiimides polymerizable functionality (due to the excess ring strain) would

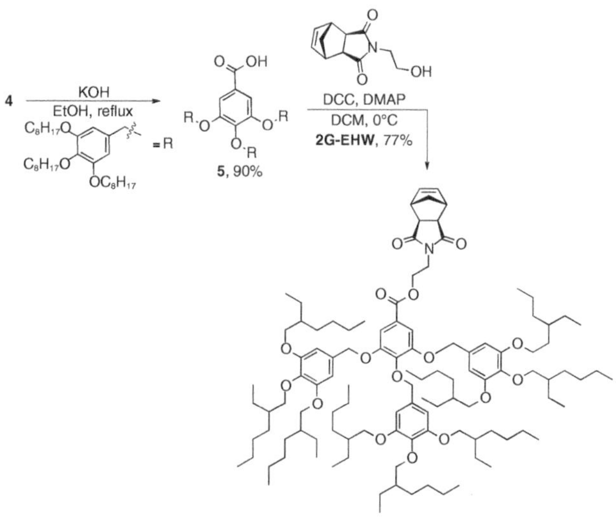

Scheme 4

Synthesis of dendronized monomer **2G-EHW**.

allow successful polymerization of second-generation wedge monomers with short linkers.

1 RESULTS AND DISCUSSION

Our approach to second-generation dendronized macromonomers was realized through the execution of standard organic transformations. Ester **4** was synthesized in a 4-step sequence starting from methyl gallate **1** (Scheme 3). A three-fold O-alkylation with 2-ethylhexyl bromide provided compound **2**, which was reduced to its corresponding benzyl alcohol (**3**) with Lithium Aluminum Hydride (LAH). The treatment of **3** with mesyl chloride followed by nucleophilic displacement of the corresponding mesylate with **1** yielded the desired ester in 68% yield over four steps.

The exposure of **4** to alkaline conditions in boiling alcoholic solvent led to the formation of the corresponding carboxylic acid **5** which was isolated in 90% yield (Scheme 4). N, N'-dicyclohexylcarbodiimide (DCC) coupling with *N*-(hydroxyethyl)-*cis*-5-norbornene-*exo*-2, 3, dicarboximide furnished the desired second-generation ethylhexyl wedge dendronized macromonomer (**2G-EHW**) in 77% yield.

The ROMP of **2G-EHW** led to the controlled synthesis of second-generation wedge-type polymers capable of being carried out over a broad range of [**2G-EHW**]:[**G3**] ratios. High conversions and remarkably low *Đ*s (entries 1-3) were achieved despite a short linker length (2 methylene units) and an extensive alkylether periphery (Tab. 1, Fig. 1).

TABLE 1

Results of the ROMP of **2G-EHW** mediated by **G3**[a]

Entry	[MM]/[G3]	Conv[b]	M_w (kDa)[c]	$Đ$ (M_w/M_n)[c]
1	200	100	4.29	1.01
2	400	99	1055	1.03
3	600	90	2086	1.17
4	800	65	2230	1.42
5	1000	0	-	-

[a] Polymerizations performed in 1 mL of THF over 2 hours at ambient temperature. [**2G-EHW**] = 90 mM.
[b] Determined by ^{1}H NMR.
[c] Determined by light scattering.

However, the reaction efficiency decreases or fails to polymerize at ratios greater than 800:1 [**2G-EHW**]:[**G3**] (Fig. 2) [23]. Nonetheless, the controlled synthesis of polymers with MW nearing 2100 kDa were efficiently achieved.

Detailed kinetic studies established a complete kinetic profile for monomer **2G-EHW**. A linear increase in MW with an increase in monomer conversion (Fig. 1a) and nearly constant *Đ* was observed during the course of polymerization; consistent with living polymerization characteristics (Fig. 3). Accordingly, for all the [**2G-EHW**]:[**G3**] ratios investigated a first-order dependence on [**2G-EHW**] was observed (Fig. 1b).

2 EXPERIMENTAL

2.1 General Considerations

$(H_2IMes)(PPh_3)_2(Cl)_2RuCHPh$ was received as a research gift from *Materia Inc.* and converted to **G3** via literature procedures [24]. All other chemicals were purchased from *Sigma Aldrich*. Solvents were purified by passage through solvent purification columns and further degassed with argon [25]. N-(hydroxyethyl)-*cis*-5-norbornene-*exo*-2,3 dicarboximide was prepared according to literature procedure [26]. All reactions were carried out in flame-dried Schlenk-type glassware on a dual-manifold Schlenk line or in a nitrogen-filled glovebox. NMR spectra were recorded on a Varian *Inova* 500 MHz spectrometer. Chemical shifts were referenced to internal solvent resonances and are reported as parts per million relative to tetramethylsilane. High resolution mass spectra were provided by the California Institute of Technology Mass Spectrometry Facility. Compound **2G-EHW** was analyzed by MALDI-TOF in reflector mode using a Voyager DE PRO time-of-flight mass spectrometer (*Applied Biosystems*). The MALDI matrix used was

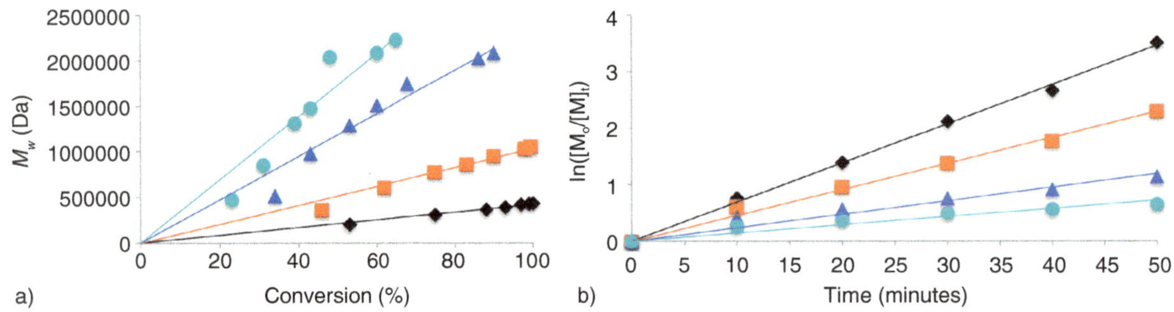

Figure 1

a) Plot of M_w as a function of **2G-EHW** conversion ([**2G-EHW**]:[**G3**] = 200 (◆), 400 (■), 600 (▲), 1000 (●). b) Semilogarithmic plots of ln([**2G-EHW**$_o$]/[**2G-EHW**$_t$]) as a function of time for the ROMP of **2G-EHW** by **G3**. Conditions: [**2G-EHW**] = 90 mM; [**G3**] = 450 μM (◆), 220 μM (■), 150 μM (▲), 110 μM (●). Polymerizations were performed in anhydrous/degassed THF at ambient temperature.

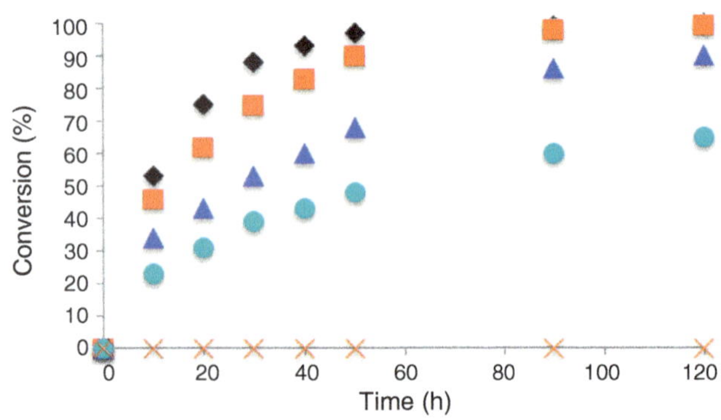

Figure 2

Kinetic data comparing the polymerization of **2G-EHW**. [**2G-EHW**]:[**G3**] ratio 200:1 (◆), 400:1 (■), 600:1 (▲), 800:1 (●), 1000:1 (X).

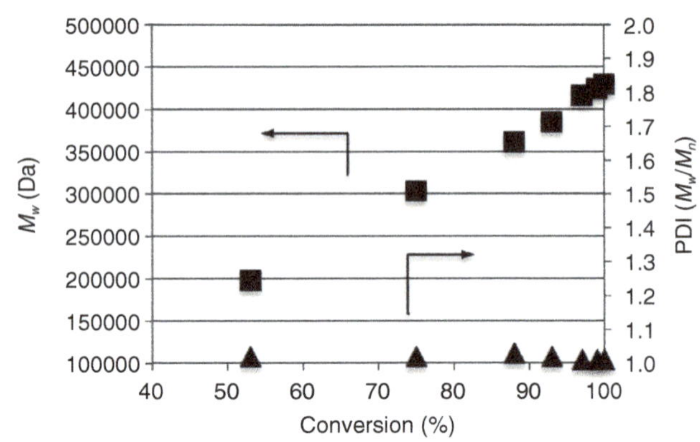

Figure 3

Plot of M_w (left) and $Đ$ (right) as a function of **2G-EHW** conversion ([**2G-EHW**]:[**G3**] = 200:1).

dithranol prepared at 10 mg/mL in THF. The instrument was externally calibrated with the manufacturer Sequazyme^{TM} Kit. Polymer MW were determined utilizing THF as the eluent by MultiAngle Light-Scattering (MALS) Gel Permeation Chromatography (GPC) using a mini-DAWN TREOS light-scattering detector, a Viscostar viscometer, and an OptilabRex refractive index detector, all from *Wyatt Technology*. An *Agilent* 1200 UV–Vis detector was also present in the detector stack. Absolute MW were determined using *dn*/*dc* values calculated by assuming 100% mass recovery of the polymer sample injected into the GPC.

2.2 Procedures

To a flame dried 500 mL round-bottom flask equipped with a magnetic stir bar was added K_2CO_3 (55.0 g, 398 mmol, 8.00 eq), anhydrous DMF (260 mL), methyl gallate (9.15 g, 49.7 mmol, 1.00 eq) and 2-ethylhexyl bromide (53.0 mL, 298 mmol, 6.00 eq). The reaction was fitted with a water-cooled condenser, placed under an inert atmosphere of argon and heated to 80°C for 14 h. The reaction was stopped by the addition of H_2O (500 mL). The mixture was transferred to a separation funnel and the aqueous layer was washed with Et_2O (3 × 300 mL). The organic layers were combined, washed with brine, dried over $MgSO_4$, vacuum filtered and concentrated under reduced pressure. The material was used crude and subjected to reduction conditions.

To a 500 mL round-bottom flask containing crude ester **2** (49.7 mmol, 1.00 eq) was added anhydrous THF (155 mL). The solution was cooled to 0°C. Once cool, LAH (3.77 g, 99.4 mmol, 2.00 eq) was carefully added. The reaction placed under an inert atmosphere of argon and allowed to stir and gradually warm to room temperature overnight. The reaction was cooled to 0°C. Once cool, the reaction was

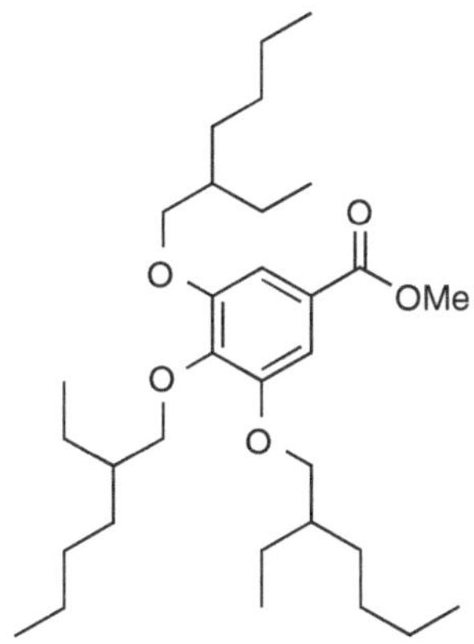

Figure 4

Methyl 3,4,5-tris((2-ethylhexyl)oxy)benzoate (2).

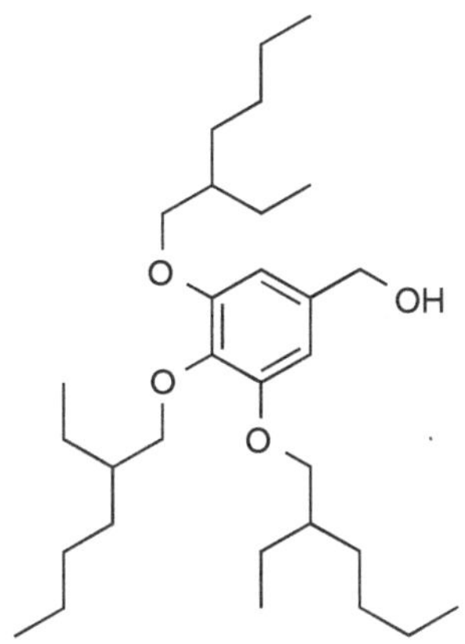

Figure 5

(3,4,5-tris((2-ethylhexyl)oxy)phenyl)methanol (3).

stopped by the addition of H_2O (20 mL) and diluted with Et_2O. A saturated aqueous solution of Rochelle's salt was added and the mixture was aggressively stirred for one hour. The two layers were separated and the aqueous layer was washed with Et_2O (3 × 300 mL). The organic layers were combined, washed with brine, dried over $MgSO_4$, vacuum filtered and concentrated under reduced pressure. The material was used crude for the next step.

To a 500 mL round-bottom flask containing crude alcohol **3** (49.7 mmol, 1.00 eq) was added anhydrous DCM (165 mL). The solution was cooled to 0°C. Once cool, distilled Et_3N (10.3 mL, 74.6 mmol, 1.50 eq) was added. MsCl (4.61 g, 59.6 mmol, 1.20 eq) was carefully added dropwise over two minutes. The reaction was placed under an inert atmosphere of argon and allowed to stir and gradually warm to room temperature overnight. The reaction was stopped by

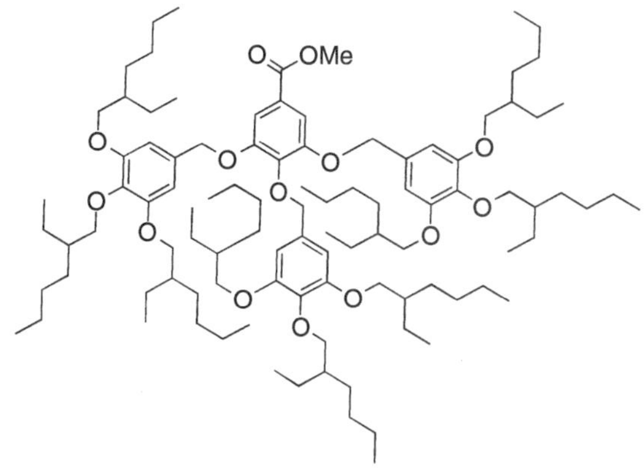

Figure 6

Methyl 3,4,5-tris((3,4,5-tris((2-ethylhexyl)oxy)benzyl)oxy) benzoate (4).

the addition of H_2O (150 mL). The two layers were separated and the aqueous layer was washed with Et_2O (3 × 200 mL). The organic layers were combined, washed with a saturated aqueous $NaHCO_3$ solution, dried over $MgSO_4$, vacuum filtered and concentrated under reduced pressure. To a separate 1 L round bottom flask was added methyl gallate (2.41 g, 13.1 mmol, 1.00 eq), K_2CO_3 (14.3 g, 104 mmol, 8.00 eq) and anhydrous DMF (300 mL). The crude mesylate (49.7 mmol, 1.00 eq) was dissolved in DMF (50 mL) and added to the reaction flask by cannula transfer. The reaction flask was equipped with a reflux condenser and heated to 80°C for 12 h then cooled to room temperature and stirred for an additional 12 h. The reaction was stopped by the addition of H_2O (300 mL) and EtOAc (300 mL). The mixture was transferred to a separation funnel and the aqueous layer was washed with Et_2O (3 × 400 mL). The organic layers were combined, washed copious amounts of H_2O, brine, dried over $MgSO_4$, vacuum filtered and concentrated under reduced pressure. The crude material was purified by flash chromatography (95:5 pentane:Et_2O then 80:20 pentane:Et_2O) on silica gel to provide the desired product as a light yellow oil (14.37 g, 68% yield over 4 steps) R_f = 0.57 (80:20 hexanes/Et_2O). ^1H NMR (500 MHz; $CDCl_3$) δ 7.36 (s, 2H), 6.67 (s, 2H), 6.65 (s, 4H), 5.14 (s, 2H), 5.07 (s, 4H), 3.86 (s, 3H), 3.84-3.76 (m, 14H), 3.72-3.66 (m, 4H), 1.75-1.25 (m, 81H), 0.96-0.85 (m, 54H); ^{13}C NMR (126 MHz, $CDCl_3$) δ 166.4, 153.5, 152.4, 142.5, 137.6, 137.5, 132.3, 131.7, 125, 109.7, 105.6, 104.8, 75.8, 75.7, 75.3, 71.6, 71.1, 70.9, 52.1, 40.6, 39.4, 30.4, 29.3, 29, 23.8, 23.7, 23.1, 23, 14.2, 14.1, 11.2, 11.1; IR (neat): 2958, 2928, 2873, 2859, 1721, 1591; HRMS-(FAB+)-m/z 1607.262 $[(M+H)^+$ requires 1607.259].

To a round-bottom flask containing **4** (9.90 g, 6.15 mmol, 1.00 eq) was added EtOH (30 mL) and KOH (3.45 g,

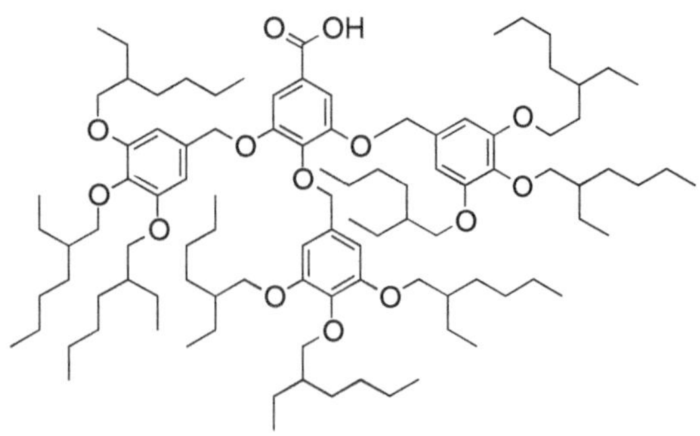

Figure 7

3-((3-((3-ethylheptyl)oxy)-4,5-bis((2-ethylhexyl)oxy)benzyl)oxy)-4,5-bis((3,4,5-tris((2-ethylhexyl)oxy)benzyl)oxy)benzoic acid (5).

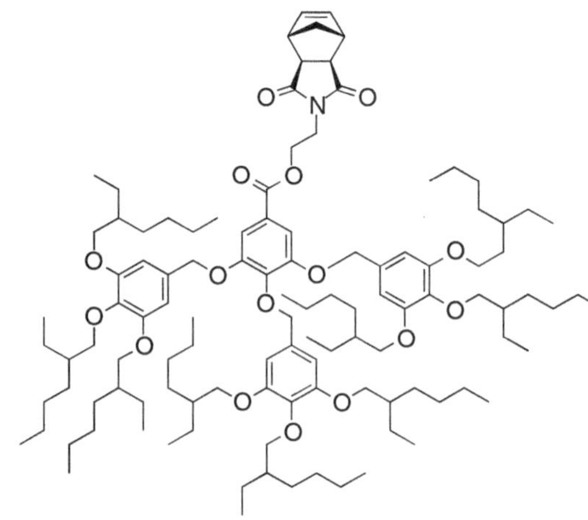

Figure 8

2-((3aR,4R,7S,7aS)-1,3-dioxo-1,3,3a,4,7,7a-hexahydro-2H-4,7-methanoisoindol-2-yl)ethyl 3-((3-((3-ethylheptyl)oxy)-4,5-bis((2-ethylhexyl)oxy)benzyl)oxy)-4,5-bis((3,4,5-tris((2-ethylhexyl)oxy)benzyl)oxy)benzoate (2G-EHW).

61.5 mmol, 10.0 eq). The flask was equipped with a water-cooled reflux condenser and heated to 80°C for 5 h. The reaction was allowed to cool to room temperature and stirred for an additional 12 h then cooled to 0°C. Once cool, conc. HCl was added until a neutral pH was achieved. The mixture was concentrated under reduced pressure then diluted with EtOAc and brine. The solution was transferred to a separation funnel and the two layers were separated. The aqueous layer was washed with EtOAc (3 × 150 mL). The organic layers were combined, washed with brine, dried over $MgSO_4$, vacuum filtered and concentrated under reduced pressure. The crude material was purified by flash chromatography (90:10 pentane:EtOAc) on silica gel to provide the desired product as a light yellow oil (8.78 g, 90% yield) R_f = 0.15 (80:20 pentane/EtOAc). ^{1}H NMR (500 MHz; $CDCl_3$) δ 7.43 (s, 2H), 6.66 (s, 2H), 6.65 (s, 4H), 5.15 (s, 2H), 5.07 (s, 4H), 3.85-3.74 (m, 14H), 3.69-3.65 (m, 4H), 1.73-1.26 (m, 81H), 0.94-0.84 (m, 54H); ^{13}C NMR (126 MHz, $CDCl_3$) δ 170.8, 153.7, 152.7, 143.5, 137.7, 132.4, 131.6, 124.1, 110.4, 105.7, 105.1, 75.98, 75.94, 75.5, 71.9, 71.3, 71.1, 40.8, 39.7, 30.7, 30.6, 29.5, 29.3, 29.2, 23.9, 23.8, 23.3, 23.2, 14.3, 14.2, 11.4, 11.3; IR (neat): 2958, 2928, 2873, 1686, 1591; HRMS-[(FAB+) m/z 1594.247 requires 1594.251)].

To a round-bottom flask containing **5** (8.78 g, 5.50 mmol, 1.00 eq) was added anhydrous DCM (150 mL) and cooled to 0°C. Once cool, N-(hydroxyethyl)-*cis*-5-norbornene-*exo*-2,3 dicarboximide (1.60 g, 6.06 mmol, 1.10 eq), DMAP (335 mg, 2.75 mmol, 0.500 eq) and N, N'-dicyclohexylcarbodiimide (1.25 g, 6.06 mmol, 1.10 eq) were added. The reaction was allowed to stir and gradually warm to room temperature for 12 h. The reaction was stopped by concentration under reduced pressure. The crude material was purified by flash chromatography (85:15 hexanes:EtOAc) on silica gel to provide the desired product as a light yellow oil (7.80 g, 77% yield) R_f = 0.37 (80:20 hexanes/EtOAc). ^{1}H NMR (500 MHz; $CDCl_3$) δ 7.33 (s, 2H), 6.67 (s, 6H), 6.24 (t, J = 2 Hz, 2H), 5.11 (s, 2H), 5.09 (s, 4H), 4.37 (t, J = 4.5 Hz, 2H), 3.89 (t, J = 5 Hz, 2H), 3.82-3.74 (m, 14H), 3.67 (dd, J = 2.5, 5.0 Hz, 4H), 3.20 (t, J = 1.5 Hz, 2H), 2.67 (d, J = 1 Hz. 2H), 1.71-1.20 (m, 81H), 0.94-0.83 (m, 56H); ^{13}C NMR (126 MHz, $CDCl_3$) δ 177.9, 165.7, 153.4, 152.6, 142.8, 137.8, 137.6, 132.5, 131.8, 124.6, 109.6, 105.5, 105, 76, 75.9, 75.5, 71.5, 71.2, 71.1, 61.9, 47.9, 45.4, 42.7, 40.7, 39.7, 37.5, 30.7, 30.6, 29.5, 29.2, 23.9, 23.8, 23.3, 23.2, 14.3, 14.2, 11.3, 11.2; HRMS: the detected mass corresponds to a radical cation with m/z 1804.3023 corresponding to $[M+Na-H]^+$ requires 1804.3043.

2.3 Synthesis of Second Generation Dendronized Homopolymer

To a 4 dram vial was added **2G-EHW** (160 mg, 0.0900 mmol, 200 eq). To the vial was added a magnetic stir bar and taken into a nitrogen filled glovebox. To the vial was added THF (1 mL). With rapid stirring, 49 μL of an appropriate concentration of **G3** in THF was quickly added *via* syringe. For kinetic analysis, a 0.05 mL aliquot of the reaction solution was taken at predetermined time intervals and injected into a 1 mL septum sealed vial containing a solution of 25 μL of ethyl vinyl ether in 0.7 mL of THF. The aliquot was analyzed by GPC to determine the MW of the polymer. After the solvent was allowed to evaporate from the vials, the polymer residue was dissolved in $CDCl_3$ and analyzed by ^{1}H NMR spectroscopy to determine the percent of monomer conversion by comparing the peaks corresponding to the wedge polymer and the unreacted monomer. The polymerization was stopped by the addition of ethyl vinyl ether (0.2 mL) and addition of methanol (25 mL). The mixture was allowed to stir for 1 h. The polymer was isolated by vacuum filtration and dried under vacuum at ambient temperature to a constant weight.

2G-EHW Homopolymer: ^{1}H NMR (500 MHz, $CDCl_3$, 25°C): δ 7.37-7.33 (m), 6.67-6.61 (m), 5.11-5.04 (m), 3.79-3.73 (m), 3.61 (bs), 1.70-1.56 (m), 1.49-1.24 (m), 0.94-0.80 (m). *dn*/*dc* value = 0.1082.

CONCLUSION

The synthesis of wedge-type dendronized polymers was achieved *via* the ruthenium-mediated ROMP of second-generation functionalized *exo*-norborene poly(ether) monomers. These substantially hindered monomers were efficiently polymerized using the highly active **G3** ruthenium carbene initiator. Well-defined ($Đ$ = 1.01-1.17) high molecular weight (2100 kDa) polymers are accessible. Furthermore, we have demonstrated that a relatively short two carbon linker length is well tolerated in the ring-opening event. This investigation has provided the foundation necessary to explore the synthesis of rigid block copolymer systems with potential for infrared reflecting photonic properties and will be disseminated in due course.

ACKNOWLEDGMENTS

This work was supported by the NIH (R01 GM031332). P.E.G is grateful to the NIH for a postdoctoral fellowship. We gratefully acknowledge support from the Camille and Henry Dreyfus Postdoctoral Program in Environmental Chemistry. The authors thank Ms. Alice B. Chang for helpful discussions.

REFERENCES

1 Grubbs R.H. (ed.) (2015) *Handbook of Metathesis*, 2nd edition, Wiley-VCH, Weinheim, Germany.

2 Hérisson P.J.-L., Chauvin Y. (1970) *Makromol. Chem.* **141**, 161-176.

3 Ivin K.J., Mol J.C. (1997) *Olefin Metathesis and Metathesis Polymerization*, Academic Press, Inc., San Diego, CA, USA.

4 Dragutan V., Streck R. (2000) *Catalytic Polymerization of Cycloolefins. Ionic, Ziegler-Natta and Ring-Opening Metathesis Polymerization*, Elsevier Science B.V., Amsterdam, The Netherlands.

5 Frenzel U., Müller B.K.M., Nuyken O. (2005) *Handbook of Polymer Synthesis*, 2nd ed., Kricheldorf H.R., Nuyken O., Swift G. (eds), Marcel Dekker, New York, NY, USA.

6 Buchmeiser M.R. (2000) *Chem. Rev.* **100**, 1565-1604.

7 Buchmeiser M.R. (2012) Ring-opening metathesis polymerization, in *Synthesis of Polymers: New Structures and Methods*, 1st ed., Schlüter A.D., Hawker C.J., Sakamoto J. (eds), Wiley–VCH, Weinheim, Germany.

8 Astruc D. (2005) *New J. Chem.* **29**, 42-56.

9 For a review of the ROMP of norbornene and its derivatives see: Floros G., Saragas N., Paraskevopoulou P., Psaroudakis N., Koinis S., Pitsikalis M., Hadjichristidis N., Merti. K., (2012) *Polymers* **4**, 1657-1673.

10 Chang A.B., Miyake G.M., Grubbs R.H. (2014) *Sequence-Controlled Polymer: Synthesis, Self-Assembly and Properties*, 1st ed., Lutz J.-F., Meyer T.Y., Ouchi M., Sawamoto M. (eds), American Chemical Society, pp. 161-188.

11 Miyake G.M., Weitekamp R.A., Grubbs R.H. (2015) *Handbook of Metathesis Vol. 3: Polymer Synthesis*, 2nd edition, Wiley-VCH, Weinheim, Germany, pp. 93-113.

12 Sveinbjörnsson B.R., Weitekamp R.A., Miyake G.M., Xia Y., Atwater H.A., Grubbs R.H. (2012) *Proc. Natl. Acad. Sci. U.S.A.* **109**, 14332-14336.

13 Miyake G.M., Weitekamp R.A., Piunova V.A., Grubbs R.H. (2012) *J. Am. Chem. Soc.* **134**, 14249-14254.

14 Miyake G.M., Piunova V.A., Weitekamp R.A., Grubbs R.H. (2012) *Angew. Chem. Int. Ed.* **51**, 11246-11248.

15 Ge J., Yin Y. (2011) *Angew. Chem. Int. Ed.* **50**, 1492-1522.

16 Galisteo-López J.F., Ibisate M., Sapienza R., Froufe-Pérez L.S., Blanco Á., López C. (2011) *Adv. Mater.* **23**, 30-69.

17 Wang J., Zhang Y., Wang S., Song Y., Jiang L. (2011) *Acc. Chem. Res.* **44**, 405-415.

18 Aguirre C.I., Reguera E., Stein A. (2010) *Adv. Funct. Mater.* **20**, 2565-2578.

19 Moon J.H., Yang S. (2010) *Chem. Rev.* **110**, 547-574.

20 Edrington A.C., Urbas A.M., DeRege P., Chen C.X., Swager T. M., Hadjichristidis N., Xenidou M., Fetters L.J., Joannopoulos J.D., Fink Y., Thomas E.L. (2001) *Adv. Mater.* **3**, 421-425.

21 Piunova V.A., Miyake G.M., Daeffler C.S., Weitekamp R.A., Grubbs R.H. (2013) *J. Am. Chem. Soc.* **135**, 15609-15616.

22 Jung H., Carberry T.P., Weck M. (2011) *Macromolecules* **44**, 9075-9083.

23 Monomer purity significantly affects the efficiency of ROMP see: Xia Y. (2015) *Macromolecules* **48**, 5656-5662.

24 Love J.A., Morgan J.P., Trnka T.M., Grubbs R.H. (2002) *Angew. Chem. Int. Ed.* **41**, 4035-4037.

25 Pangborn A.B., Giardello M.A., Grubbs R.H., Rosen R.K., Trimmer F.J. (1996) *Organometallics* **15**, 1518-1520.

26 Matson J.B., Grubbs R.H. (2008) *J. Am. Chem. Soc.* **130**, 6731-6733.

17

In Situ Generation of Molybdenum-Based Catalyst for Alkyne Metathesis: Further Developments and Mechanistic Insights

Joaquin Geng Lopez[1], Maciej Zaranek[2], Piotr Pawluc[1,2], Régis M. Gauvin[1] and André Mortreux[1]*

[1] *Univ. Lille, CNRS, Centrale Lille, ENSCL, Univ. Artois, UMR 8181 - UCCS - Unité de Catalyse et Chimie du Solide, 59000 Lille - France*
[2] *Faculty of Chemistry, Adam Mickiewicz University in Poznan, Umultowska 89b, 61-614 Poznan - Poland*
e-mail: andre.mortreux@ensc-lille.fr

* Corresponding author

Abstract — *Molybdenum-based catalysts are among the best candidates to achieve alkyne metathesis. They can be either well-defined carbynes, previously synthesized before their use, but also prepared* in situ *upon using stable molybdenum carbonyl complexes, or high oxidation state molybdenum salts that need a previous alkylation, both type of precursors being "activated" by hydroxyl-containing compounds such as phenols and silanols. This paper is presenting studies made on these systems, directed towards the knowledge of the reaction paths leading to the active species, and in particular to define the essential role of hydroxyl-containing co-catalyst in the formation of the active species, still ill-defined. From an analysis of the byproducts formed during the reaction, as well as of the initial products, reaction paths to access catalytic carbyne species is suggested, where the ligand environment consists of phenoxy (or siloxy) groups, typically required and identified to lead to alkyne metathesis in the case of well-defined catalysts.*

Résumé — Génération *in situ* de catalyseurs de métathèse des alcynes à base de molybdène : développements récents et approche mécanistique — Les catalyseurs à base de molybdène font partie des candidats les plus efficaces pour la métathèse des alcynes. Ceux-ci peuvent être soit des carbynes de molybdène préalablement synthétisés avant leur mise en œuvre, mais aussi préparés *in situ* via l'entremise de complexes à base de molybdène carbonyles, ou encore par alkylation de sels à haut degré d'oxydation, ces deux types de précurseurs nécessitant la présence de composés hydroxylés tels que des phénols ou silanols. Cette publication a pour objet de présenter des études visant à déterminer le mode de formation des espèces catalytiques impliquées dans ces deux systèmes, et en particulier du rôle du cocatalyseur hydroxylé, encore mal défini, dans la formation des espèces actives. De l'analyse des produits secondaires obtenus en cours de réaction, ainsi que des produits initiaux, on propose des voies de formation des espèces catalytiques de type carbyne, dont la particularité serait de disposer d'un environnement de ligands de type phénoxy (ou silyloxy), typiquement requis et identifiés pour favoriser le cycle catalytique relatif à la métathèse des alcynes sur des complexes carbynes bien définis.

ABBREVIATIONS

acac	Acetylacetonato
Ar	Aryl
Bu	Butyl
Et	Ethyl
GC	Gas Chromatography
$OPPh_3$	Triphenylphosphine oxide
Ph	Phenyl
$P(OPh)_3$	Triphenylphosphite
PPh_3	Triphenylphosphine
Pr	Propyl
Py	Pyridine
TOF	Turn Over Frequency

INTRODUCTION

In 1976 was held the first metathesis symposium in Mainz, where our Nobel Laureate Yves Chauvin presented some work on tungsten-based catalysts generated from the well-defined methoxyphenyl tungsten pentacarbonyl carbene complex and $TiCl_4$ for cyclopentene ring-opening polymerization [1]. During the same event, one of us presented some results related to the use of molybdenum hexacarbonyl as precursor for alkyne metathesis when associated with phenol as co-catalyst [2, 3]. Since more than four decades, tremendous efforts have been devoted to the search of new catalysts for metathesis reactions in general, in particular via the synthesis and use of well-defined, active metathesis catalysts, essentially based on ruthenium, tungsten and molybdenum. In that context, a strong interest has been devoted within the last decade to the synthesis of molecular initiators for alkyne metathesis, as revealed by recent publications in the field [4-7]. However, the use of *in situ* catalysts still remains of interest both from academic and industrial point of views, as they are generally prepared using procedures that are less sensitive to oxygen and water, whereas the synthesis and uses of well-defined and active carbenes or carbynes would require their prior synthesis and working under more specific experimental conditions.

Since our former, seminal work on molybdenum carbonyl as catalyst precursor, our interest on the alkyne metathesis reaction has been focused in the early 80s on the use of higher oxidation state molybdenum complexes, which when associated with an alkylating/reducing agent, and again a phenol as co-catalyst, have been shown to be two orders of magnitude more reactive than the former ones. This has been exemplified in a short series of papers where $MoO_2(acac)_2/AlEt_3/PhOH$ combinations were shown to be active at room temperature [8] and could be applied to functionalized alkynes [9], whereas some mechanistic insights related to the catalytic reaction have been developed [10].

The aim of this paper will be to describe further *in situ* catalytic systems based on different high oxidation state molybdenum complexes and salts, and to show the versatility of this methodology for obtaining efficient alkyne metathesis catalysts. A careful analysis of the reaction by-products will be performed on these catalysts, as well as on $Mo(CO)_6/ArOH$ systems, aiming at the knowledge of the reaction paths that could lead to the active metathesis species and their ligands environment. Examination of the structure of these by-products will allow suggesting the production of aryloxy molybdenum complexes, closely similar to the well-known carbyne complexes active in this metathesis reaction [11, 12].

1 EXPERIMENTAL SECTION

1.1 Molybdenum Complexes Precursors

$MoO_2(acac)_2$ was purchased by *Fluka*. The complexes $MoCl_2(NO)_2L_2$ (L = Py, NEt_3, PPh_3, $P(OPh)_3$, $OPPh_3$) were prepared according to the literature, via the synthesis of polymeric $[MoCl_2(NO)_2]_n$, obtained by reaction between $Mo(CO)_6$ and NOCl [13].

1.2 Catalytic Reactions

All reactions were carried out under argon in a reactor which typically consists of a round bottom glass flask of *ca.* 40 mL capacity equipped with a septum for sampling and sealed on a *ca.* 40 cm height, 2 cm diameter tube to allow condensation of the solvent during reactions conducted at high temperature[1].

In a typical experiment, the molybdenum complex (0.1 mmol) was introduced in a Schlenk tube, solubilized in 9.4 mL toluene and 0.6 mL of a molar solution of $AlEt_3$ in toluene. The alkylation process was done at room temperature during 5 min. 1 mL of this solution was then introduced into the reactor containing a 9 mL solution of 1 mmol alkyne (4-nonyne or 4-decyne) in toluene, 1 mmol phenol or silanol cocatalyst as well as 0.5 mL undecane as internal standard, previously heated at the required reaction temperature. Aliquots were taken regularly using a syringe via the septum, and immediately poured into a soda solution to stop the reaction and avoid the presence of phenol in the organic phase, which was then analyzed by GC on a *Shimadzu*

[1] This reactor model allows avoiding the presence of any trace of silicon grease, whose residual Si-OH groups are prone to act as co-catalysts in this reaction.

GC2014 apparatus equiped with a FID detector and using N_2 as carrier gas. A *Supelco* Equity-5 column (95% methylpolysiloxane + 5% phenylsiloxane, 30 m × 0.32 mm × 0.25 μm) was used starting at 100°C for 5 min, followed by a temperature program of 20°C/min for 5 min, and stabilized at 200°C. Conversions and selectivities were calculated via the internal standard method.

1.3 By-Products Production and Analysis

1.3.1 $MoO_2(acac)_2/AlEt_3/PhOH$

In the catalytic reactor described above were introduced 2.5 mmol phenol, 5 mL toluene, 1 mmol 4-octyne and 0.1 mL undecane. After complete solubilization of phenol, the catalyst was prepared as before, and all the alkylated molybdenum solution (corresponding to 0.1 mmol Mo) was introduced at room temperature; aliquots were taken from time to time to follow the evolution of the by-products formation and analyzed by GC-MS.

1.3.2 $Mo(CO)_6/ArOH$

0.05 mmol of $Mo(CO)_6$, 1 mL of toluene, 0.5 mmol of 4-octyne, 0.5 mmol of phenol or chlorophenol and 100 μL of undecane were introduced in the reactor. The mixture was stirred and heated at 110°C. Samples were taken and analyzed using the GC-MS apparatus described below.

1.3.3 GC/MS Analysis

GC-MS analyses were performed on a *Bruker* SCION-436GC – *Bruker* SCION SQ MS equipped with a 0.25 mm, 30 m-long *Agilent* DB-5 capillary column. The carrier gas was helium using a column flow of 1.0 mL/min and a temperature program of 10°C/min starting at 60°C (hold 3 min) and ending at 250°C (hold 13 min); EI: 70 eV.

2 RESULTS AND DISCUSSION

Within the frame of this paper, we will first present and compare catalytic results that have been observed on catalytic systems based on high oxidation state dioxo and dinitrosyl molybdenum/$AlEt_3$/ZOH catalysts for alkyne metathesis, with regard to the use of various and selected ligand environment. A second part will be devoted to analysis of the by-products of these reactions, as well as those observed on molybdenum hexacarbonyl/ArOH catalysts. Mechanistic considerations will be made, leading to the assumption that the co-catalyst should give rise to the formation of high oxidation state catalysts in both cases.

2.1 Dinitrosyl Molybdenum Based Catalysts

As described in a previous paper [8], the $MoCl_2(NO)_2(Py)_2/AlEt_3/PhOH$ combination is an active and selective catalyst for 4-nonyne metathesis at toluene reflux, the reaction being at equilibrium within minutes at a C≡C/Mo ratio of 100 (Eq. 1):

$$2\ C_3H_7{-}C{\equiv}C{-}C_4H_9 \rightleftharpoons C_3H_7{-}C{\equiv}C{-}C_3H_7 + C_4H_9{-}C{\equiv}C{-}C_4H_9 \tag{1}$$

The association of dinitrosyl molybdenum complexes/chloroalkylaluminum reagents to produce alkene metathesis catalysts has been known for a long time, as the first paper in the field has appeared very early in the corresponding literature, when the reaction was so called "disproportionation" [14, 15]. Of interest in the corresponding papers is that a comparison was made using molybdenum *versus* tungsten dinitrosyl complexes. This led to the conclusion that molybdenum-based catalysts are more reactive, and that the use of different ancillary, two-electron donor ligands also led to various activities. In line with these results, we performed metathesis reactions on 4-nonyne using the same procedure as described above, with different phosphorus and nitrogen ligands, using phenol as standard co-catalyst. The reactions were all run at 110°C for comparison. All reactions led to the equilibrium within minutes, with a complete selectivity into 4-octyne and 5-decyne. The results are reported in Table 1.

The use of chlorobenzene as solvent only affects the reactivity to a small extent, and a comparison of the results obtained on the different pre-catalysts leads to the conclusion that the less coordinating ligands triethylamine and triphenylphosphine oxide gave the best results. It must be emphasized that using polymeric $[Mo_2Cl_2(NO)_2]_n$ gave also good results (entry 6), but due to the poor solubility of this precursor in toluene, the activation procedure needed a much longer reaction time with the alkylating reagent $AlEt_3$, as exemplified by the evolution of the activity *versus* alkylation time at 60°C (Tab. 2).

TABLE 1
Activity of the $MoCl_2(NO)_2L_2/AlEt_3/PhOH$ for 4-nonyne metathesis

Entry	1	2	3	4	5	6
L	Pyridine	NEt_3	PPh_3	$P(OPh)_3$	$OPPh_3$	None
TOF (h^{-1})[a]	3300	4400	3950	3700[c]	4750	3600[d]
TOF (h^{-1})[b]	2150	3000	3650	2400[c]	3800	3600[d]

Conditions: T = 110°C; [C≡C]/[Mo] = 100; $[AlEt_3]/[Mo]$ = 6; PhOH/Mo = 100.
[a] Toluene as solvent.
[b] Chlorobenzene as solvent.
[c] Selectivity into metathesis products decreases to 50%.
[d] The use of a longer activation time was necessary (Tab. 2 below).

TABLE 2

Evolution of the 4-nonyne metathesis rate *versus* the activation time of $[Mo_2Cl_2(NO)_2]_n$

Activation time (min)	5	10	20	30	40	50	60
TOF (h^{-1})	57	78	200	360	385	465	516

Conditions: Same as in Table 1, toluene as solvent. $T = 60°C$.

An interesting feature to be noticed is the fact that with all catalytic systems, a further loading of alkyne was immediately metathesized to equilibrium, thus demonstrating the robustness of such systems.

2.2 Catalytic Results on the $MoO_2(acac)_2/AlEt_3/ArOH$ or ≡Si-OH Catalysts

The $MoO_2(acac)_2/AlEt_3/PhOH$ combination developed earlier [8, 10] has been the subject of some studies in this reaction. This section will first give data related to the effect of alkyne/Mo ratio, and in a second part will study the effect of the nature of the hydroxyl-containing co-catalyst on the reaction rate.

2.2.1 Effect of the Alkyne Concentration on the Reaction Rate

The evolution of the reaction rate *versus* the co-catalyst concentration has already been described, showing that the reaction was following a kinetic law in which a first order had to be applied to the phenol co-catalyst [8]. Of interest would also be to know the effect of the alkyne/Mo ratio on the activity, using constant concentration of molybdenum (10^{-3}M) and phenol (10^{-1}M). These reactions were conducted at 110°C (Tab. 3).

The evolution of the turnover frequency confirms that the reaction rate law is first order in alkyne, what was expected if considering the previous results already found at 60°C [10]. It has to be emphasized that this result corresponds to a 15-fold increase of the turnover rate upon increasing the temperature from 60 to 110°C. Such high values (40 s^{-1}) are more relevant to a "very high" activity rating on single site ethylene polymerization catalysts [16], and to the best of our knowledge, have no precedent in the field of alkyne metathesis.

Applying this catalytic system to functionalized alkynes have proved to be successful in some cases: using a substrate/PhOH/Mo ratio of 1000/100/1, the non-conjugated enyne 1-nonene-4-yne was successfully metathesized on the triple bond, the terminal double bond remaining intact, with a turnover rate of 1 680 h^{-1}. The chloro-substituted 7-chloro-3-heptyne had a reactivity similar to that of the alkyl disubstituted alkynes, affording turnover rates of 60 000 h^{-1} [9].

TABLE 3

Effect of alkyne/Mo ratio on reaction rate on the $MoO_2(acac)_2/AlEt_3/PhOH$ catalyst at 110°C

4-nonyne/ Mo	10	100	200	500	1000	1500	2000
TOF (h^{-1})	650	7000	12700	32000	62500	100250	144000

Conditions: solvent: toluene; $[Mo] = 10^{-3}M$; $[AlEt_3]/[Mo] = 6$; $[PhOH] = 10^{-1}M$; $T = 110°C$.

As compared with the $MoCl_2(NO)_2L_2$ catalyst reported above, a comparison can be made using the results observed at 110°C with a 100/100/1 (C≡C/PhOH/Mo) ratio (Tab. 3), the bis (acetylacetonato) dioxomolybdenum compound being ca 1.5 times more active than the best bis-(triphenylphosphine oxide) dichlorodinitrosylmolybdenum complex (Tab. 1, entry 5).

2.2.2 Effect of Different Co-Catalysts on the Reaction Rate

Since its discovery, the original $Mo(CO)_6$/resorcinol catalyst have been the subject of variations on the nature of the co-catalyst. Phenol itself and chlorophenol, as well as silanols and fluoro alcohols were rapidly recognized as suitable to activate the reaction, and revealed that such hydroxyl-containing compounds were essential for the reaction to proceed [17]. Following this paper, several reports aimed at the search for the best co-catalyst to be used for these *in situ* systems have appeared in the literature, for classical metathesis reactions conducted on disubstituted alkynes [18], and also for comparisons between the $MoO_2(acac)_2/AlEt_3/ArOH$ and the $Mo(CO)_6$/phenol catalytic systems applied to polymerization reaction of aromatic diynes [19]. In this paper were also reported data where different fluoro-substituted phenols were used, confirming that the use of different co-catalysts results in different reaction rates.

Our objective has then been to look for the effect of variation of the phenol/silanol (ZOH) co-catalyst on the reaction rate during metathesis of 4-decyne on alkylated $MoO_2(acac)_2/AlEt_3/ZOH$ system. For this study, we deliberately chose to conduct the reactions at 20°C, so as to obtain more accurate results in terms of comparison. These results using 4-decyne as substrate are reported in Table 4, and Figure 1 presents the kinetic profiles of the reactions.

Upon analysis of these results, it appears that although the best initial activity is achieved with *o*-fluorophenol, 3- and 4-chlorophenol gave rise to higher global turnovers, although none of the reactions is going to equilibrium. In some cases (entries 3 and 10), such a comparison is useless at this temperature, as the co-catalysts are only sparingly soluble or insoluble in toluene: any conclusion about their potential behavior as activator at higher temperature must therefore be excluded. It must be noticed however that the

TABLE 4

Initial activity and conversion of the $MoO_2(acac)_2/AlEt_3$/ZOH for 4-decyne metathesis

Entry	Co-catalyst	pKa	TOF (h^{-1})	Conversion (%)
1	2,3,4,5,6-F_5-ArOH	5.35	0	-
2	3,4-Cl_2-ArOH	8.5	53	26
3	Resorcinol[a]	9.1	0	-
4	2-F-ArOH	8.7	76	26
5	3-Cl-ArOH	8.9	59	33
6	4-Cl-ArOH	9.3	56	30.5
7	ArOH	9.9	22	19
8	2,6-$(CH_3)_2$-ArOH	10.6	0	-
9	$(C_6H_5)_3SiOH$	~ 12[b]	12.5	5.0
10	$(C_6H_5)_2Si(OH)_2$[a]	~ 12[b]	0	-
11	$(CF_3)(CH_3)_2COH$	11.8	0	-
12	$(CF_3)CH_2OH$	12.5	0	-

Conditions: T = 20°C, solvent: toluene; [C≡C]/[phenol or silanol] = 1; [C≡C]/[Mo] = 100; [$AlEt_3$]/[Mo] = 6.

[a] Insoluble in toluene.

[b] Estimated value of the pKa.

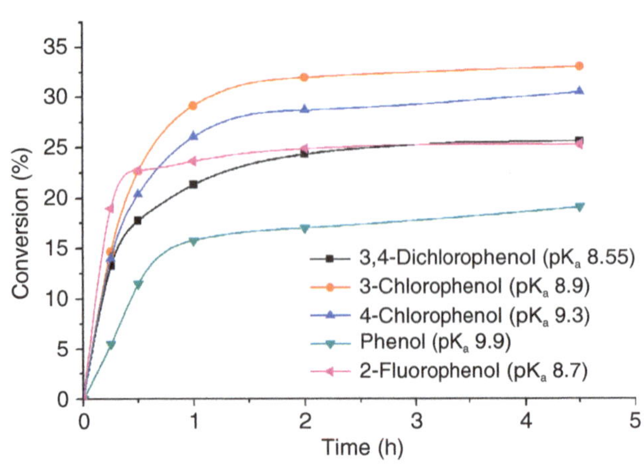

Figure 1

Reaction profiles during 4-decyne metathesis using different co-catalysts in the $MoO_2(acac)_2/AlEt_3$/Z-OH systems. See Table 4 for conditions.

more acidic halogeno-substituted phenols are much better than phenol, as found earlier on $Mo(CO)_6$/ArOH-based catalysts [17, 18], although no direct relationship between the activity and the pKa value of the co-catalysts could be found. This is similar to what was observed in several cases where a systematic study of the reactivity of *in situ* generated catalysts was conducted. These include the thermally activated $Mo(CO)_6$/ArOH catalysts during ring-closing metathesis of a functionalized diyne [18], as well the screening of alcohols and phenols for the *in situ* synthesis of aryloxy-molybdenum carbyne complex from trisamido molybdenum alkylidynes, where again the authors have found that no correlation could be found between the phenol/alcohol pKa values and the performance in ring-opening polymerization of ring-strained dibenzo[*a,e*]-8-annulene [20].

It appears from these results that the high oxidation state molybdenum/$AlEt_3$/ArOH-based catalytic systems are by far much more reactive than thermally activated $Mo(CO)_6$/PhOH, for which, using phenol as activator, a turnover frequency of 120 h^{-1} was found (*versus* 7000 h^{-1}) [8].

2.3 Mechanistic Insights

2.3.1 Previous Mechanistic Considerations

The reaction mechanism of these *in situ* catalytic systems based on molybdenum/ArOH combinations has been the subject of part of a review [21] where considerations using previous results on the $MoO_2(acac)_2AlEt_3$/PhOH combination in our group led to the conclusion that transient metallacyclobutenes might be key intermediates of the catalytic cycle [11, 12].

This hypothesis was supported by several observations which are:

- the detection of alkenes at the beginning of the reaction, arising from a reaction between an ethylidene metallacarbene initially formed during the alkylation step of the molybdenum salt by triethylaluminum;
- the fact that a first order reaction rate in phenol was established [8];
- the occurrence of an interaction via H-bonding between the proton of the hydroxyl group of the phenolic cocatalyst and the alkyne triple bond [17, 22, 23].

On the other hand, a mechanism involving the occurrence of metallacarbynes as active species and metallacylobutadiene intermediates in this metathesis has been first suggested by Katz and McGinnis [24], and established by McCullough and Schrock via the synthesis of well-defined tris-aryloxy molybdenum complexes [25].

Considering the fact that the reaction of $Mo(acac)_3$ with triethylaluminum reduces the molybdenum salt to Mo(0), as attested by the synthesis and successful use of Chatt's complex $Mo(N_2)_2(dppe)$, successfully used as precatalyst in the presence of phenol [9], we anticipated that in some way, the hydroxyl containing co-catalyst should react on molybdenum to provide higher oxidation state molybdenum species, more prone to provide active metathesis catalysts.

The question was then to know how such Mo species can be produced: the following section, dealing with a deeper examination of the by-products formed during the reaction will try to give some answers to this key problem.

2.3.2 Analysis of the By-Products Formed during Metathesis

As stated before, we had analyzed the initial by-products at the early stage of the alkyne metathesis using $MoO_2(acac)_2/AlEt_3/ArOH$ as catalyst [11, 12]. However, analyzing more carefully the GC chromatograms conducted at higher temperature revealed that small amounts of higher molecular weight products – in sub-stoichiometric quantities *versus* the catalyst – were also formed.

We were therefore interested to analyze these reaction by-products, as determination of their nature could provide some elements related to the formation of the active species.

To avoid the presence of too many products arising from the occurrence of the metathesis process, the reactions were conducted using a symmetrical alkyne, 4-octyne, and GC/MS analysis was then applied to assign their structure. Compounds **1**, **2**, **3** and **4** were formed (Fig. 2).

Analyzing their relative amounts indicates that substituted butadienic compounds **1** and **3** are produced in higher amounts than the cyclotrimer **2**, whereas this aromatic trimer becomes the major compound in the absence of phenol.

This observation prompted us to use a similar procedure aimed at the search for similar products when using our former $Mo(CO)_6/ArOH$ catalyst. Compounds **1**, **2** and **3** were also indeed produced, together with compounds **5** and **6**.

In Figure 3 are reproduced in the upper part of the GC spectrum of a reaction conducted for 4 h at toluene reflux, using 10 eq 4-octyne *versus* $Mo(CO)_6$, with and without chlorophenol.

Remarkably, the comparison of the GC chromatograms reveals several interesting features:

- one may find once more in the GC analysis **1** and **2** as the major products;
- the relative amount of aromatic trimer *versus* the butadienic compounds slightly decreases when using the phenolic reagent;
- worth to mention is also the fact that using $Mo(CO)_6$ without activator, after longer time (8 h), the tetrapropylcyclopentadienone **5** and the butylidene tetrapropylcyclopentadiene **6** were produced in larger amounts than when using the phenolic compound.

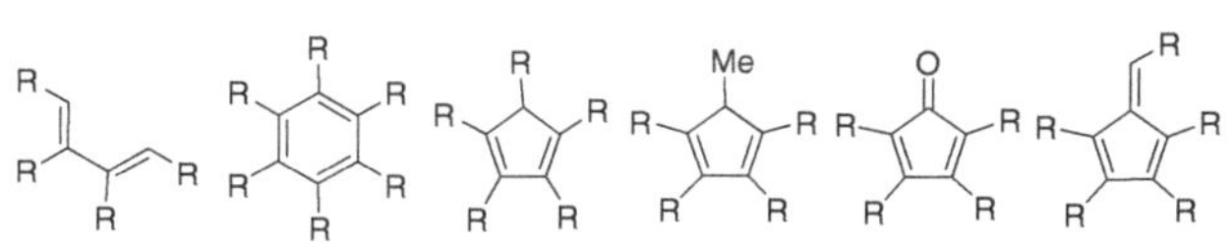

1a, 1b, 1c. (3 isomers) 2 3 4 5 6

Figure 2

Reaction by-products observed during metathesis of 4-octyne on $MoO_2(acac)_2/AlEt_3/PhOH$ and $Mo(CO)_6/ArOH$ catalytic combinations (R = Pr).

2.3.3 Discussion

The co-production of tetra-substituted butadienic compounds **1** and aromatic cyclotrimers **3** from disubstituted alkynes is scarcely described in the literature. One of these involves the use of $MoCl(CO)_2(\eta^3\text{-}C_3H_5)(NCMe)_2$, which converts 6 eq of PhC≡CPh when heated in methanol at reflux for 24 h into a mixture of hexaphenylbenzene (58%) and (*E,E*)-1,2,3,4-tetraphenylbutadiene (34%) [26]. More relevant to our concern is an attempt to establish the catalytic mechanism of the alkyne metathesis on $Mo(CO)_6/ArOH$ catalyst by Nishida *et al.* [27], who also reported the synthesis of butadienic compounds as minor products during the cross metathesis reaction between symmetrical alkyne and a dialkyne, affording mainly the aromatic [2+2+2] co-cyclization product. This reaction was done using a huge amount of "catalyst" (35 mol%), and the author suggested that the production of both compounds should arise from transient metallacyclopentadienes intermediates, which were also suggested to be key intermediates for metathesis, via a hypothetic scrambling of the alkyl groups on the metallacycle via cyclobutadienic complexes and reverse reaction [28] (Fig. 4).

The expansion of the metallacycle via insertion of a third alkyne would produce the aromatic compound after reductive elimination. In this paper however, there has been no discussion about the role of the co-catalyst in the catalytic process, neither on the molybdenum ligand environment.

From that point of view, some proposals were made by several authors, where the phenolic co-catalyst would react on molybdenum species, leading to aryloxy-molybdenum carbynes of $(ArO)_3Mo{\equiv}CR$ type [26, 29, 30]. However, in no case was reported the identification of such species, and how they could be formed from the reaction mixture consisting of $Mo(CO)_6$ and phenols.

The observation of the formation of butadienic compounds **1** in the medium is a key feature on which we then focused our attention, and isotopic labeling experiments have been performed using deuterated phenol to probe whether a protonolysis (deuterolysis) of the metallacyclopentadienic intermediates could occur. This procedure has already been used by Ibragimov *et al.* to provide evidence for the formation of aluminometallacycles in reactions involving alkynes and diethylaluminum chloride [31].

Indeed, this reaction leads to the production of partly deuterated 1,2,3,4–tetrapropylbutadiene (Fig. 5)

Our assumption and hypothesis is therefore that compounds **1** arise from a reaction between the tetra-alkyl

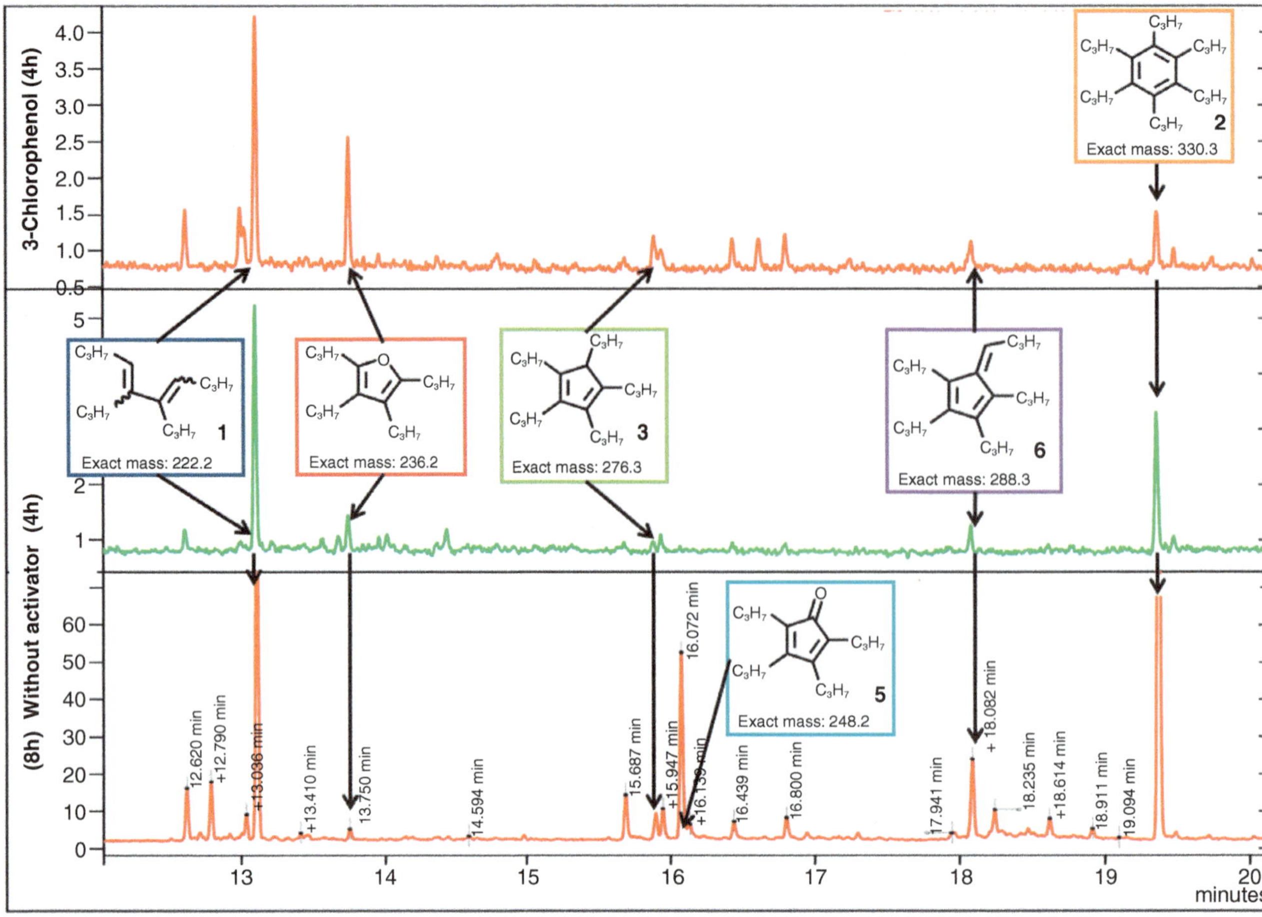

Figure 3

GC chromatogram of the higher molecular weight products obtained upon heating $Mo(CO)_6$ and 4-octyne with and without phenol at toluene reflux for 4 and 8 h.

cyclopentadienyl molybdenum intermediates and excess phenol, which in turn should also produce aryloxy-molybdenum species. Worth to be reminded at this point is the fact that the kinetic studies have shown a first order dependence of the reaction rate in phenol concentration, which may here be justified, if considering that this protonolysis reaction would be the rate determining step. Among others such as triphenylsiloxy molybdenum carbynes recently studied in depth by Heppekausen *et al.* [32], these aryloxy species are known to be ideal ligands to provide active molybdenum carbyne complexes for alkyne metathesis, as compared with tungsten-based catalysts [33]. Figure 6 is representative of the preceding hypothesis, where the production of a $(PhO)_2Mo$ transient intermediate is therefore suggested, following the preliminary molybdenacyclopentadiene complex formation.

This hypothesis does not preclude the possibility of formation the aromatic trimers, which can be produced by a further insertion of alkyne into the cyclometallated species as a competitive reaction. Of interest is also the fact that use of phenol avoids the production of the cyclopentadienone by-product: an insertion reaction of remaining CO moieties in the metallacyclopentadienyl molybdenum species would indeed lead to this ketone after reductive elimination, a process that is in competition with the protonolysis reaction. The production of tetraphenyl-substituted cyclopentadienone molybdenum complexes was reported very early by Hübel and Merenyi [34] during a reaction involving diphenylacetylene and $Mo(CO)_6$ at high temperature, where other dinuclear molybdenum cyclobutadienic complexes were synthesized [34] and well-characterized by Potenza *et al.* [35]. Notably also in the synthesis of these complexes is the fact that pentaphenylcyclopentadiene and $(Ph_5C_5)_2Mo$ were also produced, which corresponds in our case to compound **3**. Furthermore, with regard to this hypothesis, one may also consider our previous observations made on the

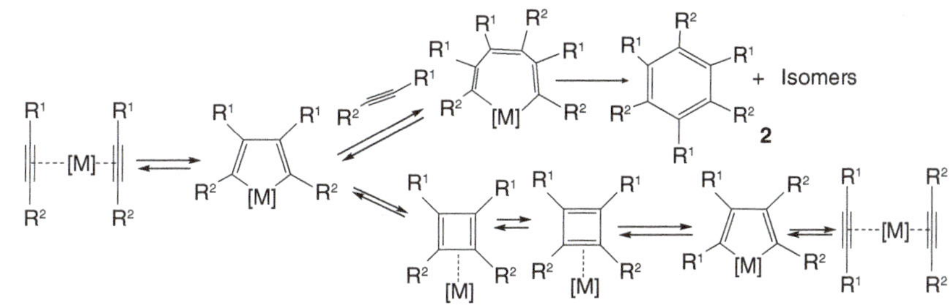

Figure 4

Metallacyclopentadiene-type mechanism for alkyne metathesis on $Mo(CO)_6$/ArOH-based catalysts [28].

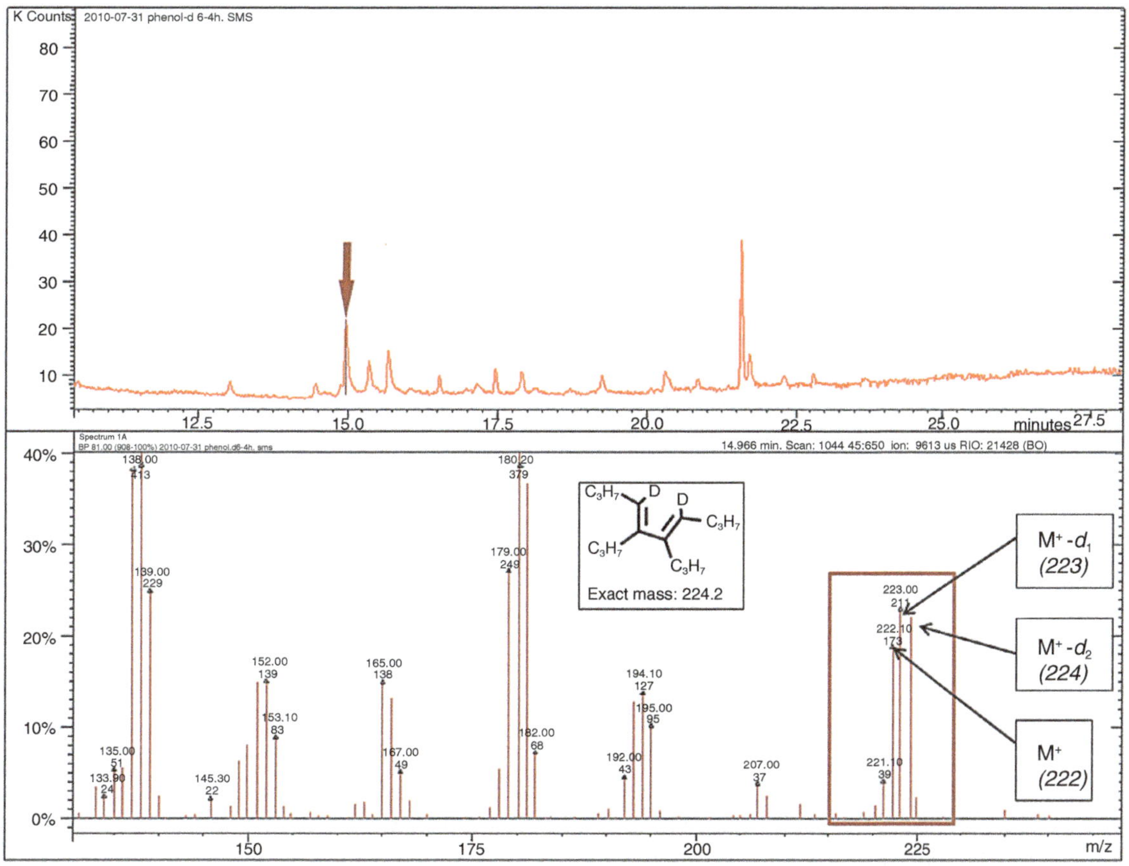

Figure 5

Mass spectrum of 1,2,3,4-tetrapropylbutadiene obtained upon heating C_6D_5OD and $Mo(CO)_6$ with 4-octyne.

reactivity of the system upon using [alkyne+$Mo(CO)_6$] mixtures previously heated at toluene reflux before phenol addition, which after rapid cooling gave catalytic activity at room temperature, whereas premixing and heating $Mo(CO)_6$ and phenol followed by alkyne addition led to a totally inactive system [2, 3, 17]. The reaction path(s) leading to the aryloxy Mo species described in Figure 6 are in total agreement with these experimental facts.

As for the presence of compound **4**, a mechanistic proposal may be suggested, which would involve an intramolecular insertion reaction of these carbenic moieties into a molybdenacyclopentadiene (Fig. 7).

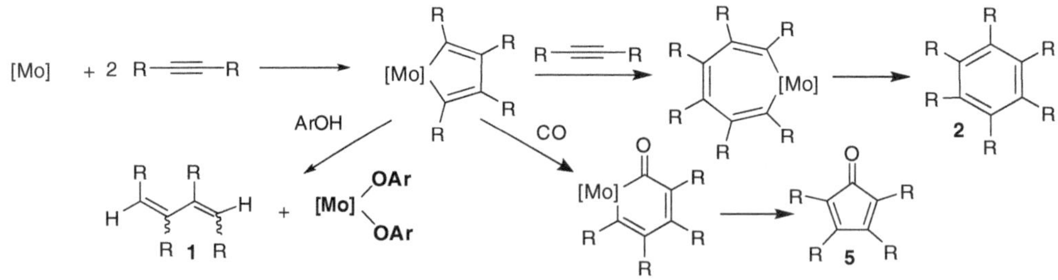

Figure 6

Suggested mechanism for the synthesis of molybdenum aryloxy species from *in situ*-generated [Mo]/PhOH/alkyne catalytic system.

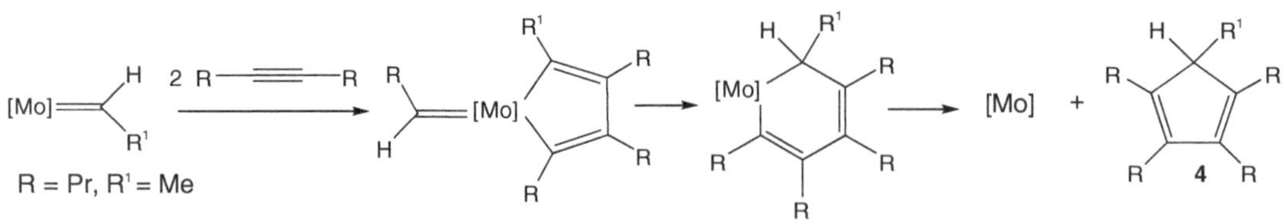

Figure 7

Proposed mechanism of formation of methyl tetrapropyl-cyclopentadiene **4**.

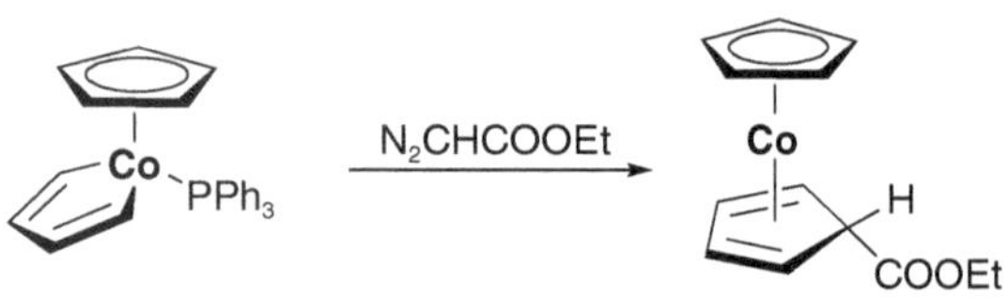

Figure 8

Synthesis of cyclopentadiene compound from the triphenyl-phosphine (cyclopentadienyl) metallacyclopentadienyl cobalt complex.

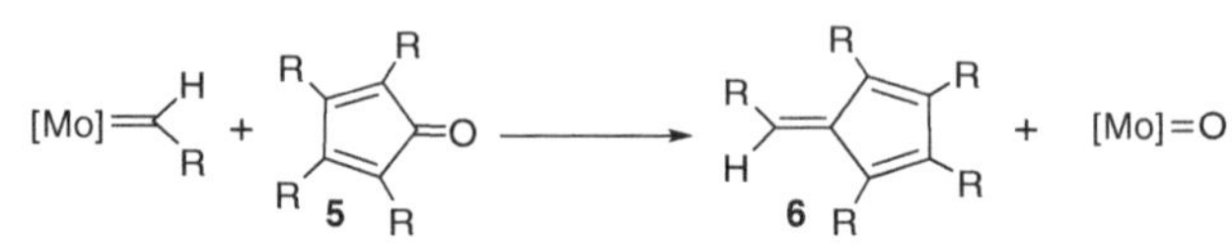

Figure 9

Plausible pathway for the formation of compound **6** as by-product.

The hypothesis of such an intramolecular reaction is supported by earlier work made on a (cyclopentadienyl) metallacylopentadienyl cobalt complex, which upon a diazacarbene addition provides the corresponding cyclopenta-diene-substituted compound (Fig. 8) [36].

Compound **6** is only present as by product when using $Mo(CO)_6$ as pre-catalyst, where due to the presence of carbon monoxide, the cyclopentadienone **5** is produced. In Figure 9 is described the way by which this ethylidene compound may be formed, via a Wittig-like reaction with a molybdenacarbene species.

If one assumes that metallacarbyne species are present in the reaction medium, this would indeed explain the presence of metallacarbenes arising from an addition reaction of phenol used in large excess, a process recognized earlier as an entry to metallacarbene species [37]. Note that this reaction path also produces an oxidized Mo=O species, which may be a termination process for metathesis.

The question then would arise to know how metallacarbynes can be involved as active species and how they are formed. As far as all attempts to detect or isolate such intermediates have failed in our hands, probably due to the extreme reactivity of these intermediates, what is following will only aim at suggesting some hypotheses on the formation of metallacarbyne species. One possibility would be to consider a further oxidative addition of phenol on the highly reactive aryloxy $(PhO)_2Mo$ to produce the unstable Mo(IV) $(PhO)_3MoH$ intermediate or $Mo(OPh)_4$. Both would dimerize and produce the $(PhO)_3Mo{\equiv}Mo(OPh)_3$ dimer, with concomitant release of dihydrogen or reduction, respectively. A metathesis reaction with the alkyne would give rise to the $(PhO)_3Mo{\equiv}C\text{-}R$ initiator of the metathesis process. In support of this assumption, such a metathesis reaction between Mo≡Mo dimers and disubstituted alkynes has indeed been reported (Fig. 10) [38], and already

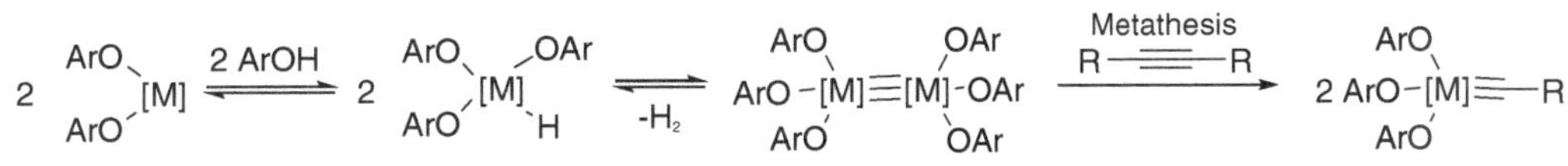

Figure 10

Putative pathway to the *in situ* production of carbyne initiators.

suggested earlier as a key reaction leading to the active species on $Mo(CO)_6$/ArOH systems [29].

On the other hand, if the activities obtained on molybdenum carbonyl-based catalysts are of the same order of those observed on tris(aryloxy)molybdenum carbynes generated *in situ* from tris amido precursors, the activities obtained on the catalytic systems based on the higher oxidation state dioxo and dinitrosyl $MoO_2(acac)_2$ and $MoCl_2(NO)_2L_2$-/AlR_3/ArOH combinations are almost two orders of magnitude higher, so that the question would be to know whether the active molybdenum species are the same in both systems and/or if the initiation process generates a similar quantity of active species. Another point worth to be considered is the fact that the alkylation procedure could give rise to another route for the metallacarbyne(s) generation.

As the role of phenol is now assigned differently from the one we proposed earlier, it seems necessary to us to revisit the experimental data that we published more than two decades ago, where a thorough GC analysis of the byproducts in the alkenes and alkyne range was performed on the $MoO_2(acac)_2$/$AlEt_3$/PhOH combination [10]. Starting from 4-nonyne as the substrate and conducting the catalytic reaction between 0 and 20°C allowed detection of 2-hexenes, 3-heptenes and concomitantly 2-heptyne and 2-hexyne as primary products before the appearance of the expected metathesis products 4-octyne and 5-decyne. That means that the initiation process is different from the one using the $Mo(CO)_6$/PhOH catalysts: the presence of these by products is consistent with the formation of not only ethylidene Mo=CHMe species via H abstraction on the $[Mo]CH_2CH_3$ moiety formed by transalkylation, but also ethylidyne [Mo≡]C-Me, responsible for the initial production of 2-hexyne and 2-heptyne via reaction with 4-nonyne. As a matter of facts, a fine tuning of the experimental conditions was necessary to achieve good results, *i.e.* the rigorous respect of the protocol related to the order of addition of the catalyst components, substrate and phenol, otherwise the reaction course deviates. These conditions are as follows:

- premixing of the molybdenum salt and triethylaluminum (1/6 ratio) at 0°C for 30 min;
- premixing of the alkyne and phenol at room temperature for 10 min;
- injection of the alkylated solution in the alkyne–phenol solution at the desired reaction temperature.

Any other protocol would give either no reaction, or polymerization if phenol is added before the introduction of alkyne.

Considering the literature devoted to the synthesis of alkylidyne molybdenum complexes, it appears that this system is closely related to the synthesis of trisalkoxymolybdenum (VI) alkylidyne complexes [39], where in a first step, using a Grignard reagent, molybdenum salts are alkylated and give stable trisalkylmolybdenum carbynes, via two successive α-H abstraction reactions. Although the alkylating reagent is here different, a possible way to phenoxymolydenum carbyne active species may therefore come from a first alkylation/alkynylation followed by protonolysis of at least one molybdenum alkyl bonds by phenol in large excess, leading to an alkyl phenoxy metallacarbyne moiety. It appears therefore that most probably, our protocol is a one pot way by which the expected carbyne species are formed. Introducing phenol before the alkyne would rather produce metallacarbenes via reaction with the phenolic reagent with the carbyne moiety. Metallacarbenes are indeed well-known to straightforwardly polymerize alkynes (Fig. 11). Worth to be mentioned here is the fact that tungsten alkylidene have been shown to rapidly interconvert with alkylidyne complexes in the presence of acids, a reaction which may also be here catalyzed by the presence of phenol [37].

Beyond these considerations which show how the alkylation process on high oxidation Mo salts results in the production of carbyne species differently from those arising from $Mo(CO)_6$ alone, another plausible and interesting alternative could be however envisaged considering data already published in the literature on such molybdenum salts/alkylaluminum devoted to their use as alkene metathesis catalysts, both on dioxo and dinitrosyl complexes. The reaction of dinitrosylmolybdenum complexes with chloroaluminum alkyls has been indeed the subject of several reports, as these were active catalysts in alkene metathesis, as already mentioned [14, 15]. Of interest is the fact that the dinitrosyl ligands were found to be still present in these reactions, in particular when using alcohols as modifiers [40, 41]. This means that in our system, the ancillary ligands could also be a combination of a dinitrosyl and aryloxy moieties, giving

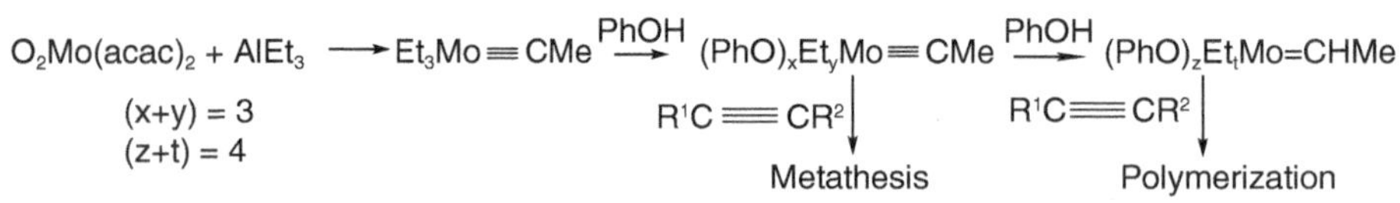

Figure 11

Mechanism proposal for the formation of metallacarbynes and metallacarbenes using the $MoO_2(acac)_2/AlEt_3/PhOH$ catalysts.

rise to much more reactive catalysts than their trisphenoxy congener. Although less documented, the same may apply when using the dioxomolybdenum precatalyst, where one may suggest that an oxo species would also be present together with one aryloxyligand, giving rise to even higher reactivities than the dinitrosyl complex. Some examples in this field have indeed shown that alike imidomolybdenum catalysts, oxo molybdenum complexes in the presence of alkylating agents give rise to efficient olefin metathesis catalysts. One early example is the use of $MoOClR_3$ arising from $MoOCl_4$ as well as mixed tungsten alkoxyalkyl oxo pre-catalysts in the presence of Lewis acids for alkene metathesis [42], and more recently that of $MoOCl_4$ in the presence of alkylating reagents such as $SnBu_4$, *n*-(BuLi) or $AlEt_3$ and EtOH, providing a nice entry for living alkyne polymerization [43].

CONCLUSION

Through this paper, one may see that the role of the phenolic and/or siloxy co-catalysts in the *in situ* synthesis of active species for alkyne metathesis is still not definitively assessed, due to the lack of spectroscopic evidence of the presence of well-defined metallacarbyne or metallacyclobutadiene complexes. However, in contrast with our former assumptions related to the role of the co-catalyst in this reaction, thanks to the analysis of by-products formed during the process, we are now proposing reaction pathways by which ancillary aryloxy or siloxy should be present in the coordination sphere of the metal. The suggested routes by which the metallacarbyne species are formed in these two systems are different, but both are using a process by which phenoxy molybdenum species are formed via protonolysis of a Mo-C bond by the phenolic co-catalyst. The fact that the high oxidation state alkylated systems are much more reactive than the molybdenum carbonyl one certainly comes from the alkylation process, which not only provides the required metal-carbon bond to be protonated, but also from its propensity to readily produce the metallacarbyne. In these *in situ* systems, we cannot know how much active species are formed from the metal precursor, neither establish a rationalized correlation between the effect of the co-catalyst structure on the activity of a given system, since these co-catalysts intervene in at least two steps in the process: the generation of the active species (thus related to pKa considerations) and afterwards as phenoxy ligands on the metallacarbynes involved in the catalytic cycle (where steric and electronic properties are key parameters). We do hope that this paper dedicated to the memory of Yves Chauvin will open the way to the discovery of more reactive catalysts, together with a further opening towards the pathways related to the formation of the active species, which has been recently considered as something relevant to mystery [44]. The present paper may have unraveled some part of this mystery, about which further studies are underway, still needed to confirm the above-described hypothesis. Indeed, although no definitive proof of the mechanistic proposal described in Figure 4 can be found in the literature, the phenoxy ligand environment may well be a key feature allowing transient and unstable cyclobutadienic complexes to be formed and responsible for the metathesis reaction to occur in these *in situ* catalytic systems.

ACKNOWLEDGMENTS

The authors thank the CNRS, the French Ministry of Higher Education and the Institut Universitaire de France for their financial support.

REFERENCES

1 Chauvin Y., Commereuc D. (1976) Tungsten carbonyl complexes activated by titanium tetrahalides as catalysts for ring-opening polymerization of cyclopentene, Book of abstracts of the *1st Metathesis Symposium*, 116-125, Mainz, Germany.

2 Mortreux A., Blanchard M. (1976) Role of phenol in the metathesis reaction of alkynes with catalysts based on $Mo(CO)_6$, Book of abstracts of the *1st Metathesis Symposium*, 106-116, Mainz, Germany.

3 Mortreux A., Blanchard M. (1974) Metathesis of alkynes on a molybdenum hexacarbonyl-resorcinol catalyst, *J. Chem. Soc. Chem. Comm.*, 786.

4 Fürstner A. (2013) Alkyne Metathesis on the Rise, *Angew. Chem. Int. Ed.* **52**, 10, 2794-2819.

5 Persich P., Llaveria J., Lhermet R., de Haro T., Stade R., Kondoh A., Fürstner A. (2013) Increasing the stuctural span of alkyne metathesis, *Chem. Eur. J.*, **19**, 39, 13047-13058.

6 Lhermet R., Fürstner A. (2014) Cross-metathesis of terminal alkynes, *Chem. Eur. J.* **20**, 41, 13188-13193.

7 Haberlag B., Freytag M., Jones P.J.,Tamm M. (2014) Tungsten and molybdenum 2,4,6-trimethylbenzylidyne complexes as robust pre-catalysts for alkyne metathesis, *Adv. Synt. Cat.* **356**, 6, 1255-1265.

8 Bencheick A., Petit M., Mortreux A., Petit F. (1982) New active and selective catalysts for homogeneous metathesis of disubstituted alkynes, *J. Mol. Catal.* **15**, 93-101.

9 Petit M., Mortreux A., Petit F. (1982) Homogeneous metathesis of functionalized alkynes, *J. Chem. Soc., Chem. Comm.*, 1385-1386.

10 Bages S., Petit M., Mortreux A., Petit F. (1990) Étude de la métathèse des hydrocarbures acétyléniques catalysée par le système $O_2Mo(acac)_2$-$AlEt_3$-PhOH, *J. Mol Catal.* **59**, L25-L29.

11 Part of this work has been presented in the *XVIIIth International Symposium on Metathesis (ISOM XVIII)*, Leipzig, 2009.

12 Part of this work has been presented in the *XIXth International Symposium on Homogeneous Catalysis (ISHC XIX)*, Ottawa, 2014.

13 Cotton F.A., Johnson B.G.F. (1964) Dichlorodinitrosylmolybdenum, dichlorodinitrosyltungsten, and some of their derivatives, *Inorg. Chem.* **3**, 1609-1612.

14 Zuech E.A. (1968) Homogeneous catalysts for olefin disproportionation, *Chem. Commun.*, 1182.

15 Zuech E.A., Hughes W.B., Kubicek D.H., Kittleman E.T. (1968) Homogeneous catalysts for olefin disproportionation from nitrosyl molybdenum and tungsten compounds, *J. Am. Chem. Soc.* **92**, 528-531.

16 Britovsek G., Gibson V.C., Wass D.F. (1999) The search for new-generation olefin polymerization catalysts: life beyond metallocenes, *Angew. Chem. Int. Ed.* **38**, 428-447.

17 Mortreux A., Delgrange J.C., Blanchard M., Lubochinsky B. (1977) Rôle du phénol dans la réaction de métathèse des hydrocarbures acétyléniques sur les catalyseurs à base de $Mo(CO)_6$, *J. Mol. Catal.* **2**, 73-82.

18 Sashuc V., Ignatowska J., Grela K. (2004) A fine-tuned molybdenum hexacarbonyl/phenol initiator for alkyne metathesis, *J. Org. Chem.* **69**, 7748-7751.

19 Bly R.K., Dyke K.M., Bunz U.H.F (2005) A study of molybdenum catalysts in the polymerization of 2,5 didodecyl-1,4-dipropynylbenzene, *J. Organomet. Chem.* **690**, 825-829.

20 Fisher F.R., Nuckolls C. (2010) Design of living ring-opening living alkyne metathesis, *Angew. Chem. Int. Ed.* **49**, 7257-7260.

21 Coutelier O., Mortreux A. (2006) Alkyne metathesis catalysts: scope and future, *J. Mol. Catal. A: Chemical* **254**, 96-104.

22 Van Huong P. (1963) Influence des solvants sur le spectre infrarouge des carbures acétyléniques. Propriétés acides et basiques de ces carbures, *Revue de l'IFP* **XVIII**, 9, 37-51.

23 Yoshida Z., Ishibe N., Ozoe H. (1972) Hydrogen bonding of phenol with acetylenes and allenes, *J. Am. Chem. Soc.* **94**, 14, 4948-4952.

24 Katz T.J., McGinnis J. (1975) The mechanism of the olefin metathesis reaction, *J. Am. Chem. Soc.* **97**, 1592.

25 McCullough L.G., Schrock R.R. (1984) Metathesis of acetylenes by molybdenum (VI) alkylidyne complexes, *J. Am. Chem. Soc.* **106**, 4067-4068.

26 Goodyear J.W., Hemingway C.W., Harrington R.W., Wiseman M.R., Brisdon B.J. (2002) The crystal structure of $[Mo(NCS)(CO)_2(\eta^3\text{-}C_3H_5)(NCMe)_2]\cdot MeCN$ and the reactions of $\{Mo(CO)_2(\eta^3\text{-}C_3H_5)+\}$ containing species with symmetric alkynes, *J. Organomet. Chem.* **664**, 1-2, 176-181.

27 Nishida M., Shiga H., Mori M. (1998) [2+2+2] cocyclisation using $Mo(CO)_6$-*p*-ClPhOH, *J. Org. Chem.* **63**, 8606-8608.

28 Kaneta N., Hirai T., Mori M. (1995) Reaction of alkyne having hydroxyphenyl group with $Mo(CO)_6$, *Chem. Lett.*, 627-628.

29 Huc V., Veihofen R., Martin-Jimenez I., Oulié P., Lepetit C., Lavigne G., Chauvin R. (2003) Ether-induced rate enhancement of Mo-catalyzed metathesis under mild conditions, *New J. Chem.* **27**, 10, 1412-1414.

30 Kloppenburg L., Song D., Bunz U.W.F. (1998) Alkyne metathesis with simple catalyst systems: Poly(*p*-phenyleethylenylene)s, *J. Am. Chem. Soc.* **120**, 7973-7974.

31 Ibragimov A.G., Khafizova L.O., Gilfanova G.N., Yakupova L.R., Borisova A.L., Dzhemilev U.M. (2003) Synthesis and transformation of metallacycles -31- Catalysts based on cobalt complexes in reactions of trialkyl-and alkylhaloalanes with olefins, allenes and acetylenes, *Russ. Chem. Bull., Int. Ed.* **52**, 2434-2439.

32 Heppekausen J., Stade R., Kondoh A., Seidel G., Goddard R., Fürstner A. (2012) Optimized synthesis, structural investigations, ligand tuning and synthetic evaluation of silyloxy-based alkyne metathesis catalysts, *Chem. Eur. J.* **18**, 10981-10299.

33 Zhu J., Jia G., Lin Z. (2006) Theoretical investigation of alkyne metathesis catalyzed by W/Mo alkylidyne complexes, *Organometallics* **25**, 7, 1812-1819.

34 Hübel W., Merenyi R.J. (1964) Über Organometall-Komplexe, XIV, Organomolybdän-Verbindungen, *J. Organomet. Chem.* **2**, 213-221.

35 Potenza J.A., Johnson R.J., Chirico R., Efraty A. (1977) Molecular structure of a complex with a Mo=Mo bond and a bridging cyclopentadienone group, *Inorg. Chem.* **16**, 9 2354-2359.

36 O'Connor J.M., Pu L., Uhrhammer R., Johnson J.A. (1989) A new mode of carbene reactivity: coupling with two alkynes to generate highly substituted cyclopentadiene products, *J. Am. Chem. Soc.* **111**, 1889-1991.

37 Freudenberger J.H., Schrock R.R. (1985) Preparation of di-*tert*-butoxytungsten (VI) alkylidene complexes by protonation of tri-*tert*-butoxytungsten (VI) alkylidyne complexes, *Organometallics* **4**, 1937-1944.

38 Strutz H., Schrock R.R. (1984) Metathesis of molybdenum-molybdenum triple bond with acetylenes to give alkylidyne complexes, *Organometallics* **3**, 1600-1601.

39 Mc Cullough L.G., Schrock R.R., Dewan J.C., Murdzek J.C. (1985) Preparation of trialkoxymolybdenum(VI) alkylidyne complexes, their reactions with acetylenes, and the X-Ray structure of $Mo[C_3CMe_3)_2][OCH(CF_3)_2]_2(C_5H_5N)_2$, *J. Am. Chem. Soc.* **107**, 5987-5998.

40 Keller A., Szterenberg L. (1989) Dialkoxydinitrosylmolybdenum compounds as olefin metathesis precatalysts: catalyst formation and electronic structure, *J. Mol. Catal.* **57**, 207-220.

41 Keller A. (1991) Dinitrosylalkylidene complexes containing η-alkylalkoxy ligand: synthesis, spectroscopic characterization and metathesis activity, *J. Organomet. Chem.* **407**, 237-246.

42 Kress J., Wesolek M., Le Ny J.P., Osborn J.A. (1981) Molecular Complexes for Efficient Metathesis of Olefins. The Oxo-ligand as a Catalyst-Cocatalyst Bridge and the Nature of the Active Species, *J. Chem. Soc., Chem. Comm.* 1039-1040.

43 Hayano S., Masuda T., Ji H., Advincula R.C. (2000) On the role of the catalysts components in the living metathesis polymerization of substituted acetylenes by $MoOCl_4$-based systems, *Polymer Bulletin* **44**, 129-136.

44 Schrock R.R. (2013) Alkyne metathesis by molybdenum and tungsten alkylidyne complexes, *Chem. Commun.* **49**, 5529-5531.

PERMISSIONS

 Every chapter published in this book has been scrutinized by our experts. Their significance has been extensively debated. The topics covered herein carry significant findings which will fuel the growth of the discipline. They may even be implemented as practical applications or may be referred to as a beginning point for another development.

The contributors of this book come from diverse backgrounds, making this book a truly international effort. This book will bring forth new frontiers with its revolutionizing research information and detailed analysis of the nascent developments around the world.

We would like to thank all the contributing authors for lending their expertise to make the book truly unique. They have played a crucial role in the development of this book. Without their invaluable contributions this book wouldn't have been possible. They have made vital efforts to compile up to date information on the varied aspects of this subject to make this book a valuable addition to the collection of many professionals and students.

This book was conceptualized with the vision of imparting up-to-date information and advanced data in this field. To ensure the same, a matchless editorial board was set up. Every individual on the board went through rigorous rounds of assessment to prove their worth. After which they invested a large part of their time researching and compiling the most relevant data for our readers.

The editorial board has been involved in producing this book since its inception. They have spent rigorous hours researching and exploring the diverse topics which have resulted in the successful publishing of this book. They have passed on their knowledge of decades through this book. To expedite this challenging task, the publisher supported the team at every step. A small team of assistant editors was also appointed to further simplify the editing procedure and attain best results for the readers.

Apart from the editorial board, the designing team has also invested a significant amount of their time in understanding the subject and creating the most relevant covers. They scrutinized every image to scout for the most suitable representation of the subject and create an appropriate cover for the book.

The publishing team has been an ardent support to the editorial, designing and production team. Their endless efforts to recruit the best for this project, has resulted in the accomplishment of this book. They are a veteran in the field of academics and their pool of knowledge is as vast as their experience in printing. Their expertise and guidance has proved useful at every step. Their uncompromising quality standards have made this book an exceptional effort. Their encouragement from time to time has been an inspiration for everyone.

The publisher and the editorial board hope that this book will prove to be a valuable piece of knowledge for researchers, students, practitioners and scholars across the globe.

LIST OF CONTRIBUTORS

C. Ghotbi and V. Taghikhani
Department of Chemical and Petroleum Engineering, Sharif University of Technology, Tehran - Iran

T. Jafari Behbahani
Department of Chemical and Petroleum Engineering, Sharif University of Technology, Tehran - Iran
Research Institute of Petroleum Industry (RIPI), Tehran – Iran

A. Shahrabadi
Research Institute of Petroleum Industry (RIPI), Tehran – Iran

Andreas A. Danopoulos
Laboratoire de Chimie de Coordination, Institut de Chimie (UMR 7177 CNRS), Université de Strasbourg, 4 rue Blaise Pascal, 67081 Strasbourg Cedex - France
Institute for Advanced Study (USIAS), Université de Strasbourg, 5 allée du Général Rouvillois, 67083 Strasbourg Cedex - France

Pierre Braunstein
Laboratoire de Chimie de Coordination, Institut de Chimie (UMR 7177 CNRS), Université de Strasbourg, 4 rue Blaise Pascal, 67081 Strasbourg Cedex – France

Tom Bosserez, Jan Rongé and Johan Martens
KU Leuven, Centre for Surface Chemistry and Catalysis, Kasteelpark Arenberg 23, Post box 2461, 3001 Leuven - Belgium

Jan van Humbeeck
KU Leuven, Materials Engineering Department (MTM), Kasteelpark Arenberg 44, Post box 2450, 3001 Leuven - Belgium

Sophia Haussener
École Polytechnique Fédérale de Lausanne (EPFL), Laboratory of Renewable Energy Science and Engineering, Station 9, 1015 Lausanne – Switzerland

B. Auffray and B. Garcia
IFP Energies nouvelles, 1 et 4 avenue de Bois Préau, 92852 Rueil-Malmaison - France

C.-P. Lienemann and L. Sorbier
IFP Energies nouvelles, Rond-point de l'échangeur de Solaize, BP 3, 69360 Solaize - France

A. Cerepi
EA 4592 G&E, ENSEGID, Université de Bordeaux, 1 allée F. Daguin, 33607 Pessac Cedex – France

Teresa Pinto, Kai Szeto, Nesrine Oueslati, Véronique Dufaud and Frédéric Lefebvre
Laboratory of Chemistry, Catalysis, Polymers and Processes (C2P2), University of Lyon 1, CNRS, CPE, UMR 5265, 43 Bd du 11 Novembre 1918, 69616 Villeurbanne - France

Nadine Essayem
Institut de Recherches sur la Catalyse et l'Environnement de Lyon (IRCELyon), University of Lyon 1, CNRS, UMR 5256, 2 Avenue Albert Einstein, 69626 Villeurbanne – France

V. Kumar Saw, A. Mandal and S. Laik
Gas Hydrate Laboratory, Department of Petroleum Engineering, Indian School of Mines, Dhanbad, 826004 - India

G. Udayabhanu
Department of Applied Chemistry, Indian School of Mines, Dhanbad, 826004 – India

Guilherme B. Machado and Antonio H.M. da F.T. da Silva
Petróleo Brasileiro S.A., Petrobras/Research and Development Center, Av. Horácio Macedo, 950, Cidade Universitária, Ilha do Fundão, CEP: 21941-915, Rio de Janeiro, RJ - Brazil

José E.M. Barros
Department of Mechanical Engineering, Universidade Federal de Minas Gerais, UFMG, Av. Antonio Carlos, 6627, Campos Universitário, Pampulha, CEP: 31270901, Belo Horizonte, MG - Brazil

Sérgio L. Braga and Carlos Valois M. Braga
Department of Mechanical Engineering, Pontifícia Universidade Católica do Rio de Janeiro, PUC-Rio, rua Marquês de São Vicente, 225, Gávea, CEP: 22453-900, Rio de Janeiro, RJ - Brazil

Ellen C.Z. de Azevedo
Consultant, rua Dias da Cruz, 673, Casa 10/101, Meier, CEP: 20720-011, Rio de Janeiro, RJ – Brazil

H. Liu and H. Zhang
MOE Key Laboratory of Petroleum Engineering in China University of Petroleum, Beijing 102249 - China

J. Wang
MOE Key Laboratory of Petroleum Engineering in China University of Petroleum, Beijing 102249 - China

Department of Petroleum & Geosystems Engineering, The University of Texas at Austin, Austin 78712 - USA

K. Sepehrnoori
Department of Petroleum & Geosystems Engineering, The University of Texas at Austin, Austin 78712 - USA

G. Zhang
CNOOC Tianjin Branch Company, Tianjin 300452 - China

P. Liu
Sinopec Shengli Oilfield Branch Company, Shandong Dongying 257000 – China

Elena Balantseva, Anna Maria Ferrari and Gloria Berlier
Università di Torino, Dipartimento di Chimica and NIS (Nanostructured Interfaces and Surfaces) Centre, Via P. Giuria 7, 10125 Torino - Italy

Bruno Camino
Department of Chemistry, Imperial College London, South Kensington London SW7 2AZ - United Kingdom

R. Passalacqua, G. Centi and S. Perathoner
University of Messina and INSTM/CASPE, Department of Electronic Engineering, Industrial Chemistry and Engineering, V.le F. Stagno d'Alcontres 31, 98166 Messina – Italy

Florian Allouche, Victor Mougel, Wolfram Grüning and Christophe Copéret
Department of Chemistry and Applied Biosciences, ETH Zürich, Vladimir Prelog Weg 1-5, 8093 Zürich – Switzerland

Pierre H. Dixneuf, Christian Bruneau and Cédric Fischmeister
Institut des Sciences Chimiques de Rennes, UMR 6226 CNRS-Université de Rennes, "Organometallics: Materials and Catalysis", Centre for Catalysis and Green Chemistry, Campus de Beaulieu, 35042 Rennes – France

Raju Dey, Manoja K. Samantaray, Emmanuel Callens, Ali Hamieh, Edy Abou-hamad, Santosh Kavitake and Jean-Marie Basset
King Abdullah University of Science & Technology, KAUST Catalysis Center (KCC), Thuwal 23955-6900 - Saudi Arabia

Abdul-Hamid M. Emwas
Imaging and Characterization Core Lab, King Abdullah University of Science and Technology (KAUST), Thuwal 23955-6900 - Saudi Arabia

Margarida F. Costa Gomes
Clermont Université, Université Blaise Pascal, Institut de Chimie de Clermont-Ferrand, UMR 6296 CNRS, Équipe Thermodynamique et Interactions Moléculaires, ICCF-TIM, BP 80026, 63171 Aubière - France

Leila Moura
Clermont Université, Université Blaise Pascal, Institut de Chimie de Clermont-Ferrand, UMR 6296 CNRS, Équipe Thermodynamique et Interactions Moléculaires, ICCF-TIM, BP 80026, 63171 Aubière - France
Université de Lyon, UMR 5265 CNRS, Université de Lyon 1-ESCPE Lyon, C2P2, Équipe Chimie Organométallique de Surface, 69616 Villeurbanne – France

Catherine C. Santini
Université de Lyon, UMR 5265 CNRS, Université de Lyon 1-ESCPE Lyon, C2P2, Équipe Chimie Organométallique de Surface, 69616 Villeurbanne – France

Pablo E. Guzmán and Robert H. Grubbs
Arnold and Mabel Beckman Laboratories for Chemical Synthesis, Division of Chemistry and Chemical Engineering, California Institute of Technology, Pasadena, California 91125 - United States

Victoria A. Piunova
IBM Almaden Research Center, 650 Harry Road, San Jose California 95134 - United States

Garret M. Miyake
Department of Chemistry and Biochemistry, Materials Science and Engineering Program, University of Colorado, Boulder, Colorado 80309 - United States

Joaquin Geng Lopez, Régis M. Gauvin and André Mortreux
Univ. Lille, CNRS, Centrale Lille, ENSCL, Univ. Artois, UMR 8181 - UCCS - Unité de Catalyse et Chimie du Solide, 59000 Lille - France

Piotr Pawluc
Univ. Lille, CNRS, Centrale Lille, ENSCL, Univ. Artois, UMR 8181 - UCCS - Unité de Catalyse et Chimie du Solide, 59000 Lille - France
Faculty of Chemistry, Adam Mickiewicz University in Poznan, Umultowska 89b, 61-614 Poznan – Poland

Maciej Zaranek
Faculty of Chemistry, Adam Mickiewicz University in Poznan, Umultowska 89b, 61-614 Poznan – Poland

Index